Fracture Mechanics of Ceramics

Volume 3

Flaws and Testing

Volume 1	Concepts, Flaws, and Fractography
Volume 2	Microstructure, Materials, and Applications
Volume 3	Flaws and Testing
Volume 4	Crack Growth and Microstructure

Fracture Mechanics of Ceramics

Volume 3

Flaws and Testing

Edited by R. C. Bradt

Department of Materials Science and Engineering
Ceramic Science and Engineering Section
Pennsylvania State University
University Park, Pennsylvania

D. P. H. Hasselman

Department of Materials Engineering
Virginia Polytechnic Institute and State University
Blacksburg, Virginia

and F. F. Lange

Science Center
Rockwell International
Thousand Oaks, California

SPRINGER SCIENCE+BUSINESS MEDIA, LLC

Library of Congress Cataloging in Publication Data

Symposium on the Fracture Mechanics of Ceramics, Pennsylvania State University, 1973.
Fracture mechanics of ceramics.

Includes bibliographical references.
CONTENTS: v. 1. Concepts, flaws, and fractography. – v. 2. Microstructure, materials, and applications. – v. 3. Flaws and testing. – v. 4. Crack growth and microstructure.
1. Ceramics – Fracture – Congresses. I. Bradt, Richard Carl, 1938- ed. II. Hasselman, D. P. H., 1931- ed. III. Lange, F. F., 1939- ed. IV. Pennsylvania. State University. V. Symposium on the Fracture Mechanics of Ceramics, Pennsylvania State University. 1977. VI. Title.
TA430.S97 1973 620.1'4 73-20399
ISBN 978-1-4615-7019-6 ISBN 978-1-4615-7017-2 (eBook)
DOI 10.1007/978-1-4615-7017-2

Originally published by Plenum Press, New York in 1978
Softcover reprint of the hardcover 1st edition 1978

PREFACE

These volumes, 3 and 4, of Fracture Mechanics of Ceramics constitute the proceedings of an international symposium on the fracture mechanics of ceramics held at the Pennsylvania State University, University Park, PA on July 27, 28, and 29, 1977. Volumes 1 and 2 were published previously as the proceedings of a symposium of the same name held July 11, 12, and 13, 1973, also at Penn State.

All four volumes published to date concentrate on the fracture aspects of the mechanical behavior of brittle ceramics in terms of the characteristics of cracks.

The program chairmen gratefully acknowledge the financial assistance for the symposium provided by the Office of Naval Research, the Energy Research and Development Administration, and the Army Research Office. Without their support the quality and magnitude of this conference simply would not have been possible.

Numerous individuals contributed to the success of the conference, but unfortunately they cannot all be listed here. However the program chairmen would especially like to recognize the contributions of Penn State Conference Coordinator, Mr. Ronald Avillion, whose expertise in planning and organization was indispensable; Dr. Fred R. Matson for his interesting after dinner speech; and Drs. A. M. Diness, J. C. Hurt, and D. W. Readey for their encouragement and valuable suggestions regarding the program.

Finally, we wish to also thank our joint secretaries for the patience and help in bringing these proceedings to press.

University Park, PA — R. C. Bradt

Blacksburg, VA — D. P. H. Hassleman

Thousand Oaks, CA — F. F. Lange

July, 1977

CONTENTS OF VOLUME 3

CONTENTS OF VOLUME 4

FUNDAMENTALS OF THE STATISTICAL THEORY OF FRACTURE

S. B. Batdorf

Materials Department
School of Engineering and Applied Science
University of California, Los Angeles, California 90024

ABSTRACT

The first important study of fracture statistics was that of Weibull. His work was based on the tacit assumption that only the component of stress normal to a crack plane contributes to its fracture, and on the use of simple analytical formulas for failure probability. Recent progress in short-term fracture includes the use of more refined fracture criteria and a search for better distribution functions for the frequency of cracks, based on microstructural considerations. Use of the critical value of strain energy release rate as a fracture criterion leads to improved agreement with experiment. Consideration is also given to the statistics of fracture in static fatigue and in dynamic fracture.

INTRODUCTION

Ceramic materials are generally brittle, and nominally identical specimens of a brittle material may exhibit large variations in fracture stress, especially if the specimens are small. When brittle materials are employed in practical structures, the designer must be able to assure himself of an acceptably low probability of failure during service. Typically, on the basis of laboratory data on a limited number of specimens uniformly loaded in simple tension or pure bending, the probability of failure must be calculated for structural members of different sizes and shapes and under completely different loading conditions. The tool for accomplishing this is fracture statistics.

The first important contribution to this subject was made by Weibull in 1939 [1,2], and his theories are still the basis for most calculations in this field. He exploited the analogy between a stressed brittle structure and a loaded chain, which breaks when the strength of its weakest link is exceeded. Before discussing weakest link theory [WLT], it is appropriate to enquire into its range of applicability.

Clearly, the fracture of a single fiber is similar to the breaking of a chain. The fiber may be regarded as a linear array of very short elements, and the fracture of any element causes failure of the fiber as a whole. Similarly, when an isolated crack in an elastic body is loaded normal to its plane, it will become unstable and grow catastrophically, causing fracture. The fracture stress of the entire body is that of the weakest crack. On the other hand, there are many situations to which WLT does not apply. For instance, in a bundle of fibers the first fiber failure does not ordinarily cause failure in the bundle as a whole; rather, failure is the result of damage accumulation. Similarly, in some stress states (e.g. pure compression) a crack in an elastic body will generally grow in a direction which results in crack arrest. Here, also, failure is a result of progressive damage, and WLT does not apply.

Ideally, fracture statistics should be based on a proper consideration of three elements — extreme value statistics, fracture mechanics, and material microstructure. Weibull's theory is based almost exclusively on the first element. In recent years, progress has been made in incorporating the other two elements into weakest link theory. The purpose of the present paper is to outline the current status of fracture statistics and indicate some directions of expected future progress. In the effort to give a clear and concise description of the field, historical perspective has been somewhat slighted. The primary emphasis is on the underlying physical concepts rather than computational techniques. It is hoped that those whose contributions may have been omitted or underemphasized will understand and forgive.

WEAKEST LINK THEORY (WLT)

We first derive the fundamental equation of weakest link theory. Let it be assumed that a stressed solid can fail due to any of a number of independent and mutually exclusive mechanisms or causes, each involving infinitesimal probability of failure $(\Delta P_f)_i$. The probability that the i'th mechanism will not cause failure is $(P_S)_i = 1 - (\Delta P_f)_i$. The overall probability of survival is the product of the individual probabilities of survival, i.e.,

$$P_S = \prod_i \left(P_S\right)_i = \prod_i \left\{1 - \left(\Delta P_f\right)_i\right\}$$

$$\cong \prod_i \exp\left[-\left(\Delta P_f\right)_i\right] = \exp\left[-\sum_i \left(\Delta P_f\right)_i\right] \quad . \tag{1}$$

The sum of the individual probabilities of failure appearing in the final equality above was called by Weibull the "risk of rupture" and was given the symbol B.

In evaluating B, Weibull added together the probabilities of failure of all the elements of volume ΔV in the entire body. The probability of failure of the i'th element ΔV_i in simple tension σ, for example, is

$$\left(\Delta P_f\right)_i = n(\sigma)\ \Delta V_i \quad , \tag{2}$$

where $n(\sigma)$ is the number of flaws per unit volume with a strength less than σ. If $n(\sigma)$ is less than unity, it can be regarded as the probability that such a flaw will occur in a unit volume. Thus, the probability of failure is given by

$$P_f = 1 - P_S = 1 - \exp\left[-\int dV\ n(\sigma)\right] \quad . \tag{3}$$

For a uniformly stressed body

$$P_f(\sigma) = 1 - \exp\left[-V\ n(\sigma)\right] \quad , \tag{4}$$

or

$$n(\sigma) = \frac{1}{V} \ln P_S^{-1} \quad . \tag{5}$$

If $P_f(\sigma)$ is determined by testing a number of specimens in simple tension, Eq. (5) can be used to obtain $n(\sigma)$ and Eq. (3) can then be used to evaluate the probability of failure of a body of arbitrary size and shape and nonuniformly distributed tensile stress. By assuming, as Weibull did, that compressive stresses do not contribute to fracture, the above procedure can also be used to calculate the probability of failure in pure bending, or indeed any combination of uniaxial tension and compression. A limitation to this procedure is that in ignoring fracture due to compression, it does not adequately account for failure in all stress states.

Weibull also gave a procedure for calculating P_f for polyaxial stress states when the failure statistics for simple tension are known. Basically, his procedure is to calculate B by averaging the tensile stress in all directions [1,3]. This is intuitively plausible, but not rigorous, and it has been questioned by some investigators. Barnett et al. [4] have formulated a number of alternative procedures they considered equally plausible. Presumably, in part at least, because of doubts concerning Weibull's procedure, some investigators [5,6,7] have chosen to use an approximation in which it is assumed that with respect to fracture, the principal stresses act independently, i.e.,

$$P_S(\sigma_1, \sigma_2, \sigma_3) = P_S(\sigma_1)\, P_S(\sigma_2)\, P_S(\sigma_3) \quad , \tag{6}$$

where σ_1, σ_2, σ_3 are the principal stresses.

There is some evidence that Weibull had second thoughts concerning the validity of his recommended procedure, since he wrote in 1966 [8]: "Another problem of a more theoretical nature will be to deduce the effect of bi- and triaxiality on the distribution functions of one-dimensional stresses. If the principal stresses are acting independently of each other, and it seems that such materials may exist, then the problem may be soluble along lines previously sketched. In other cases the solution is very intricate and will certainly require close examination of the physical behavior of the material in question."

WLT FOR POLYAXIAL STRESS CONDITIONS

To put weakest link theory for polyaxial stress states on a firm physical foundation, we will make the explicit assumption that the flaws responsible for fracture are microcracks in the material. We will further assume that the cracks do not interact, and that each crack has a critical stress σ_c defined as the remote tensile stress applied normal to the crack plane which will cause fracture. Fracture under combined stresses occurs when the effective stress σ_e acting on the crack is equal to σ_c. The effective stress is some function of the applied stresses at the location of the crack, and its precise form depends on the fracture criterion employed. The search for fracture criteria leads us to fracture mechanics. Before discussing particular fracture criteria, however, we formulate the theory of fracture statistics for an arbitrary fracture criterion by leaving σ_e unspecified.

The potential causes of failure are the individual cracks. For purposes of analysis, it is convenient to group the cracks according to location, the applied stress state, and crack critical stress. We assume that the stress state varies slowly so that within a volume element ΔV all cracks will be subject to the same macroscopic stress. We also assume that the material is macroscopically homogeneous, so that a function $N(\sigma_c)$ can be defined as the number of cracks per unit volume having a critical stress equal to or less than σ_c. The probability that a crack having a critical stress in the range σ_c to $\sigma_c + d\sigma_c$ exists in volume element ΔV is the $\Delta V\,[dN(\sigma_c)/d\sigma_c]\,d\sigma_c$.

If such a crack is actually present, the probability that it will fracture depends on its orientation, the stress state, and the fracture criterion. We assume there is a solid angle Ω such that fracture of a crack will occur if and only if its normal lies within Ω. This means that if the normal lies within Ω, $\sigma_e > \sigma_c$, where σ_e is the effective stress corresponding to the fracture criterion selected. If the cracks are randomly oriented, the probability that a crack will fracture under the applied stress Σ is $\Omega(\Sigma,\sigma_c)/4\pi$.

Now the probability of failure due to a crack in the critical stress range $d\sigma_c$ located in volume element ΔV is the product of the above probabilities, i.e.,

$$\left(\Delta P_f\right)_i = \left(\Delta V \frac{dN(\sigma_c)}{d\sigma_c}\, d\sigma_c\right)\left(\frac{\Omega(\Sigma,\sigma_c)}{4\pi}\right) \quad . \tag{7}$$

Substituting Eq. (2) into Eq. (1), and changing sums into integrals, we obtain

$$P_S = \exp\left[-\int dV \int d\sigma_c \frac{dN}{d\sigma_c}\frac{\Omega}{4\pi}\right] \quad . \tag{8}$$

We note in passing that $N(\sigma_c)$ is independent of stress state and depends only on the material. Also, since σ_c is defined as the stress which causes fracture when applied normal to the crack plane, $N(\sigma_c)$ can be converted directly into crack size distribution when K_{Ic} is known. In these respects it differs from Weibull's n and the g(S) of McClintock and others [9,10,11]. These functions represent the number of cracks per unit volume that will be fractured by a particular applied stress state, and they depend on the stress ratio. More precisely, they are essentially equivalent to the second integral in Eq. (8). The function g(S) is the same as N for the particular case of hydrostatic tension, because then Ω is 4π or zero depending on the stress level.

In simple tension and equibiaxial tension, analytical expressions can be found for Ω for some fracture criteria, at least, and used together with Eq. (3) to evaluate P_S. In the general case, we find Ω by integrating $d\Omega$ over the range in which $\sigma_e > \sigma_c$. One way of accomplishing this is to integrate over the entire angular range but include a suitable operator H in the integral:

$$P_S = \exp\left[-\iiint dV\, d\Omega\, d\sigma_c\, H(\sigma_e,\sigma_c)\, \frac{dN}{d\sigma_c}\right] \quad , \tag{9}$$

where

$$H(\sigma_e,\sigma_c) = 1 \text{ when } \sigma_e > \sigma_c$$
$$= 0 \text{ where } \sigma_e < \sigma_c \quad . \tag{10}$$

We can now carry out the integral over σ_c first, with the result

$$P_S = \exp\left[-\iint dV\, d\Omega\, N(\sigma_e)\right] \quad . \tag{11}$$

The effective stress causing fracture is a function of both the component of stress normal to the crack plane, σ_n, and the shear stress τ parallel to the crack plane. An approximation frequently employed is the assumption that the cracks are shear-insensitive, i.e., that $\sigma_e = \sigma_n$. Using this approximation, Eq. (11) becomes equivalent to Weibull's rule for polyaxial stress states [1], and Eq. (8) reduces to the equation of Batdorf and Crose [12]. Eq. (8) and Eq. (11) are equivalent formulations of the same theory. The physical justification of the former is the more readily apparent, but the latter is more convenient for computational purposes.

RELATION TO GRIFFITH THEORY

In 1924, Griffith [13] published a theory for the fracture of solids under biaxial stress conditions. He assumed the presence of a large number of identical cavities in the form of nearly flat elliptical cylinders (Griffith cracks) with their axes normal to the plane of the stresses, but otherwise randomly oriented. He calculated the maximum peripheral tensile stress on the free surface of such a crack as a function of crack orientation, and assumed fracture would occur whenever this maximum value exceeded the intrinsic strength of the material.

This fracture criterion led to the failure envelopes shown in Fig. 1. In the tension-tension quadrant and most of the tension-compression quadrant, failure obeys a maximum tensile stress criterion. Fracture initiates in a crack whose plane is normal to the largest principal tensile stress. Thus, only the normal stress σ_n acts on the crack plane, and tensile-type fractures result. In the portion of the tension-compression quadrant near the compression axis, material strength is locally exceeded when the crack normal is inclined at somewhat more than 45° to the axis of maximum

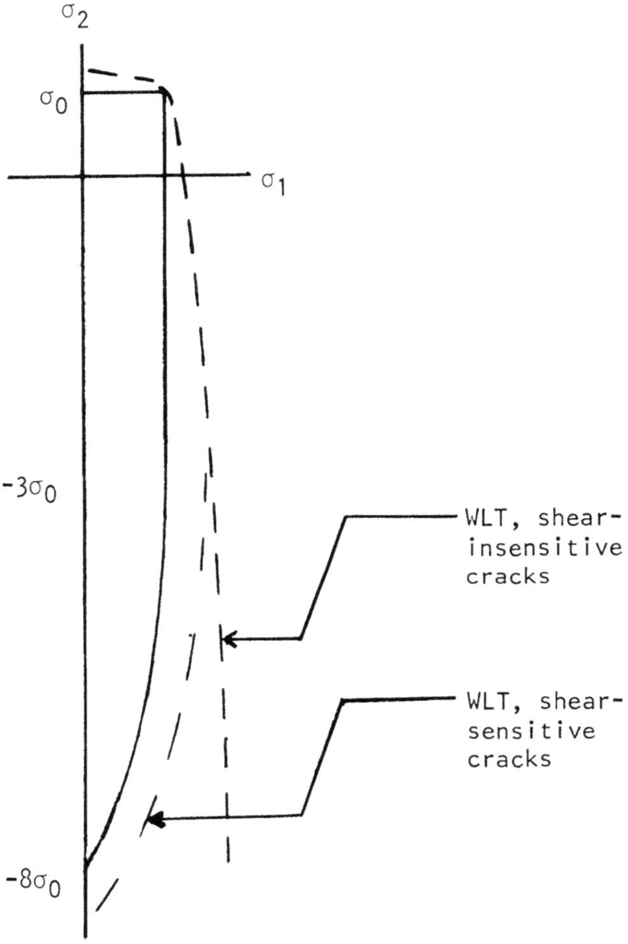

Fig. 1. Comparison of Fracture Theories for Biaxial Stresses

tension. Here both σ_n and τ contribute to failure and the resulting oblique fracture is often called a shear-type failure.

In any crack-based statistical theory it is most unlikely that the weakest crack will be oriented at exactly the angle of maximum vulnerability, so Griffith's curves must be regarded as the lower limit for fracture, i.e., they represent the contour for $P_f(\sigma_1,\sigma_2) <<< 1$. A typical $P_f(\sigma_1,\sigma_2) = 0.5$ contour, for the case of shear-insensitive cracks, is shown as a dashed curve. This curve covers both quadrants almost completely, but does not provide for failure due to pure compression. In the case of shear-sensitive cracks, the entire $P_f(\sigma_1,\sigma_2) = 0.5$ contour predicted by WLT will fall slightly outside of Griffith's failure envelope as shown schematically in Fig. 1. For high compressive stresses, the results differ significantly from those for shear-insensitive cracks. However, as noted earlier, in this region WLT does not apply. Thus, Weibull theory and the more refined theories to be discussed later are limited to cases in which no principal compressive stress exceeds the maximum principal tensile stress in absolute value by more than a factor of about 3. This limitation is probably not too serious from a practical point of view because the percent dispersion in fracture stress is generally much less in compression than in tension [14,15].

CRACK DENSITY FUNCTIONS

The derivative of the crack density function, $N'(\sigma_c)$, gives the number of cracks per unit volume per unit range in critical stress. One might at first be inclined to expect this function to be Gaussian or near-Gaussian. In actuality it turns out that Gaussian distributions are incompatible with WLT. This is readily demonstrated as follows.

We have seen that any WLT must take the form

$$P_S(\sigma) = - \exp\left[- f(\sigma)\right] \quad , \tag{12}$$

whence

$$P_f'(\sigma) = - P_S'(\sigma) = f'(\sigma) \exp\left[- f(\sigma)\right] \quad . \tag{13}$$

If the distribution is Gaussian,

$$P_f'(\sigma) = A \exp\left[- a\,(\sigma - b)^2\right] \quad . \tag{14}$$

Eq. (13) and Eq. (14) are incompatible, because if $f'(\sigma)$ is a constant, $f(\sigma)$ cannot be a quadratic function. Another reason for excluding the Gaussian distribution is that it implies a nonvanishing probability of tensile failure when the applied stress is compression.

Weibull introduced two distribution functions

$$P_f(\sigma) = 1 - \exp\left[- V k\sigma^m\right] \tag{15}$$

and

$$P_f(\sigma) = 1 - \exp\left[- Vk (\sigma - \sigma_u)^m\right] \quad (\sigma > \sigma_u) \tag{16a}$$

$$= 0 \qquad (\sigma < \sigma_u) \quad . \tag{16b}$$

Each may be regarded as a skewed Gaussian distribution with a skewness, which can be either positive or negative, determined by the value of m. The 2-parameter form allows failure to occur at any positive value of the tensile stress, while the 3-parameter form implies that fracture cannot occur for $\sigma < \sigma_u$.

The function $P_f(\sigma)$ is conventionally determined by conducting N tests and numbering the observed fracture stresses $\sigma_1 \ldots \sigma_N$ in ascending order. It is then usually assumed that

$$P_f(\sigma_j) = \frac{j}{N + 1} \quad . \tag{17}$$

A more sophisticated statistical treatment [16] leads to the conclusion that

$$P_f(\sigma_j) = \frac{j - 0.3}{N + 0.4} \quad . \tag{18}$$

The difference between Eq. (17) and Eq. (18) becomes insignificant for large values of N and for simplicity we shall use Eq. (17).

A simple technique for determining the parameters m, σ_u, and k is to write Eq. (16a) in the form

$$\ell n\, \ell n\, (1 - P_f)^{-1} = \ell n\, Vk + m\, \ell n\, (\sigma - \sigma_u) \quad . \tag{19}$$

Next, $\ell n\, \ell n\, (1 - P_f)^{-1}$ is plotted against $\ell n\, (\sigma - \sigma_u)$ for various assumed values of σ_u. The value of σ_u adopted is that for which the N data points most nearly lie on a straight line. The slope of the line chosen is m, and Vk is the value of $\ln (1 - P_f)^{-1}$ for $\sigma = \sigma_u + 1$. When Weibull's 2-parameter form is used, σ_u is arbitrarily taken to be zero.

Before passing on, we note that a least squares fit in $\ln \ln (1 - P_f)^{-1}$ vs. $\ln (\sigma - \sigma_u)$ space is in general not a least squares fit in P_f vs. σ space. Consequently, the Weibull form found in the manner described is not the best possible fit to the data. However, if the scatter in the data is not large, it will be adequate for most practical purposes.

Fracture statistics are relatively simple when Weibull's 2-parameter form is used. If the probability of failure in simple tension is

$$P_f = 1 - \exp (- V k_T \sigma^m) \quad , \tag{20}$$

then that in bending can be shown to be [1]

$$P_f = 1 - \exp [- V k_B \sigma^m)] \quad , \tag{21}$$

where k_B/k_T depends on the cross section. For a rectangular cross section

$$k_B = k_T/(2m + 2) \quad . \tag{22}$$

Under the uniform principal stresses σ_1, σ_2, σ_3, the probability of failure is given by

$$P_f = 1 - \exp \left[- V k \left(\frac{\sigma_2}{\sigma_1}, \frac{\sigma_3}{\sigma_1} \right) \sigma_1{}^m \right] \quad . \tag{23}$$

The constant $k (\sigma_2/\sigma_1, \sigma_3/\sigma_1)$ is an analytical function of the stress ratios [17]. Many other cases are treated in [18].

Weibull's 3-parameter representation has a similar universal applicability when attention is limited to uniaxial stress problems. In going from uniaxial to polyaxial stress states, however, the functional form changes [19], i.e. it is not possible to express both uniaxial and polyaxial stress data in the form $P_f = 1 - \exp [- V k (\sigma - \sigma_u)^m]$. To assist those desiring to use Weibull's 3-parameter representation in conjunction with his theory for polyaxial stress states, Dukes [20] has carried out parametric calculations with the aid of a computer.

Weakest link theory implies a volume effect that is sometimes misapplied. Consider, for instance, the situation illustrated schematically in Fig. 2. Let us assume specimens were tested, so we know $P_f(\sigma)$ over the range $0.1 \lesssim P_f \lesssim 0.9$. The known region is shown as a solid curve, and the dotted extension is an extrapolation of the data, accomplished by assuming that the mathematical function used to represent the test data continues to be valid outside the range in which it was tested. All that WLT can tell us is that a specimen 10 times larger will have 10 times the risk of rupture over the stress range tested. It does not answer the generally more interesting question, what stresses correspond to

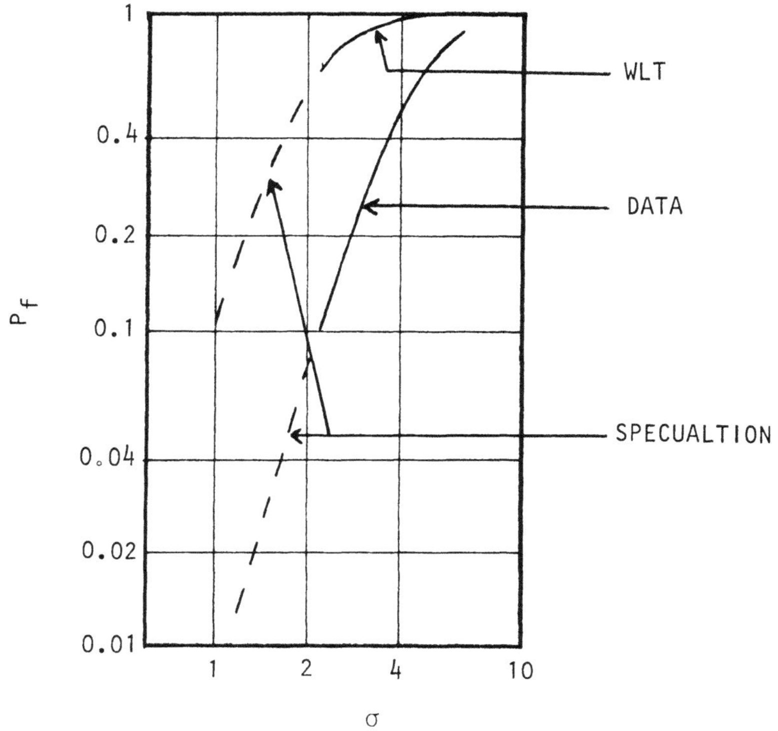

Fig. 2. Data, WLT inference concerning volume effect, and speculative extrapolations.

probabilities of failure of the larger specimen in the probability range tested, i.e., $0.1 \lesssim P_f \lesssim 0.9$, unless the validity of the extrapolation to lower stress levels can be established.

This raises the question whether the Weibull distribution functions have a fundamental significance that might justify confidence in such extrapolation. Weibull made no such claim; he regarded them only as convenient mathematical tools having wide applicability [21]. It is true that they coincide with the third asymptotic form of extreme value theory. However, the founders of this theory did not assert that all distributions must approach a limiting form, but only that if a distribution does approach a limiting form, it must be one of the three identified in the theory of extreme values [22,23]. Basically, the limiting form results when a distribution can be represented as a power series. When this happens, only the leading term need be considered for large samples (i.e. large specimens with very many cracks). We shall show later, however, that microstructural considerations suggest that such a power series is not possible, and that as a result no limiting form exists for fracture.

FRACTURE CRITERIA

The direct stress applied normal to a crack plane, σ_n, results in a very high local stress at the root of the crack, while direct stresses in the plane of the crack do not. The shear stress τ applied parallel to the crack plane also results in very high local stresses. Consequently, the effective stress σ_e causing fracture is a function of both σ_n and τ.

The statistical analysis of fracture is appreciably simplified, however, by the assumption that $\sigma_e = \sigma_n$, i.e. that the cracks are shear-insensitive. In this case the properties of the cracks are completely characterized by σ_c, the remote stress normal to the crack plane that will cause fracture. There is then no need to specify the size or shape of the crack or Poisson's ratio. And because any given crack is weakest when oriented normal to the largest principal tensile stress, as pointed out in the section "Relation to Griffith Theory," the approximation $\sigma_e = \sigma_n$ is not a bad one. This approximation was explicitly made by Batdorf and Crose [24], and is implicit in Weibull's treatment of polyaxial stress states [1].

The fracture criterion used by Griffith to take into account the effects of shear is the assumption that fracture occurs when the local tensile stress at some point on the crack surface exceeds the intrinsic strength of the material. Many authors have discussed the stress distribution around cavities of various types under various loading conditions. For present purposes, the most concise and general treatment is probably that of Mirandy and Paul [25] who give an explicit formula for the maximum stress on the surface of an ellipsoid with semiaxes a > b >> c under arbitrary applied loads. They find that

$$\sigma_{max} = \frac{b}{c}\,\frac{1}{E}\left\{\sigma_n + \sqrt{\sigma_n^2 + \tau^2 F}\right\} \quad , \tag{24}$$

where E is an elliptic integral that depends on b/a and F is a complicated function of the geometry of the crack and its orientation relative to the shear stress. We have defined the effective stress to be σ_n in the absence of shear, and some function of σ_n and τ when shear is present. Accordingly, we conclude that for a material containing cracks of only a single plan form (i.e. a fixed value of b/a)

$$\sigma_e = \frac{1}{2}\left\{\sigma_n + \sqrt{\sigma_n^2 + \tau^2 F}\right\} \quad . \tag{25}$$

Using the formulas in [25], it can be shown that for a Griffith crack (G.C.) with its axis normal to the applied shear stress

$$\sigma_e = \frac{1}{2}\left\{\sigma_n + \sqrt{\sigma_n^2 + \tau^2}\right\} \quad \text{(G.C.)} \quad , \tag{26}$$

while for a penny-shaped crack (P.S.C.)

$$\sigma_e = \frac{1}{2}\left\{\sigma_n + \sqrt{\sigma_n^2 + \tau^2/(1 - 0.5\nu)^2}\right\} \text{(P.S.C.)} \quad , \tag{27}$$

where ν is Poisson's ratio. The latter, possibly less familiar result, is also given in [26].

There are, however, grounds for doubting that maximum local tensile stress represents an acceptable fracture criterion. Consider, for instance, an ellipsoidal cavity with principal axes a = 4b loaded in tension parallel to the c axis. The maximum local stress occurs in the equatorial plane, and it is uniform around the entire circumference. Thus, according to the maximum local tensile stress criterion, all points on the equatorial belt are equally likely to fracture and the crack should advance in all directions in its own plane. On the other hand, the region which is above some fixed fraction of the maximum stress is four times as wide at the ends of the b axis as at the ends of the a axis. Thus, if stressed volume is important, growth should start first at the ends of the b axis. Similarly, the stress intensity factor is largest at the ends of the b axis, where it is twice as large as at the ends of the a axis. If stress intensity governs fracture, the b axis should extend first, and the crack should circularize. It is generally agreed that this is what actually happens.

Another fracture criterion is the assumption that fracture occurs when the elastic strain energy released when the crack grows is equal to the energy stored in the newly created free surfaces, i.e. when the strain energy release rate reaches some critical value. The strain energy release rates are well-established for a Griffith crack or penny-shaped crack growing in its own plane. For the former [27]

$$G = \frac{1 - \nu^2}{E} \left[K_I^2 + K_{II}^2\right] , \tag{28}$$

where

E = elastic modulus

$$K_I = \sigma \sqrt{\pi a} \tag{29}$$

$$K_{II} = \tau \sqrt{\pi a} . \tag{30}$$

Thus,

$$\sigma_e = \sqrt{\sigma_n^2 + \tau^2} \quad (G.C.) . \tag{31}$$

For the latter [28]

$$G = \frac{(1 - \nu^2)}{E} \left[K_I^2 + K_{II}^2 + \left(\frac{1}{1 - \nu^2}\right) K_{III}^2\right] , \tag{32}$$

where

$$K_I = 2\sigma_n \sqrt{a/\pi} \tag{33}$$

$$K_{II} = 4 \tau \sqrt{a/\pi} \sin \gamma/(2 - \nu) \tag{34}$$

$$K_{III} = 4 \tau \sqrt{a/\pi} \cos \gamma (1 - \nu)/(2 - \nu) . \tag{35}$$

The minimum value of G occurs at the points on the crack periphery where only modes I and II are involved, and here

$$\sigma_e = \sqrt{\sigma_n^2 + \tau^2/(1 - 0.5\nu)^2} \quad \text{(P.S.C.)} \quad . \tag{36}$$

Unfortunately, under the combined action of σ_n and τ, cracks do not extend in their own plane so the above expressions are not strictly correct. There is at present no consensus regarding the proper fracture criterion for cracks extending out of their plane. Swedlow [29] listed nearly thirty differing treatments of this subject that had appeared by 1975, and others have been proposed since that time. We will not attempt to employ any of these theories in this review. Instead, we will use the equations just derived noting, however, that the true value of σ_e must be smaller than that found for coplanar extension. That is because the lowest instability mode is the mode that actually occurs.

Some idea of the differences between the various fracture criteria discussed herein can be obtained by assuming the material obeys the Weibull 2-parameter form [Eq. (15)], and plotting the ratio of the risk of rupture under equibiaxial tension to that for uniaxial tension vs. the parameter m. This is done in Fig. 3 which is adapted from [17] and [30]. The ratio is largest for shear insensitive cracks. The remaining curves are for the criteria in Eqs. (26), (27), (31), and (36) respectively. We note that the ratio $k_{eq\ bias}/k_{uniax}$ increases with m for all criteria, and is smallest for the fracture criteria based on strain energy release rate. Also included in the comparison is the assumption that with respect to fracture, the principal stresses behave independently. This leads to a constant ratio of 2. The comparison suggests that for $m \gtrsim 3$ the independence assumption will lead to unconservative estimates of the statistics of failure in biaxial tension. We note in passing that for the low values of P_f desired in most structural applications, the failure probability ratio is the same as the k ratio.

Experimental data on uniaxial and equibiaxial bending of alumina plates were recently obtained by Sines and Giovan [30]. In Fig. 4 the equibiaxial data are approximated by a Weibull 2-parameter curve. The uniaxial results were computed for shear-insensitive cracks and also for the fracture criteria listed in Eq. (27) and Eq. (36). As would be anticipated on the basis of the preceding discussion, the agreement is best using Eq. (36). For details of the analysis, see [30].

The relative merits of various fracture criteria can also be tested in other ways. For instance, Petrovic and Mendiratta [32] tested the variation of tensile fracture stress of controlled surface cracks with crack angle. The conclusion from this experiment is that Eqs. (31) and (36) are in reasonably good agreement with experiment, and are the best of the criteria listed in this paper.

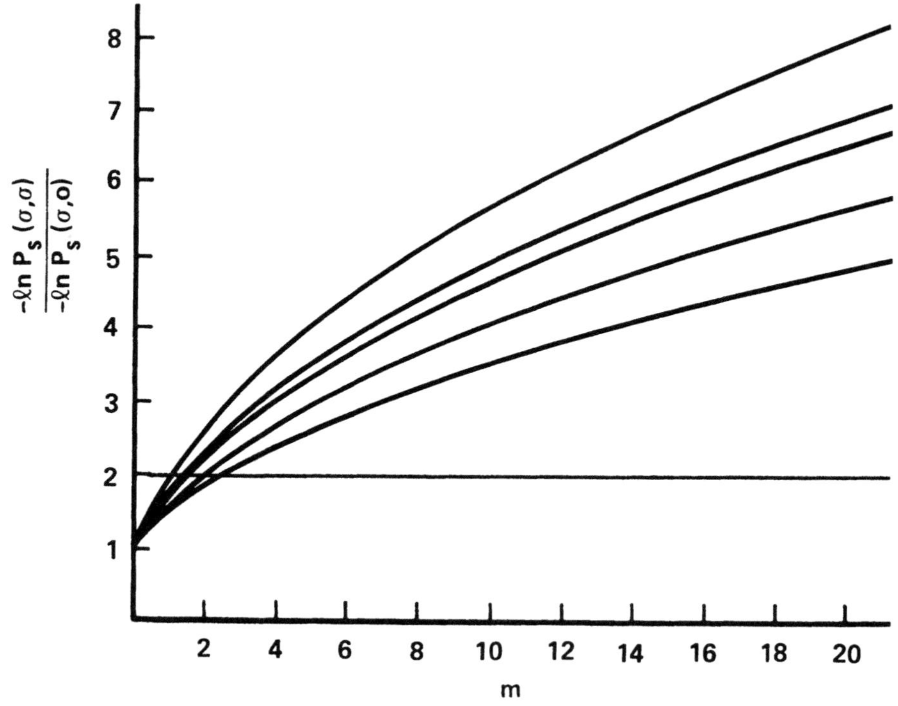

Fig. 3. Relation Between Failure Probability Under Equibiaxial Loading and That for Uniaxial Loading as a Function of Weibull Parameter m.

WLT FOR ANISOTROPIC MATERIALS

Anisotropic brittle materials may be anisotropic with respect to their elastic properties, their fracture properties, or both. The elastic anisotropy will influence the stresses resulting from a given load application, but will not directly affect the calculation of probability of failure, which depends only on the stress state and crack distribution.

Two different approximate techniques have been proposed for determining the fracture statistics of certain types of polygraphite that are isotropic in one plane but have a lower fracture stress in the direction normal to the plane of isotropy. In one approach, the cracks were assumed randomly oriented, but they were given critical stresses that varied with orientation [31]. In the other, the critical stresses were assumed unaffected by orientation, but cracks were given a preferred orientation [32]. It is likely that there is actually both a preferred orientation and a variation

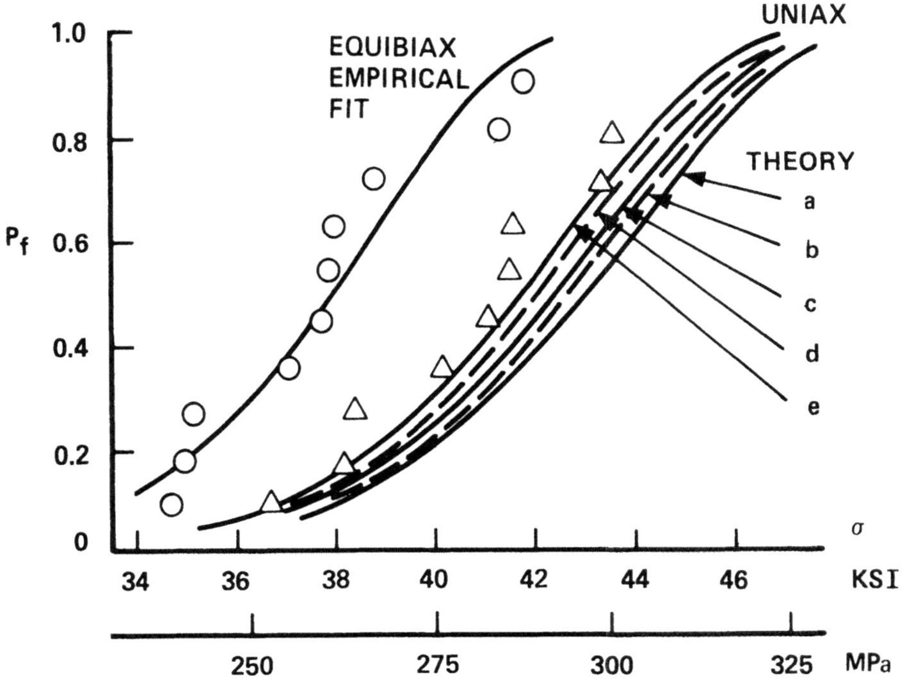

Fig. 4. Theory Compared with Data of Giovan and Sines

in crack strength with orientation, but a theory incorporating both would be considerably more complex and not really needed, at least for graphite. Both theories that attribute the entire anisotropy to only one effect appear to be in satisfactory agreement with the available data.

WLT FOR SURFACE DISTRIBUTED CRACKS

In his first paper on the statistics of fracture, Weibull included a discussion of surface-distributed flaws [1]. In this treatment, the only change was to replace $B = \int n_1(\sigma)\, dV$ by $B = \int n_2(\sigma)\, dA$ where A represents area, and n_2 might or might not be the same as n_1. Thus the analogy to the theory for volume distribution flaws is very close.

In a more recent analysis the fracture statistics of surface distributed cracks have been worked out on the assumption that crack planes are always normal to the material surface [35]. As a result, the orientation of a crack is specified by a single angle, whereas for volume distributed cracks two angles are required. The analogue to Eq. (8) applicable to surface distributed cracks of the type under discussion is

$$P_s = \exp\left[-\iint dA\, d\sigma_c \frac{\omega}{\pi} \frac{dN}{d\sigma_c}\right] \tag{37}$$

A rather surprising relation exists between the fracture statistics for volume distributed cracks and the fracture statistics for surface distributed cracks. Let us assume that uniaxial data are used to determine the fracture statistics for specimens in uniform biaxial tension. It has been shown [36] that the result is the same for both theories, i.e., it does not matter whether we use Eq. (8) or Eq. (37). This is quite a convenience for those who like to do their own calculating. Finding the biaxial stress statistics for Weibull's 3-parameter form using Eq. (8) generally requires the use of a large digital computer. But using Eq. (37) the solution can readily be found using a simple programmable hand computer such as the HP-25 or SR-56.

There are materials having surface distributed cracks but no interior cracks, e.g. glass, but probably no materials with volume distributed cracks, but no surface cracks. It is known from fracture mechanics that a crack near the surface is generally weaker than an identical crack far from the surface. This complicates the interpretation of laboratory data on small specimens and the application of the results to larger specimens. It becomes necessary to separate out the surface and volume contributions to the fracture statistics. An analysis of this type has recently been carried out by Rebholz and Teter [37].

MICROSTRUCTURAL CONSIDERATIONS

So far our discussion of fracture statistics has been based on a combination of extreme value theory and fracture mechanics. The crack density function had to be inferred from test data, and it was pointed out that extrapolation of this function to stresses outside the region bounded by the highest and lowest observed fracture stresses is unwarranted. Thus without additional sources of information it is not possible to obtain reliable estimates of the fracture behavior of very much larger specimens or the stresses at which extremely low probabilities of failure can be assured.

To resolve this difficulty, McClintock [23] recently proposed a theory for the crack density function based on microstructural considerations. He assumed that cracks are random aggregations of imperfectly bonded pairs of grains. If the probability that two adjacent grains are unbonded is q, the probability that two such unbonded pairs are side by side is q^2, etc. In this manner the statistics of crack size can be found, and using the value of the critical stress intensity factor for the material this can be converted into statistics of critical stress. From this in turn one can obtain the statistics of failure. The material model employed by McClintock was a 2-dimensional one in which all grains were identical rectangles and all cracks were normal to the applied tensile stress. It was shown that the probability of a crack being longer than a is exp $[-a/\lambda]$, and that the resulting fracture statistics do not approach an asymptotic form for large specimens.

This theory was similar in many ways to one published earlier by Fisher and Holloman [38]. These authors considered randomly oriented, penny-shaped cracks in a 3-dimensional elastic solid. They made the judgmentally-based assumption that the probability of a crack having a radius larger than r is exp $[-r/\lambda]$. Using this size distribution they used fracture mechanics to obtain a critical stress distribution. Employing the assumption $\sigma_e = \sigma_n$, they then worked out the fracture statistics. The relation of their work to Weibull's and other WLT's escaped general attention because they retained the binomial form in their calculations rather than converting to exponentials [see Eq. (1)].

Batdorf [39] applied the McClintock concept of the origin of cracks to a consideration of randomly oriented penny-shaped cracks in a 3-dimensional polycrystal. He found that the probability of a crack having a radius greater than r is exp $[-r^2/\lambda^2]$. The fracture criterion employed was that expressed in Eq. (31).

Among the principal findings of this theory are the following:

(1) The total number of cracks is finite rather than infinite as in Weibull theory. The most probable critical stress is $s_c \simeq 1$ where s is a reduced stress to be discussed later. Weibull's 2- and 3-parameter forms imply that the number of cracks with a given fracture stress increases monotonically with stress; thus the most probable fracture stress is infinity.

(2) The probability of failure in simple tension is given by

$$P_f(s) = 1 - \exp\left[-N_oV \int_o^s \left(1 - \frac{s_c}{s}\right) \frac{4}{s_c^5} \exp\left(-1/s_c^4\right) d\, s_c\right] \qquad (38)$$

where N_oV is the total number of cracks. For a sufficiently small probability of failure

$$P_f(s) \cong (NV_o/4)s^4 \exp(-1/s^4) \tag{39}$$

(3) Examining Eq. (39) we see that $P_f(s) = 0$ for $s = 0$, and $P_f(s) > 0$ when $s > 0$. However, $P_f(s)$ cannot be expressed as a power series in s because at $s = 0$, P_f and all its derivatives vanish. It turns out that over a finite range of P_f such as $0.01 \leq P_f \leq 0.99$, the prediction of this theory can be fitted very closely using Weibull's 3-parameter form. However, m and σ_u are both very weak functions of the volume instead of being volume independent as in Weibull theory. There is some experimental confirmation of the predicted direction of change of Weibull parameters with volume change [39].

(4) The reduced stress is related to the actual stress through the equation

$$s = 2\sigma A_o^{0.25}/K_{Ic}\,\pi^3 \left[\ln(q^{-1})\right]^{0.25} \tag{40}$$

where

A_o is grain cross sectional area

K_{Ic} is the critical stress infinity factor for mode I

q is the probability that adjacent grains will be unbonded.

Knowledge of how the fracture statistics depend on grain size, effectiveness of intergranular bonding, and critical stress intensity factor should be very helpful in determining what changes in processing variables will effect desired improvements in material properties. Results qualitatively similar to those just discussed are obtained with McClintock's theory and that of Fisher and Holloman, except that the latter considered only a structureless elastic solid, and therefore contained no analogue to Eq. (40).

More recently McClintock has refined his 2-dimensional model by using regular hexagons for grains and assuming grain boundaries have a Weibull-type strength distribution [40]. Such a model assumes no cracks are present initially, but as the load increases, cracks are created, grow, coalesce, and eventually cause the load capacity of the specimen to pass its peak and decline.

Another theory in which cracks are created by the loading process has been devised for polygraphites [41]. This theory is based on a consideration of the low tensile strength of a graphite crystal in the c-direction. The grains are assumed to be randomly oriented, so that here and there a chance aggregation of grains of nearly the same orientation are arrayed in the shape of a penny. Such an array opens up and acts like a crack when the applied stress exceeds the capacity of all of its constituent grains.

Two-dimensional models are generally of qualitative rather than quantitative value, but they help guide our thinking. Three-dimensional models are obviously preferable in principle, but their complexity requires greater development time and effort. At present, simplifying assumptions are used whose influence on the accuracy of predictions is somewhat uncertain. However, statistical theories of fracture incorporating extreme value theory, fracture mechanics and microstructural considerations are as yet in their infancy. The present writer believes that such theories offer the greatest long range promise for future progress.

STATIC FATIGUE

Up to this point we have discussed primarily idealized short term brittle fracture in which preexisting cracks are unaffected by increasing stress until their strength is exceeded, at which time they expand suddenly and fracture the specimen or structural part. Some attention was devoted to situations in which cracks are created and grow as the stress increases. We now consider fracture resulting from subcritical crack growth — slow growth of cracks with passage of time, even at constant stress. Whereas the objective in short-term fracture is to predict the probability of failure in a given stress state, in static fatigue it is to predict the time to reach a given probability of failure when the stress state is specified.

It has been determined by a number of investigators conducting mode I crack propagation studies that for a given system [material, temperature, environment, etc.] there is a unique relationship between crack velocity and the crack tip intensity factor K_I [42,43]. This is usually expressed in the form

$$v = A\, K_I^n \tag{41}$$

If a is the crack length

$$v = da/dt \tag{42}$$

and

$$K_I = \sigma Y \sqrt{a} \tag{43}$$

where Y is a geometrical factor. Combining these relations with Eq. (41) it is readily shown [44] that the time to failure is given by

$$T = 2\left[K_{Ii}^{2-n} - K_{Ic}^{2-n}\right] \left[(n - 2)\,\sigma^2\, AY^2\right] \qquad (44)$$

where K_{Ii} is the initial value of K_I.

The problem now is how to go from a laboratory situation with an artificial crack of prescribed size growing in mode I to the general situation in service, in which an unknown distribution of randomly oriented natural cracks undergoes mixed mode growth in an arbitrary stress state. Such a goal has yet to be achieved. In fact it is beyond the state of the art in fracture mechanics, since it involves a knowledge of the growth rate and eventual critical stress of a non-planar crack. However, significant progress is being made.

Most theoretical work in this area depends on three basic assumptions or limitations (1) the stress state is uniform simple tension, (2) crack planes are assumed to be normal to the applied stress both for calculating crack size and for determining crack velocity, (3) the short-term fracture statistics are adequately described by Weibull's 2-parameter form.

Davidge, McLarin, and Tappin [44] have used these assumptions to develop strength-probability-time (SPT) relations. An SPT diagram for alumina is shown in Fig. 5. Such a diagram can be used to find the stress corresponding to an acceptable probability of failure during the design life of a structural element.

Evans and Wiederhorn [45] have shown how the statistics of failure are affected by prior proof testing. The minimum time to failure is found by noting that no crack longer than the critical length for the proof test can be present in any of the surviving specimens. They then use Weibull statistics and Eq. (43) to obtain the actual distribution of crack size in the surviving specimens. From this they solve for the probability of fracture as a function of stress and time under load. Results for soda-lime glass in water are shown in Fig. 6. The comparison between theory and experiment suggests that the former may be conservative by between a half and a whole order of magnitude. Discrepancies of this sort may be due in part to errors in determining material parameters and propagation constants. The manner in which these affect the statistics of fracture in static fatigue have recently been analyzed in some detail [46,47].

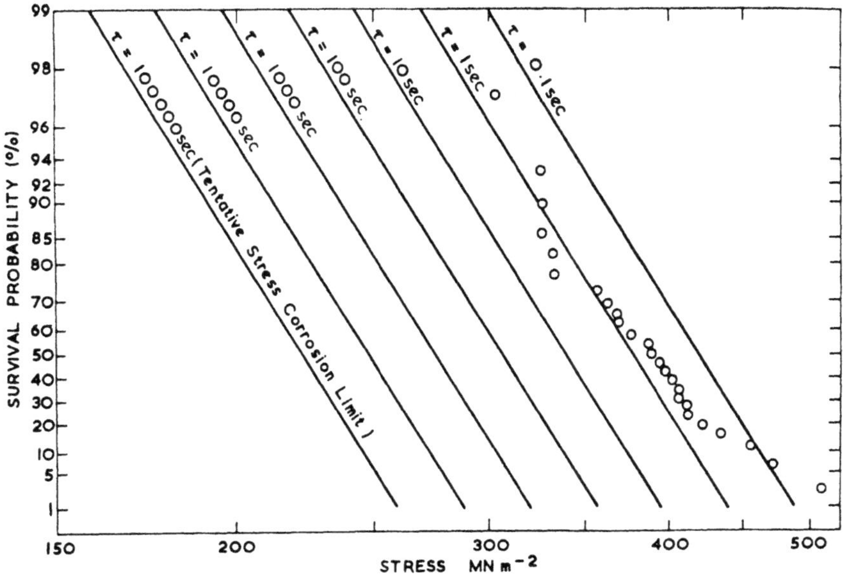

Fig. 5. SPT Diagram for Alumina [44]

DYNAMIC FRACTURE

In dynamic fracture studies, an intense stress pulse of very short duration passes through the material. Many cracks grow. After passage of the stress wave, the specimen may still be in one piece but in damaged condition, or it may be fragmented into a few or many pieces.

In treating internal crack damage, Curran, Shockey and Seaman [47,48] have developed a fracture model that takes into account three aspects of the fracture process not usually included in treatments of short-term fracture: (a) nucleation of cracks as a function of stress; (b) growth of cracks as a function of stress, time, fracture toughness, and initial crack size; (c) decrease in strength and stress attenuation with increasing damage.

The nucleation rate is assumed to have the form

$$\dot{N} = N_o \exp\left[\left(\sigma - \sigma_{no}\right)/\sigma_1\right] \tag{45}$$

where σ_{no} is the threshold stress for nucleation, and N_o and σ_1 constants. The cracks are assumed to be nucleated with a distribution of radii given by

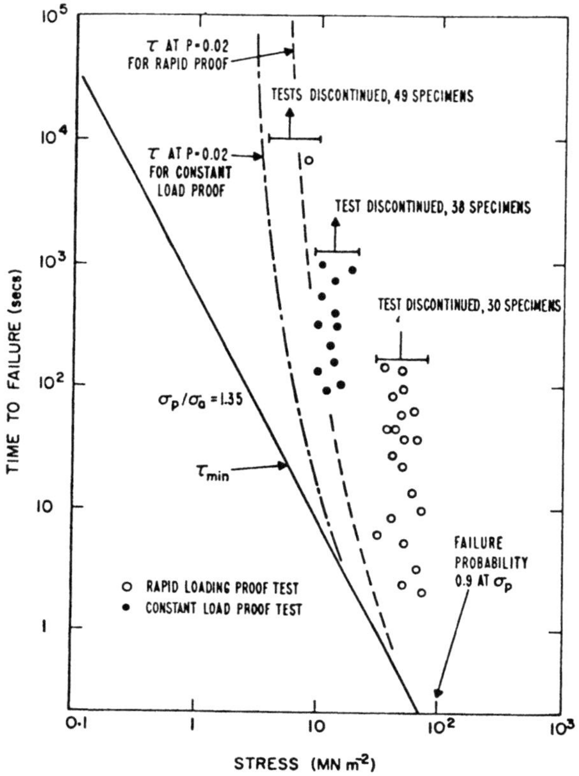

Fig. 6. A Comparison of Failure Times After Proof Testing [45]

$$\Delta N(r) = \Delta N_o \exp\left[- r/r_1\right] \tag{46}$$

where ΔN_o is the total number/cm^3 nucleated in the time interval t to t + Δt, ΔN is the number having a radius greater than r, and r_1 is a constant. The growth of cracks was tested in many ways, the most successful being the assumption that

$$\dot{r} = r\,(\sigma - \sigma_{go})/4\eta \tag{47}$$

where η is the crack tip viscosity. The threshold stress for crack growth is assumed to be given by

$$\sigma_{go} = 0.5\, K_{Ic}\, \sqrt{\pi/r} \tag{48}$$

where K_{Ic} is the plane strain fracture toughness.

As the stress wave progresses through a material, energy is extracted in creating and enlarging cracks, so the pulse is attenuated. The calculation of the number and size of cracks as a function of time and location, their effects on the stiffness and damping of the material, and the effects of these in turn on the stress pulse propagation is a strongly coupled and involved process requiring a computer code. Accordingly we must expect to be limited to numerical results for the particular cases chosen for investigation. However the number and size distributions of cracks in damaged specimens and fragment size in fragmented specimens are in reasonable agreement with experiment.

In the case of extremely short stress pulses the diffraction of the wave by cracks and delay times in the initiation of crack growth become important. These and other refinements in the theory have been discussed by Kalphoff and Seaman [50] and by Vardar and Finnie [51].

CONCLUDING DISCUSSION

It appears from the foregoing that Weibull's theory for uniaxial stress states is essentially correct as it stands, except for the limitations that it applies only to tensile type fractures, and that caution must be exercised in making predictions implying a knowledge of $n(\sigma)$ outside of the stress range in which it has been established by experiment. We have shown that Weibull's treatment of polyaxial stress states implies the assumption that cracks are shear insensitive, i.e., $\sigma_e = \sigma_n$. It is felt that there is sufficient evidence of the superiority of the fracture criterion of Eq. (31) to warrant its use in conjunction with either Eq. (8) or Eq. (11) for future treatments of polyaxial stress states, and abandonment of Weibull's treatment and the assumption of independence of principal stresses. More research is needed, however, in the areas of crack interaction and shear-type fractures, where weakest link theory does not apply.

Theories including due consideration of extreme value theory, fracture mechanics and material microstructure have only recently been introduced into the literature. Although complex, they offer the greatest long range promise and should be developed much more completely.

Much progress has recently been made in the statistical treatment of static fatigue, including the effects of proof testing. The accuracy of predictions is impaired somewhat, however, by the tacit assumption that the critical crack is normal to the applied stress. Also little has been done to analyze time to failure under polyaxial stress conditions.

In dynamic fracture, considerable success has been achieved in accounting for experimental data on crack damage and fragmentation in the case of materials that have been studied fairly thoroughly. There are so many material parameters, however, that optimizing their values to match theory to experiments involving all of them simultaneously may not result in reliable values for each parameter. This in turn makes transfer of the knowledge gained to untested materials difficult.

We conclude from all this that a lot of progress has been made in statistical theory of fracture, and that useful techniques exist to guide designers in their consideration of short-term, long duration, and dynamic fracture conditions. It is also evident that a lot of work remains to be done.

ACKNOWLEDGEMENT

The support of the writer's efforts in Fracture Statistics by the Office of Naval Research and the Air Force Office of Scientific Research is gratefully acknowledged. The technical monitor was Dr. Nicholas Perrone of ONR.

REFERENCES

1. W. Weibull, "A Statistical Theory of the Strength of Materials," Ingeniorsvetenskapakademiens. Handlinger Nr. 151, 1939.

2. W. Weibull, "The Phenomenon of Rupture in Solids," Ingeniorsvetenskapakademiens. Handlingar Nr. 153, 1939.

3. D.M. Shur, "Statistical Criteria of the Danger of Material Fracture in a Complex Stress State," Mashinovedenie, No. 1, 197, pp. 51-58.

4. R.A. Barnett et al, "Fracture of Brittle Materials Under Transient Mechanical and Thermal Loading," U.S. Air Force Flight Dynamics Laboratory, Report AFFDL-TR-66-220, March 1966.

5. A.M. Freudenthal, "Statistical Approach to Brittle Fracture," Chap 6 of Vol. II, Fracture, edited by H. Liebowitz, Academic Press, 1968.

6. J. Margetson, "A Statistical Theory of Brittle Failure for an Anisotropic Structure Subjected to a Multiaxial Stress State," AIAA Paper 76-632, AIAA/SAE 12th Propulsion Conf., Palo Alto, Ca., July 1976.

7. P. Stanley, A.D. Sivill, and H. Fessler, "Applications of the Four Function Weibull Equation to Design of Brittle Components," Proc. of Int. Symp. on Frac. Mech. of Ceramics, Univ. Park, Pa., July 27-29, 1977.

8. W. Weibull, "A Survey of Statistical Effects in the Field of Material Failure," Applied Mechanics Surveys, ed. by H.N. Abramson, H. Liebowitz, J.M. Crowley and S. Juhasz, Spartan Books, Washington, D.C., 1966.

9. F.A. McClintock and A. Argon, "Mechanical Behavior of Materials," Addison-Wesley Publishing Co., Reading, Mass., 1966.

10. J.R. Matthews, F.A. McClintock, and W.J. Shack, "Statistical Determination of Surface Flow Density in Brittle Materials," J. of Am. Ceramic Soc., Vol. 59, No. 7-8, July-Aug. 1976, p. 304-308.

11. A.G. Evans, "A General Approach for the Statistical Analysis of Multiaxial Fracture," to be published.

12. S.B. Batdorf and G.J. Crose, "A Statistical Theory for the Fracture of Brittle Structures Subjected to Nonuniform Polyaxial Stresses," J. Appl. Mech., Vol. 41, No. 2, 459-64, June, 1974.

13. A.A. Griffith, "The Theory of Rupture," Proc. 1st Intl. Cong. Appl. Mech., Delft, 55-63.

14. Marc Adams and G. Sines, "Determination of the Biaxial Compressive Strength of a Sintered Alumina Ceramic," J. Am. Ceramic Soc., Vol. 59, No. 7-8, July-Aug., 1976, p. 300-304.

15. M. Adams, "Strength of Brittle Ceramics in Compressive Stress States," Ph.D. Thesis, UCLA, Aug., 1975.

16. L.G. Johnson, "Theory + Techniques of Variation Research," Elsevier Pub. Co., Amsterdam, 1964.

17. S.B. Batdorf, "Some Approximate Treatments of Fracture Statistics for Polyaxial Tension," Int. J. of Fracture, Vol. 12, No. 1, Feb. 1977, pp. 5-11.

18. N.A. Weil and I.M. Daniel, J. Am. Cer. Soc., 47:6, 1964.

19. O. Vardar and I. Finnie, Int. J. of Fracture, 11, 1975, 495-508.

20. W.H. Dukes, Handbook of Brittle Material Design Technology, AGARDograph No. 152 (Dec., 1970).

21. W. Weibull, "A Statistical Distribution Function of Wide Applicability," J. App. Mech. 18, 293-297, 1951.

22. E.J. Gumbel, "Statistics of Extremes," Columbia Univ. Press, 1958.

23. F. A. McClintock, "Statistics of Brittle Fracture," in Vol. 1, Fracture Mechanics of Ceramics, Ed. by R.C. Bradt, D.P.H. Hasselman, and F.F. Lange, Plenum Press, 1973.

24. S.B. Batdorf and J.G. Crose, "A Statistical Theory for the Fracture of Brittle Structures Subjected to Nonuniform Polyaxial Stresses," J. App. Mech., Vol. 41, No. 2, 459-64, June, 1974.

25. B. Paul and L. Mirandy, "An Improved Fracture Criterion for Three Dimensional Stress States," Trans. ASME 98, 159, 1975.

26. H. Koide, "Fracture Initiation in Brittle Polycrystalline Material Such as Rock," Proc., Intl. Conf. on Mech. Beh. of Mtls, Vol. IV, Soc. of Materials Science, Kyoto, Japan, 455, 1972.

27. P.C. Paris and G.C. Sih, "Stress Analysis of Cracks," Am. Soc. Test Mat. Special Technical Report STP 381, p. 30- , 1965.

28. M.K. Kassir and G.C. Sih, J. Appl. Mech., 33, (1966), p. 601.

29. Swedlow, "Criteria for Growth of the Angled Crack," Am. Soc. Test. Mat. Publication, ASTM STP 601, 1976, p. 506-520.

30. S.B. Batdorf and H.L. Heinisch, Jr., "Weakest Link Theory Reformulated for Arbitrary Fracture Criterion," UCLA Rpt, ENG-7728, Ap. 1977.

31. M. Giovan and G. Sines, to be published.

32. J.J. Petrovic and M.G. Mendiratta, "Mixed Mode Fracture from Controlled Surface Flaws in Hot-Pressed Si_3N_4," J. Am. Cer. Soc., Vol. 59, No. 3-4, March-April 1976, p. 163-167.

33. S.B. Batdorf, "A Statistical Theory for the Fracture of Transversely Isotropic Brittle Materials of Moderate Anisotropy," Air Force Report No. SAMSO-TR-73-361, Nov. 1973.

34. J.G. Crose, J.D. Buch, E.Y. Robinson, "A Fracture Criterion for Anisotropic Graphites in Polyaxial Stress States," Proc. 12th Conf. on Carbon, 1975, p. 131.

35. S.B. Batdorf and H.L. Heinisch, Jr., "Fracture Statistics of Brittle Materials with Surface Cracks," UCLA Rpt. No. UCLA-ENG-7703, Jan. 1977.

36. S.B. Batdorf and D.J. Chang, "On the Relation Between the Fracture Statistics of Volume Distributed and Surface Distributed Cracks," UCLA Rpt. No. UCLA-ENG-7723, Feb. 1977.

37. M.J. Rebholz and R.D. Teter, "Thermal Structural Reliability Analysis of Reentry Vehicle Nosetips," Lockheed Missile and Space Co., Rpt LMSC-D436937, June, 1976.

38. J.C. Fisher and J.H. Holloman, "A Statistical Theory of Fracture," Trans. Am. Inst. Min. Met. Eng. 171, Aug. 1947, p. 546.

39. S.B. Batdorf, "Fracture Statistics of Brittle Materials with Intergranular Cracks," Nuclear Eng. and Design 35, (1975), 349-360.

40. F.A. McClintock and Howard J. Mayson, "Principal Stress Effects on Brittle Crack Statistics," ASME Appl. Mech. Div., Vol. 16, "Effects of Voids on Material Deformation."

41. J.D. Buch, "Mechanical Behavior Model for Graphites," Special Technical Publication 605, ASTM, 1976, p. 124-144.

42. S.M. Wiederhorn, J. Am. Ceram. Soc., 50 (1967) 407.

43. A.G. Evans, Intl. J. Frac., 9 (1973) 267.

44. R.W. Davidge, J.R. McLaren, G. Tappin, "Strength-Probability-Time (SPT) Relationships in Ceramics," J. Mat. Sci., 8 (1973) 1699-1705.

45. A.G. Evans and S.M. Wiederhorn, "Proof Testing of Ceramic Materials — an Analytical Basis for Failure Prediction," Int. J. Frac., Vol. 10, No. 3, Sept. 1974, 379-392.

46. S.M. Wiederhorn, E.R. Fuller, J. Mandel, and A.G. Evans, "An Error Analysis of Failure Prediction Technique Derived from Fracture Mechanics," J. Am. Ceram. Soc., 59 [9-10] 403-411 (1976).

47. Dawn F. Jacobs and J.E. Ritter, Jr., "Uncertainty in Minimum Lifetime Predictions," J. Am. Ceram. Soc., Vol. 59, No. 11-12, Nov.-Dec. 1976, 481-487.

48. D.R. Curran, D.A. Shockey, and L. Seaman, "Dynamic Fracture Criteria for a Polycarbonate," J. App. Phys. Vol. 44, No. 9, Sept. 1973, 4025-4038.

49. D.R. Curran, L. Seaman, and D.A. Shockey, "Dynamic Failure of Solids," Physics Today, Vol. 30, No. 1, Jan. 1977, p. 46 ff.

50. J.F. Kalphoff and D.A. Shockey, "Instability of Cracks Under Impulsive Loads," J. App. Physics Vol. 48 No. 3, 1977 p. 986-993.

51. O. Vardar and I. Finnie, "The Prediction of Fracture in Brittle Solids Subjected to Very Short Duration Tensile Stresses," to be published in Intl. J. Frac.

A GENERAL APPROACH FOR THE STATISTICAL ANALYSIS OF FRACTURE

A. G. Evans

Science Center, Rockwell International

Thousand Oaks, California 91360

ABSTRACT

A generalised approach for analysing fracture statistics under uniaxial or multiaxial states of stress is described. The approach does not require prior assumptions about the form of the strength distribution function, and allows multiaxial fracture probabilities to be derived from uniaxial data.

1.0 INTRODUCTION

The statistical analysis of fracture in brittle materials has generally been based on the Poisson postulates.[1] These postulates enable the probability $\delta\phi$ that a flaw, with a strength S or smaller, exists in a small volume element δV to be expressed in terms of $g(S)dS$, the number of flaws per unit volume with a strength between S and S+dS. Thereafter, if the flaws are assumed to be non-interacting and fracture occurs by the direct extension of pre-existing flaws, $\delta\phi$ can be equated to the probability of fracture of the element δV, at the stress S. The probability of fracture Φ of a body of volume V can then be determined from the product of the survival probabilities $(1 - \delta\phi)$ of each element. In the limit, $V/\delta V \to \infty$,[2,3]

$$1 - \Phi(S) = \exp\left[- \int_V dV \int_o^S g(S)dS\right] \quad (1)$$

The primary objective of a statistical analysis of fracture strength is to derive g(S) from data obtained using well-defined specimen geometries. Subsequently, the derived g(S) can be used to estimate the fracture probability of components fabricated from the same material. The initial approach to this problem, pioneered by Weibull,[4,5] was to assume a reasonable (asymptotic) functional form for g(S), then to integrate eqn. (1) to allow the distribution parameters to be derived from fracture probability data. However, real flaw strength distributions are not necessarily characterized by Weibull's distribution, and a less arbitrary approach to statistical analysis is preferred. Such an approach has recently been proposed by Matthews et al[6], who demonstrate that by introducing the sample stress distribution, equation (1) can be manipulated (in certain cases) to obtain g(S) in terms of $\Phi(S)$, without assuming _a priori_ that g(S) has a specific functional form. Matthews et al analyzed the one-dimensional uniaxial case,

$$S = S_m (1 - z/\ell)^k \tag{2}$$

(where S_m is the peak sample stress, ℓ is a length dimension in the z direction, and k is a constant), showed that for surface flaw control;[6]

$$g(S_m) = \frac{|k|}{2b\ell} [(1 + 1/k)\ \xi'(S_m) + S_m\ \xi''(S_m)] \tag{3}$$

where b is the sample dimension normal to ℓ, and

$$\xi(S_m) = -\ \ell n\ [1 - \Phi(S_m)] \tag{4}$$

However, the flaw strength distribution g(S) is stress state dependent (i.e., it differs for uniaxial, biaxial or triaxial tension). Stress state effects have previously been predicted by simply evaluating the stress in any orientation from the principal stresses, incorporating this into a specific g(S)dS function (the Weibull function[4]), and integrating over stress and volume to obtain the fracture probability.[7] This approach does not recognize the flaws as being unique entities operated on by the multiaxial stress, and does not, therefore, represent a fundamental way of treating the multiaxial effect. Recently Batdorf and Crose[8] have suggested a greatly superior approach that allows the flaws to be considered in a fundamental manner. However, their choice of a crack extension criterion - a critical normal stress criterion - is not the most appropriate,[9,10] and their approach (which invokes a Taylor series for g(S)dS to obtain the final result) is mathematically complex.

In this paper, a generalised approach for evaluating the stress state dependence of g(S) is presented, using a pertinent crack extension criterion; then expressions for obtaining g(S) from fracture data are derived, for the important uniaxial and biaxial specimen geometries used in ceramic strength testing. Finally, fracture data for silicon nitride and alumina are examined, to illustrate the approach and to demonstrate the insights that can be gained.

2.0 CRACK EXTENSION CRITERIA

Several criteria for crack extension under multiaxial states of stress have been proposed. These criteria generally concur that fracture occurs when some combination of the modes I, II and III stress intensity factors attains a critical value and that this critical value is a material dependent parameter. Since the fracture initiating flaws in brittle materials are irregularly shaped defects,[2] such as inclusions, pores, surface grains (often as sites for machining flaws) etc., the simplest flaw geometry that incorporates an adequate set of flaw variables is the penny-shaped flaw. A recent study of the angular dependence of fracture from semi-circular surface cracks[9,10] has shown that, for brittle materials, crack extension occurs when the maximum coplanar strain energy release rate G reaches a critical value, G_c. The coplanar strain energy release rate for plane strain conditions is simply,[11]

$$\frac{EG}{(1-\nu^2)} = K_I^2 + K_{II}^2 + \frac{1}{(1-\nu)} K_{III}^2 \qquad (5)$$

where E is Young's modulus. Hence, the crack extension criterion of present interest can be derived by obtaining the maximum strain energy release rate and equating this to the critical value G_c to obtain

$$\frac{\pi E G_c}{4(1-\nu^2)a} = \sigma_n^2 + \frac{\tau_m^2}{(2-\nu)^2} \qquad (6)$$

3.0 MULTIAXIAL FRACTURE STRESS RELATIONS

Flaw orientation effects on fracture can be most readily visualized by allowing all flaws with normals occurring in the angular range θ to θ + dθ to occupy a pair of small volume elements, dθ (Fig. 1a). Then, if the length of these elements is large enough that flaws contained within them fully sample the flaw size distribution, these elements can be used to construct volumes that represent both the size and orientation distributions of the flaws.

For randomly oriented flaws, there are equal numbers of flaws in each orientation θ and size range a to a+da. All volume elements are thus of equal size, and a spherical volume pertains (Fig. 1a).

Fracture under multiaxial states of stress can be evaluated by superimposing the principal stress axes onto the flaw orientation volume using the crack extension relation (eqn. 6) to obtain the fracture strength distribution for each element, and finally summing over all volume elements. The three basic tensile fracture condi-

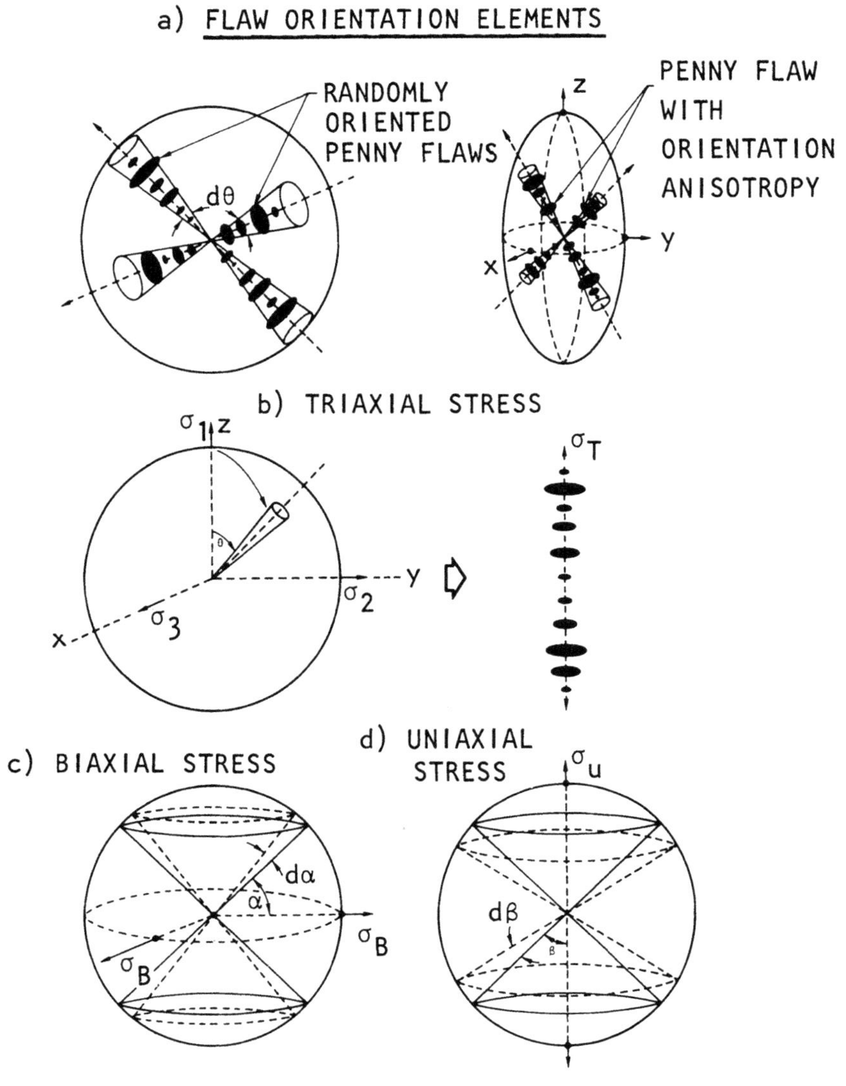

Fig. 1 The orientation relations used for developing the multiaxial theory.

tions which span the range of practical importance in the tension/tension sector (equi-triaxial, equi-biaxial and uniaxial) are considered in detail. For simplicity of presentation, the analysis is confined to randomly oriented flaws.

3.1 Equi-triaxial Fracture

For equi-triaxial tension, the principal stress axes can be positioned at any inclination. Hence, for all flaw orientations the applied stress σ_T can be considered to act in a direction normal to the crack plane (the in-plane shear stresses are then necessarily equal to zero). Thus, for randomly oriented flaws, the spherical volume of Fig. 1a can be reduced to a column (Fig. 2b), wherein all orientation effects have been relinquished. The equi-triaxial strength, S_T, is thus obtained from eqn. (6) as

$$S_T^2 = \frac{E\,G_c}{4(1-\nu^2)a} \tag{7}$$

The distribution of equi-triaxial strengths, $g(S_T)dS_T$ (equal to the number of flaws per unit volume with a strength between S_T and $S_T + dS_T$) is thus uniquely related to the distribution in flaw radii $g(a)da$ through the critical strain energy release rate G_c. It is therefore the most fundamental descriptor of the statistics of fracture in a material containing randomly oriented flaws. However, there are no existing techniques for measuring the strengths of brittle materials in the equi-triaxial condition, and this basic distribution must usually be inferred from other fracture data.

3.2 Equi-biaxial Tension

For equi-biaxial tension, the stress axes can be positioned at any inclination within the stressed plane. Thus, symmetry requires that the normal and shear stresses be constant anywhere within a sectorial element, inclined at an angle α to this plane (Fig. 1c), such that all of the cracks with normals lying within one of these elements experiences the same stress state. The stress S_B needed to propagate cracks located in these elements can be derived directly from eqn. (6), by substituting for $\sigma_n(\alpha, S_B)$ and $\tau_m(\alpha, S_B)$, to obtain;

$$S_B^2 = S_T^2/[\cos^2\alpha(\cos^2\alpha + \sin^2\alpha/(2-\nu)^2)] \tag{8}$$

The transformation from the triaxial to the biaxial distribution can now be achieved by considering the extension of flaws contained in the sectorial elements. The volume ΔV of each sectorial element

$\Delta\alpha$, in a sphere of unit volume, is;

$$\Delta V = \frac{\cos\alpha \ \Delta\alpha}{2} \tag{9}$$

The number of flaws per element with a strength between S_T and $S_T + \Delta S_T$ is thus

$$\left| g(S_T)\Delta S_T \right|_\alpha = \frac{\cos\alpha}{2} \Delta\alpha \ g(S_T)\Delta S_T \tag{10}$$

If the elements are small enough that the biaxial strengths of the flaws within the elements are uniquely related to their triaxial strengths (through eqn. (8)), the biaxial strength distribution for each element can be derived directly from the triaxial distribution. Thereafter, the complete biaxial distribution, $g(S)_B \Delta S_B$, can be obtained by summation over all elements at constant stress.

This process is most readily visualized by considering discrete angular elements. The biaxial frequency function for each element is obtained by a translation of each part of a triaxial strength histogram along both the density and strength axes, as illustrated in Fig. 2a. Then, restructuring the elemental biaxial strength histogram into histograms based on a uniform strength range ΔS_B (Fig. 2b), the complete biaxial distribution $g(S_B)\Delta S_B$ can be constructed by summation over all elemental histograms at each ΔS_B (Fig. 2b). The inverse transformation, from $g(S_B)\Delta S_B$ to $g(S_T)\Delta S_T$, can also be performed, by an iterative procedure involving the inverse logic.

A convenient analytic expression relating the biaxial and triaxial strengths is;[12]

$$g(S_B) = \int_0^{\pi/2} g[S_T(S_B,\alpha)] \left[\cos\alpha \sqrt{\cos^2\alpha + \sin^2\alpha/(2-\nu)^2}\right] \cos\alpha \ d\alpha \tag{11}$$

However, a functional relation for $g(S_T)$ is needed to complete the analysis. Applying the versatile power function;

$$g(S_T) = \left(\frac{S_T - S^*}{S_o}\right)^m \equiv \left(\frac{S_T}{S_o}\right)^m = (1 - S^*/S_T)^m \tag{12}$$

the biaxial distribution becomes;[12]

$$g(S_B) = \left(\frac{S_B}{S_o}\right)^m I_B \ (m, S^*/S_B, \nu) \tag{13}$$

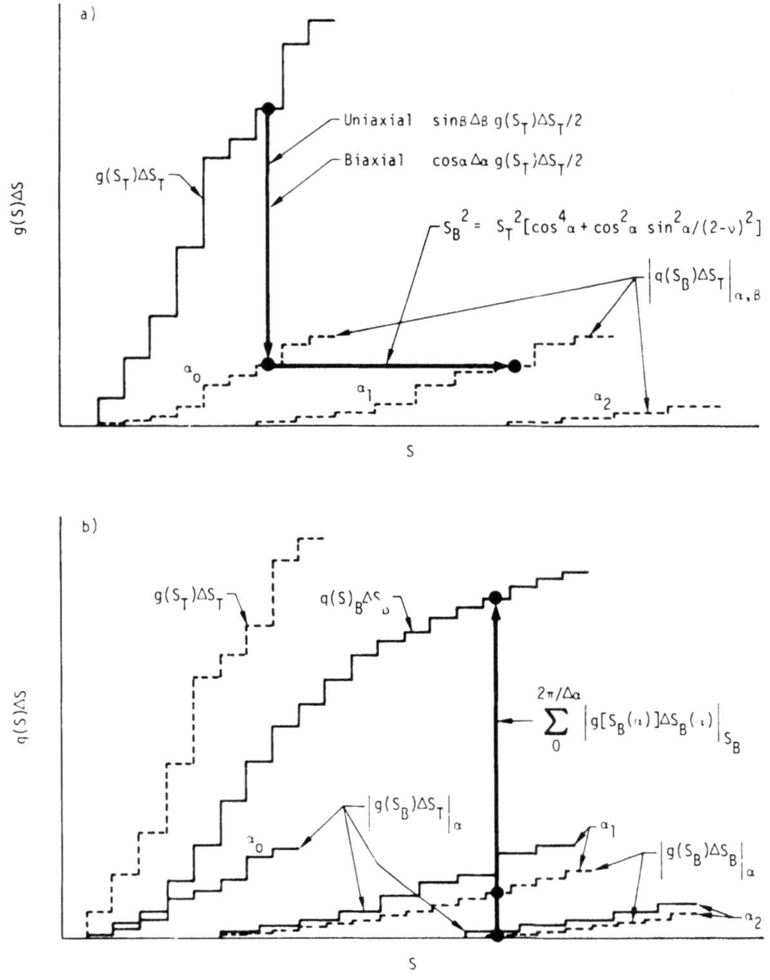

Fig. 2 The discrete element (histogram) approach for correlating strength distributions in different stress states. (a) The development of elemental histograms by translating along the distribution and strength axes, (b) the restructuring of the elemental histograms into histograms of constant ΔS followed by summation at constant stress to obtain the final distribution.

3.3 Uniaxial Tension

For uniaxial tension, the flaws that experience equivalent states of stress also have normals that occupy sectorial elements (Fig. 1 d). The uniaxial strength, S_u, of flaws in one of these elements is identically equal to that for the biaxial element. However, the volume of the stressed element in terms of the pertinent orientation ($\beta = 90 - \alpha$) is now, for a sphere of unit volume;

$$\Delta V = \frac{\sin\beta \ \Delta \beta}{2} \tag{14}$$

The transformation from the triaxial to the uniaxial distribution is thus quite similar to that described above for the biaxial distribution (Fig. 3), except that the elemental density is;

$$\left| g(S_B) \ \Delta S_B \right|_\beta = \frac{\sin\beta}{2} \ g(S_T)\Delta S_T \ \Delta\beta \tag{15}$$

The equivalent analytic expression for the uniaxial strength distribution is;[12]

$$g(S_u) = I_u \ (m, S^*/S_u, \nu)\left(\frac{S_u}{S_o}\right)^m \tag{16}$$

Values of the biaxial/uniaxial and triaxial/uniaxial strength ratios at equivalent fracture probabilities, derived from eqns. (12), (13) and (16) are plotted in Figs. 3, 4.

4.0 FLAW DENSITY RELATIONS

4.1 Uniaxial Strength Tests

The uniaxial tests that are most widely used to obtain fracture strength data for ceramics[2] are illustrated in Fig. 5; these include the three and four point flexural tests, the tensile test and the expanded ring test.[13] Because the derivations of flaw density relations from fracture strength data are uniquely dependent on the specimen geometry,[14] separate expressions are required for each method.

a) <u>Tensile Test</u>. For volume flaws;

$$g(S_m) = \frac{\xi'(S_m)}{V} \tag{17}$$

where V is the volume of the gauge section. The equivalent expres-

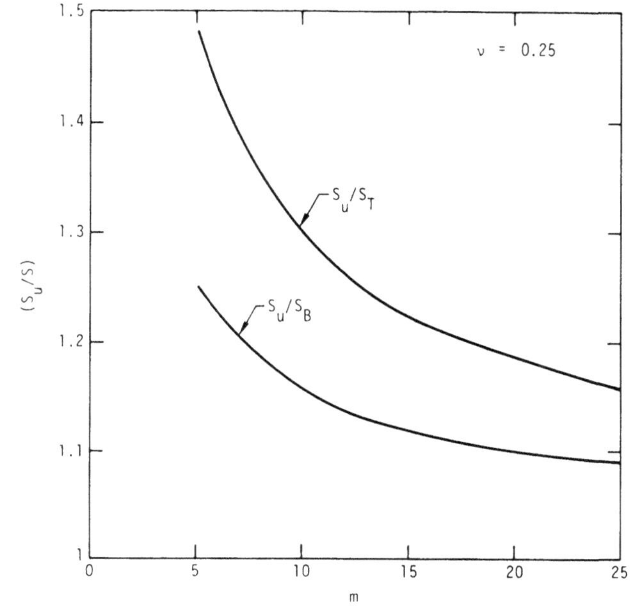
1.5
1.4
1.3
1.2
1.1
1
0
5
10
15
20
25
m
(S_u/S)
ν = 0.25
S_u/S_T
S_u/S_B

Fig. 3

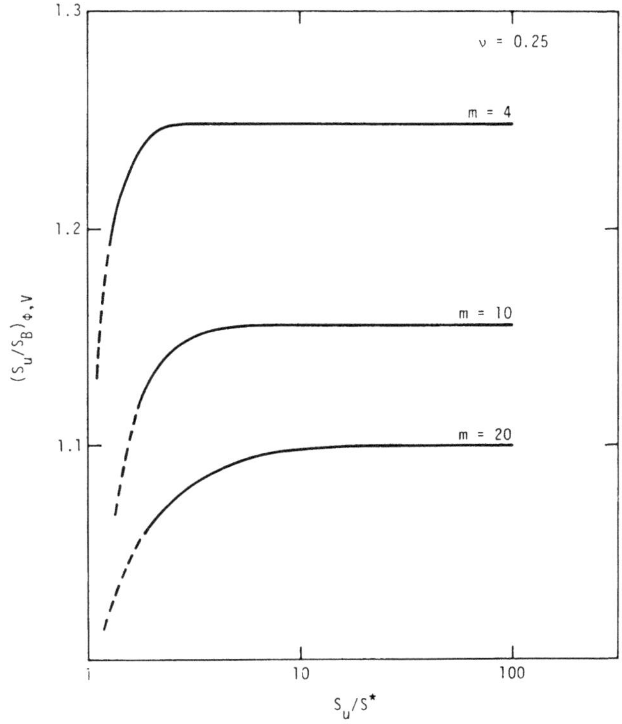
1.3
1.2
1.1
1
10
100
S_u/S*
(S_u/S_B)_φ,ν
ν = 0.25
m = 4
m = 10
m = 20

Fig. 4

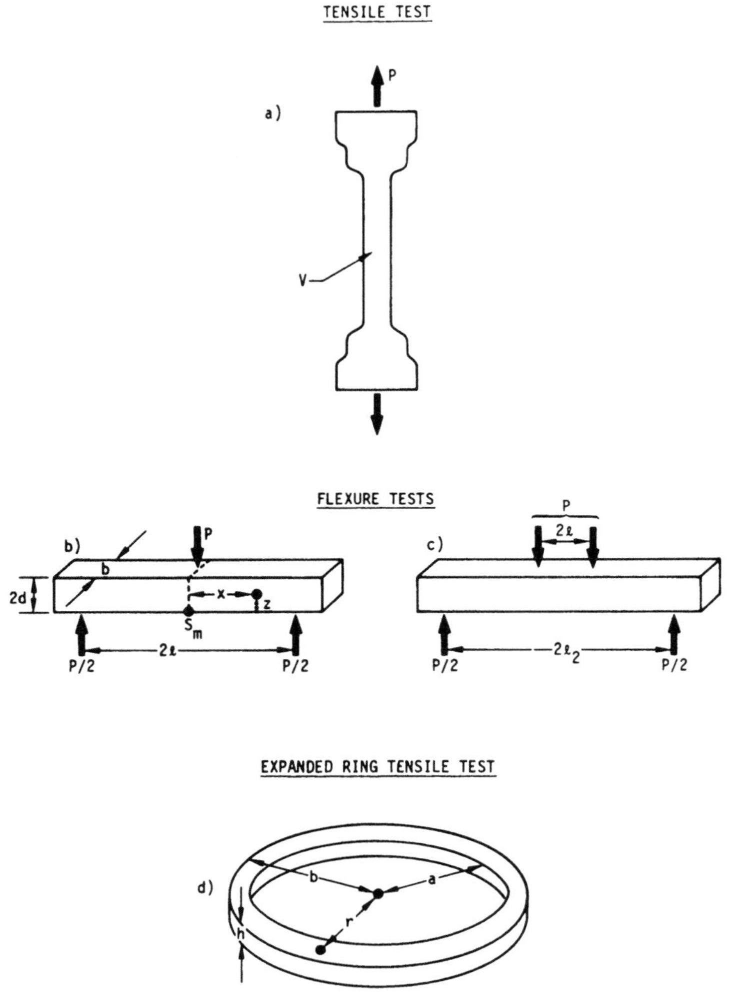

Fig. 5 Uniaxial strength testing techniques for ceramics, (a) tensile test, (b) three point flexure, (c) four point flexure, (d) expanded ring tensile test.

sion for surface flaws is;

$$g(S_m) = \frac{\xi'(S_m)}{A} \tag{18}$$

where A is the surface area of the gauge section.

b) Three Point Flexure. For volume flaws;

$$g(S_m) = \frac{1}{2b\ell d}\left[4\,\xi'(S_m) + 5S_m\xi''(S_m) + S_m^{\,2}\,\xi'''(S_m)\right] \tag{19}$$

where b is the sample width, 2d is the thickness and 2ℓ the distance between the outer loading points.

For surface flaws;

$$g(S_m) = \frac{1}{2b\ell}\,[2\xi'(S_m) + S_m\,\xi''(S_m)] \tag{20}$$

c) Four Point Flexure. For four point flexure, only fractures that occur between the inner loading points (separation $2\ell_1$) can be analysed. For this constraint on data acceptance, the flaw density function for volume flaws is;

$$g(S_m) = \frac{1}{2b\ell_1 d}\,[2\xi'(S_m) + S_m\xi''(S_m)] \tag{21}$$

For surface flaws;

$$g(S_m) = \frac{\xi'}{2b\ell_1}\,(S_m) \tag{22}$$

d) Expanded Ring Test. For volume flaws the best approximation is;

$$g(S_m) \approx \frac{2(b/a)^2}{\pi a^2 h[1+(b/a)^3]}\,(2\xi' + \xi''\,S_m) \tag{23}$$

where h is the sample thickness, b is the outer radius and a is the inner radius. For surface flaws, the best approximation is;

$$g(S_m) \approx \frac{\xi'}{\pi b\,[h(1+a/b) + b\,(1-(a/b)^2)]} \tag{24}$$

4.2 Biaxial Strength Tests

A number of relatively simple flexural tests are available for biaxial testing in the tension/tension quadrant.[15] These tests use disc samples supported near the disc periphery and loaded over a region around the disc center. The most convenient of these tests for statistical analysis is the concentric ring test, wherin the disc is supported uniformly around its periphery and loaded over a concentric circle, radius r_o. In this case, the tensile stress is equibiaxial and uniform over the zone r_o (except for a small, <0.5%, increase opposite the loading circle[16]) and varies linearly through the thickness. Hence, when fractures outside the zone r_o are excluded, the expression for $g(S_m)$ is;

$$g(S_m) = \frac{1}{\pi r_o^2 d}\left[2\xi'(S_m) + S_m \xi''(S_m)\right] \quad (25)$$

for volume flaws, and

$$g(S_m) = \frac{\xi'(S_m)}{\pi r_o^2} \quad (26)$$

for surface flaws, where 2d is the sample thickness.

The stresses in the zone outside r_o are relatively complex and deviate from the equibiaxial condition. It is not possible, therefore, to derive a meaningful $g(S_m)$ from test data that include the fractures originating in the outer zone. Similarly, the interpretation of data obtained on other biaxial test geometries is dubious, because the complex stress fields are only crudely approximated by the functions that yield tractable expressions for $g(S_m)$.

5.0 DATA ANALYSIS AND INTERPRETATION

5.1 Analysis

The procedures for data analysis are illustrated by examining extensive data obtained for two commercial hot pressed silicon nitrides (NC 132 and HS 130). These two materials differ primarily in their inclusion contents; specifically, the NC 132 material has a substantially reduced size distribution of the inclusions that act as sites for volume controlled fracture.

The HS 130 material was tested using the three point flexure and expanded ring methods, while the NC 132 material was tested in four point flexure. A total of 90 specimens were used for ring testing, and 1000 specimens each were used for four point and three point flexure testing.

The data analysis procedure firstly requires that the test data be ordered, and then that the cumulative probability of fracture $\Phi(S_m)$ be determined as a function of the peak stress, S_m.

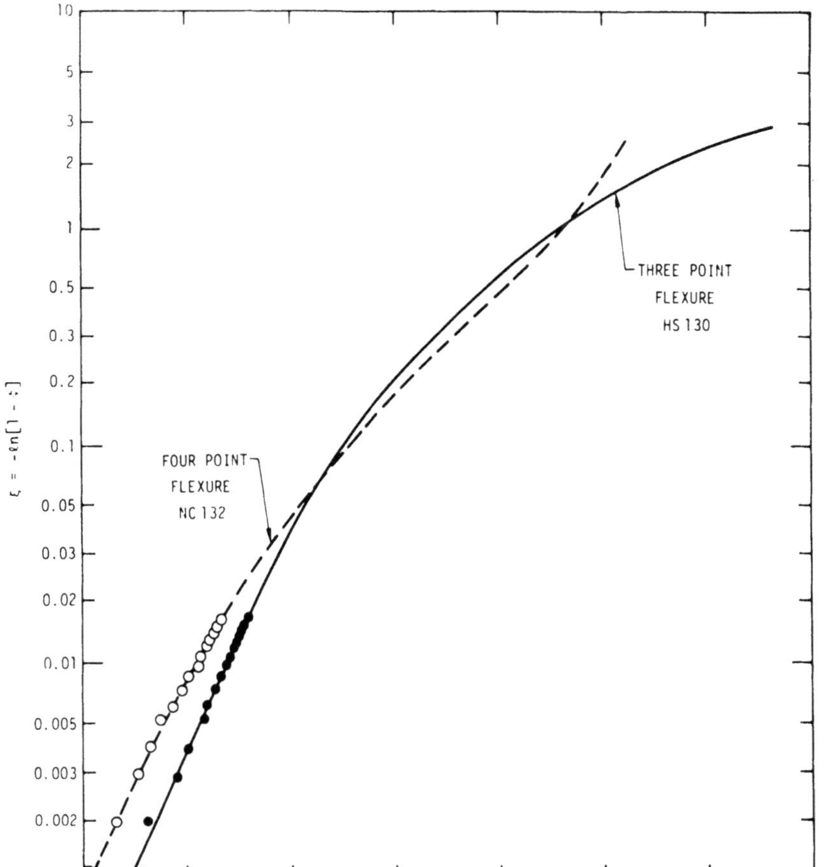

Fig. 6 Relations between the probability factor ξ and the peak stress at fracture, S_m, for three point flexure data on HS 130, and four point flexure data on NC 132.

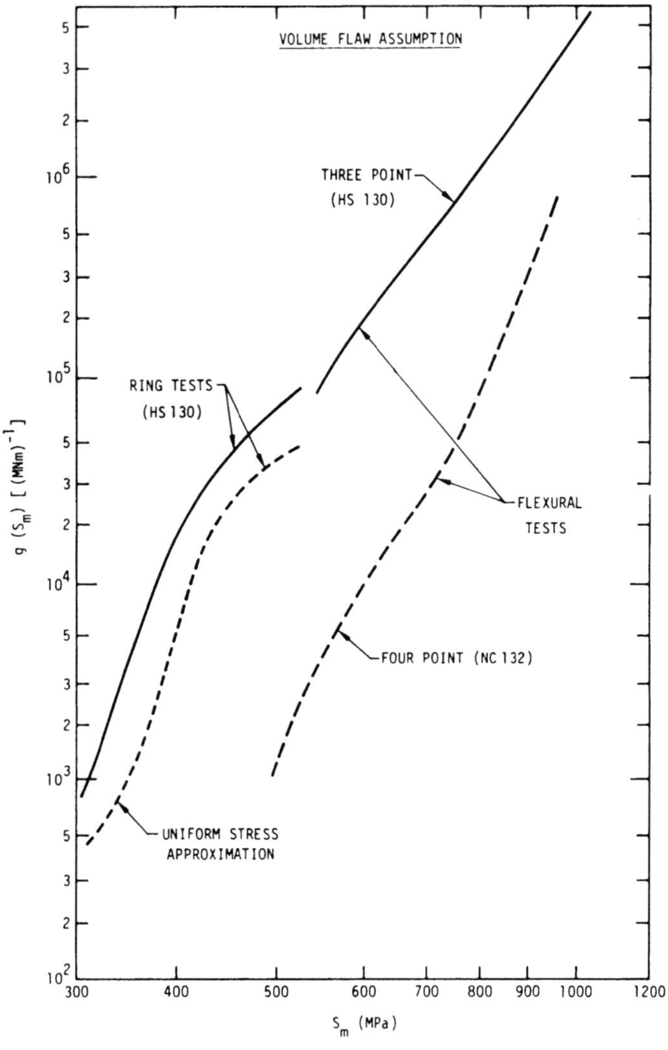

Fig. 7 The strength distribution function, $g(S_m)$, as a function of the strength, S_m, derived assuming volume flaw control.

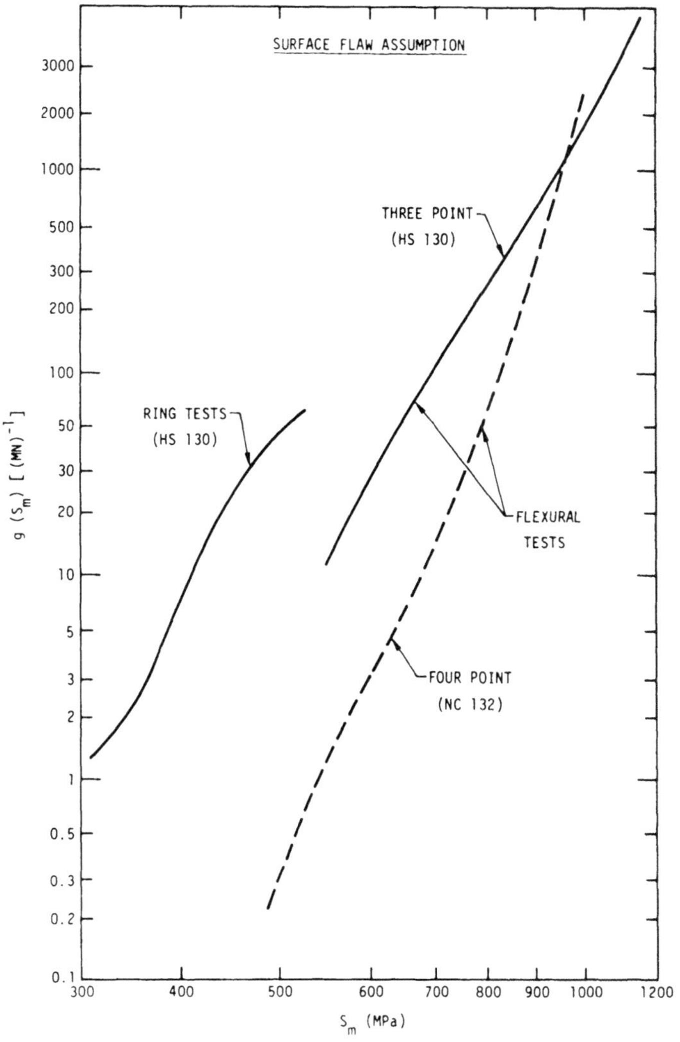

Fig. 8 The strength distribution function $g(S_m)$ as a function of the strength, S_m, derived assuming surface flaw control.

The quantity $\xi(\equiv -\ell n(1-\Phi))$ is then evaluated and plotted as a function of S_m (e.g. Fig. 6). Localized peaks in $\xi(S_m)$ are not usually significant, and polynomial fits that tend to smooth out such local oscillations, but retain the general shape of the curve, are generally preferred for subsequent analysis. Thereafter, the derivatives of $\xi(S_m)$ which determine $g(S_m)$ can be deduced directly from the polynomials.

The resultant strength distributions obtained from the three data sets, assuming volume flaw control or surface flaw control are plotted in Figs. 7 and 8.

5.2 Stress State Correlations

Uniaxial and equi-biaxial fracture data for conditions of equivalent volume have recently been obtained for alumina by Giovan and Sines,[17] using four point flexure and concentric ring test techniques (Fig. 9). These data are used both to demonstrate the method of data analysis and as a basis for comparison with the theory. The strength distribution functions are derived from the test data by assuming surface flaw control and by excluding fractures that initiated outside the uniformly stressed central zones. The $\xi(S)$ and $g(S)$ curves derived from the fracture data can be adequately described by a power function, as usually obtains for a relatively small sample. The exponent, m, is similar for both the uniaxial and biaxial distributions, consistent with theoretical expectations, and has a least squares value for the uniaxial data of 14.3. This m value is used to obtain the uniaxial/equi-biaxial and uniaxial/equi-triaxial strength ratios, at constant volume and fracture probability. Then, the equi-biaxial and equi-triaxial strengths are derived from the least squares fit of the uniaxial strength data, and compared with the measured equi-biaxial strengths (Fig. 9). The correlations is good, indicating the essential validity of the present analysis.

5.3 Interpretation

Strength distribution curves obtained using an adequate number of samples (Figs. 7, 8) contain important implications about the nature of the strength controlling defects and about the size dependence of the fracture strength. For example, the relatively good correspondence in $g(S_m)$ obtained for the HS 130 using the ring and flexure test data and a volume flaw assumption, implies that fracture in this material is primarily volume flaw controlled at the strength levels wherein $g(S_m)$ has been correlated (~ 450-650 MPa). Volume flaw control at larger or smaller strengths is not necessarily implied by such a correlation, although fractographic observations

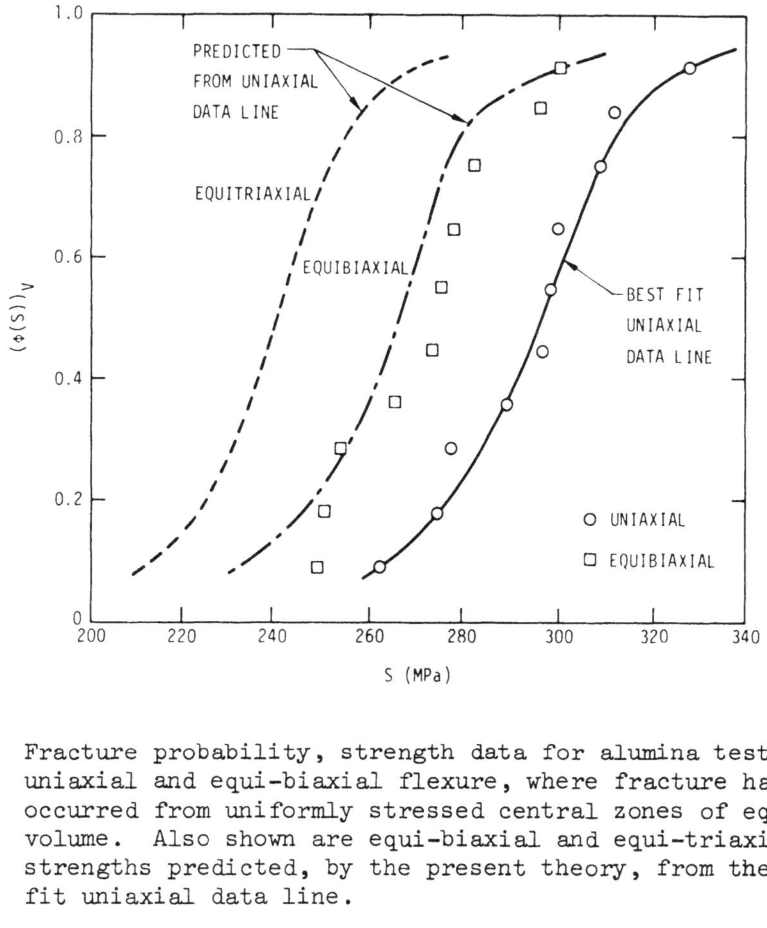

Fig. 9 Fracture probability, strength data for alumina tested in uniaxial and equi-biaxial flexure, where fracture has occurred from uniformly stressed central zones of equivalent volume. Also shown are equi-biaxial and equi-triaxial strengths predicted, by the present theory, from the best fit uniaxial data line.

suggest (as might be anticipated) that fracture is still volume flaw controlled at strength levels below 450 MPa. Conversely, the approximate correspondence of the surface flaw $g(S_m)$ curves (Fig. 8) for HS 130 and NC 132 at high strength levels ($\gtrsim$ 900 MPa) suggests that this is a regime of surface flaw control, because (as noted above) these materials differ principally in their content of volume distributed defects.

The occurrence of surface flaw controlled fracture at high strength levels and volume flaw controlled fracture at lower strength levels is expected to be typical of many structural ceramic materials. Also, it is quite likely that more than one population of volume flaws (and perhaps surface flaws) exists over the complete strength range of practical concern. This highlights the futility of using data obtained in one strength range for predicting fracture in an entirely different strength range. It is vital, therefore, for predicting the fracture probabilities of ceramic components that strength distributions be obtained in the pertinent strength range. Usually, this will require tensile data obtained on test specimens with stressed volumes equal to a significant fraction of the component volume.

CONCLUSION

A general approach for treating fracture statistics has been described. The strength of the approach is the ability to derive strength distribution functions from fracture data, and to interrelate the uniaxial and multiaxial forms of these functions. The approach also allows insights to be gained about the fracture sources, especially when multimodal distributions occur or when data from more than one test geometry are available.

ACKNOWLEDGEMENTS

The author gratefully acknowledges the financial support of the Rockwell International Independent Research and Development Program.

REFERENCES

1. B. W. Lindgren and G. W. McGrath, Introduction to Probability and Statistics (McMillan, N.Y.) 1959.

2. A. G. Evans and T. G. Langdon, Progress in Materials Science <u>21</u> (1976) 171.

3. A. S. Argon and F. A. McClintock, Mechanical Behavior of Materials, Addison Wesley (Reading, Mass.) 1966.

4. W. Weibull, Proc. Roy. Swedish Inst. Eng. Res. No. 153 (1939).

5. W. Weibull, Jnl. Appl. Mech., 18 (1951) 293.

6. J. R. Matthews, F. A. McClintock and W. J. Shack, Jnl. Amer. Ceram. Soc., 59 (1976) 304.

7. W. H. Dukes, Handbook of Brittle Materials Resign Technology, AGARD - AG 152-71 (Feb. 1971).

8. S. B. Batdorf and J. G. Crose, Jnl. Appl. Mech. (1974) 41.

9. J. J. Petrovic and M. G. Mendiratta, Jnl. Amer. Ceram. Soc. 59 (1976) 163.

10. J. J. Petrovic and M. G. Mendiratta, Jnl. Amer. Ceram. Soc., to be published.

11. B. R. Lawn and T. R. Wilshaw, Fracture of Brittle Solids (Cambridge Univ. Press) 1976.

12. A. G. Evans, Jnl. Amer. Ceram. Soc., in press.

13. R. Sedlacek and F. A. Halden, Rev. Sci. Instn. 33 (1962) 298.

14. A. G. Evans and R. L. Jones, Jnl. Amer. Ceram. Soc., in press.

15. J. B. Wachtman, W. Capps and J. Mandel, Jnl. Materials, 7 (1972) 188.

16. G. Sines, Private communication.

17. M. Giovan and G. Sines, to be published.

APPLICATIONS OF THE FOUR FUNCTION WEIBULL EQUATION IN THE DESIGN OF BRITTLE COMPONENTS

P. Stanley*, A. D. Sivill+ and H. Fessler*

* Department of Mechanical Engineering, University of Nottingham, England

+ Lucas Service U.K., Aston, Birmingham, England

SUMMARY

The equation gives the failure probability of a component in terms of the unit strength and variability (characterised by the Weibull modulus) of the material, the size and shape of the component and the type and magnitude of the loads acting on it. The shape of the component and the type of loading determine the form of the stress distribution; this, in turn, determines an important characteristic quantity, the stress-volume integral, from which the failure probability for a specified load is quickly and easily evaluated.

Stress volume integrals for a cylinder subjected to thermal loading are dervied from calculated stresses and discussed. The use of finite element calculations in the determination of stress-volume integrals is also described. A nomograph is presented to aid the evaluation of failure probabilities.

INTRODUCTION

Since the strength of nominally identical brittle components varies significantly it is necessary to base design studies on probability of failure calculations; the four-function Weibull equation (1) has been developed for this purpose. The development of the equation for the determination of the probability of failure of a brittle component of complex shape subjected to a general three-dimensional stress system has been described elsewhere (1) and its use has been shown to be valid in a particular case (2).

The stress-dependent function in the equation is the stress-volume integral $\Sigma(V)$ and the evaluation of the integral is the crucial step in the calculations of failure probability. In general, the evaluation requires knowledge of the principal stresses throughout the component; these can be determined directly from standard "strength of materials" or theory of elasticity solutions in simple cases, or experimentally by frozen-stress photoelasticity, or numerically by, for example, finite element computations. Stress-volume integrals for mechanical loadings have been determined analytically (1) and experimentally (3). This paper gives analytical results for thermal loading of simple cylinders and deals with some features of finite element stress analysis when used to determine stress-volume integrals.

Notation

a	unit surface area
A	total surface area
E	Young's modulus
H	step function
l	length
m	Weibull modulus
P_f	probability of failure
r, R	radius
t	time
T	temperature
v	unit volume
V	total volume
α	coefficient of linear expansion
ν	Poisson's ratio
σ	normal stress
$\bar{\sigma}_{fa}$	mean failure stress of unit surface area of the material under uniform uniaxial tension
$\bar{\sigma}_{fv}$	mean failure stress of unit volume of the material under uniform uniaxial tension
$\Sigma(A)$	stress-area integral
$\Sigma(V)$	stress-volume integral

Suffices

i	inner
nom	nominal
o	outer
r	radial
z	axial
θ	hoop

THE FOUR FUNCTION WEIBULL EQUATION

The failure probability of a brittle component depends on

i) the strength of the material
ii) the variability of strength of the material
iii) the shape of the component
iv) the size of the component
v) the type and magnitude of the loads applied to the component.

The effects of these several factors are incorporated in equation (1):

$$P_f = 1 - \exp\left\{ - \left(\frac{1}{m}\,!\right)^m \left(\frac{\sigma_{nom}}{\bar{\sigma}_{fv}}\right)^m \frac{V}{v} \Sigma(V) \right\} \qquad (1)$$

where P_f is the failure probability
m is the Weibull modulus
$\frac{1}{m}\,!$ is the gamma function of $(\frac{1}{m} + 1)$
V is the component volume
v is the unit volume
σ_{nom} is a nominal stress proportional to the applied load
$\bar{\sigma}_{fv}$ is the unit volume strength of the material (see Notation)
$\Sigma(V)$ is the stress-volume integral.

The equation contains four non-dimensional functions and introduces two material parameters m, the Weibull modulus, which characterises the variability in the strength of the material, and $\bar{\sigma}_{fv}$, the mean failure stress of unit volume of the material under uniform uniaxial tension. The first function $[(1/m)!]^m$ is associated with variability only; it is tabulated in reference (4) and has been plotted in references (1) and (2); it varies from 0.66 for m = 5 to 0.54 for m = 20. The second function $(\sigma_{nom}/\bar{\sigma}_{fv})^m$ combines the effects of material strength and applied load, and is the only function affected by the magnitude of the load. The use of the mean unit strength in this function is important in that it allows the size-dependence of the strength of brittle materials to be conveniently accounted for by the inclusion of the third function V/v. The determination of $\bar{\sigma}_{fv}$ from the results of fracture tests on beam specimens is discussed in reference (5). The effects of component shape and type of loading are represented in the fourth function, the stress-volume integral, $\Sigma(V)$, which is a volume integral of the three principal stresses in the form

$$\Sigma(V) = \int_V \left\{ \left(\frac{\sigma_1}{\sigma_{nom} H(\sigma_1)} \right)^m + \left(\frac{\sigma_2}{\sigma_{nom} H(\sigma_2)} \right)^m + \left(\frac{\sigma_3}{\sigma_{nom} H(\sigma_3)} \right)^m \right\} \frac{dV}{V} \qquad (2)$$

where σ_1, σ_2, σ_3 are the three principal stresses at a point in the component

$H(\sigma)$ is a step function defined to account for the greater strength of the material under compressive loading.

If $\sigma \geqslant 0$, $H(\sigma) = 1$, if $\sigma < 0$, $H(\sigma) = -8$.

The non-dimensional nature of $\Sigma(V)$, with respect to both stress and volume, is an important feature.

Equation (1) is arranged to ease the task of the designer. The stress-volume integral has some similarities to the stress concentration factor frequently used in conventional design calculations. Both are non-dimensional, depend on shape of component and type of loading but are independent of size of component and magnitude of loads. Both have to be determined by specialists in stress analysis. When the material properties and loads are known the probability of failure of a brittle component can be calculated manually from the stress-volume integral, just as the factor of safety for a metallic component can be derived from the stress concentration factor. The nominal stress, σ_{nom}, in $\Sigma(V)$ is chosen by the stress analyst to be a simple function of load.

Assumptions

The following assumptions, relating to both strength and strength variability, are made in developing equation (1):

1) The material contains many flaws of different severity. When the stresses at a particular flaw become critical for that particular flaw, a crack is initiated and propagates without further increase in load, resulting in fracture (i.e. the material is completely brittle).

2) The treatment is essentially a linear elastic one. Non-linear and time-dependent stress-strain characteristics are not allowed for. In the examples given below it is assumed that material properties are independent of temperature. (The effects of the temperature-dependence of material properties have been studied and will be described elsewhere (6).)

3) The material is isotropic and statistically homogeneous, i.e. the likelihood of finding a flaw of a given severity within an arbitrarily small volume of the material is constant throughout the body.

4) The severity of a flaw is independent of its position whether on the surface or within the volume of the component; this is referred to as the "volume flaw" assumption.

(Alternatively it may be assumed that surface flaws are so much more severe than interior flaws that all fractures initiate at the surface of a component; this is referred to as the "area flaw" assumption - see below.)

5) The Weibull postulate with zero "threshold stress" adequately describes the variation of fracture strength of similar components. The introduction of a finite threshold stress may become worthwhile as work in this field develops.

6) The stress-volume integral must incorporate an assumption or "criterion" regarding failure under multiaxial stresses. In the present work an "independent action" failure criterion has been used which assumes that failure due to one principal stress is independent of the presence of the other two. This failure criterion also takes account of the much greater compressive strength of a ceramic material compared with its tensile strength. Like the zero threshold stress assumption the independent action criterion can be modified if subsequent developments warrant it.

"Area Flaw" Failure Criterion

Using the alternative assumption that all failures originate at the surface of the component, equation (1) may be replaced by

$$P_f = 1 - \exp\left\{ - \left(\frac{1}{m}\,!\right)^m \left(\frac{\sigma_{nom}}{\bar{\sigma}_{fa}}\right)^m \frac{A}{a}\, \Sigma(A) \right\} \qquad (3)$$

where A is the component surface area
a is unit surface area
$\bar{\sigma}_{fa}$ is the unit area strength of the material (see Notation)
$\Sigma(A)$ is the stress-area integral given by

$$\Sigma(A) = \int_A \left\{ \left(\frac{\sigma_1}{\sigma_{nom} H(\sigma_1)} \right)^m + \left(\frac{\sigma_2}{\sigma_{nom} H(\sigma_2)} \right)^m + \left(\frac{\sigma_3}{\sigma_{nom} H(\sigma_3)} \right)^m \right\} \frac{dA}{A} \quad (4)$$

HOLLOW CYLINDERS WITH STEADY RADIAL TEMPERATURE GRADIENT

A long, hollow open-ended cylinder length l, with inner (R_i) and outer (R_o) surfaces at different uniform temperatures, has been considered. The theoretical elastic stresses (7) have been taken to apply over the entire length of the cylinder. For steady-state heat flow the temperature (T) at a radius r is given in terms of the internal (T_i) and external (T_o) temperatures by the expression,

$$T = \frac{(T_i - T_o)}{\log_e (R_o/R_i)} \log_e (R_o/r) \quad (5)$$

The nominal stress was taken as the absolute value of $\alpha E(T_i - T_o)/(1 - \nu)$ and the stress-volume integral was evaluated by numerical integration. A hundred concentric cylindrical elements of equal radial thickness were used in the computation and the ratio of compressive to tensile strength was taken as 8 throughout. The two cases of $T_i > T_o$ and $T_i < T_o$ were considered separately and the values of $\Sigma(V)$ versus R_i/R_o are shown in Figs. 1a and 1b respectively, for four values of Weibull modulus.

Stress-area integrals were also evaluated (neglecting the cylinder end surfaces) for $T_i > T_o$ and $T_i < T_o$. (Since the hoop and longitudinal stresses are constant on each surface, the integrals were determined analytically.) The values of $\Sigma(A)$ versus R_i/R_o are shown in Figs. 2a and 2b.

Throughout this work the material properties have been taken to be independent of temperature and the results presented are only valid under these conditions.

The difference in values of $\Sigma(V)$ (and $\Sigma(A)$) for the two temperature conditions is a consequence of the greater strength of the material under compressive stresses. With a value of 8 for the strength ratio the tensile stresses dominate the integrals even when the maximum compressive stresses are numerically greater than the maximum tensile stresses. Consequently, since the tensile stresses for the $T_i < T_o$ condition are greater than those for the $T_i > T_o$ condition, the integrals for the former condition are greater than those for the latter. This dominating effect of the tensile stresses would persist for smaller values of the strength

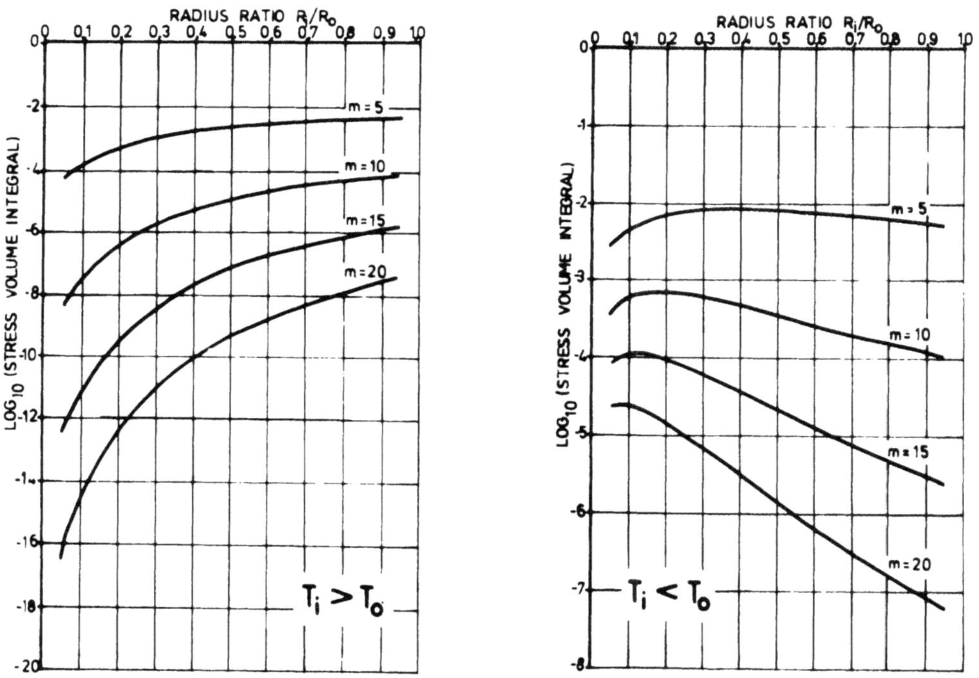

Fig. 1 Stress-volume integrals for hollow cylinders under steady, radial, thermal gradients.
Fig. 1a $T_i > T_o$. Fig. 1b $T_i < T_o$.

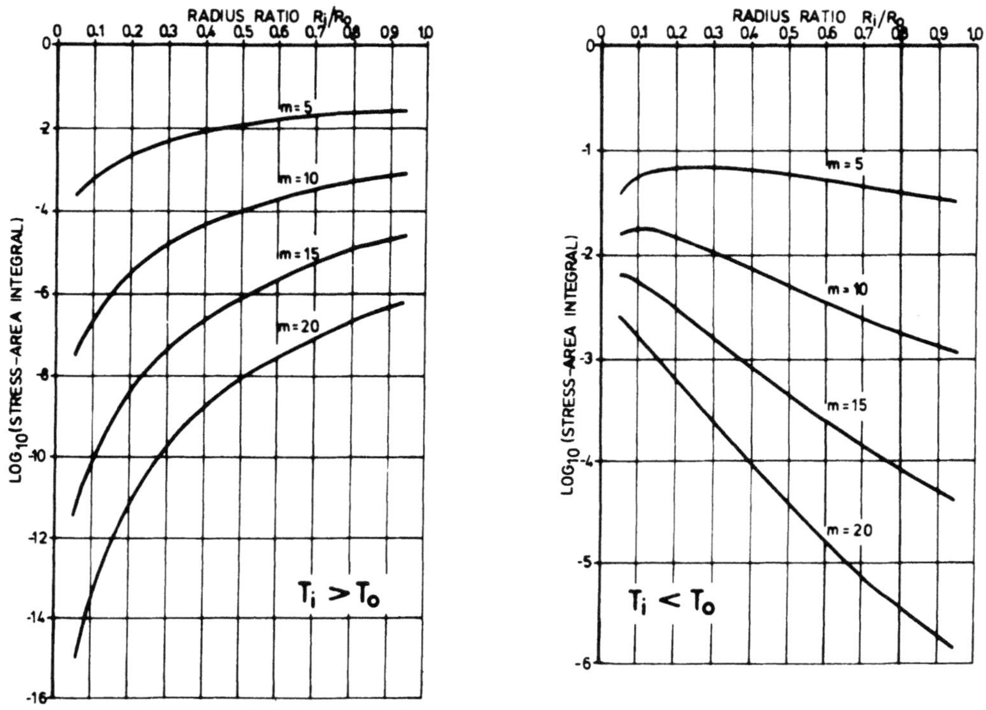

Fig. 2 Stress-area integrals for hollow cylinders under steady, radial thermal gradients.
Fig. 2a $T_i > T_o$. Fig. 2b $T_i < T_o$.

ratio down to at least 2. For all conditions the stress integrals decrease with increasing values of m.

When $T_i < T_o$ both $\Sigma(V)$ and $\Sigma(A)$ pass through a maximum as R_i/R_o decreases. The maximum stress increases with decreasing R_i/R_o, though, since it occurs at the bore, the relative volume over which it acts decreases; the combination of the two effects causes the maximum in the stress integrals.

FINITE ELEMENT CALCULATIONS

The finite element method is widely accepted as the most versatile numerical method of stress analysis. The method requires the "idealisation" of the component or structure into a number of elements and the specification of the appropriate boundary conditions. The stresses at a large number of discrete points in the component are calculated from this data; the stresses obtained are necessarily approximate and considerable experience and judgement are usually required if the method is to be used successfully.

The numerical procedures involved in the method are long and may be complex; a number of computing schemes have been devised to facilitate these procedures. Such a scheme, called PAFEC 70+ (8) has been developed in the Department of Mechanical Engineering at Nottingham University and was used in this work. The extension of this scheme to permit the direct evaluation of stress integrals (equations (2) and (4)) presented no difficulties. This was done for a particular two-dimensional finite element, and the resulting program was tested on a curved two-dimensional cantilever beam for which an analytical stress solution is available.

Procedure

In the PAFEC 70+ scheme (8) the curvilinear, 8-node, isoparametric, in-plane finite element is transformed to a square to enable the integral of the strain energy to be evaluated. The principal stresses are then calculated at sixteen points within the element; the points are positioned at "Gauss points" to facilitate the integration process. (Gaussian integration (9) is a form of numerical integration which has gained wide acceptance for finite element purposes. An nth order Gauss process fits the value of a function at N points (called Gauss points) to a polynomial of order (2N - 1). This function is then integrated analytically.) The values of the two principal stresses σ_1 and σ_2 were determined at each of these positions and the quantity $(\sigma_1{}^m + \sigma_2{}^m)dV$ was calculated for the transposed element by 4th order Gaussian integration. The product of this and the Jacobian of the transformation gave the integral for the actual element. A summation over all elements resulted in the integral for the body, from which the stress-volume integral was calculated using the nominal stress and component volume.

Results

A thick curved bar ($R_o/R_i = 1.5$) loaded as a cantilever as shown in Fig. 3 was chosen to prove the program and to study the dependence of the accuracy of the stress-volume integral on the number of elements. This "component" has significant, non-uniform stress gradients. There is an analytical stress solution from which the exact stress-volume integral was calculated by numerical integration. The six different meshes used in the investigation are shown in Fig. 3. After the 3-element idealisation the mesh was refined by equal sub-division as indicated, the refinement being concentrated in the region of highest tensile stress. In each case the stress-volume integral was calculated for the bar using four values of Weibull modulus; the results are given in Fig. 4 as multiples of the theoretical value. (For this form of presentation the choice of nominal stress is immaterial.)

As the number of elements increases, the computed stress-volume integrals converge to the exact value for all Weibull moduli. The larger the m value, the more predominant is the contribution of the maximum stress to the stress-volume integral and therefore the smaller are the elements required in the high stress regions for a satisfactory estimate of the integral. There is, for the higher m values, a significant over-estimate, (over 6% for $m = 20$), before the true value is obtained. The risk of underestimating the stress-volume integral (and consequently the probability of failure) as a result of using an insufficiently refined idealisation is evident. Instead of trying to formulate guide lines for the number of elements to be used for particular stress gradients in components for which the stress-volume integral is not known, it is suggested that the mesh be refined in the high stress regions until a satisfactory convergence is obtained in successive evaluations of the integral. The type of element used is particularly important in determining the rate of convergence.

Errors in the computed stresses can result from incorrect positioning of the mid-side nodes of isoparametric elements (10), the usual effect being to increase the stresses locally. When stress-integrals are calculated from these stresses this error is "magnified" m times and can lead to seriously incorrect $\Sigma(V)$ values. This dependence is so critical that even "round-off" error in nodal positioning may be detected as an increased stress-integral. Table 1 gives an indication of the likely magnitude of the effect produced by defining nodal positions in the cantilever structure to 3 significant figures and to 5 significant figures. The 24 element mesh shown in Fig. 3 was used in this part of the work.

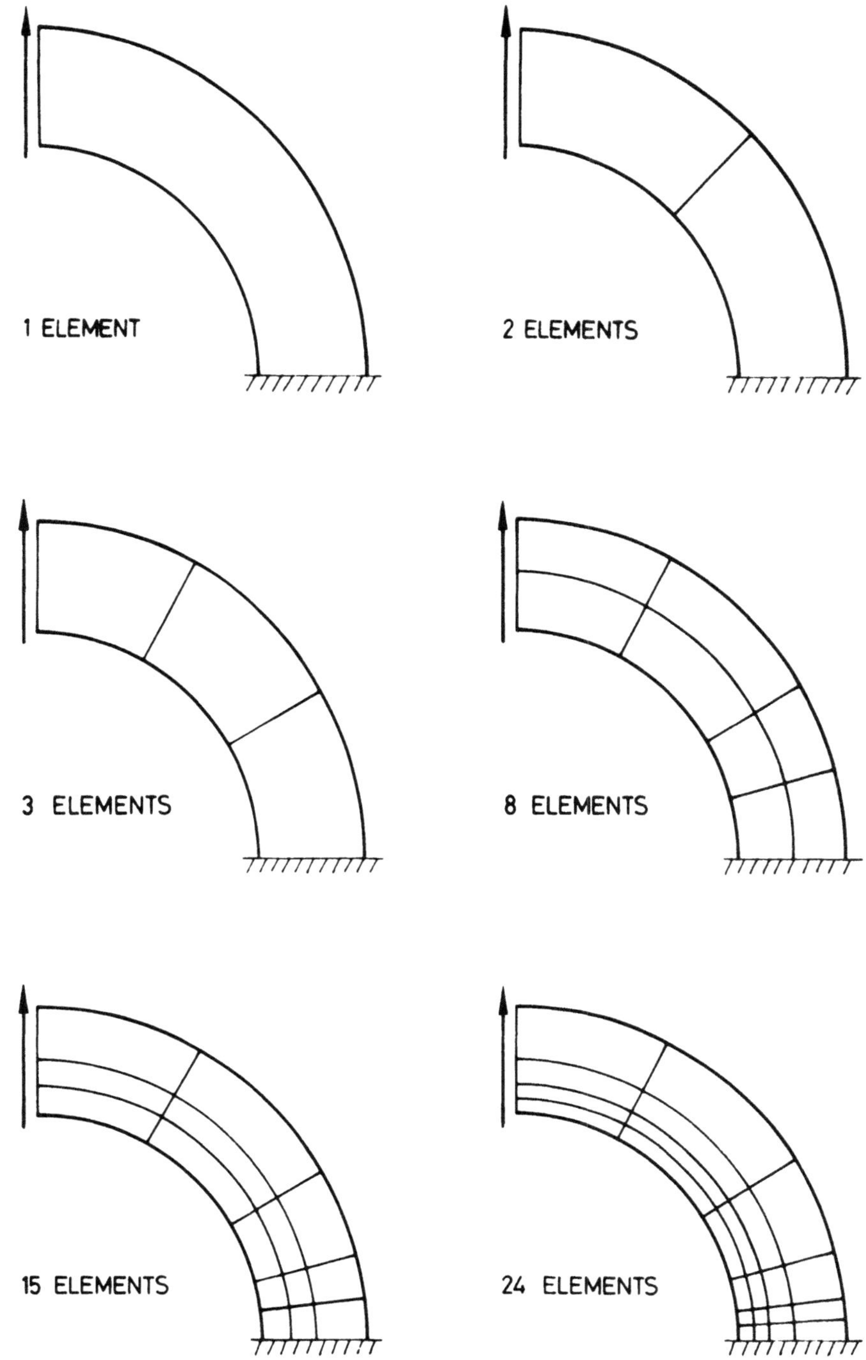

Fig. 3 Loading and meshes used in finite element computations of thick curved bar $R_o/R_i = 1.5$.

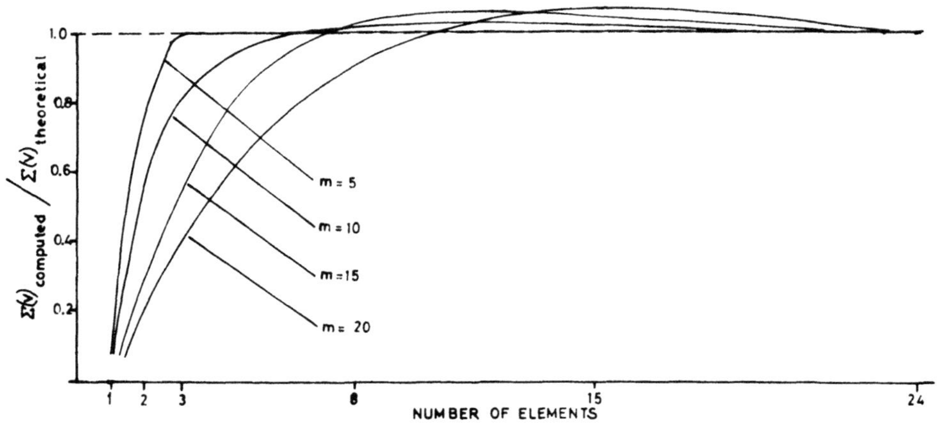

Fig. 4 Accuracy of stress-volume integrals for curved cantilever (R_o/R_i = 1.5) from different finite element computations.

Table 1 Errors in stress-volume integrals caused by round-off (24 element mesh)

Weibull Modulus m	Error in Σ(V) Relative to the Exact Value	
	Nodes to 3 Sig. Figs.	Nodes to 5 Sig. Figs.
5	-	-
10	+3%	-
15	+9%	+3%
20	+13%	+4%

Note: - indicates less than 1%

An aid to prediction

A nomograph to evaluate equation (1) has been prepared which is of general validity and which provides a visual impression of the dependence of P_f on the separate non-dimensional variables. The chart is valid for any component and any material. A failure probability estimate can be quickly determined by progressing around the chart (Fig. 5), choosing the appropriate value from a family of curves at each stage starting with $\sigma_{nom}/\bar{\sigma}_{fv}$ as indicated by the arrows.

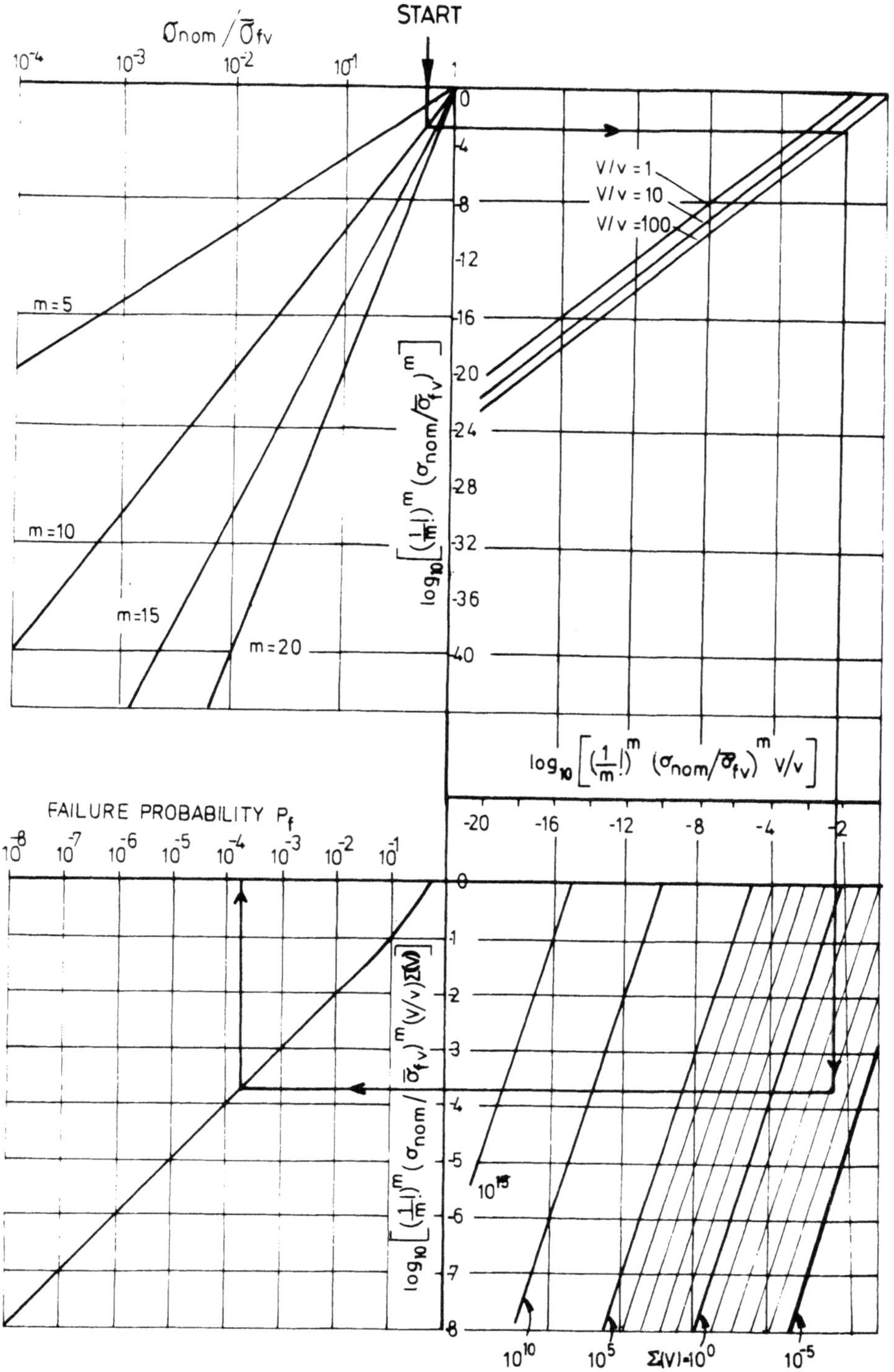

Fig. 5 A nomograph formulation of the four-function Weibull equation

DISCUSSION

The six assumptions made here are reasonable and convenient but the four-function equation could be used with other assumptions. Some of the alternatives require further comment.

The four-function equation is based on a Weibull failure distribution, which has been shown (11) to be a satisfactory representation of scatter. Other probability distributions could, in principle, be used to predict component failure probability but it is considered unlikely that their use would offer any advantages over the present method.

The positions of critical flaws are very difficult to determine. With statistical homogeneity and isotropy the assumptions that the critical flaw may occur anywhere (leading to $\Sigma(V)$) and that it will only occur at the surface (leading to $\Sigma(A)$) are the extreme possibilities. It seems likely that surface flaws will be more severe than interior flaws of the same size but the complications of a combined criterion are not justified at present.

The "independent action" assumption used in the evaluation of the stress integrals is simple and predictions based on it have been shown to conform with experimental results (2).

The stress-volume and stress-area integrals apply to any component of the particular shape subjected to the particular form of loading, i.e. they are independent of size, material and the specific value of load. Using analytical stress analysis it is only possible to deal with complete cylinders. In extending the work to cases of cylinders with holes or other irregularities, and with non-uniform temperature patterns, numerical stress analysis is essential. Some relevant work is being carried out at present by one of the authors. The results presented here can be useful in preliminary analysis of simple tubular components.

The nomograph is generally applicable for any calculation based on the volume flaw assumption. A similar diagram could of course be constructed for the area flaw assumption. The nomograph is useful for rapid evaluation of failure probability, especially when the effects of changes in any one of the relevant quantities has to be assessed during design.

The work involved in hand-calculation of stress integrals is usually prohibitive. The finite element program modifications and developments described are natural extensions of the finite element method for brittle material requirements and comprise a powerful and convenient means of evaluating failure probability. However, accurate knowledge of stresses in the highly stressed regions is essential in this work since stress values are raised to a high power. It is particularly important in the automatic calculation of stress-integrals to examine the stress solution for discontinuities and evidence of unsatisfactory mesh configurations. A greater number of significant figures should not be taken as indicating a higher degree of accuracy.

The dependence of the accuracy of the stress-volume integral on the size and shape of the element used in its evaluation will be a feature specific to the element and an assessment of this dependence is considered to be an important preliminary in any major numerical study of failure probability. It is an unfortunate feature of the 8-node isoparametric element used in the present work that the mid-side nodes must be precisely positioned. The sensitivity of the result to this detail is such that 'round-off' error can significantly affect the computed stress-integral value. The effect can be minimised by defining nodal coordinates to as many significant figures as possible.

Instead of the finite element method photoelastic methods can be used for the detailed stress analysis (e.g. reference (3)). Once a system is established, the finite element method is likely to be less laborious than a photoelastic one because the input for the stress integral determination is automatically produced. However, the photoelastic method need not be as laborious as appears because, for material of good consistency (say m values greater than 10), the contributions to the stress integrals from regions of low stresses are negligible so that it is often unnecessary to carry out a photoelastic analysis of the whole component.

ACKNOWLEDGEMENTS

This work has been carried out with the support of the Procurement Executive, Ministry of Defence. The authors wish to thank Dr. D. J. Godfrey. Dr. R. D. Henshell and the other members of the PAFEC team (now PAFEC Ltd.) provided valuable assistance with the finite element work.

REFERENCES

1. Stanley, P., Fessler, H. and Sivill, A.D., "An Engineer's Approach to the Prediction of Failure Probability of Brittle Components", Proc. Brit. Ceram. Soc., 1973, Vol. 22, p. 453.

2. Stanley, P., Sivill, A.D. and Fessler, H., "The Application and Confirmation of a Predictive Technique for the Fracture of Brittle Components", Fifth Int. Conf. on Experimental Stress Analysis, Udine, Italy, 1974.

3. Stanley, P., Sivill, A.D. and Fessler, H., "A Design Exercise for a Ceramic Turbine Disc subjected to Centrifugal Stresses", Submitted to J. Strain Analysis.

4. Abramovitz, M. and Stegun, I.A., "Handbook of Mathematical Functions", N.B.S. publication AMS 55.

5. Stanley, P., Sivill, A.D. and Fessler, H., "The Unit Strength Concept in the Interpretation of Beam Test Results for Brittle Materials", 1976, Proc. I.Mech.E., Vol. 190, p. 585.

6. Stanley, P. and Chau, F.S., to be published.

7. Timoshenko, S.P. and Goodier, J.N., "Theory of Elasticity", 1970, McGraw-Hill, 3rd Ed.

8. Henshell, R.D. (Ed.) "PAFEC 70+", Mechanical Engineering Department, Nottingham University, 1972.

9. Zienkiewicz, O.C., "The Finite Element Method in Engineering Science", McGraw-Hill, London, 1971.

10. Bond, T.J., "Some Considerations of the Finite Element Method in Stress Analysis", Ph.D. Thesis, Nottingham University, 1972.

11. Sivill, A.D., "The Development and Application of a Statistical Basis for the Design of Brittle Components", Ph.D. Thesis, Nottingham University, 1974.

MULTIPLE FLAW FRACTURE MECHANICS MODEL FOR CERAMICS

H. A. Nied and K. Arin

General Electric Company

Schenectady, New York

ABSTRACT

A multiple flaw fracture mechanics model applicable to ceramic materials is formulated which includes the interaction between a distribution of flaws located on a critical fracture plane. The opening mode stress intensity factor K of this system of flaws, modeled as an array of circular cracks, is determined in terms of the area fraction, the remote uniaxial stress field and the flaw characteristic size. The resulting mixed boundary value problem is formulated in terms of the Papkovich-Neuber functions represented by appropriate series expansions which satisfy the Laplace equations. By imposing the boundary conditions, the problem is reduced to a set of dual series equations which can be rewritten as a Fredholm integral equation of the second kind. This integral equation is then simply solved using known numerical techniques. Finally, the stress intensity factor K is obtained as the strength of the singularity of the near tip stress field by a limiting process. When this solution, expressed in terms of the local defect area fraction, is compared to the known plane strain solution for the stress intensity factor of collinear cracks, it can be shown that the simpler 2 dimensional plane strain solution is valid over a range of defect area fractions of interest in ceramics. Accordingly, a simpler engineering solution based on Koiter's solution modified by an appropriate constant is proposed for estimating the strength of ceramics.

The results of the analysis show that the opening mode stress intensity factor and the material strength is strongly influenced by the degree of local defect area fraction.

INTRODUCTION

Generally, ceramics are inherently hard and brittle, tending to resist deformation when subjected to applied loads. Ceramics are not homogeneous materials but are composed of a complex multiphase composite of individual grains and pores. The fabrication process introduces many internal microflaws and defects into the structure other than simple surface defects. These flaws and defects produce structural discontinuities in the material. Fractography of fracture surfaces has revealed that microstructural pores and void clusters are indeed frequent sources of failure in ceramics [1].

It is not the purpose of this paper to survey the evolution and development of various fracture mechanics models applicable to brittle materials such as ceramics, but to cite several popular models. A fracture model was proposed by Griffith [2] in his 1920 paper which included the presence of internal defects in brittle materials. Using an energy balance theory he was able to relate the strength of the material in terms of the characteristic flaw size modeled as a sharp crack and fracture energy required to create a unit amount of fracture surface. This early work formed the foundation for the field of linear elastic fracture mechanics. Interesting fracture mechanics models representing void or pore type flaws in ceramics have been devised using Bowie's solution [3] for the stress intensity factor at the tip of a radial crack emanating from a circular hole. In the adaptation of that model to ceramic materials, one or two radial intergranular boundary cracks are assumed to emanate from a circular hole representative of a single, isolated pore as shown in Figure (1). Rice [1] and Evans [4] describe the application of this model for determining the strength of ceramics, using the assumption that the ratio of the characteristic grain size to the effective pore diameter is equal or greater than one. Rice [1] has characterized pore defects in some ceramics consisting of a microstructure having protrusions into the cavity similar in shape to stalagtites and stalagmites. This type of defect generally contains sharp cusps emanating from the cavity which coincide with grain boundaries. A solution for the stress intensity factor obtained by Gresco [5] for a four cusp hypocycloid (Fig. (2)) subjected to a uniaxially applied remote load can provide a suitable fracture mechanics model for flaws of this sort. If the remotely applied load is orthogonal to the surface passing through any two cusps, the opening mode stress intensity factor K is only 1.0206 K for a Griffith type crack of equal size. The general inadequacy of these fracture mechanics models are that they are isolated, two dimensional tunnel type flaws which do not take into account flaw interaction associated with large flaw volume fractions. Evans and Langdon [4] have presented the results for the interaction effect based on the well known solution obtained by Koiter [6] for an array

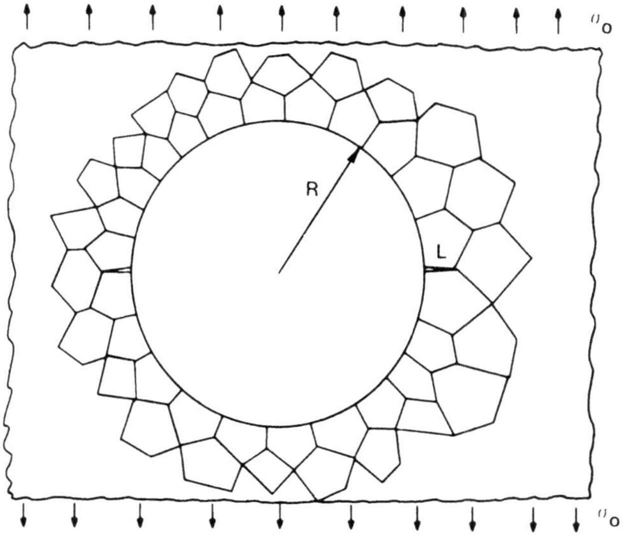

Figure 1. Fracture mechanics model of a pore defect in ceramics based on Bowie's solution.

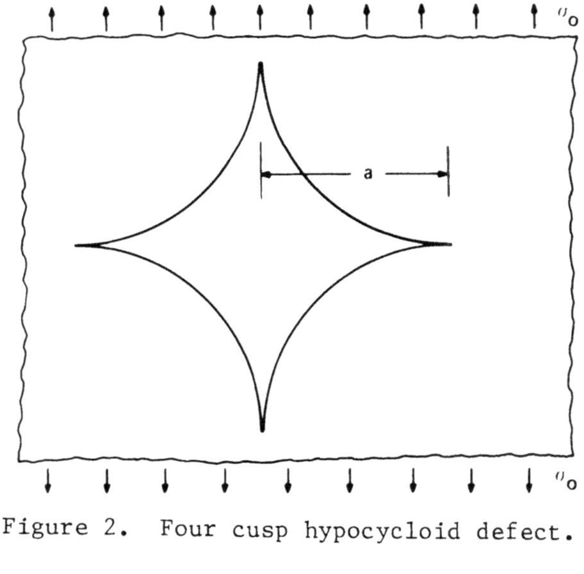

Figure 2. Four cusp hypocycloid defect.

of collinear cracks shown in Figure (3). Koiter's solution for the stress intensity factor is given by:

$$K = \left[\frac{2b}{\pi a}\tan\left(\frac{\pi a}{2b}\right)\right]^{1/2}\sigma_o\sqrt{\pi a} \tag{1}$$

Based on this solution the interaction effect becomes appreciable when flaw separation distances become less than the flaw length. This solution provides an analytical tool for predicting the coalescence of flaws in a porous region. However, this model which is a two-dimensional solution tends to underestimate the strength of ceramic materials since the flaws are modeled as tunnel cracks. In this paper, a multiple flaw fracture mechanics model is presented which consists of a planar array of interacting circular cracks. Later it is shown that the two dimensional solution of Koiter can be used as an approximate solution for a planar array of flaws when modified by a constant determined from the three dimensional solution.

MULTIPLE FLAW MODEL

Consider a ceramic material containing a three dimensional distribution of pore or cavity type flaws. Many reaction bonded structural ceramics such as silicon nitride possess high remaining porosity which influence the strength. A model for this type of defect distribution can be visualized by passing a plane cutting through the region of maximum porosity normal to the uniaxially applied load which then becomes the critical stress plane. The microstructure of each pore or cavity is assumed to consist of many crack sites randomly emanating from the internal surface of the nearly spherical void along large grain boundaries or cusps. In addition, it is assumed that the trace of probable fracture surface of each individual pore will project as circular cracks on the plane of critical stress as shown in Figure (4). The technique for interpreting and quantitatively measuring the distribution of the trace of these flaws are contained in reference [7]. The degree of porosity can be measured as the local area fraction which is the area ratio of the void to the void-free material. Statistical models can be employed to prescribe the characteristic size of the flaw and separation distance in terms of the mean and deviation values. The two major features of this fracture model are the assumptions that (1) the fracture surface will pass through the region of maximum porosity and (2) the microstructural singularities of the cavity produce an effect equivalent to a penny shaped crack. The net result is that material defect system is reduced to an array of interacting circular cracks shown in Figure (5). The entire domain can be represented by an array of hexagon enscribed penny shaped cracks which cover the critical plane. A fracture mechanics model

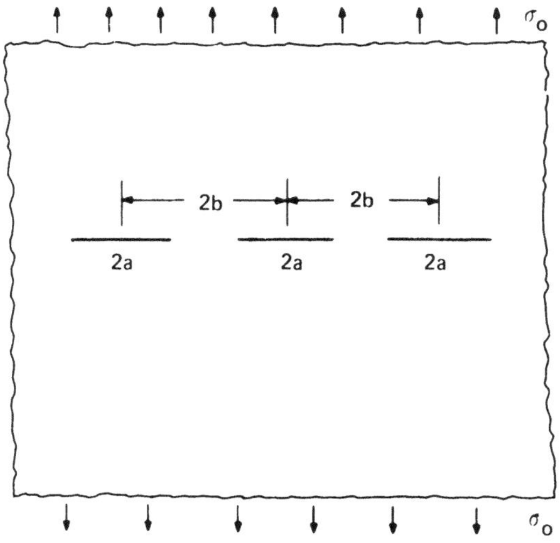

Figure 3. Array of collinear cracks.

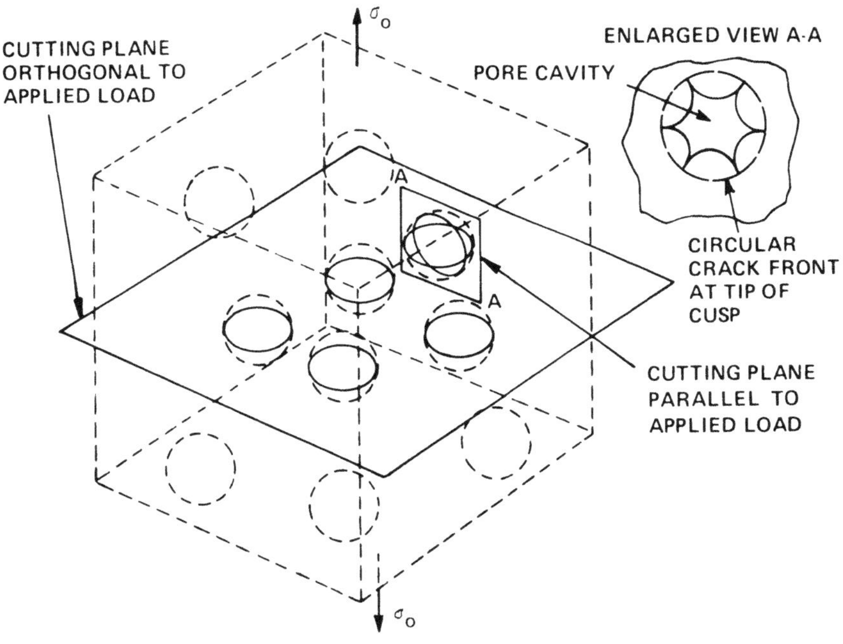

Figure 4. Trace of flaws on plane of critical stress.

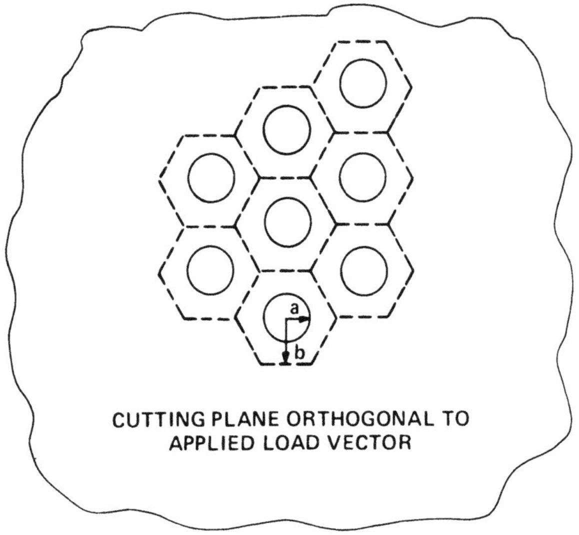

Figure 5. Planar array of circular cracks.

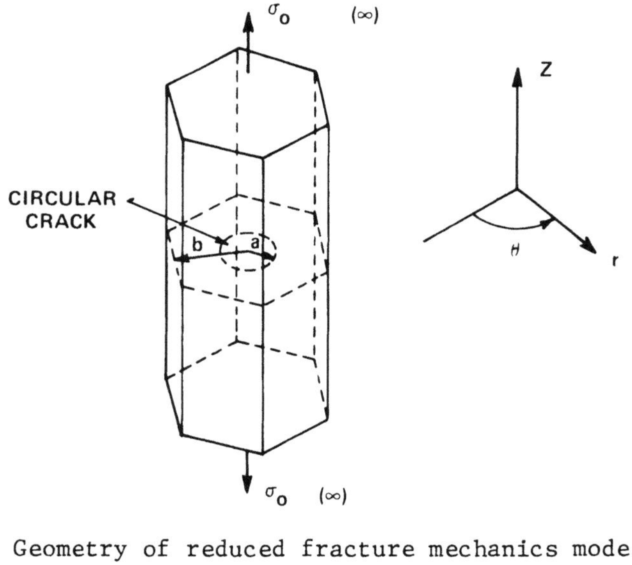

Figure 6. Geometry of reduced fracture mechanics model for an array of interacting circular cracks.

can be constructed from this array by the removal of a single hexagonal bar containing the circular crack and imposing the proper boundary conditions on the lateral surfaces. Figure (6) shows the geometry for a suitable three dimensional model. The fracture mechanics solution to be sought for this model must require that the lateral hexagonal surface be free of shear stress and supported in such a way that the displacement vector normal to the surface must vanish. In addition, the crack surfaces are traction free. The displacement vectors and stress tensors are functions of the space variable r, z and θ for a cylindrical coordinate system. The spatial dependency in θ is due to the hexagonal outer surface. A solution to this boundary value problem is currently being worked on since it will provide an exact solution to the material model throughout the continuum. However, in order to obtain a first order approximation that retains the feature of radial crack interaction, a further simplification of this material model can be introduced by assuming an axisymmetric model which will eliminate the solution dependency on θ. Figure (7) shows the modified geometry.

This model resembles the material model used by Hashin [8] to determine the bounds of the elastic modulii for a composite material consisting of reinforcement fibers of radii "a" instead of the trace of circular cracks for this model. The final configuration of the material model can be constructed by removing a single element from the array in Figure (7). Hence, the corresponding mixed boundary value problem has been reduced to the solution of a circular cylinder containing an interior penny shaped crack normal to its axis which is subjected to a uniform remote stress field σ_0 shown in Figure (8).

SOLUTION OF THE BOUNDARY VALUE PROBLEM

For this elasticity problem we assume that the cylindrical lateral surface is free of shear stress and is supported in such a way that the radial component of the displacement vector vanishes on the surface, in addition, to the traction free conditions on the crack surface. The effect of a penny shaped crack on the stress distribution in a long circular cylinder under tension has been treated by Sneddon [11].

The Navier governing equations in terms of displacements for an isotropic, elastic material written in the cylindrical coordinate system for this axisymmetric problem is given by:

$$(\lambda+\mu)\frac{\partial}{\partial r}\left[\frac{1}{r}\frac{\partial}{\partial r}(ru_r) + \frac{\partial u_z}{\partial z}\right] + \mu\left[\nabla^2 u_r - \frac{u_r}{r^2}\right] = 0 \quad (2)$$

$$(\lambda+\mu)\frac{\partial}{\partial z}\left[\frac{1}{r}\frac{\partial}{\partial r}(ru_r) + \frac{\partial u_z}{\partial z}\right] + \mu\left[\nabla^2 u_z\right] = 0 \quad (3)$$

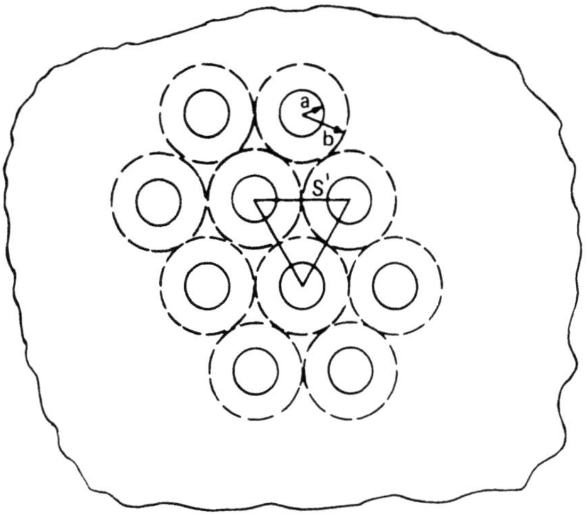

Figure 7. Planar array of circular cracks for an isotropic void model.

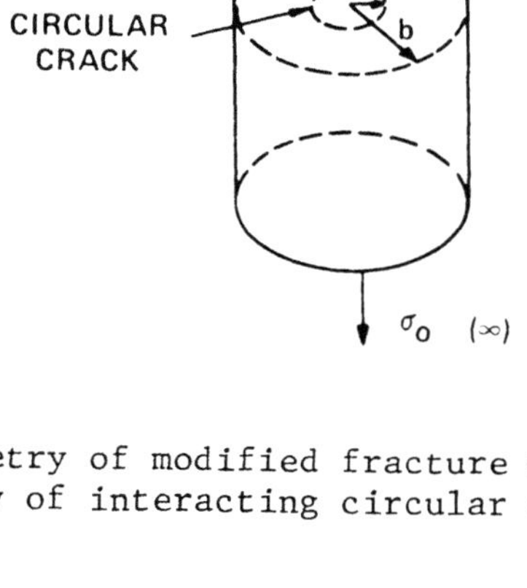

Figure 8. Geometry of modified fracture mechanics model for an array of interacting circular cracks.

where λ and μ are Lamé elastic constants, u_r and u_z are the displacement vectors in the radial and axial direction respectively.

Due to the linearity of the equations, this problem can be recast into two component problems by using the principle of superposition. Accordingly, the solution of each component problem shown in Figure (9) will be sought with the complete solution provided by the superposition of the two component solutions.

Component Problem I

The solution to this axisymmetric elasticity problem is given by:

$$u_r = u_\theta = 0, \quad u_z = \frac{(1+\nu)(1-2\nu)}{(1-\nu)} \frac{\sigma_0}{E} z \tag{4}$$

The strain tensor is:

$$\varepsilon_r = \varepsilon_\theta = 0, \quad \varepsilon_z = \frac{(1+\nu)(1-2\nu)}{(1-\nu)} \frac{\sigma_0}{E} \tag{5}$$

where the components of the shear tensor vanish.

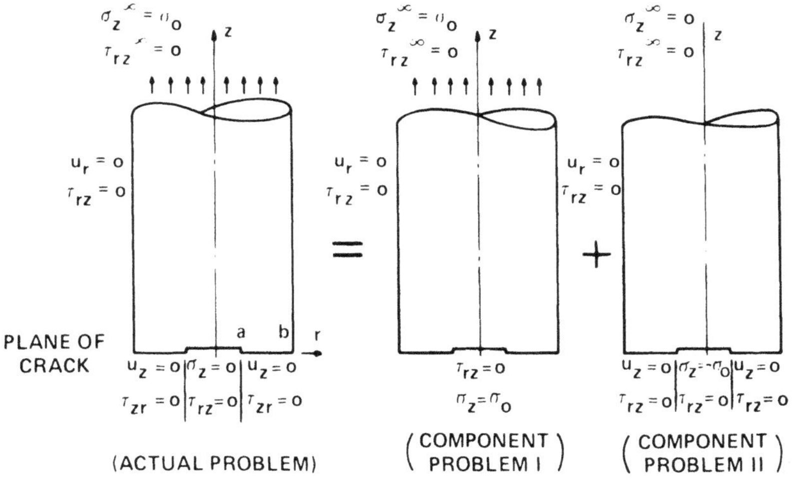

Figure 9. Boundary conditions of problem by superposition for actual and component problems.

$$\varepsilon_{r\theta} = \varepsilon_{\theta z} = \varepsilon_{rz} = 0 \tag{6}$$

$$\tau_{r\theta} = \tau_{\theta z} = \tau_{rz} = 0 \tag{7}$$

The stress solutions are:

$$\sigma_\theta = \sigma_r = \frac{\nu}{1-\nu}\,\sigma_0, \; \sigma_z = \sigma_0 \tag{8}$$

where ν and E are the elastic constants for Poisson's ratio and Young's modulus, respectively.

Component Problem II

Using the formulation from [9], the governing equations, displacements and stresses are written in terms of the Papkovich-Neuber functions ϕ and ψ which satisfy the Laplace equations:

$$\nabla^2\phi = \nabla^2\psi = 0 \tag{9}$$

$$2\mu u_r = -\frac{\partial\phi}{\partial r} - z\,\frac{\partial\psi}{\partial r} \tag{10}$$

$$2\mu u_z = \kappa\psi - \frac{\partial\phi}{\partial z} - z\,\frac{\partial\psi}{\partial z} \tag{11}$$

$$\sigma_r = -\frac{\partial^2\phi}{\partial r^2} + 2\nu\,\frac{\partial\psi}{\partial z} - z\,\frac{\partial^2\psi}{\partial r^2} \tag{12}$$

$$\sigma_z = 2(1-\nu)\,\frac{\partial\psi}{\partial z} - \frac{\partial^2\phi}{\partial z^2} - z\,\frac{\partial^2\psi}{\partial z^2} \tag{13}$$

$$\tau_{rz} = \frac{\partial}{\partial r}\left[-\frac{\partial\phi}{\partial z} - z\,\frac{\partial\psi}{\partial z} + (1-2\nu)\psi\right] \tag{14}$$

where the material constant κ is given by

$$\kappa = 3 - 4\nu \tag{15}$$

By choosing a suitable series representation

$$\phi = A_0 + \sum_{n=1}^{\infty} A_n e^{-k_n z} J_0(k_n r) \tag{16}$$

$$\psi = B_0 + \sum_{n=1}^{\infty} B_n e^{-k_n z} J_0(k_n r) \tag{17}$$

containing an exponential term in z and the radial dependency through the Bessel function of zero order J_0, the unknown coefficients A_k and B_k are determined by imposing the following boundary conditions:

$$\tau_{zr}(r,0) = 0 \qquad 0 \leq r \leq b \tag{18}$$

$$\sigma_z(r,0) = f(r), \text{ prescribed } 0 \leq r < a \tag{19}$$

Mixed conditions (19)–(20)

$$u_z(r,0) = 0 \qquad a < r \leq b \tag{20}$$

$$u_r(b,z) = 0 \tag{21}$$

$$\tau_{rz}(b,z) = 0 \tag{22}$$

also

$$(\tau,\sigma) \rightarrow 0 \text{ as } z \rightarrow \infty \tag{23}$$

$$(\tau,\sigma) \text{ finite as } r \rightarrow 0 \tag{24}$$

From (21) and (22) one obtains the functional condition

$$J_1(k_n b) = 0 \qquad n = 1,2,... \tag{25}$$

and from (18) the relationship between the constants

$$A_n = -\frac{(1-2\nu)}{k_n} B_n \qquad n = 1,2,... \tag{26}$$

The mixed boundary conditions (19) and (20) can then be written as:

$$-\sigma_z(r,0) = \sum_{n=1}^{\infty} B_n k_n J_0(k_n r) = -f(r), \quad 0 \leq r < a \tag{27}$$

$$\frac{\mu}{1-\nu} u_z(r,0) = \frac{\kappa B_0}{2(1-\nu)} + \sum_{n=1}^{\infty} B_n J_0(k_n r) = 0, \; a < r \leq b \tag{28}$$

Following a method similar to Sneddon [10] by rewriting these equations in terms of non-dimensional variables, inverting equation (28) and substituting into equation (27), the problem is reduced to the following set of equations:

$$-\frac{\kappa B_0}{4b(k-\nu)} = \int_0^{a/b} h(\tau)\tau d\tau \tag{29}$$

$$B_n = -\frac{2}{k_n J_0^2(k_n b)} \int_0^{a/b} h(\tau)\ \sin(k_n b\tau)d\tau \tag{30}$$

where the function $h(\tau)$ is the solution to the Fredholm equation:

$$h(\tau) - \int_0^{a/b} L(\tau,u)\ h(u)du = \chi(\tau), \qquad 0 \le \tau < \frac{a}{b} \tag{31}$$

The kernel in this integral equation is given by:

$$L(\tau,u) = \left(\frac{2}{\pi}\right)^{5/2} \int_0^{\infty} \frac{K_1(y)}{I_1(y)} \sinh(\tau y)\sinh(uy)dy \tag{32}$$

where K_1 and I_1 are modified Bessel functions. The load function may be written as:

$$\chi(\tau) = \frac{2}{\pi} \int_0^{\tau} \frac{uF(u)du}{\sqrt{\tau^2-u^2}}, \qquad F(u) = f(bu) \tag{33}$$

In the special case of a uniform crack surface pressure $f(bu) = -\sigma_0$, (33) becomes:

$$\chi(\tau) = -\frac{2\sigma_0}{\pi}\tau \tag{34}$$

In general, the Fredholm equation (31) can be conveniently solved by any standard numerical method. If collocation is used, this equation written in a discretized form becomes:

$$h(\tau_i) - \sum_{k=1}^{N} L(\tau_i,\tau_k)h(\tau_k)\Delta\tau_k = \chi(\tau_i),\ i = 1, \ldots, N \tag{35}$$

from which $h(\tau_k)$ is obtained

The stress intensity factor can be obtained from the following stress expression:

$$\sigma_z(b\rho,0) = \sigma_z(r,0) = -\sum_{n=1}^{\infty} B_n k_n\ J_0(k_n r) \tag{36}$$

$$= 2\int_0^{a/b} h(\tau)d\tau \sum_{n=1}^{\infty} \frac{J_0(k_n b\rho)\sin(k_n b\tau)}{J_0^2(k_n b)} \tag{37}$$

where $\rho = r/b$.

The series in (37) can be evaluated by the method described in [10]. By considering the dominant part as $\rho \to c$, the stress intensity factor K is determined by the following limiting process where $c = a/b$.

$$K\sqrt{\frac{c}{a}} = \lim_{\rho \to c} \sqrt{2\pi(\rho - c)}\,\sigma_z(\rho b, 0) \tag{38}$$

The resulting solution is:

$$K = -\frac{h(c)}{c}\,\sigma_o\sqrt{\pi a} \tag{39}$$

RESULTS AND CONCLUSION

The solution to determine the stress intensity factor for the original problem is completely provided by the solution obtained from Component Problem II. The results of that solution are shown in Figure (10) as the normalized stress intensity factor as a function of the crack to cylinder radii ratio (a/b). In the limit as the ratio a/b approaches zero, the solution reduces to that for a penny shaped crack in an infinite medium. At this point it is convenient to express the results in terms of the localized area fraction which is a suitable material parameter providing a measure of the degree of porosity on the critical fracture plane. Figure (11) is a plot of the normalized stress intensity factor as a function of the localized crack area fraction A_f. In this figure both the solution for the interaction of an array of circular cracks and an array of collinear cracks according to solution (1) are compared. It is seen that the stress intensity factor for the two dimensional solution is higher than that for the three dimensional solution. This implies that the critical strength of the material determined by the two dimensional solution will underestimate the strength based on a 3-D criterion.

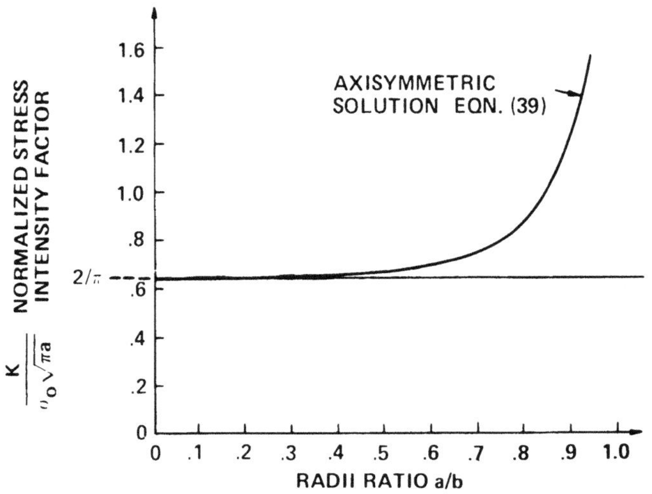

Figure 10. Normalized stress intensity factor ($K/\sigma\sqrt{\pi a}$) versus crack to cylinder radii ratio (a/b).

The range of applicability of the simpler two dimensional solution can be examined by introducing the parameter S defined by

$$S = \frac{\sigma_{0(cr)}^{(\infty)} - \sigma_{0(cr)}}{\sigma_{0(cr)}^{(\infty)}} \times 100 \tag{40}$$

which is a measure of the percentage decrease in the critical stress compared to the value of that for an isolated crack in an infinite medium. This parameter provides a suitable means to compare the solutions of the multiple flaw model of an array of circular cracks to a representation for the array of 2 dimensional collinear cracks obtained from Koiter's solution [6]. The critical strength decrease is plotted in Figure (12). The critical stress decrease for the

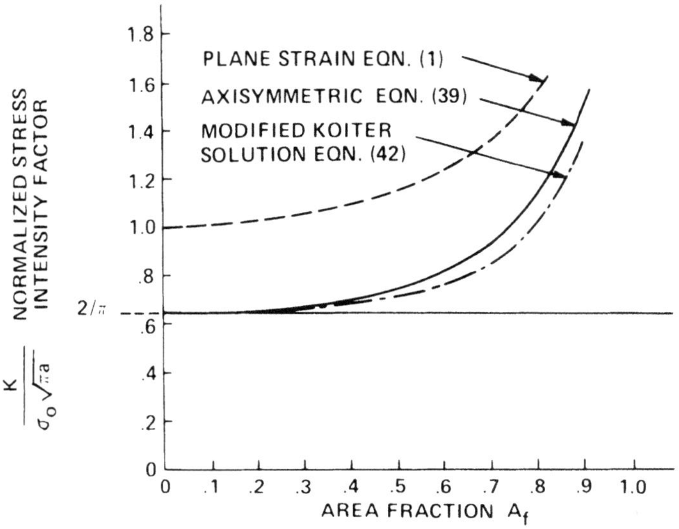

Figure 11. Normalized stress intensity factor $(K/\sigma\sqrt{\pi a})$ versus crack area fraction.

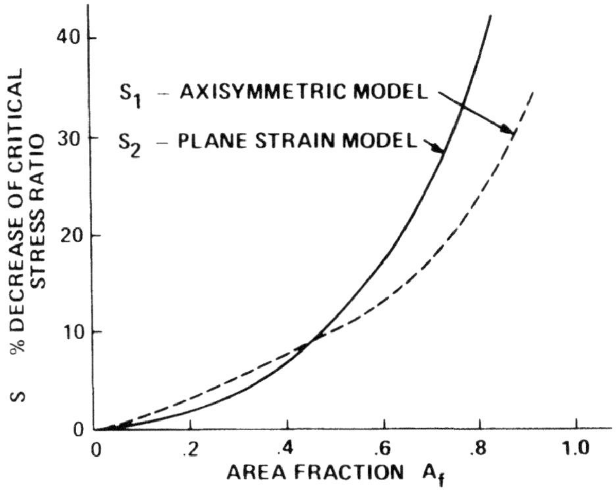

Figure 12. Percentage decrease in critical stress versus flaw area fraction.

array of circular cracks based on an axisymmetric model is shown by curve S_1. Curve S_2 is a plot based on Koiter's solution where the area fraction A_f is the ratio of the crack length to the spacing distance for a uniform thickness strip. The cross over between these 2 curves occurs at an area fraction of about 45%. This result implies that the strength of the material or the stress intensity factor for an array of flaws would provide a conservative value when the 2-dimensional plane strain solution of Koiter is used as a model in this range of defect area fraction. A better approximation is immediately obtained when Koiter's solution is modified by a factor $2/\pi$ which arises directly from the 3-D solution. Hence, the solution given by equation (1) can provide a good engineering estimate to the 3-dimensional flaw interaction problem when the following modified expression is used.

$$K = \frac{2}{\pi}\left[\frac{2b}{\pi a}\tan\left(\frac{\pi a}{2b}\right)\right]^{1/2}\sigma_o\sqrt{\pi a} \qquad (41)$$

This equation can be rewritten in terms of the area fraction

$$K = \frac{2}{\pi}\left[\frac{1}{A_f}\tan\left(\frac{\pi}{2}A_f\right)\right]^{1/2}\sigma_0\sqrt{2a} \qquad (42)$$

from which an expression for the critical stress is easily obtained.

$$\sigma_{cr} = \frac{\pi}{2}\frac{K_{cr}}{\sqrt{\frac{2a}{A_f}\tan\left(\frac{\pi}{2}A_f\right)}} \qquad (43)$$

where K_{cr} is the critical stress intensity factor which is a material property obtained by test.

SUMMARY

Under certain simplifying assumptions, a multiple flaw material model can be reduced to a planar array of interacting circular cracks which is applicable to ceramics having a distribution of pore type defects. Using a fracture mechanics formulation, the stress intensity factor for a single element of the crack array can be determined. This provides a solution for the strength of the material in terms of the defect area fraction. It was shown that the two dimensional solution of Koiter modified by a constant term provides a good approximation to the more complex three dimensional solution. See Figure (11).

REFERENCES

1. R. W. Rice, "Fracture Mechanics of Ceramics," V. 1, R. C. Bradt, ed., Plenum Press, pp. 323-345, (1973).

2. A. A. Griffith, Phil. Trans. Roy. Soc. (London) Ser. A, 221, pp. 163, (1920).

3. O. L. Bowie, J. Math. and Phys., 35, 60-71, (1956).

4. A. G. Evans and T. G. Langdon, "Structural Ceramics," Prog. Matl. Sci., V. 21, Pergamon Press., (1976).

5. A. P. Gresco, Koncentlatsla Naprazszhenii, V. 1, pp. 104, (1965).

6. W. T. Koiter, Ingenieur Arch., V. 28, pp. 168-172, (1959).

7. E. E. Underwood, "Quantitative Stereology," Addison-Wesley, (1970).

8. Z. Hashin, "Ceramic Microstructure," R. M. Fulrath, ed., John Wiley and Sons, Chap. 14, pp. 313-341, (1966).

9. K. Arin and F. Erdogan, Intl. J. Eng. Sci., 9, 213, (1971).

10. I. N. Sneddon, "Mixed Boundary Value Problems in Potential Theory," North Holland Publishing Co., (1966).

11. I. N. Sneddon and M. Lowengrub, "Crack Problems in the Classical Theory of Elasticity," John Wiley and Sons, (1969).

ANALYSIS OF MICROVOIDS IN Si_3N_4 CERAMICS BY SMALL ANGLE NEUTRON SCATTERING

P. PIZZI

FIAT Research Center

Orbassano TURIN (ITALY)

1. INTRODUCTION

The quality and mechanical properties of ceramics are greatly determined by pores or inclusion concentration and dimensions (1-3). The minimum thermal or mechanical stress to initiate crack propagation is, in fact, a function of crack length and density (4,5). The knowledge of these values with non destructive measurements is possible by the use of the small angle neutron scattering technique (SANS). In fact the small angle neutron scattering is particularly useful to examine "inhomogeneties" greater than ~20 Å (6). In the term "inhomogeneties" we summarize all compositional or positional atomic density fluctuations arising from defects like dislocations, microvoids or different phases in materials.

Moreover the high penetration of neutrons in materials allows to obtain such information in a non destructive way on components of different thicknesses, and to follow the degeneration processes due to creep, thermal fatigue or others (7,8). In particular the use of neutrons of high wavelength allows to avoid the Bragg scattering contribution of the crystalline lattice and to evidence the "inhomogeneties" ranging between 20 and 10^5Å. By using the SANS device placed at the Galileo reactor (9) the neutron scattering from silicon nitride at different densities and treatments has been measured; some comparisons with S.E.M. measurements have been also performed but the purpose of the work was essentially oriented to demonstrate and evaluate the performance of the SANS technique with c eramics materials. In fact the "inhomogeneties" dimension present in Si_3N_4 as microvoids are usually in the range from a few hundreds of Angstroms to tenths of micrometers. This large extension of scattering center dimensions gives rise to the superposition of multiple neutron refraction from large voids to the usual diffraction phenomena for smaller voids (10). Therefore the interpretation of the scattering effects is particularly important to obtain the desired non destructive information on the silicon nitride microstructure. Before showing the obtained results on the hot pressed and reaction bonded silicon nitride of different densities and submitted to thermal fatigue a brief summary of the neutron scattering theory will be given.

2. THEORY

The purpose of the "small angle neutron scattering" (SANS) is to furnish information about dimension, volume concentration and possibly shape of the "inhomogeneties" present in the materials. Such type of information are generally obtained by comparison between the angular distribution of the scattered neutrons and the scattering functions of scattering center models.

Usually for SANS one should notice what follows (11):

- It is not possible to distinguish between "particle scattering" and scattering from void of the same shape and size.
- It is difficult to discriminate between shape and dimensional polidispersion effects.
- In the case of particles or voids of dimension a one has to consider the following phase difference between the impinging and diffracted waves:

$$\phi = \frac{2\pi\ a}{\lambda}\ (1 - n') = \Delta\eta . a . \lambda \tag{1}$$

where: λ = neutron wavelength
n' = refraction index for neutrons
$\Delta\eta$ = scattering length density between the matrix and the scattering centers.

Generally we have the following two cases:
a) diffraction for $\phi << 1$
b) refraction for $\phi >> 1$

The first case is usually well known and discussed in literature (11) while the second one is not generally considered in details and will be partially treated here (12).

It has been demonstrated that in the case of refraction ($\phi >> 1$) the scattering cross-section $d\sigma/d\Omega$ for spherical particles of radius R is (12) described by:

$$\frac{d\sigma}{d\Omega} \cong \frac{\pi R^2 \Delta\eta^2 \lambda^4}{(\Delta\eta^2\lambda^4 + \pi^2\theta^2)^2} \tag{2}$$

where θ is the scattering angle as shown in the schematic picture of the measuring geometry of Figure 1. In this figure are also reported the collimation slits D1, D2 and the multidetector plane perpendicular to the neutron beam axis.

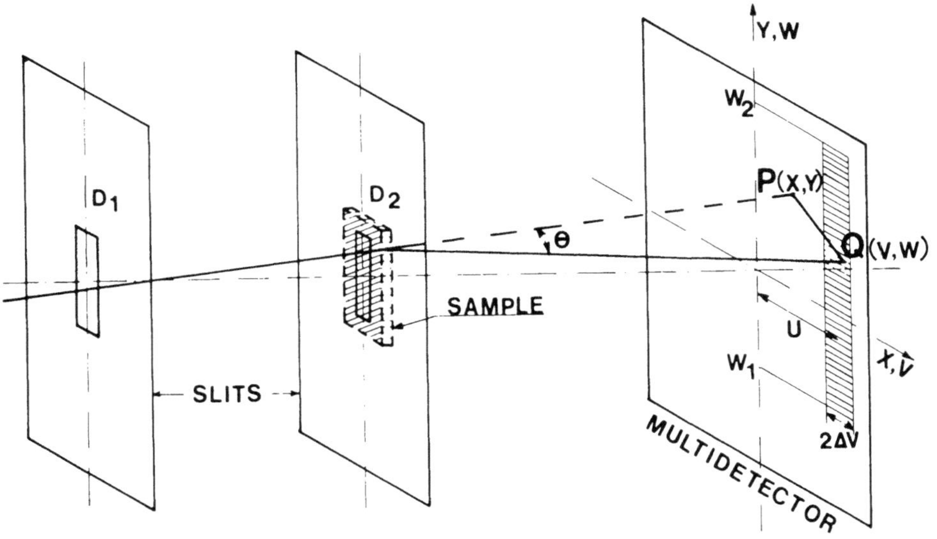

Figure 1. Schematic of Apparatus

If the scattering is multiple, as is usual in porous media, the scattered neutron intensity distribution I(θ) is furnished by the following relationship:

$$I(\theta) = I_o \exp(- \Sigma_a . D) . W(\theta,n) \qquad (3)$$

where: I_o is the direct beam

Σ_a is the absorption cross section

D is the thickness of the sample

and W(θ,n) is the Snyder-Scott distribution function, that is (13):

$$W(\theta,n) = \frac{1}{n} \int_0^{\infty} \cos \theta.s.\exp \left(n(q(s) - 1)\right) ds \qquad (4)$$

where:

$$q(S) = \int_{-\infty}^{+\infty} \exp \left(i\ \theta\ S\right)\ p(\theta)\ d\theta$$

with

n = number of refractions

p(θ)= probability to have a certain scattering in the angle θ.

For $\theta > \theta_{min}$ the above distribution is approximated by the following:

$$W(\theta,n) = \frac{\Delta\eta^2.n}{8\pi^2} \cdot \frac{\lambda^4}{\theta^3} \left\{1 - 3n \frac{\Delta\eta^2\lambda^4}{2\pi^2\theta^2} \left(0,89 + \ln \frac{\Delta\eta\lambda^2}{2\pi\theta}\right)\right\} \qquad (5)$$

While at low scattering angles ($\theta < \theta_{min}$) the gaussian distribution represent very well the scattered neutron intensity function. From a measurement of beam broadening Δθ it is then possible to obtain the number of multiple refractions n, that is:

$$\Delta\theta^2 - \Delta\theta_o^2 = \alpha.n. \left(\frac{\Delta\eta\lambda^2}{2\pi}\right)^2 . \left\{\beta + \ln \left(\frac{\gamma}{\Delta\eta\lambda^2}\right)\right\} \qquad (6)$$

where α, β, γ are constants depending on the shape of the particles and the measuring geometry and $\Delta\theta_o$, Δθ are the beam width measured at one-half of the maximum intensity.

Now if we suppose, in the case of ceramics, a uniform distribution of N voids/cm^3 of single volume V_p, inside a larger grain volume V_g we can write:

$$V_g \cdot \rho_p = V_g \cdot \rho - V_p \cdot \rho \qquad (7)$$

where: ρ_p = apparent density of the sample

ρ = true density

From the relation (7) one can deduce:

$$V_p = \frac{1}{N} \cdot \frac{\rho - \rho_p}{\rho} \tag{8}$$

If we consider, moreover, that the number of refractions n is proportional to:

$$n = N \cdot \sigma_R \cdot D \tag{9}$$

with σ_R the single refraction cross-section equal to $2\pi R^2$ one /12/:

$$V_p = \sigma_R \cdot \frac{D}{n} \cdot \left(\frac{\rho - \rho_p}{\rho}\right) \tag{10}$$

In the particular case of spherical particles the relation (10) furnishes for the average radius $\bar{R}$:

$$\bar{R} = \frac{3}{2} \cdot \frac{D}{n} \cdot \frac{\rho - \rho_p}{\rho} \tag{11}$$

3. EXPERIMENTAL RESULTS

The small angle neutron scattering technique has been applied to Si_3N_4 ceramics of different densities and different manufacturing procedures. Essentially hot pressed and reaction bonded Si_3N_4 of densities from 2,28 to 3,2 g/cm^3 have been measured. For hot pressed Si_3N_4 submitted to thermal fatigue the scattering cross-section and microvoids evolution have been measured at various cycle numbers. In some cases the results have been compared with scanning electron microscopy observations.

3.1. SANS on Si_3N_4 of different densities

Samples of Si_3N_4 ceramics, hot pressed and reaction bonded, of about 0,8 cm of thickness have been measured at two different wavelengths of 8 and 10 Å. The relative neutron intensity vs. the square of the scattering vector $\bar{K} = \frac{2\pi\theta}{\lambda}$ is shown in Fig. 2, where a large effect on densities is noticeable.

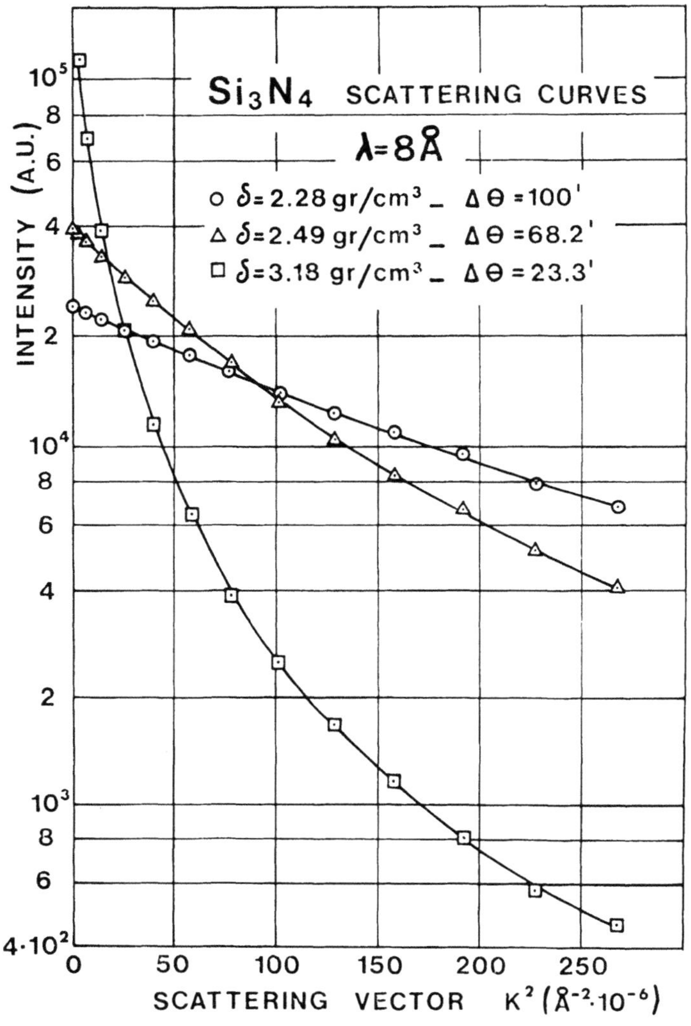

Figure 2. Si_3N_4 Scattering Curves

To evaluate and discuss the results, in Fig. 3 the SANS measurements for δ = 2,46 and 3,18 g/cm^3 at 8 and 10 Å have been shown. In the same figure the scanning electron microscopy for hot pressed Si_3N_4 show visible micropores also inside the grains. Differences in the scattered intensities at the two different wavelengths are due to multiple refraction effects as will be discussed later. Analogous measurements have been performed for hot pressed Si_3N_4 of small different densities as shown in Fig. 4.

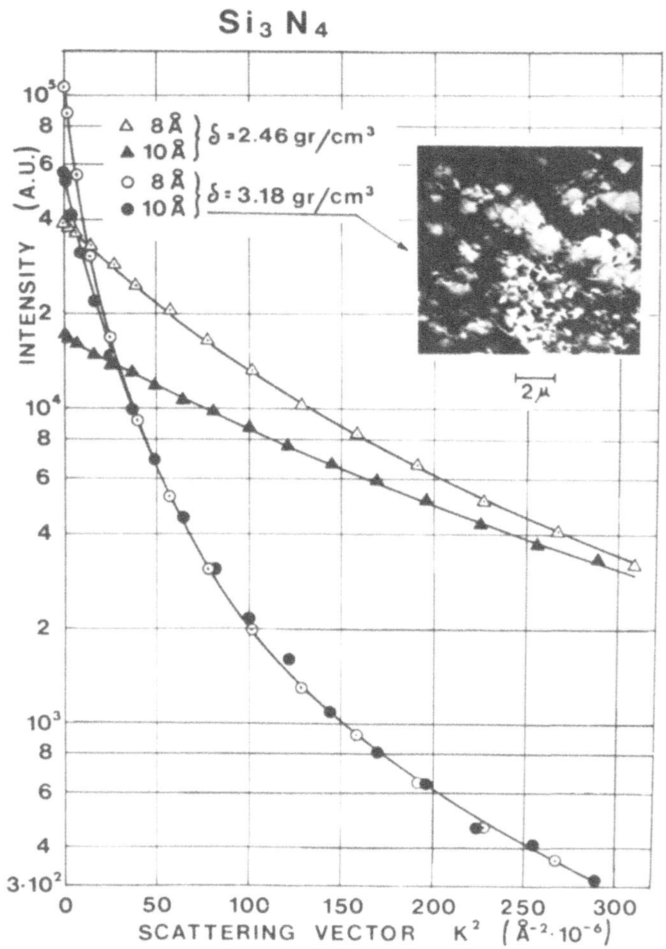

Figure 3. Wavelength Effects on Scattering Intensities

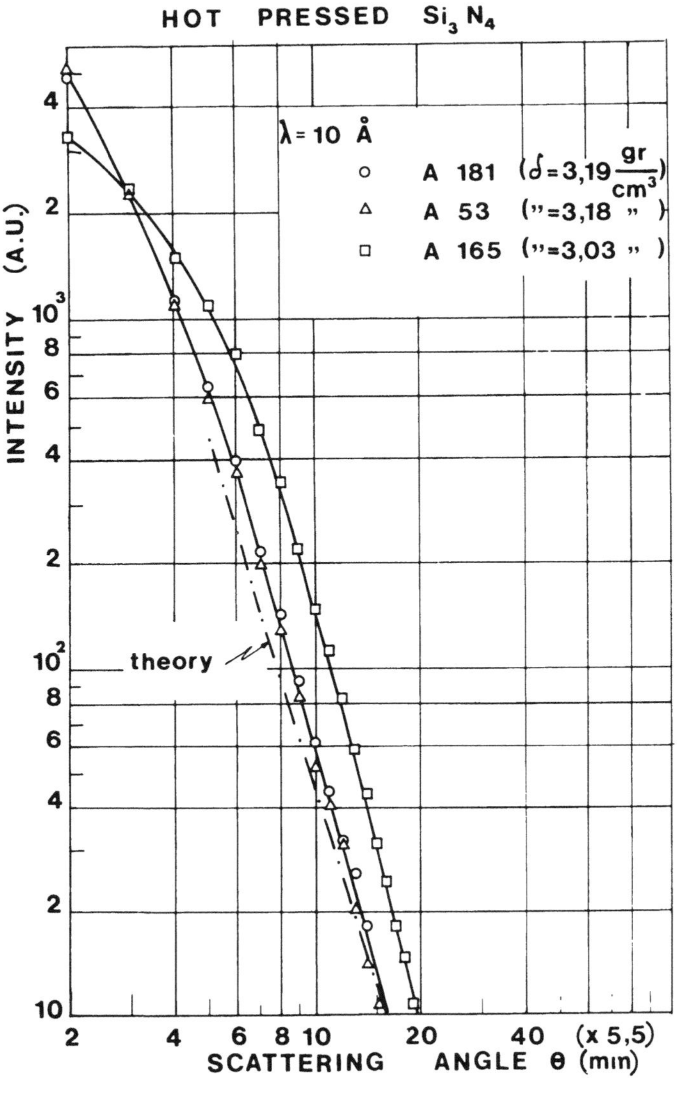

Figure 4. Scattering, Compared with Theoretical

Small density differences give large scattering effects; in particular the presence of large microvoids is clearly visible from multiple refraction effects at different wavelengths. Slit and wavelength resolution corrections are performed on the curves shown in Fig. 4 where for comparison a curve obtained for n = 50 calculated on the basis of the foregiven theory (equation 5) is also shown.

3.2. Effects of thermal fatigue on hot pressed Si_3N_4 by SANS

Hot pressed silicon nitride specimens of parallelepiped shape have been submitted to thermal fatigue cycles between room temperature and 1300°C every four minutes. In the hot pressed samples is also present 5% of MgO and the apparent initial density is 3, 18 gr/cm^3.

Measurements by SANS have been performed at two different wavelengths of 7,5 and 10 Å. An example of the obtained scattering curves for the initial samples and after 2400 cycles is shown in Fig. 5 and in Table I is reported the beam broadening at different cycle numbers. The scattering function is essentially due to refraction effects from larger voids, in fact, the differences in scattering at low angles (Table I) for two wavelengths and comparison with theory suggest this interpretation of SANS curves. From the beam broadening $\Delta\theta$ and using (6) the number of refractions has been calculated. The refraction density, that is, the number of refractions per unit length is reported in Fig. 6 in function of thermal fatigue cycles.

By supposing microvoids of spherical shape from (11) it is possible to deduce an initial value for the radius $\bar{R}_o$ equal to $\sim$2,7μm. This value is consistent with the refraction effect; in fact, if we keep in account a value of the scattering length density $\Delta\eta$ for Si_3N_4 of 8,6.10 cm, from (1) one obtains for the phase difference ϕ a value of about 5 at 10 Å.

The increase of the refraction density (n/D), during thermal fatigue, is connected by (9) with the number of scattering voids N and with the refraction cross section σ_R. If we suppose that the most important effect is the increase in σ_R due to the increase in the size of microvoids, then it is possible to calculate the increase of the average radius $\bar{R}$ by the relation $\sigma_R = 2.\pi.R^{-2}$

The evolution of the radius $\bar{R}$ in function of the cycle numbers is shown in Fig. 7. It is interesting to observe an initial large increase in the size and a second stage with a constant subcritical radius rate increase, as will be discussed later in more detail.

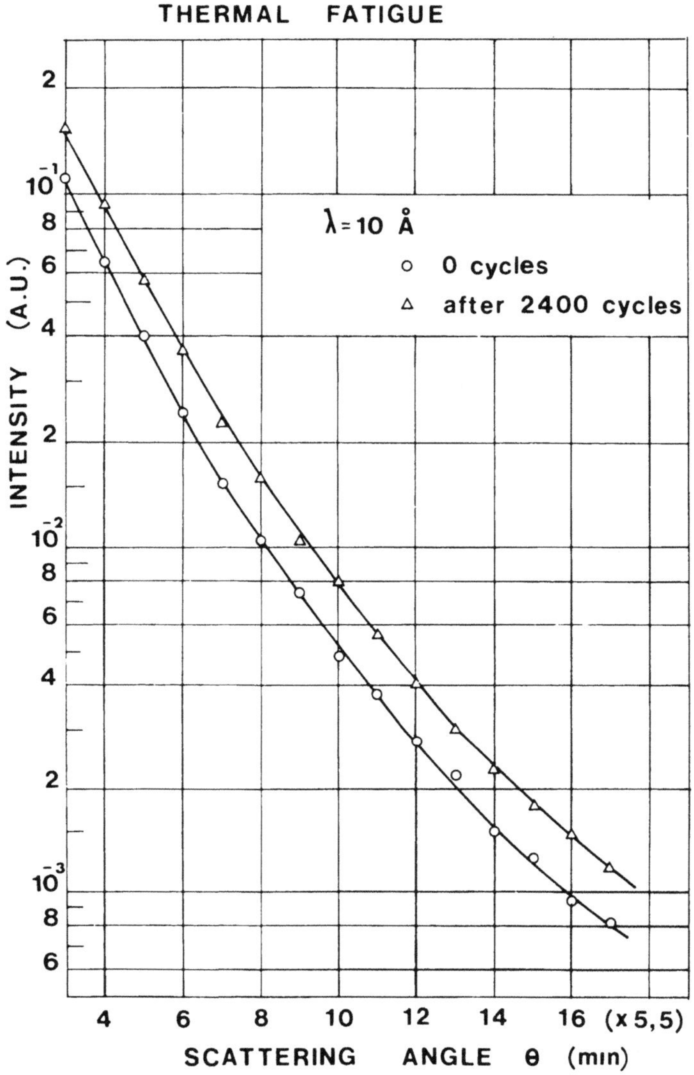

Figure 5. Thermal Fatigue

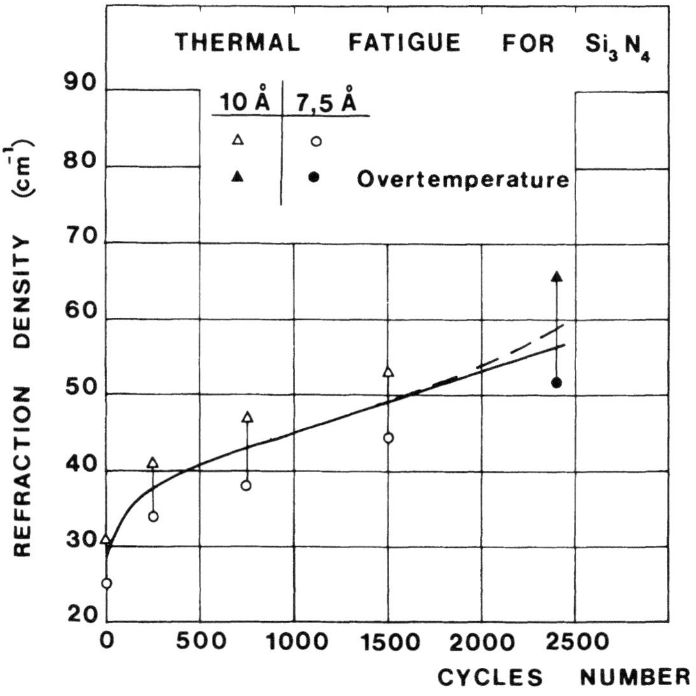

Figure 6. Refraction vs. Number of Cycles

TABLE I

Cycles Nr	Beam broadening $\Delta\theta$ [10^{-3} rad] λ = 7,5Å	λ = 10Å	Direct beam $\Delta\theta_o$ [10^{-3} rad]
0	3,3	5,1	2,2
250 (top)	3,6	5,9	2,2
250 (tip)	3,6	5,7	2,2
750	3,7	6,3	2,2
1500	3,9	6,6	2,2
2400	4,2	7,4	2,2

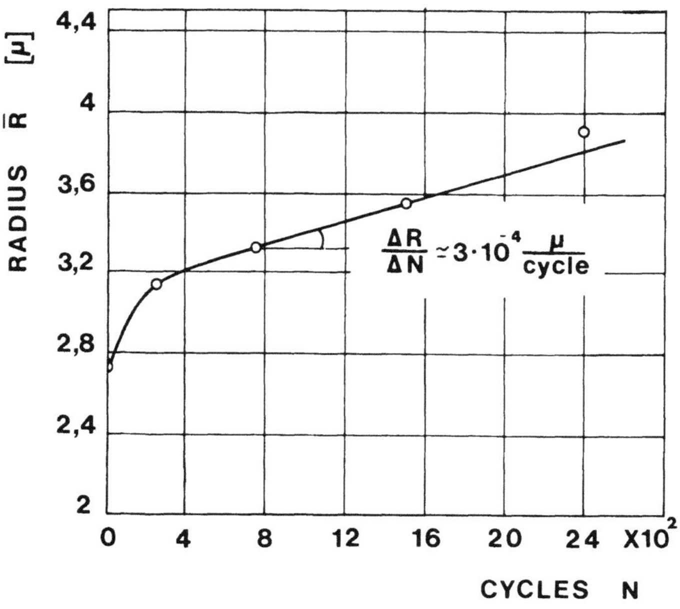

Figure 7. Microvoid Evolution vs. Thermal Fatigue

To avoid interferences from oxidation effects in some cases the samples have been measured with and without the presence of the oxide layer; no appreciable differences have been observed.

4. DISCUSSION

In the case of reaction bonded Si_3N_4 the presence of large microvoids is easily visible by SANS measurements and is practically the most important effect covering all other diffraction effects. For the two specimens having 2,46 and 2,28 gr/cm^3 densities, the obtained values for the average radius of microvoids are 5 and 3 μm. The high density of these voids give quite a gaussian distribution of the scattered intensity also at large angles as visible in Fig. 2 and 3. It should be, then, very difficult to observe other microstructural effects due to the presence of other phases or heterogeneities.

In the case of the hot pressed silicon nitride submitted to thermal fatigue it is important to point out the following:

a) The presence in some cases of a small number of microvoids of large dimensions is very important for the high contribution to the scattering as in the case of the specimens submitted to thermal fatigue.

b) The evolution of microvoid dimensions during cycling is consistent with thermal fatigue theories /14,15/. In these theories the growth is due to the diffusion of vacancies from the bulk or along the grains to the tip of the original voids. The growth law is in general of the following type /14/:

$$\frac{da}{dt} = \alpha \quad 1 - \exp\,(\beta - \gamma a) \tag{12}$$

where: α, β, γ are constants depending on the properties of the material.

There is a subcritical growing of the dimensions a of the microvoid dimensions.

This behaviour is confirmed by the SANS measurements shown in Fig. 7, where after an initial high growth rate, the rate of subcritical growing decreases and is equal to $\sim 3.10^{-4}$ μm/cycle.

The initial growth rate is equivalent to $\sim 2.10^{-3}$ μm cycle. These values are in agreement with the values of Amman et al. that for Si_3N_4 thermally cycled from about 1300°C to 40°C the growth rate was found to be 10^{-3} μm/cycle /15/.

c) The possibility to follow the degeneration processes in Si_3N_4 by SANS has been demonstrated by the preliminary experiences here reported. It remains to complete essentially what follows:

- to study further the theory to interpret completely the superposition of refraction and diffraction phenomena and to analyze the shape and dimensional distribution effects.

- to extend the applications of SANS both to thermal fatigue at different temperatures and characteristics of Si_3N_4 and to other degeneration phenomena like creep. In fact if we consider the high costs involved in these experiences where usually it is necessary to have a statistical approach, the use of SANS allows to decrease the number of specimens and consequently the costs.

- to examine the possibility to control components like turbine blades or others because of the N.D. characteristic of SANS it is of great importance also to develop other N.D. methods.

- to keep in account and examine in detail the small contribution to SANS eventually given by Bragg scattering for larger crystallographic planes of α and β hexagonal structures.

ACKNOWLEDGMENTS

The author is grateful to FIAT management for permission to publish and their stimulating effort to the development and applications of the technique, and in particular for the contribution of Dr. H. Walther. This work has been performed in collaboration with E. Ansaldi, B. Giorgi, F. Iazzi, U. La Malfa, A. Rastaldo, P. L. Tarditi for SANS experiments and sample preparations. Particular thanks are due to P. Cortese for the collaboration in the theory and the important discussions. This work has been performed with the collaboration of the CAMEN Center for SANS measurements.

REFERENCES

1. D. P. H. Hasselman, J. Amer. Ceram. Soc. vol. 52, n. 11, Nov. 1969.
2. H. R. Baumgartner, D. W. Richerson, Symposium on Fracture Mechanics of Ceramics, Penn State 1973.
3. R. W. Davidge, Symposium on Fracture Mechanics of Ceramics, Penn State 1973 - Bradt, Hasselman, Lange, Plenum Press, N.Y.
4. J. W. Edington, D. T. Rowcliffe, J. L. Henshall, Powder Metallurgy International vol. 7, n. 2, 1975.
5. Ram Kossowsky, J. of Mat. Sci. 8 (1973) 1603-1615.
6. W. Schmatz, T. Springer, J. Schelten, K. Ibel, J. of Appl. Cryst. (1974) 7, 96.
7. G. Cherubini, A. Olivi, (CAMEN); P. Pizzi, H. Walther, (FIAT), La Metallurgia Italiana vol. LXVII, n. 6, Giugno 75.
8. P. Pizzi, H. Walther, P. Cortese, 8th World Conf. on N.D. Testing - CANNES September 1976 - 3L7.
9. G. P. Galotto, P. Pizzi, H. Walther, (CRF): V. Angelastro, N. Cerullo, G. Cherubini, (CAMEN), Nucl. Instr. and Methods, 134 (1976) 369-378.
10. A. Guinier, G. Fournet, John Wiley, New York, 1955.
11. C. G. Shull, L. C. Roess, J. of Appl. Phys. vol. 18, March 1947.
12. R. J. Weiss, Phys. Rev. vol. 83, n. 2, July 15, 1951.
13. H. S. Snyder, W. T. Scott, Phys. Rev. vol. 76, n. 2, July 15, 1949.
14. R. Dutton, Symposium on Fracture Mechanics of Ceramics, vol. 2, Penn State, July 1973.
15. C. L. Ammann, J. E. Doherty, C. G. Nessler, Mat. Science and Engineering 22 (1976) 15-22.

CRACK BLUNTING IN SINTERED SiC

C. A. Johnson

General Electric Corporate Research and Development

Schenectady, New York

INTRODUCTION

Submicron powder of β-SiC with small additions of boron and carbon (approximately 0.5 w/o of each) can be densified to greater than 98% of theoretical by solid state sintering in a reducing atmosphere at approximately 2100°C.[1] The consolidated material is composed of β and/or α-SiC,[2-4] depending on impurities and sintering conditions, with isolated inclusions of carbon and B_4C. The resulting high temperature properties are excellent in terms of creep, oxidation, and stress rupture,[3-5] probably due to the lack of any continuous grain boundary phase. These excellent properties combined with the availability of shaping techniques such as slip casting and injection molding qualify boron-doped sintered SiC as one of the most promising ceramics for structural applications in oxidizing environments at temperatures of 1400°C and above.

The dimensional stability of SiC during sintering is sufficient to preshape a complex body to over-sized dimensions in the green state and reproducibly sinter to final density with dimensional tolerances which are satisfactory for most applications. For some applications such as bearing and attachment surfaces, however, dimensions must be defined more accurately, thus requiring a final machining operation in the densified state. Diamond grinding of brittle ceramics such as sintered SiC introduces surface damage to the body which, depending on the relative severity of the machining flaws, generally reduces the fracture strength. It was observed in the present study that much of the strength degradation induced by diamond grinding in sintered SiC could be restored by refiring the machined body in vacuum at temperatures approaching the original sintering temperature. The actual mechanism of strength recovery is believed to be a crack "blunting" mechansim through the chemical dissociation and removal of a small amount of SiC from all free surfaces including the walls of defects which intercept the surface (such as machining flaws).

The equilibrium partial pressure of vapor species above SiC plus graphite has been studied as a function of temperature by Drowart and DeMaria[6] using a mass spectrometer to analyze the vapor effusing from a Knudsen cell. For temperatures below 2100°C, they reported that Si was the predominate vapor species with only minor amounts of other molecules such as SiC_2, Si_2C, Si_2, Si_3C, Si_2C_2, SiC, Si_2C_3, and Si_3. The equilibrium pressure of silicon at 2100°C was reported at approximately 10^{-3} atmospheres which is sufficiently high to allow measurable rates of weight loss due to dissociation of SiC in most graphite furnaces. The dissociated silicon vapor is transported to cooler parts of the furnace where it recombines with carbon insulation, etc. The dissociated carbon generally remains on the original surface and is readily observed on all specimens sintered under low total furnace pressures. The kinetics of the dissociation reaction are strongly influenced by the total pressure of inert gas above the SiC. This pressure dependence arises because the rate of dissociation is generally limited by the rate of vapor transport away from the surface which, in turn, is dependent upon the total gas pressure. Low pressure therefore tends to enhance the vapor transport process, increase the rate of SiC dissociation, and more effectively "blunt" any surface flaws.

The experimental work supporting the crack blunting hypothesis and characterizing the rate of SiC dissociation falls into three catagories: strength testing to demonstrate the improvement in fracture strength due to crack blunting; direct observation of thermal shock cracks which were "blunted"; and weight loss measurements to characterize the rate of material removal as a function of temperature, time and total inert gas pressure. The experimental results of these three areas are discussed in the following three sections.

STRENGTH TESTING

The first experimental observation of crack blunting in sintered SiC was in a study of machining and subsequent heat treatment on fracture strength. A series of 53 specimens were die-pressed in a 0.2 x 2.0 in. double-acting die at 4 kpsi (28 MPa) from a batch of SiC powder which had been screened through an 8 μm screen as a dispersion in benzene to remove foreign material and large agglomerates, spray frozen to prevent segregation of binders, etc., and freeze dried to remove the benzene[4]. The specimens were sintered in a carbon resistance furnace* at approximately 2080°C for 20 minutes re-

*Astro Industries, Santa Barbara, Ca.

sulting in final cross sectional dimensions of 0.165 x 0.165 in., a final density of approximately 95 percent of theoretical and a microstructure composed primarily of 3 to 5 μm grains of β-SiC. Of these specimens, 33 were tested in the as-sintered state at room temperature in a sphere-loaded three-point bend fixture[4] with a 1.5 in. span resulting in an average strength of 94.3 kpsi (650 MPa). A second group of 10 specimens were diamond ground (320 grit) parallel to the specimen length removing 0.0005 in. per pass for a total of 0.005 in. per side. This treatment decreased the average strength to 66.2 kpsi (456 MPa), apparently due to severe machining induced flaws. A third group of 10 specimens were ground in an identical manner but, prior to testing, they were refired in vacuum (0.3 to 0.4 torr of helium) at the original sintering temperature for 20 minutes. This treatment increased the strength from 66.2 to 90.3 kpsi (456 to 623 MPa).

Refiring at or very near the original sintering temperature risks both a small change in dimensions due to additional densification and an increase in the largest grain size due to exaggerated grain growth during any β to α transformation. Lower temperatures, on the other hand, will be less effective at crack blunting since the rate of dissociation decreases with decreasing temperature. It was therefore decided to study the strength of machined specimens as a function of refiring temperature up to the original sintering temperature. A series of 98 specimens of SiC were pressed and sintered in the same manner as those described above. They were diamond ground to 0.150 in. square cross section and were refired in vacuum (0.3-0.4 torr helium) in groups of 12 at six different temperatures with the remaining 26 tested without refiring to estimate the "as-machined" strength. They were tested in three-point bending as described above, and the results are plotted on Figure 1 as fracture strength vs. refiring temperature. The specimens receiving no refiring are plotted as specimens refired at room temperature. Each data point represents an average strength and the corresponding error bar includes the 95 percent confidence interval on the average as calculated from Student's t-distribution. The strength of the as-machined specimens averaged approximately 81 kpsi (560 MPa) and, as expected, was the lowest strength. The highest average of approximately 96 kpsi (660 MPa) was measured for those specimens refired at the original sintering temperature.

Due to the rather large degree of scatter, a dashed curve was drawn through the data of Figure 1 as a suggested relationship between strength and refiring temperature. The sigmoidal shape reflects the limiting effects which are expected at very high and very low refiring temperatures. At progressively higher temperatures the strength will no longer be controlled by machining flaws but will become limited by internal defects which are not affected by the refiring treatment. At progressively lower temperatures there is negligible dissociation of SiC and no crack blunting.

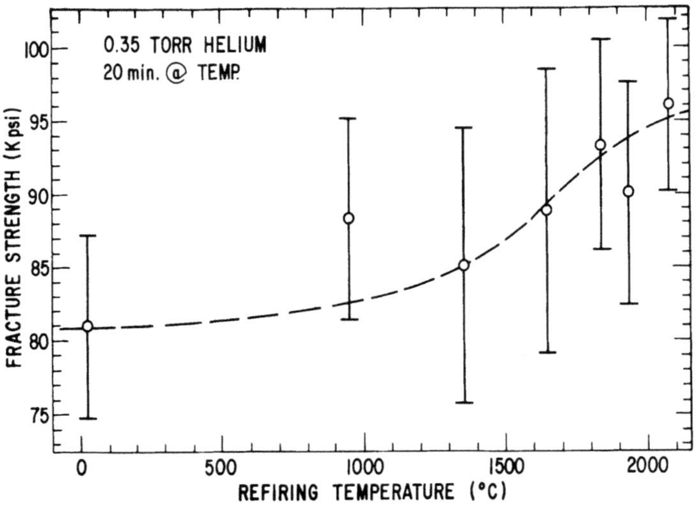

Figure 1. Effect of refiring temperature on the bend strength of diamond ground sintered SiC.

The data included on Figure 1 suggest that the greatest degree of strength recovery in machined specimens requires a refiring temperature no lower than the original sintering temperature. It is expected, however, that the increased benefits of refiring at approximately 100°C below the sintering temperature will compensate for the small sacrifice in strength by maintaining both the original microstructure and the machined dimensions.

The reduction in severity of surface flaws through the dissociation of SiC is not limited to machined surfaces but can also be used to minimize the severity of surface fabrication defects during sintering. The sintering time and temperature are defined by the final density and microstructure which are desired. The principal remaining variable which allows control of the amount of dissociation is the total pressure of inert gas in the sintering furnace. In order to characterize the dependence of fracture strength on furnace pressure during sintering, a series of 150 specimens were die-pressed as described above. The bars were sintered in ten groups of fifteen in a carbon resistance furnace at pressures ranging from 0.35 to 760 torr of helium. Pressure was controlled with a cartesian diver pressure control with a small, continuous flow of helium. All specimens were heated to temperature at 1000°C per hour and held at 2085°C for 30 minutes, resulting in 94 to 95 percent of theoretical density. They were tested in the "as-sintered" state (no machining, etc.) as described above, and the results are plotted on Figure 2 as fracture strength vs. furnace pressure. Again each data point represents the average strength and the corresponding error bar includes the 95 percent confidence interval on the average. The data points of the highest and lowest furnace pressures each represent two groups (30 specimens).

As shown in Figure 2, the lowest furnace pressures yielded an average strength of approximately 105 kpsi (720 MPa) and the highest pressures yielded approximately 80 kpsi (550 MPa). The sigmoidal shape of the dashed curve drawn through the data reflects the expected limiting strengths at both high and low pressures. As furnace pressure is decreased, the "blunting" of surface flaws by material removal can only increase the fracture strength to the point where failure is controlled by internal flaws. At lower furnace pressures, the average strength is expected to be constant. The location of fracture origins supports this reasoning with an increasing proportion of failures observed from subsurface flaws as the furnace pressure was decreased. A lower limiting strength might also be expected at high furnace pressure where the kinetics of the dissociation reaction are reduced to the point of little or no material removal. This is also supported qualitatively by the observation of no free carbon on the surfaces of bodies sintered in one atmosphere of helium, whereas similar firing conditions under vacuum produce an obvious coating of carbon on all surfaces.

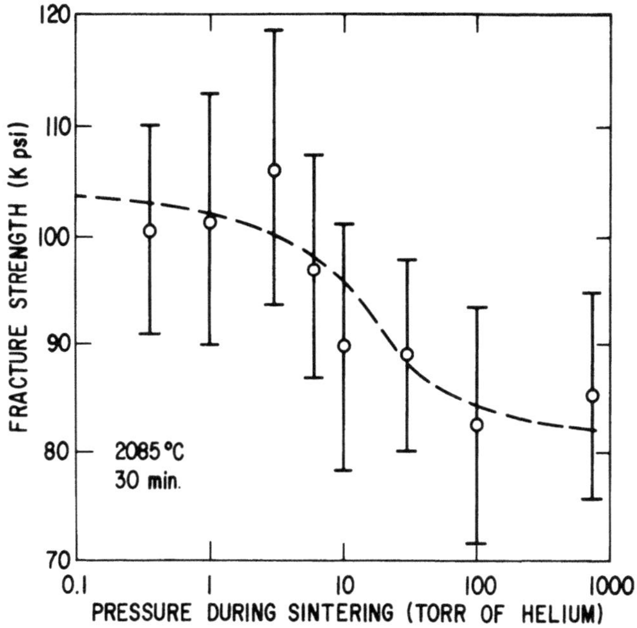

Figure 2. Effect of furnace pressure during sintering on the bend strength of SiC tested in the as-sintered condition.

DIRECT OBSERVATION OF CRACK BLUNTING

The "blunting" or "rounding" of the tips of sharp cracks was easily resolved with both optical and scanning electron microscopy. The material used for these experiments was boron-doped hot-pressed SiC which is described in more detail in the next section. Polished tablets measuring 1.0 x 0.5 x 0.050 in. were heated to 600°C in air and quenched into water at room temperature. This thermal shock treatment resulted in several fragments of the original tablet, some of which contained arrested thermal shock cracks. Figure 3 is an optical micrograph of the tip of such a thermal shock crack illustrating that it is indeed sharp and that the precise location of the end of the crack is difficult to resolve at this magnification.

Fragments containing arrested thermal shock cracks were refired at 1955°C for 30 minutes with a furnace pressure of 0.8 torr of helium. The shape of the resulting blunted crack on one fragment was observed optically at progressively greater depths by carefully polishing SiC from the surface. Figures 4a-c are micrographs of three different sections, all at the same magnification as Figure 3 (different thermal shock crack). Figure 4a was photographed after the surface carbon was removed by polishing with 1 μm diamond abrasive. Very little SiC was removed, therefore this plane was defined as the fired surface of the fragment. From the micrograph, the radius of the crack tip appears to be roughly 5 μm which is orders of magnitude larger than the radius of the sharp crack of Figure 3. The new crack extending from near the tip of the blunted crack on Figures 4b and c was introduced during specimen mounting after Figure 4a was photographed. Figure 4b was photographed after approximately 25 μm of SiC had been removed. The crack tip radius still appears to be roughly 4-5 μm. Figure 4c was photographed after an additional 13 μm for a total of approximately 38 μm had

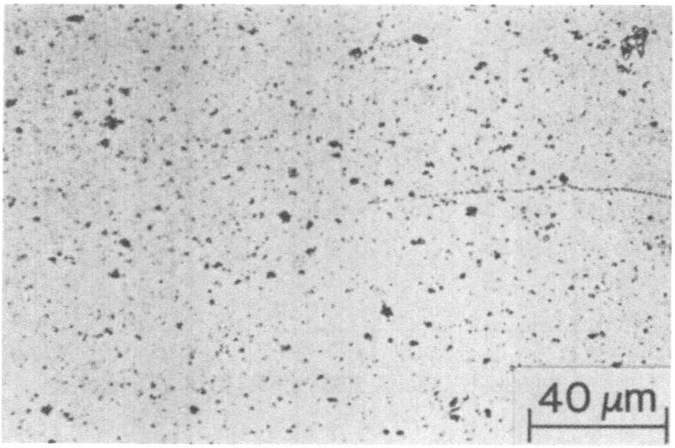

Figure 3. Optical micrograph of the tip of a thermal shock crack.

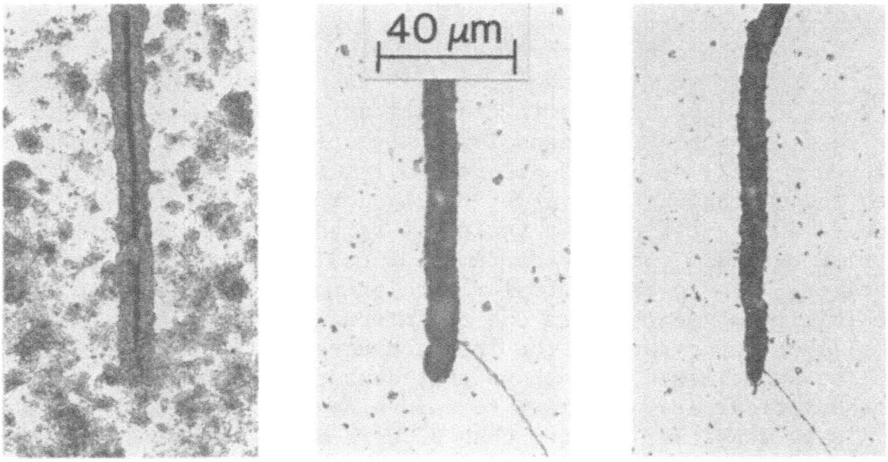

(a) (b) (c)

Figure 4. Optical micrographs of a "blunted" thermal shock crack at 0 (a), 25 (b) and 38 µm (c) from the fired surface.

been removed. At this depth, the crack tip radius was reduced to roughly 3 µm. Polishing continued up to a total depth of approximately 85 µm where the crack tip radius could still be resolved.

A similar fragment with a blunted crack was placed in a furnace for one hour at 700°C in air to oxidize the carbon. Figure 5 is an SEM micrograph of the tip of the crack, again illustrating the blunted shape of a previously sharp crack after a refiring treatment in vacuum.

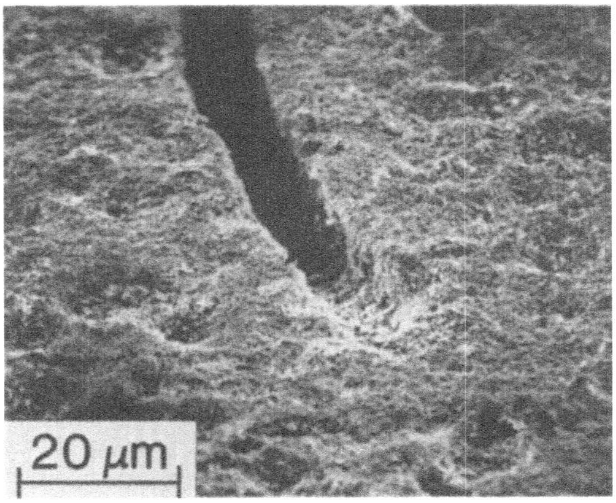

Figure 5. SEM micrograph of a "blunted" thermal shock crack.

WEIGHT LOSS EXPERIMENTS

One of the most attractive features of the refiring treatment for the repair of machining damage is the dimensional stability during refiring, thus maintaining any critical dimensions defined during machining. The dissociation reaction, however, necessarily removes a small amount of SiC from all surfaces and slightly modifies the dimensions. In order to characterize the thickness of SiC removed during refiring, a series of weight loss experiments were carried out as a function of temperature, time and furnace pressure. Specimens with no open porosity were preferred to eliminate the possibility of weight loss due to internal dissociation and transport of silicon vapor through the interconnected network of porosity. Earlier experiments have shown that densities of at least 97 percent of theoretical are necessary to assure no open porosity[4]. Sintering of SiC to densities greater than 97 percent of theoretical generally leads to β to α transformation and exaggerated grain growth[2], therefore the specimens for this study were hot-pressed from boron-doped powder. One hundred grams of screened and freeze-dried powder was hot-pressed in a 2.0 in. diameter graphite die for one hour at 1900°C and 10 kpsi (69 MPa). The resulting billet had a density of approximately 99.5 percent of theoretical, a microstructure composed of β-SiC, and a grain size of less than one micron. Tablets measuring 1.0 x 0.5 x 0.050 in. were cut and ground from this billet. They were cleaned and weighed to the nearest 0.1 mg and were positioned vertically during refiring in a carbon resistance furnace. The specimens were heated to temperature at 1000°C per hour while furnace pressure was maintained with a cartesian diver pressure control with a small, continuous flow of helium.

The amount of SiC removed by dissociation was calculated by two methods. The first assumed that the weight loss during refiring was exclusively due to loss of silicon vapor with no carbon transport, thus defining the weight of SiC dissociated. In the second method, the carbon was removed by oxidation at 700°C for one hour. The total weight loss provided a more direct measurement of the SiC weight loss. The first method is in error if some carbon-containing vapor species (SiC_2, Si_2C, etc.) contribute to the weight loss. Based on the very low equilibrium pressures of such species, this error should be minimal. A third method involving the direct measurement of thickness change was not sufficiently sensitive to yield meaningful results.

The results of these experiments are included on Figures 6-8 where thickness and weight per area of SiC removed by dissociation are plotted versus temperature, time, and pressure respectively. Figure 6 is plotted as a function of reciprocal temperature and includes data measured at six different temperatures from 1670 to 2050°C, all for 30 minutes at temperature and 0.8 torr of helium. Not plotted are the results of an experiment at 1065°C which resulted

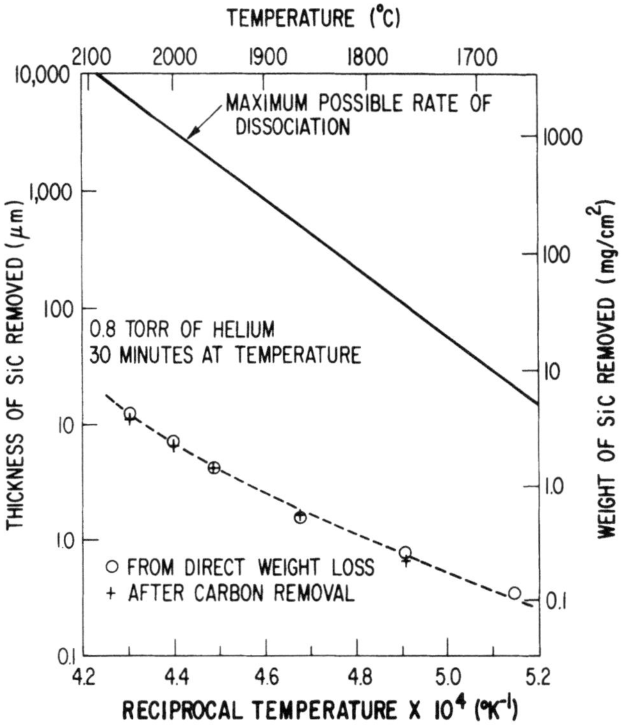

Figure 6. Temperature dependence of the observed and the maximum possible rate of SiC dissociation.

in no resolvable weight loss. The 4 to 5 μm of SiC removed by refiring at 1955°C is acceptably small for maintaining most dimensions. In cases where these tolerances are unacceptable (±0.0002 in.), the body can be machined over-sized by the appropriate amount.

Also included in Figure 6 is the temperature dependence of the maximum possible rate of material removal by dissociation as calculated by assuming Hertz-Langmuir evaporation into a perfect vacuum[7]. This was calculated from the relationship:

$$J = P\left(\frac{M}{2\pi RT}\right)^{\frac{1}{2}}$$

where J is the flux in mass per unit area, P and M are the equilibrium pressure (from Drowart and DeMaria)[6] and the molecular weight of the vapor species, R is the gas constant and T is absolute temperature. Although the temperature dependences are nearly equivalent, the measured rate of dissociation is approximately 2-2.5 orders of magnitude slower than the calculated maximum. A variety of factors

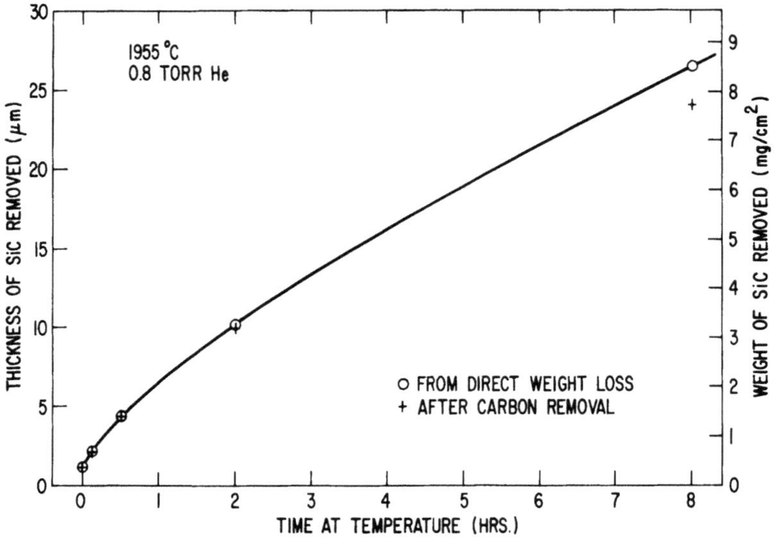

Figure 7. Time dependence of SiC dissociation.

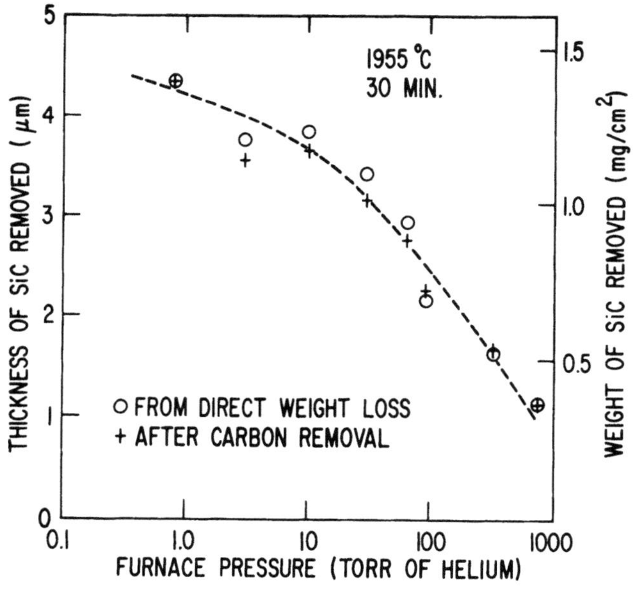

Figure 8. Furnace pressure dependence of SiC dissociation.

may have contributed to this difference including an evaporation constant significantly less than unity, control of the rate by diffusion through the progressively thicker layer of surface carbon, or control of the rate by diffusion of silicon away from the specimen in a less than perfect vacuum. Regardless of the source, the slower than maximum rate of dissociation has important implications for sintering. From Figure 6, 30 minutes at 2100°C could result in greater than one centimeter of SiC dissociated from all free surfaces during sintering instead of the 20 μm or so predicted on Figure 6 and actually observed during sintering in vacuum.

Figure 7 includes the measured time dependence of weight loss at 1955°C and 0.8 torr of helium with total times at temperature from zero to eight hours. The finite weight loss at zero time is due to dissociation while approaching temperature at 1000°C per hour and while furnace cooling afterwards. If the rate of dissociation is controlled by diffusion of silicon vapor through the permeable surface layer of carbon, the time dependence should be approximately parabolic. On the other hand, if the rate is controlled by vapor diffusion away from the specimen, the time dependence should be linear. The behavior observed on Figure 7 is between linear and parabolic suggesting that both may contribute to the time dependence of dissociation rate.

Figure 8 includes the measured weight loss as a function of furnace pressure from 0.8 to 760 torr of helium, all for 30 minutes at 1955°C. As expected, lower pressures of inert gas allow faster rates of dissociation.

DISCUSSION

Integrating the data of the previous three sections leads to several points of discussion:

1. By assuming a penny-shaped crack geometry and using the critical stress intensity factor for sintered SiC, one can calculate that the depth of a machining flaw necessary to cause failure at 80 kpsi (550 MPa) is approximately 10 μm. The increase in strength observed after refiring can therefore be caused by two possible strengthening mechanisms. The first is simply a decrease in flaw size by removal of a uniform layer of material from the specimen surface through dissociation. As shown in Figures 6-8, many refiring conditions can remove 5-10 μm of SiC, thus supporting this hypothesis. Data from Figure 1, however, show that refiring at 1650°C for 20 minutes at a pressure of 0.35 torr can increase the fracture strength from approximately 80 to 90 kpsi (550 to 620 MPa). The best estimate of the thickness of SiC removed by this refiring treatment according to the data of Figures 6-8 is only 0.3 to 0.4 μm. The observed increase in strength cannot be rationalized through purely

a decreased flaw length model. The second possible strengthening mechanism is crack blunting. For a constant length defect under a uniform applied stress, the maximum stress at the tip is related to the radius of curvature at the flaw tip through an inverse square root relationship. An increase in fracture strength of 20 percent only requires an increase in the crack tip radius of 40 to 50 percent. Even a small amount of dissociation at the crack tip could increase the radius sufficiently to increase the average strength, thus supporting crack blunting as the strengthening mechanism. The difference in apparent crack tip radius from Figure 3 to Figures 4 and 5 is several orders of magnitude, but the effective increase in curvature is much smaller due to localized faceting, pitting, etc. at the tip of the blunted crack.

2. The widths of the blunted thermal shock cracks of Figures 4a and 5 are approximately 8 μm which is consistent with the 4.5 μm loss of SiC calculated from weight loss experiments under the same refiring conditions (each wall of the crack receded approximately 4 μm). It is also interesting to note from Figures 4 and 5 that the walls of the blunted cracks are nearly parallel to a depth that is much greater than the width. After refiring at 1955°C for 30 minutes at 0.8 torr, a severe machining flaw with a maximum depth of 25 μm should have a crack tip with approximately the appearance of Figure 4b.

3. If one assumes that the level of strength in Figure 1 is merely a function of the amount of SiC dissociated from the surface and is independent of the specific conditions leading to that dissociation, then the information available in Figures 6-8 can provide a semi-quantitative prediction of the strength resulting from other refiring conditions. Qualitatively, lower furnace pressures and longer times at temperatures during the refiring treatment should result in equivalent amounts of dissociation at lower temperatures, thus shifting the curve on Figure 1 to the left.

4. Material with blunted cracks should exhibit peculiar subcritical crack growth behavior. As a sharp crack initiates at the tip of a blunted crack and begins to propagate, the stress intensity factor at the tip will increase initially at a faster rate than the square root of the crack length. After the sharp crack has propagated a distance equal to several widths of the blunted crack, all benefits of the refiring operation will have disappeared. When stress rupture data of such a refired material is plotted as fracture stress versus time to failure, the relationship will include an unusually large negative slope at short times to failure. At longer times to failure, the behavior will return to normal.

CONCLUSIONS

Surface dissociation of SiC at high temperatures and low pressures can be used to increase the crack tip radius or "blunt" potential strength limiting surface defects. The mechanism can be beneficial both in reducing the severity of surface fabrication defects during sintering and in reducing the severity of machining induced surface damage during a refiring treatment. The refiring treatment recovers a large degree of the strength loss incurred during machining and maintains any critical dimensional tolerances defined during the machining operation.

ACKNOWLEDGMENTS

The author would like to thank Drs. R. A. Giddings and S. Prochazka for many helpful discussions. This work was supported in part by the Naval Air Systems Command.

REFERENCES

1. S. Prochazka, "Sintering of Silicon Carbide," p. 239, Ceramics for High Performance Applications, Ed. by Burke, Gorum and Katz, Brook Hill, Mass., 1974.
2. C. A. Johnson and S. Prochazka, "Microstructures of Sintered SiC," Ceramic Microstructures '76, Ed. by Fulrath and Pask, 1977.
3. S. Prochazka, R. A. Giddings, and C. A. Johnson, "Investigation of Ceramics for High-Temperature Turbine Vanes," Final Report, Contract N62269-74-C-0255, G.E. Report No. SRD-74-123, Nov. 1974.
4. S. Prochazka, C. A. Johnson and R. A. Giddings, "Investigation of Ceramics for High-Temperature Turbine Components," Final Report, Contract N62269-75-C-0122, G.E. Report No. SRD-75-126, Dec. 1975.
5. C. A. Johnson and S. Prochazka, "Investigation of Ceramics for High-Temperature Turbine Components," Final Report, Contract N62269-76-C-0243, 1977.
6. J. Drowart and G. DeMaria, "Thermodynamic Study of the Binary System Carbon-Silica Using a Mass Spectrometer," p.16, Silicon Carbide-A High-Temperature Semiconductor, Ed. by O'Connor and Smiltens, Pergamon Press, N.Y., 1960.
7. A. W. Searcy, "The Reactions of High-Temperature Materials," p.57, Chemical and Mechanical Behavior of Inorganic Materials, Ed. by Searcy, Ragone and Colombo, John Wiley and Sons, Inc., N.Y., 1970.

ALTERATION OF FLAW SIZES AND K_{Ic}'S OF Si_3N_4 CERAMICS BY MOLTEN SALT EXPOSURE

W. C. Bourne and R. E. Tressler

Ceramic Science Section, Materials Science Department

The Pennsylvania State University, University Park, Pa.

INTRODUCTION

Silicon nitride ceramics are being considered for structural applications at elevated temperatures in corrosive environments including exposure to gaseous and molten salts (1). Although silicon nitride is generally considered to be corrosion resistant, relatively little effort has been expended to document in detail the effects of molten salt exposure on the mechanical behavior of silicon nitride ceramics. There have been several studies of the effects of oxidation in gaseous ambients on the strengths of hot-pressed and reaction sintered Si_3N_4. For example, Evans and Davidge (2) have shown that oxidation at 1400°C can markedly improve the high temperature strength of reaction sintered Si_3N_4 as long as the samples are not cycled to room temperature after oxidation and before testing. These results have been partially explained by flaw healing upon oxidation and microcracking (due to $\beta \rightarrow \alpha$-cristobalite transformation) upon cooling. Most studies of the effects of oxidation on the strengths of the hot-pressed varieties of Si_3N_4 have concluded that, when the starting strengths are limited by microstructurally controlled flaws, prolonged oxidation results in strength reduction (3). In some instances preferential oxidation (resulting in pitting) of certain regions of the microstructure has been identified as the cause of major strength losses (4).

Tressler et al. (5) have shown that the extent and rate of molten salt corrosion depends on the concentration of free oxide ions in the molten salt in NaCl-Na_2SO_4 binary mixtures. McKee and Chatterji (6) have presented a qualitative framework in which to evaluate the "active" to "passive" oxidation of SiC in gaseous and molten salt environments in which the oxygen potential at the SiC surface is the critical parameter. Similar arguments are suggested to obtain for Si_3N_4 materials. Thus, the healing or accentuating of surface flaws can be generally understood in terms of the corrosion process at the surface of the ceramics.

As pointed out by Tressler et al. (5), significant penetration of contaminant cations and anions from molten salt or gaseous agents into Si_3N_4 ceramics can occur upon extended exposure at elevated temperatures. Since the time-dependent mechanical properties of these materials are controlled by grain boundary processes and the contaminant penetration is expected to occur via the grain boundaries, one may well expect alteration of these properties of these materials after extended exposure to contaminants at service temperatures.

This paper reports on a study of the effects of molten salt exposure on the strength limiting parameters (K_{Ic} and flaw size) of MgO-Si_3N_4 hot pressed material* and reaction sintered Si_3N_4**.

EXPERIMENTAL PROCEDURE

The three major elements in the experimental program were as follows:

(1) Determine the fracture stress as a function of final surface machining to establish a surface machining condition that would give the maximum fracture stress for these materials, presumably, controlled by microstructure related flaws rather than machining induced flaws (7).

(2) Expose test bars to an environment which causes active corrosion and one which caused only contaminant penetration (NaCl-Na_2SO_4 eutectic and NaCl, respectively) for times sufficient to allow penetration of the contaminant species to depths several times greater than the critical flaws sizes in the unexposed materials.

(3) Measure the fracture stress and critical stress intensity factor of exposed and unexposed samples at room temperature and 1200°C to establish the changes in the strength limiting parameters (K_{Ic} and critical flaw sizes).

Specifically, test bars (0.20 cm x 0.60 cm x 3.8 cm) were final machined perpendicular to the final tensile axis using 120, 220, 400, 600, and 1200 grit diamond wheels all under identical machine conditions. In all cases 500 μm of the tensile surface was removed in the final grinding process to assure the removal of rough machining damage from the parallel facing operation. The sample edges adjacent to the tensile surface of reaction sintered samples were bevelled by

* NC-132, The Norton Co., Worcester, Mass.

** NC-350, The Norton Co., Worcester, Mass.

hand using 14 μm silicon carbide on a polishing wheel. All mechanical testing was done in four-point bending with upper and lower spans of 0.792 cm and 2.38 cm, respectively, and a crosshead speed of 0.05 cm/min. The final machining conditions selected from this evaluation (see Figures 1 and 2 of Results and Discussion) were 1200 grit diamond for the hot-pressed Si_3N_4 and 600 grit for the reaction sintered Si_3N_4.

The tests specimens were exposed to the molten salts at 1000°C by totally immersing the specimens (4-6 at a time) in 60-80 gm batches of the eutectic composition (37% NaCl and 63% Na_2SO_4) and pure NaCl in an atmosphere of still air. The hot-pressed samples were exposed for 144 hours to assure penetration of approximately four times the initial flaw size of the untreated material (5). The reaction sintered material was treated for 50 hours after which microprobe analyses showed significant amounts of sodium, sulfur, and chlorine at 300 μm from the surface of the material, several times the critical flaw size of the untreated material.

Strength measurements were made in four point bend as described above in air at room temperature and in air at 1200°C in a platinum wound resistance heated tube furnace in which the sample residence time before testing was three minutes.

Controlled flaws were introduced into untreated and treated samples perpendicular to the tensile axis using a Knoop indenter with a 500 gm load. The samples were then broken in four point bend as described above, the flaw dimensions were measured using an optical microscope and/or SEM photographs, and K_{Ic} was calculated from these dimensions and the failure stresses using the relationship given by Petrovic et al. (8).

Qualitative analyses for Na, S, and Cl were performed on samples as described in the Results and Discussion using a non-dispersive x-ray flourescence attachment* on the SEM**.

RESULTS AND DISCUSSION

Surface Machining Effects on Strength

In Figures 1 and 2 the results of the surface machining grit size vs. strength study are presented in the form of the Griffith equation assuming failure from surface flaws proportional to the size of the machining grit. In the case of the hot-pressed material the data points for the coarse grit machining fall on a straight line

* X-ray analyzer, Kevex-ray, Burlingame, Calif.

** Stereoscan 600, Cambridge, Monsey, N.Y.

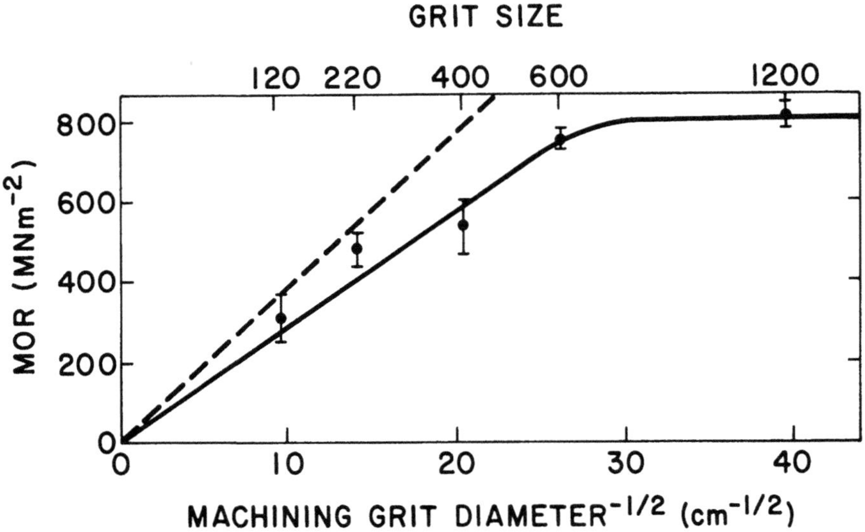

Figure 1. The plot of strength vs. surface finish for hot-pressed Si_3N_4. The dotted line shows the relation predicted by the Griffith equation using the average experimental K_{Ic}.

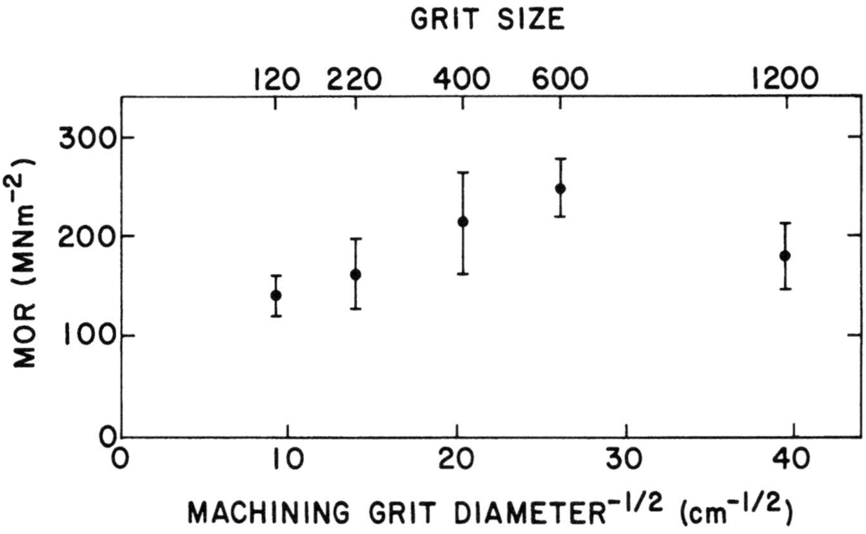

Figure 2. The plot of strength vs. surface finish for reaction sintered Si_3N_4.

which extrapolates through the origin confirming failure initiated by surface flaws generated by machining which are proportional to the machining grit size. The 600 grit machining flaws are apparently slightly larger than the flaws which are caused by microstructural features as revealed by the 1200 grit diamond machined specimens. Using the K_{Ic} value reported in Table I below and assuming a half-penny shaped surface flaw, the dotted line in Figure 1 results from the Griffith equation. Obviously the smaller grit sizes inflict more severe flaws as a fraction of their size than do the coarser grit sizes as found previously for SiC (7).

The reaction sintered material (density = 2.48 g/cm^3, 20% porosity) showed a less straightforward result in which the data do not fall in a straight line through the origin. Apparently, there is an interaction of the machining grit with the pores in the structure such that the flaws are not directly proportional to the grit size with the proportionality constant independent of grit size. Thus the surface machining operation which gave the highest strength (600 grit) was chosen for the molten salt exposure experiments. For the hot pressed Si_3N_4 the 1200 grit finish was chosen for the specimens which were exposed to the molten salt.

Effects of Molten Salt Exposure on σ_f and K_{Ic}

Hot Pressed Si_3N_4. The results of the strength and K_{Ic} tests at room temperature and 1200°C for untreated and treated specimens in the Na_2SO_4-NaCl eutectic are summarized in Table I(a) and for NaCl in Table I(b). The samples which were annealed in the eutectic salt were washed in hot H_2O in an ultrasonic cleaner for 15 min. after which the surface showed well-defined grain facets and grain boundaries. They also showed pitted areas caused by preferential oxidation by the eutectic salt. After a similar washing treatment, the samples from the NaCl treatment were covered with a glassy layer of variable thickness. To facilitate accurate alignment in the test fixture this layer was removed by a 15 min. etching treatment in 10% HF at room temperature. After this treatment, the surfaces of these specimens appeared similar to those exposed to the eutectic salt, however, they showed no areas of pitting due to preferential oxidation.

Considering first the room temperature results, the molten salt exposure caused significant strength reductions in both cases, being more severe in the more corrosive eutectic salt. The K_{Ic} measurements using the controlled flaw technique showed significant reductions in the K_{Ic} values in the contaminated layers of both types of specimens. The K_{Ic} value for the untreated material is slightly higher but within the range of scatter of K_{Ic} values reported for this material using other measurement techniques (9). The reduction

Table I(a)

Experimental Values for Strength and K_{Ic} at 25°C and 1200°C for Hot Pressed Si_3N_4, Untreated and Treated for 144 hrs. at 1000°C in Na_2SO_4-NaCl Eutectic

	K_{Ic} $(MNm^{-3/2})$[1]	σ_f (MNm^{-2})[2]	$C_{avg.}$ (μm)[3]
Untreated			
25°C	4.93±0.31[4]	826±31.7[4]	22
1200°C	5.95±1.51	529±51.8	79
Treated			
25°C	4.18±0.47	Unpitted 426 (1 meas)	66
		Pitted 308±104	126
1200°C	4.61±1.33	246±25.5	219

Table I(b)

Hot Pressed Si_3N_4, Treated for 144 hrs. at 1000°C in NaCl

	K_{Ic} $(MN/m^{3/2})$	σ_f (MN/m^2)	C(μm)
Treated 100% NaCl 25°C	3.99±0.48	540±44.4	34.3
1200°C	3.86±0.56	373±61.1	67.3

1. By controlled flaw technique, 3-5 tests.

2. In four-point bending, 5-6 tests.

3. Calculated from $C = \dfrac{[K_{Ic}(E_K)^2]}{1.25\pi\sigma_f^2}$

4. ± one standard deviation

in K_{Ic} after NaCl exposure is particularly large. Using these K_{Ic} values and the average strengths, the critical flaw sizes were calculated from the Griffith relation for half-penny shaped surface flaws. For the samples exposed to the eutectic salt the flaw size was increased several fold, depending on whether or not failure initiated from the corrosion pits which were caused by selective attack of certain regions of the microstructure. The flaw size in the NaCl treated specimens was increased by 50%; and, combined with the 20% reduction in K_{Ic}, caused a large reduction in the strength.

The data from the 1200°C tests showed similar trends with the treated specimens having lower K_{Ic} values than the untreated materials. The K_{Ic} value for the original material showed a significant increase from 25°C to 1200°C and nearly the same magnitude as reported by others (9). The strength values were again considerably lower (30% lower for the NaCl treated specimens and 50% lower for the eutectic treated specimens) than the original material at 1200°C. As noted in previous studies, the strength of the original material decreases significantly from 25°C to 1200°C even though the K_{Ic} value at 1200°C is apparently larger. This result has been attributed to subcritical crack growth during the strength test (10). From the calculated critical flaw sizes at 1200°C vs. 25°C one can conclude that the original flaw sizes have nearly tripled in length provided the measured K_{Ic} is applicable to the fracture process in the strength test. Mendiratta _et al_. (11) have shown that the measured K_{Ic} value is strongly dependent on crack velocity in hot pressed Si_3N_4 at 1300°C.

The strengths of the contaminated specimens at 1200°C followed the same trends as at 25°C in that the eutectic salt caused more drastic increases in flaw severity than the NaCl, but the NaCl caused larger decreases in K_{Ic} than the eutectic salt. In fact, with the original glassy coating intact it is possible that the NaCl treated specimens would not show flaw size increases relative to the original machined specimens (12). However, the decrease in K_{Ic} relative to the original material is significant, particularly at 1200°C where the contaminated region of the NaCl treated material has a K_{Ic} value one-third lower.

Reaction Sintered Si_3N_4. The molten salt exposure effects on K_{Ic} and strength of reaction sintered Si_3N_4 are detailed in Tables II(a) and II(b). The samples were prepared for testing by the same process as were the hot pressed specimens. The values of K_{Ic} measured here by the controlled flaw technique are lower by at least a factor of two than reported by other investigators (13,14). The error arises from the difficulty in measuring the dimensions of the critical flaw in this porous material. The microcrack caused by the indentation links up with the adjacent pores to form a critical flaw which is surely underestimated by measuring the original microcrack.

Table II(a)

Experimental Values for Strength and K_{Ic} at 25°C and 1200°C for Reaction Sintered Si_3N_4, Untreated and Treated for 144 hrs. at 1000°C in Na_2SO_4-NaCl Eutectic

	Apparent K_{Ic} $(MNm^{-3/2})$[1]	σ_f (MNm^{-2})[2]
Untreated		
25°C	0.876±0.05[3]	246±29.8[3]
1200°C	1.07±0.38	256±50.3
Treated		
25°C	0.778±0.22	110±10.7
1200°C	2.08±0.23	135±15.6

Table II(b)

	Apparent K_{Ic} $(MNm^{-3/2})$	σ_f (MNm^{-2})
Treated 100% NaCl		
25°C	1.36±0.20	108±15.0
1200°C	1.12±0.21	118±15.5

1. By controlled flaw technique, 3-5 tests.

2. In four-point bending 5-6 tests.

3. ± one standard deviation

Therefore, the apparent K_{Ic} values reported here can only be used to judge the relative toughnesses of the contaminated zone in the reaction sintered Si_3N_4. For this reason critical flaw sizes were not calculated and tabulated at 25°C. The molten salt exposure (both the eutectic and NaCl) caused strength reductions of ~50% while having very little affect on the apparent fracture toughness except for a reproducible increase in the case of the NaCl treated material. Thus, the major affect of the salt exposure was to drastically increase the surface flaw sizes. In the case of the NaCl treated material the critical flaw sizes may not have been as severely affected if the glassy surface layer had not been etched away, thereby exposing the accentuated flaws in the surface of the Si_3N_4.

The 1200°C tests revealed strengths which were comparable to the room temperature strengths in the three groups of specimens, that is to say the 50% strength reduction due to exposure persists at 1200°C. However, in the case of the eutectic treated material the K_{Ic} value at 1200°C was more than double the value at 25°C. This result suggests that a glassy phase with low viscosity has permeated the pore structure and/or grain boundaries causing much higher energy absorption during fracture but also allowing considerable slow crack growth which results in a strength which is not much higher than the room temperature strength despite the dramatically increased K_{Ic}. This phenomenon was not observed in the NaCl treated specimens where the free oxygen ion activity in the melt was much lower than in the eutectic melt. There should be a measureable difference in the strain rate dependence of the strength of these two different groups of salt treated materials if the scenario suggested above is correct.

Qualitative Analyses for Contaminant Ions

To test the hypothesis that the affected zone was caused by grain boundary penetration of the ionic species in the melts, three types of specimens were analyzed: (1) fracture surfaces of untreated strength specimens, (2) fracture surfaces of treated strength specimens, (3) polished sections in a plane perpendicular to the exposed surfaces. In Figure 3 the display of the x-ray fluorescence counts as a function of x-ray energy are shown for these three types of specimens with the latter two from the eutectic salt exposure. In Figure 3(a) from the untreated specimen the signal for Na, Cl, and S are lost in the background. In Figure 3(b), the spectrum from the fracture surface of the treated sample (intergranular fracture) the signals for Na, Cl, and S are quite strong compared to the spectrum from the same distance from the surface in the polished section (Figure 3(c)). This result demonstrates that the contaminants are concentrated in the grain boundary phase,

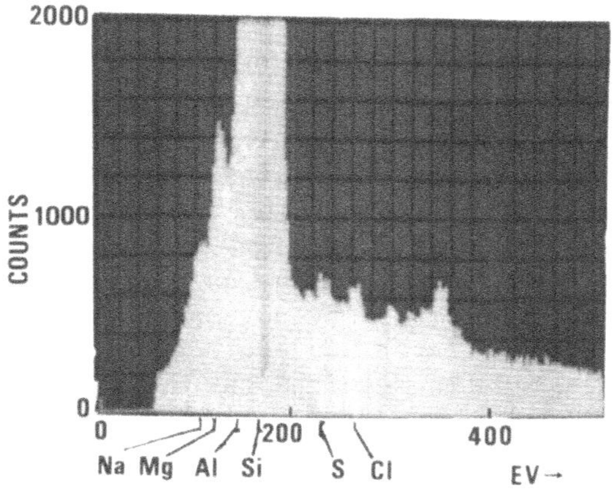

Figure 3(a). Non-dispersive x-ray spectrum from the fracture surface of untreated, hot-pressed Si_3N_4. Spot size ~80 μm, centered ~40 μm from original surface. 2000 counts = full scale, 2.5 min. count time.

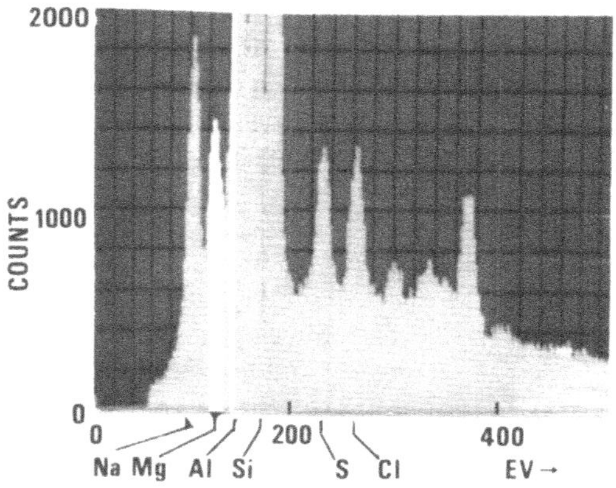

Figure 3(b). Spectrum from fracture surface of sample of hot-pressed Si_3N_4 treated in eutectic salt for 144 hrs. at 1000°C. Same location and conditions as 3(a).

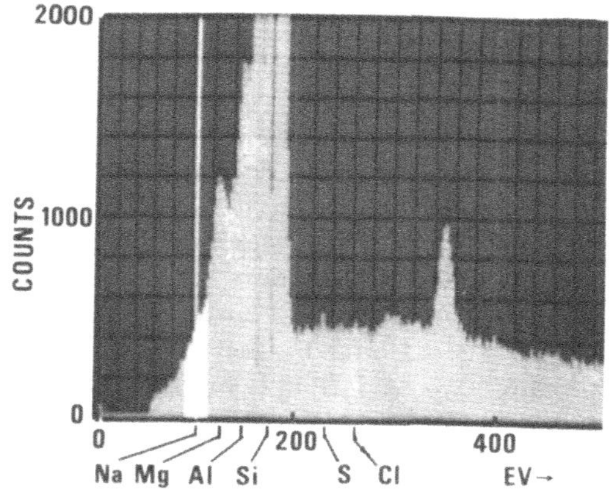

Figure 3(c). Spectrum from polished surface perpendicular to original sample surface for hot-pressed Si_3N_4 treated as in 3(b). Same location and conditions as in 3(a).

presumably, a complex silicate glass.

The analogous experiments for the NaCl treated hot pressed Si_3N_4 resulted in similar conclusions with Na and Cl being the contaminants. The results for the reaction sintered material are not as straightforward since the polished sections intersected many pores which contained the contaminant species; thus, the x-ray spectrum for the polished section and the fracture surface are not as markedly different as in the case of the hot pressed material.

CONCLUSIONS

The experiments on molten salt exposure effects on the strength limiting parameters for hot-pressed and reaction-sintered Si_3N_4 ceramics have shown that for relatively short exposure times the critical flaw sizes are substantially increased and the K_{Ic} values of the surface region are measurably reduced in the case of the hot-pressed material. The more oxidizing eutectic mixture of $NaCl-Na_2SO_4$ caused greater flaw size increases and lesser K_{Ic} decreases than the less oxidizing NaCl melt. The results at 1200°C show evidence for subcritical crack growth which is similar to that observed for untreated hot-pressed Si_3N_4.

The apparent K_{Ic} values for the sintered material are not much affected by the molten salt exposure except for the material exposed to the eutectic mixture in which the apparent K_{Ic} at 1200°C was double that of the untreated material. The strengths of these materials were dramatically reduced by molten salt exposure apparently due to increased surface flaw sizes.

Qualitative analyses of the affected zones of the test specimens indicated that the contaminants diffuse along the grain boundaries thus modifying the grain boundary controlled properties.

These results indicate the potential for significant degradation of the mechanical properties of Si_3N_4 after exposure to sodium containing oxidizing and relatively non-oxidizing molten salts.

ACKNOWLEDGEMENTS

The authors acknowledge the support of the U.S. Army Research Office for this work.

REFERENCES

1. See for example Ceramics for High Performance Applications, eds. J. J. Burke, A. E. Gorum and R. N. Katz, Brook Hill Publishing Co., Chestnut Hill, MA, 1974.

2. A. G. Evans and R. W. Davidge, J. Mater. Sci., 5, 314 (1970).

3. S. C. Singhal, presented at the NATO Advanced Study Institute on Nitrogen Ceramics, Canterbury, England, August, 1976, to be published.

4. S. W. Freiman, J. J. Mecholsky and R. Munson, presented at the 79th Annual Meeting of the American Ceramic Society, Chicago, Illinois, April, 1977.

5. R. E. Tressler, M. D. Meiser and T. Yonushonis, J. Amer. Ceram. Soc., 59, 278 (1976).

6. D. W. McKee and D. Chatterji, J. Amer. Ceram. Soc., 59, 441 (1976).

7. D. C. Cranmer, R. E. Tressler and R. C. Bradt, J. Amer. Ceram. Soc., 60, 230 (1977).

8. J. J. Petrovic, L. A. Jacobson, P. K. Talty and A. K. Vasudevan, J. Amer. Ceram. Soc., 58, 113 (1975).

9. A. G. Evans and S. M. Weiderhorn, J. Mater. Sci., 9, 270 (1974).

10. F. F. Lange, J. Amer. Ceram. Soc., 57, 84 (1974).

11. M. G. Mendiratta, J. Wimmer, and I. Bransky, J. Mater. Sci., 12, 212 (1977).

12. F. F. Lange, J. Amer. Ceram. Soc., 53, 290 (1970).

13. J. T. Barnby and R. A. Taylor in Special Ceramics 5, eds. P. Popper, The British Ceramic Research Association, Stoke-on-Trent, England, 1972.

14. R. W. Davidge in Proceedings for Ref. 3.

WEIBULL PARAMETERS AND THE STRENGTH OF LONG GLASS FIBERS

William D. Scott and A. Gaddipati*

Ceramic Engineering Division
University of Washington
Seattle, Washington 98195

ABSTRACT

Theoretical bimodal Weibull distributions were generated and various shapes of the resulting "linear" Weibull plots are shown. Data for the strength of 119 specimens of coated silica glass fibers 1m long are presented and a technique for extracting two distributions from the sample is described. Three methods of estimating Weibull parameters are compared for this experimental distribution.

INTRODUCTION

Recent advances in the use of glass fibers as optical waveguides and the use of brittle ceramics in structural applications has prompted renewed interest in the problem of design reliability of brittle materials. In the case of optical waveguides, the problem is one of predicting the probability of failure, or the minimum extreme strength, of a one kilometer glass fiber from mechanical strength measurements on much shorter samples. The nature of brittle failure, the need to include a significant size parameter in scaling up the test by a factor of 1000 or more, and

---

*Now Research Ceramist, Quartz and Chemical Products Division, General Electric Company, Cleveland, OH 44110.

This work was supported by the Naval Ocean Systems Center, Defense Advanced Research Projects Agency, and by NASA Grant NGL 48-002-004 to the University of Washington.

the need to predict the population behavior from a limited sample all require the application of statistical methods of design reliability.

The purpose of this paper is to bring to the reader's attention some published techniques for estimating statistical parameters, to explore the effect of parameter variation on the statistical distribution functions and to report some results on the strength of coated silica glass fibers.

STATISTICS OF THE STRENGTH OF BRITTLE MATERIALS

The mechanical failure of a brittle material is caused by the propagation of a microscopic crack or Griffith flaw. In the case of glass fibers, these flaws are a few micrometers or less in size and occur primarily on the surface of the material. The flaw size, the number of flaws and their frequency of occurrence may be described by statistical distributions. The general problem is one of predicting the minimum extreme value of strength resulting from a sample (flaw) population of a given size. The exact solution for extreme value distributions has been given by Gumbel (1,2), and Epstein (3,4) has published extreme value formulations for a number of different statistical distributions. One type of extreme value model which has wide applicability in failure statistics of many kinds and brittle failure in particular is the Weibull model (5,6).

The Weibull cumulative distribution function (CDF) relating the probability of failure, $F(x)$ to the strength, x is

$$F(x) = 1 - \exp\{-(x/\sigma)^{\lambda}\} \qquad (1)$$

where σ and λ are the scale parameter and the shape parameter respectively.* The corresponding probability density function (PDF) defined as $dF(x)/dx$ is

$$f(x) = (\lambda/\sigma)(x/\sigma)^{\lambda-1} \exp\{-(x/\sigma)^{\lambda}\} \qquad (2)$$

$$0<x,\sigma,\lambda$$

Transposing and taking the logarithm twice transforms Eq. 1 into

$$\ln\ln(1/(1-F(x))) = \lambda\ln x - \lambda\ln\sigma \qquad (3)$$

which is a linear equation with slope λ.

*The notation and nomenclature will follow that of K.V.Bury, Statistical Models in Applied Science, Ref. 7, where the derivation and properties of this and other models are given in detail.

The effect of specimen size on observed strength is actually observed as the variation of the extreme value of flaw severity when the flaw sample size is increased, i.e. the number of flaws "sampled" in a given individual strength test. Following Bury (7) the sample size enters the Weibull CDF as

$$F(x) = 1 - \exp\{-(L/L_1)(x/\sigma)^{\lambda}\} \qquad (4)$$

where L_1 is a unit length (or area or volume) containing N_1 flaws with associated contant σ_1. L is some greater length with the same number of flaws per unit area and $\sigma = \sigma_1/N_1{}^{1/\lambda}$.
At constant probability of failure, e.g. 0.5, the relation of size to failure stress is

$$x_1/x_2 = (L_2/L_1)^{1/\lambda} \qquad (5)$$

This was originally established by Weibull (5) on empirical grounds and utilized recently by Tarliyal and Kalish (8), Kalish et al.(9), Kurkjian et al. (10), and Mauer et al. (11).

ESTIMATION OF THE WEIBULL PARAMETERS σ and λ

1. Linear Plotting

The most common technique for estimating the parameters σ and λ is to plot the observed data in the linear form of Eq. 3. The expected value of F(x) is obtained from the ordered data where

$$F(x) = i/(n+1) \qquad (6)$$

in which i is the ith order of failure and n is the total number of specimens in the sample. In addition to initial estimates of σ and λ, this plot also gives information as to the suitability of the proposed model. Substantial systematic curvature in what should be a linear plot indicates the need for a third parameter in the Weibull function (7) or, as will be shown later, a bimodal or multi-modal population.

2. Maximum Likelihood Method

A second method called Maximum Likelihood (ML) minimizes the variance of the parameters σ and λ. Thoman, Bain and Antle (12) analyzed this technique for the Weibull distribution and published confidence intervals for the parameters and unbiasing factors for the shape parameter as a function of sample (test) size. The Maximum Likelihood value of λ is found by an interative procedure and the expression given by Thoman (Ref. 12, Sec. 5) converges to within $\pm$ 0.001 in about four iterations using starting values from the linear plot of Eq. 3. The ML estimate of σ is then found directly.

3. Non-Linear Least Squares Estimation

Davies (13) proposed using direct non-linear least squares curve fit to the CDF, Eq. 1 as an improvement in some situations over linear fits to Eq. 3 or maximum likelihood estimates. Non-linear curve fitting programs are not generally available, however, a well documented public computer program called NLWOOD* has been described by Daniel and Wood (14). It is an iterative routine which will evaluate σ and λ directly from Eq. 1. While the program does not minimize the squared residuals, it permits a user selected criteria of relative change in fitted parameters and/or sum of squared residuals as a test of fit. It provides confidence intervals on the parameters, sum of the squared residuals and useful graphical information on the individual residuals. A comparison of the three methods for estimation of Weibull parameters has also been given by Heavens and Murgatroyd (17).

WEIBULL PARAMETERS AND THE RESULTING DISTRIBUTING FUNCTIONS

Theoretical Weibull distributions were generated by assuming values for σ and λ, and a sample size of 60 and calculating the corresponding strength using Equations 1 and 6. Plots were made of Equations 1, 2 and 3 for the assumed parameters to observe the effect of systematic parameter variations on the shapes and locations of the various curves. In addition, the effect of bimodal distributions on the shape of "linear" plots was studied by merging two sets of calculated strength values, reordering, and replotting the merged set as if it was a single unknown sample. The Weibull distributions selected were similar to distributions observed in our laboratory for the strength of 7 and 10mm glass rods in 3-point bending.

Figure 1 shows the effect of varying σ with constant λ, (Table I). This is essentially the size effect for a given population where the smallest specimen size is represented by curve D and the largest size by curve A, nearly 2200 times larger than D. (An extrapolation of 50cm fiber tests to 1km is a size effect of 2000) Although the Weibull parameter λ determines the slope of the linear plot (Fig. 1c) it is clearly incorrect to say that λ determines the "steepness" of the CDF (Fig. 1a) or the

---

*The source program, FORTRAN listing and User's Manual is available from the SHARE Library (Number 360D-13.6.007), for IBM, and the VIM Library (Number G2-CAL-NLWOOD), for CDC.

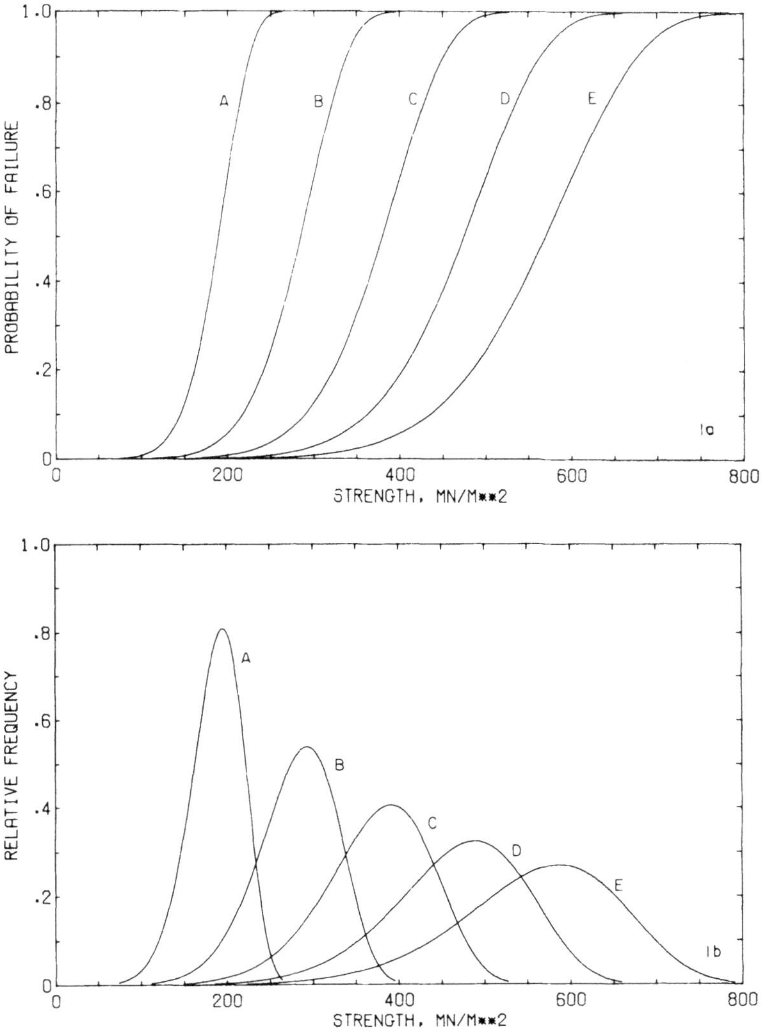

Figs. la,lb. Calculated Weibull CDF and PDF curves with λ constant and σ increasing left to right. See Table I. This is equivalent to change in specimen size (e.g., length) in a constant population in which A specimens are nearly 2200 times longer than E specimens. The PDF curves are scaled to equal areas.

dispersion of the PDF (Fig. lb). These two features which relate to the variability of a set of measurements around the mode or average are a function of both the shape parameter and scale parameters and only in the case where the scale parameter (or the more directly measured, average strength) is the same will the shape parameter alone determine the "steepness" of the CDF.

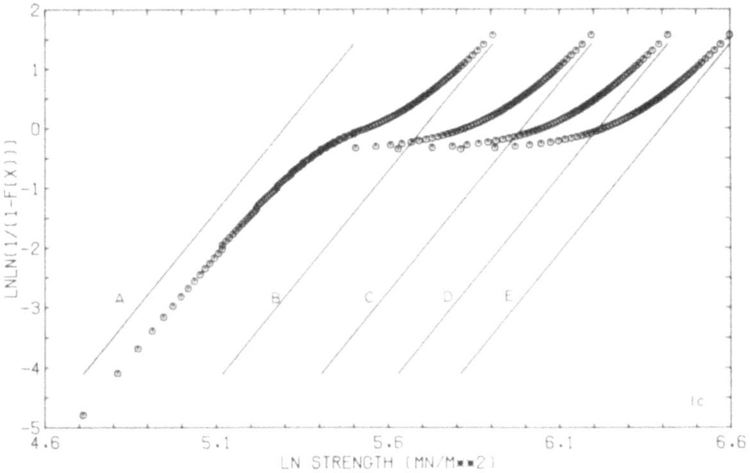

Fig. 1c. Linear forms of the Weibull functions of 1a and 1b. The data points were calculated from two individual functions and then merged, reordered and plotted as if coming from a single sample.

TABLE I

PARAMETERS USED IN CALCULATING WEIBULL FUNCTIONS

		Curve				
		A	B	C	D	E
Fig. 1	σ	200	300	400	500	600
	λ	7.0	7.0	7.0	7.0	7.0
Fig. 2	σ	200	300	400	500	600
	λ	5.0	7.0	9.0	11.0	13.0
Fig. 3	σ	400	400	400	400	500
	λ	7	6	5	4	3

The points in Figure 1c are the merged, calculated values from curves A-B, A-C, A-D and A-E, i.e. sampling values which might arise if the real populations were an unknown bimodal distribution. As seen in Figure 1c, λ for lower strength distribution (A) would be fairly accurately determined from the slope of the low strength points. However, in $\ln\sigma$ which is the value on the abscissa where the linear plot intersects the horizontal line of $\ln \ln (\phi) = 0$, would have substantial error. Virtually no accurate information can be obtained from this plot for the higher distribution of the pair.

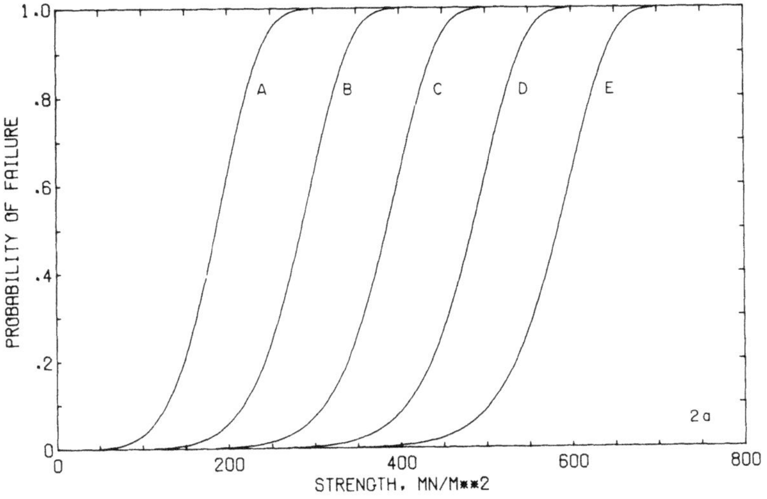

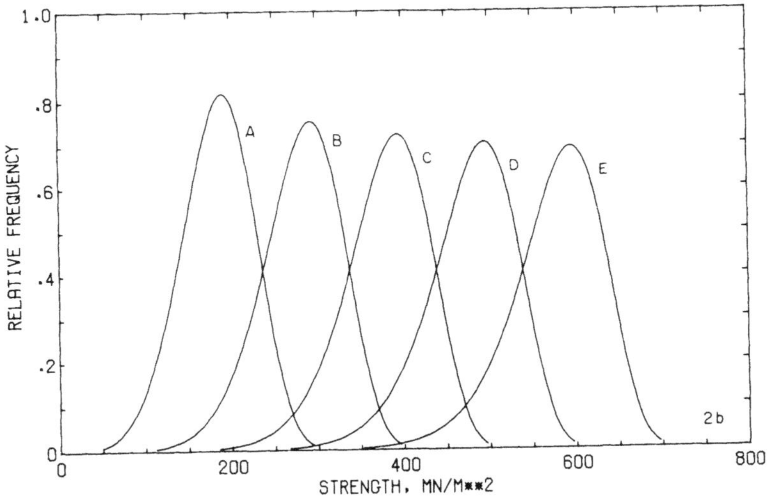

Figs. 2a,2b. Calculated Weibull CDF and PDF curves with increasing λ and σ (Table I).

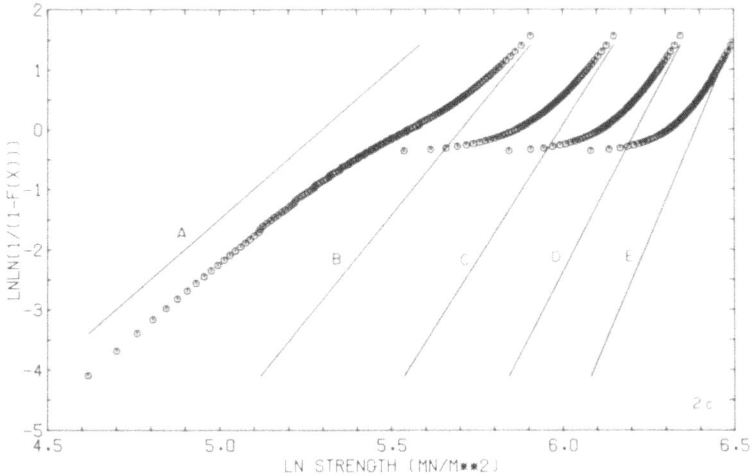

Fig. 2c. Linear forms of the Weibull functions of 2a and 2b. The data points are merged samples as described in Fig. 1c.

Figure 2 is an example where both σ and λ increase (Table I). This produces PDF curves with very similar central portions but different tails skewed to the left. Again errors in estimating σ from the merged samples will arise if the lower strength population is assumed to coincide with the data points. This error occurs in the lower part of Figures 1c and 2c because the order plotting function for each of the true populations is i/(n+1) while for the merged population it was i/(2n+1). As pointed out by Tariyal and Kalish (8) to separate bimodal distributions, one should know the fraction of observations belonging to each distribution. With this information and only the lower part of the strength data, e.g. 30 of the 60 values, the order plotting would be correct and more accurate estimates of σ and λ would result.

Curves A through D in Figures 3a and 3b have constant σ and decreasing λ. Curve E has higher σ and lower λ (Table I). This set was selected as an example of bimodal distributions with substantial overlap. The shape of the merged points in Figs. 3c,d, and e is similar to that observed for the strength of glass fibers in the present work and by Mauer (15).

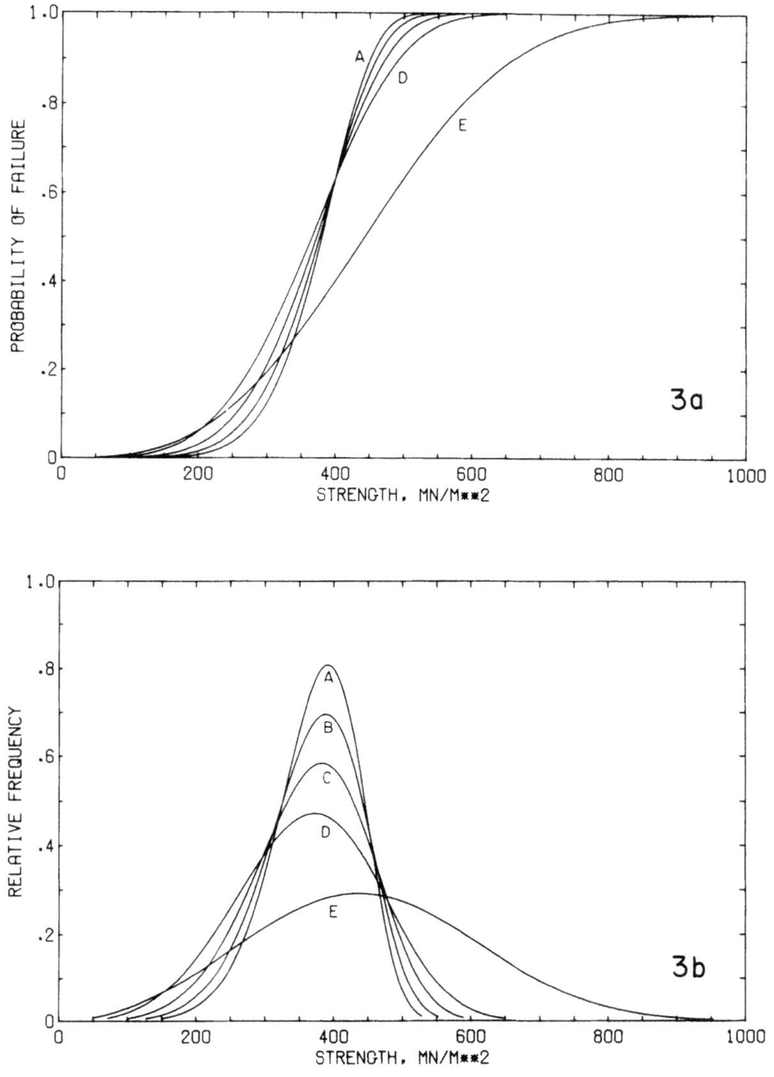

Figs. 3a,3b. Calculated Weibull CDF and PDF curves. A through D have constant σ and decreasing λ. Curve E has higher σ and lowest λ. (See Table I)

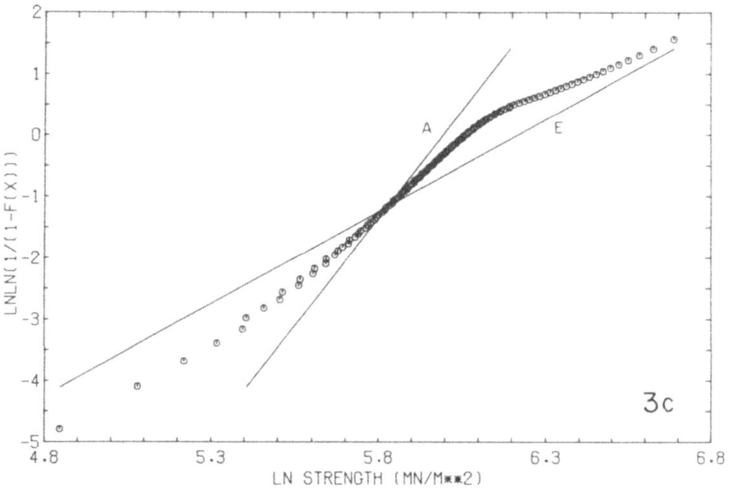

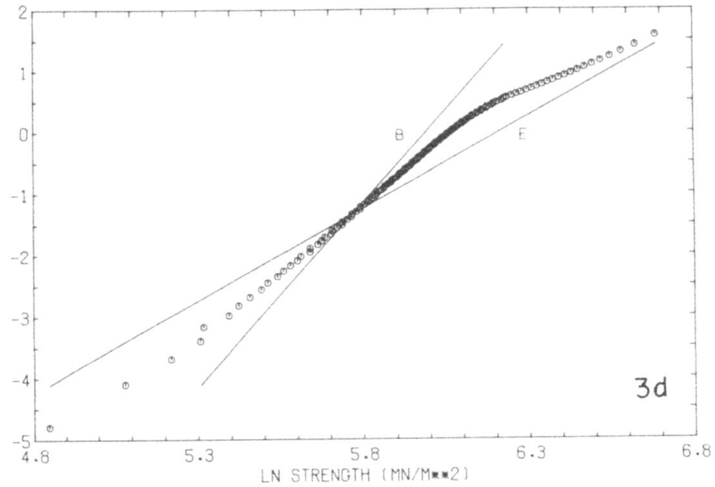

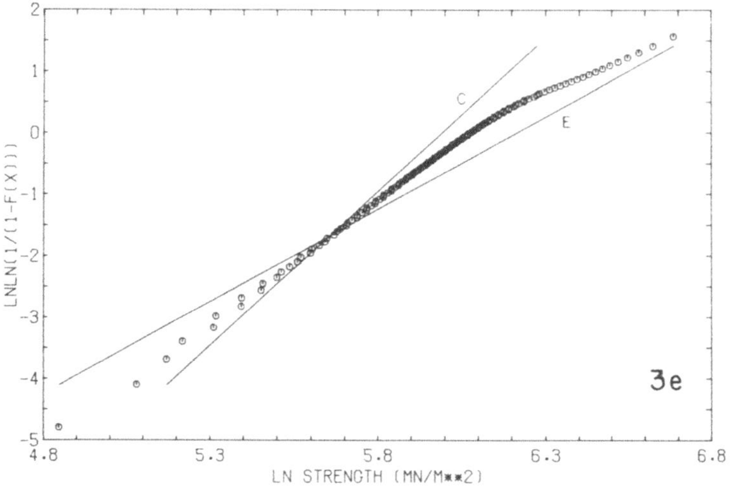

Figs. 3c, 3d, 3e. Linear forms of the Weibull functions of 3a and 3b. The data points are merged as described in 1c.

STRENGTH OF SILICA GLASS FIBERS COATED WITH POLYETHYLENE

1. Experimental procedure

Silica glass fibers 125 to 139 μm diameter were pulled from high purity commercial silica rods in a nitrogen protected graphite-sucecptor induction furnace. The furnace temperature was 2000°C, the drawing rate was 1 to 3 m/sec. Up to 600 m were drawn in one continuous filament. The fibers was protected from mechanical damage by coating with polyethylene which was heated to 325°C in an annular container surrounding the fiber and was extruded around the fiber by nitrogen or argon pressure.

Tensile strength tests were done by attaching polyethylene grips to aligned fibers using molten polyethylene as cement and pulling the fibers in tension in an Instron testing machine. It was very difficult to measure strengths greater than 3.5 GN/m^2 because the polyethylene would shear out of the grip pads. Wrapping the fiber around 3 cm roller grips was more successful, but it was still not possible to measure very high strengths of short 10 cm fibers.

One large set of 119 specimens, 106 cm long were measured using the Instron machine. A few tests were also made on longer fibers up to 10 m long using a constant load rate apparatus in which long fibers were suspended vertically and load (water) was added to a container suspended at the bottom.

2. Results and Discussion

Figure 4 shows the direct CDF plot for the 119 specimens. The dashed line is the curve calculated using the ML estimates of the Weibull Parameters, and the solid line was obtained using parameters from the non-linear direct fit. The deviation at the higher strength is very similar to that seen by Davies (13) in silicon nitride. Figure 5 shows the "linear" plot of this data assuming Equations 3 and 6.

One possible correction for the curvature in Figure 5 is the introduction of a location parameter μ in the Weibull CDF as

$$F(x,\mu,\alpha) = 1 - \exp\{-(\frac{x-\mu}{\sigma})^{\lambda}\} \qquad (7)$$

This corresponds to a minimum strength, μ, below which failure does not occur. A third parameter was introduced in steps of 35 MN/m^2 from zero to the minimum observed strength of 248 MN/m^2, and the resulting CDF curve calculated at each trial. Although the ML CDF fit (as well as the non-linear direct fit) improves slightly as judged by a reduction in sum of squared residuals, the curvature

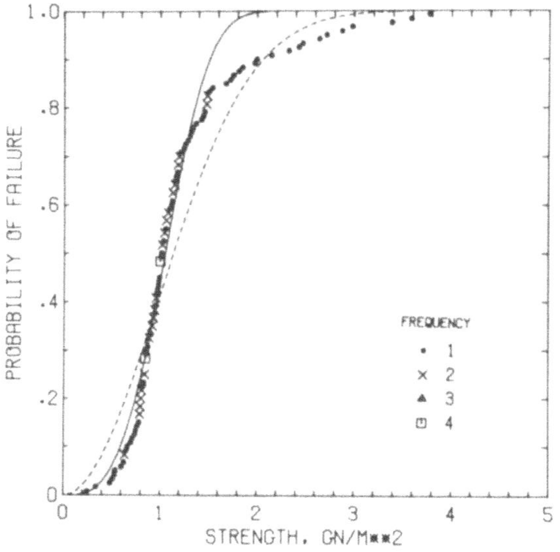

Fig. 4. Direct CDF plot for 119 coated silica glass fibers 106 cm long. The dashed line is the Weibull function calculated from parameters estimated by the Maximum Likelihood Method, the solid line from the non-linear direct fit.

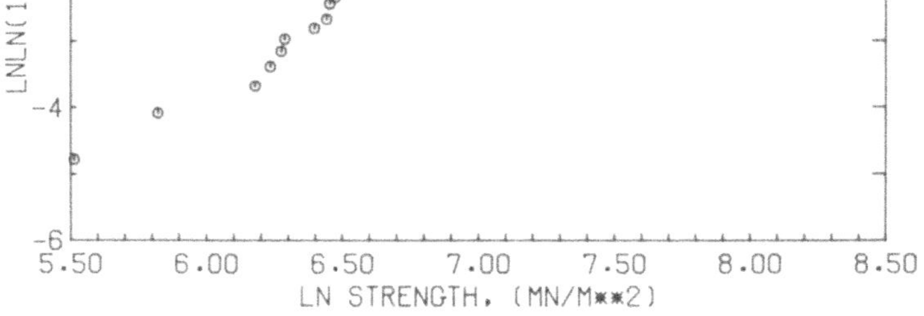

Fig. 5. Strength of 119 coated silica glass fibers plotted according to Eq. 3.

of the linear plot or the fit at the upper end of the CDF was not significantly improved.

Based on Fig. 3 and other similar trials, the curvature in Fig. 5 was assumed to arise from an overlapping bimodal distribution. The parameters of the two distributions were estimated as follows: Based on straight line portions at the ends of the curve of Fig. 5, the population totals were estimated to be 80 specimens for the lower portion and 39 for the upper portion. These sample sets will be called Set I and Set II respectively. Using these population sub-sets and assuming the lower 40 and upper 30 data points of the total set to belong to each of the individual distributions alone, the 30 data points were re-ordered against their respective sub-set population. Weibull parameters were estimated using either straight line probability plotting or non-linear curve fitting to this partial data. Maximum likelihood estimates could not be used because the 30 points are not the entire sample.

Two pairs of Weibull parameters σ and λ were estimated from these plots. Theoretical sets of 80 and 39 data points were then calculated using these parameters, and two theoretical sets, i and ii were merged and re-ordered to correspond to the original experimental observation. The ordered positions of individual points belonging to Set i and Set ii were determined in the merged population. For example, in the merged 119 theoretical points, the following positions were occupied by points from Set ii: Position No. 1,4,8,14,21,30,39,49,59,67,75,82,87,91,93-to-119. These same positions were then selected from the ordered experimental data and assigned to Set II with the remaining assigned to Set I. Weibull parameters were then recalculated for the experimental data and the results are given in Table II and Fig. 6. There is a large variation in the parameters estimated by the three different techniques for the full set of 119 specimens. The estimation techniques are much more consistent for the assumed subset.

One can now calculate the influence of these two distributions on the strength of long fibers, e.g. 1000 m. Using Eq. 4, the failure probability for each distribution at 1km and various stresses is obtained. The total failure probability is then

$$P_{Total} = 2\ P_I/3 + P_{II}/3$$

assuming the fraction of the populations from each distribution remains 2/3 and 1/3. The results are given in Fig. 8 which shows the calculated size effect on strength using Eq. 1 and the joint probability of two distributions. Data points for the experimental results for fiber lengths from 1.06 to 11.93 m are also shown.

TABLE II

RESULTS OF VARIOUS METHODS OF WEIBULL PARAMETER ESTIMATION

	Full Data Set, 119				Partial Data, Set I				Partial Data, Set II			
Method	Sample Size	Sigma	Lambda	95% C.L.	Sample Size	Sigma	Lambda	95% C.L.	Sample Size	Sigma	Lambda	95% C.L.
NLLS	119	1170	3.29	3.10 3.49	40 of 80	1005	4.71	4.21 5.21	30 of 39	1916	2.14	1.86 1.97
NLLS	-	-	-	-	80 of 80	1017	5.08	4.94 5.23	39 of 39	1910	2.05	1.94 2.15
LLS	119	1338	2.54	-	80 of 80	1030	4.70	-	39 of 39	1954	2.01	-
ML	119	1359	2.03	1.79 2.26	80 of 80	1029	4.79	4.08 5.45	39 of 39	1942	2.05	1.54 2.50

NLLS = Non Linear Least Squares. LLS = Linear Least Squares. ML = Max. Likelihood.

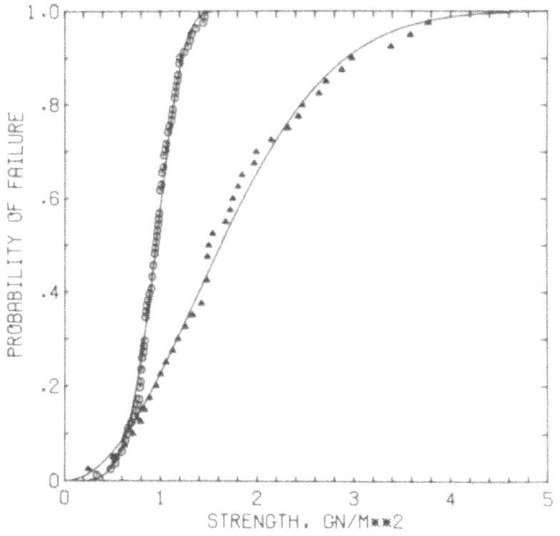

Fig. 6. Weibull CDF curves for assumed bimodal population from the samples of Figs. 4 and 5 (See Table II)

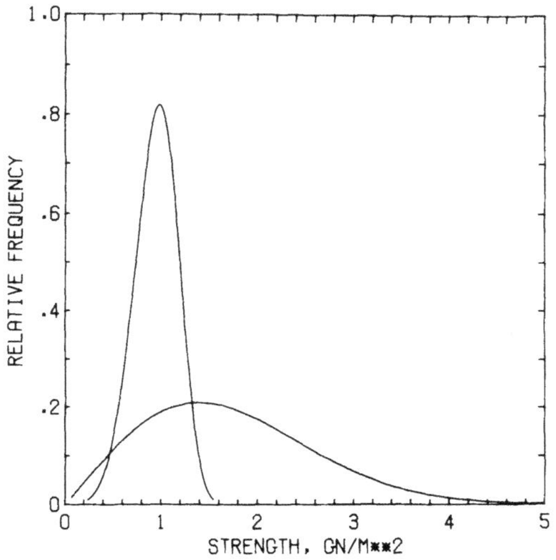

Fig. 7. Weibull PDF curves corresponding to Fig. 6. The curves are scaled to equal areas.

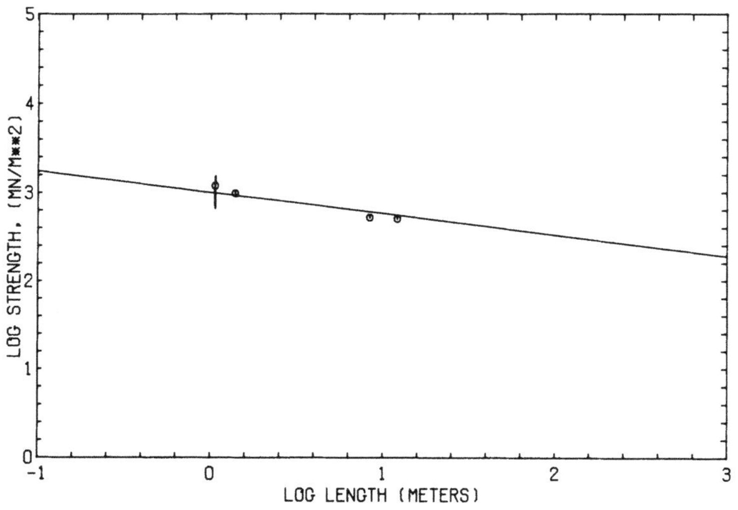

Fig. 8. Average strength of experimental fibers and calculated 0.5 probability from the joint distribution.

In calculating the contribution to failure from the two distributions it was found that the tail of the low λ distribution dominated at long lengths. It is interesting that the data from which this distribution was extracted was in the higher strength portion of the original experimental measurement.

CONCLUSIONS

As in any statistical modeling, the foregoing analysis is an attempt to define functions which represent the underlying phenomenon. The fact that one can obtain functions which appear to do this is not a confirmation of the nature of the underlying phenomenon, but with hypothesis testing, only places some confidence measurement on the validity of the assumed model.

In mechanical property testing of brittle materials where bimodal distributions are suspected, independent physical confirmation of different failure modes and measurement of the fraction belonging to each distribution should be attempted. Direct fractographic analysis which would indicate different types of flaw origins would be useful. However, it would be very tedious and perhaps impossible for large numbers of high strength fibers. The analytical techniques recently published by Matthews et al. (16) may be useful in defining different independent populations of flaw origins.

Maximum Likelihood method of estimation of Weibull parameters is the most attractive since no ordering of the data is required, the convergence is very rapid, and the form of the expression given by Thoman et al. (12) easily adapted to a computer. Thoman et al. have also provided the analysis of the inferences on the Weibull parameters. However, the ML method can not give correct results if the underlying distribution is not in fact a single Weibull distribution. For this reason, probability plotting should be employed as a check on the assumption of the statistical model and to furnish a first estimate of λ needed for the ML iteration. When the estimates obtained by least squares, non-linear direct fit and Maximum Likelihood are widely different, it is an indication that the assumption of a single modal Weibull distribution is incorrect.

REFERENCES

1. E. J. Gumbel, Statistics of Extremes, Columbia University Press, (1958).

2. E. J. Gumbel, "Statistical Theory of Extreme Values and Some Practical Applications," Nat'l Bur. Stats. Appl. Math. Series, No. 33, (Feb. 1954).

3. Benjamin Epstein, "Application of the Theory of Extreme Values in Fracture Problems," Amer. Statistical Association J., pp 403-12, (Sept. 1948).

4. Benjamin Epstein, "Statistical Aspects of Fracture Problems," J. Appl. Phys, 19, [2], (1948).

5. W. Weibull, "A Statistical Theory of the Strength of Materials," Proc. Royal Swedish Inst. for Eng. Res., No 151 (1939).

6. W. Weibull, "A Statistical Distribution Function of Wide Applicability," J. Appl. Mech., p. 293-97, (Sept. 1951).

7. K. V. Bury, Statistical Models in Applied Science, Wiley,(1976)

8. B. K. Tariyal and D. Kalish, "Application of Weibull-type Analysis to the Strength of Optical Fibers," Mat. Sci and Engr., 27, 69-71 (1977).

9. D. Kalish, B. K. Tariyal, and R. O. Pickwick, "Strength and Gage Length Extrapolations in Optical Fibers," Amer. Ceram. Soc. Bulletin, 56 [5], 491 (1977).

10. C. R. Kurkjian, R. V. Albarino, J. T. Krause, H. N. Vazirani, F. V. DiMarcello, S. Torza, and H. Schonhorn, "Strength of 0.04-50-M Lengths of Coated Fured Silica Fibers," Appl. Phys. Let., 28, [10], 588-90 (1976).

11. R. D. Mauer, R. A. Miller, D. D. Smith, J. C. Trondsen, "Optimization of Optical Waveguides-Strength Studies," Office of Naval Research, March 1974, AD777118.

12. D. R. Thoman, Lee J. Bain, C. E. Antle, "Inferences on the Parameters of the Weibull Distribution," Technometrics, 11, [3], 445-60, (1969).

13. D. G. S. Davies, "The Statistical Approach to Engineering Design in Ceramics," Proc. Brit. Ceram. Soc., p. 429-52, (1973).

14. Cuthbert Daniel and Fred S. Wood, Fitting Equations to Data, Wiley-Interscience, (1971).

15. Robert D. Mauer, Strength of Optical Waveguides, Applied Physics Letters, 27 [4] 220-21, (1975).

16. J. R. Matthews, F. A. McClintock and W. J. Shack, "Statistical Determination of Surface Flaw Density in Brittle Materials," J. Amer. Ceram. Soc., 59 [7-8], 304-8, (1976).

17. J. W. Heavens and P. N. Murgatroyd, "Analysis of Brittle Fracture Stress Statistics," J. Amer. Ceram. Soc., 53 [9], 503-5, (1970).

SURFACE FLAWS AND THE MECHANICAL BEHAVIOR OF GLASS OPTICAL FIBERS

William E. Snowden

General Electric Co., Nela Park

Cleveland, Ohio 44112

ABSTRACT

The strength and static fatigue behavior of glass optical waveguide fibers were investigated. Bimodal distributions of surface flaws were generally observed, and several types of defects were identified. Consideration of strength - length aspects of these fibers indicated that the strength of 1 km lengths is in the range of 300-400 MN/m^2. Static fatigue results suggested a stress corrosion process with a characteristic constant $n \simeq 26$, but data analysis was complicated by the presence of a wide range of critical flaw sizes in the test populations.

INTRODUCTION

The susceptibility of glass to surface damage, resulting in a significant degradation in strength, poses a major problem in the development of communications systems using glass optical fibers. For these optical waveguide (OWG) fibers, surface flaws are generated during drawing and reeling processes. Damage occurs in varying degress depending primarily on drawing conditions, which can lead to contamination by foreign particles, and the effectiveness of the fiber coating in reducing surface abrasion. Strength data for OWG fibers, usually obtained using specimens of relatively short length, typically exhibit high variability. It is thus difficult to predict with confidence the probabilities of survival of very long lengths during installation and in service. Problems of mechanical performance are further complicated by stress corrosion and static fatigue, where slow growth of small surface flaws over long periods of time may limit safe service stresses to relatively low levels. The potential worth of OWG fibers for a wide range of communications applications, however, provides urgent incentive for intensive study of strength and static fatigue phenomena.

Strength measurements for several types of optical fibers and coating systems have been described in the literature.[1-4] For polymer-coated furnace drawn fibers ranging in length from 0.04 m to 20 m, individual strength values of from less than 500 MN/m^2 to about 7000 MN/m^2 have been obtained. For any given set of these fibers, variability in strength has been high, and bimodal distributions of flaws have generally been observed. For coated fused silica fibers drawn using laser heating,[5] an average tensile strength for 20 m specimens of 5250 MN/m^2 has been reported, and failures were found to occur from a single mode of flaws of exceptionally uniform size.[4] Strength data have been analyzed using the Weibull[6,7] weakest-link concept of brittle fracture, and attempts to predict the strength of very long (1 km) lengths of fiber have been described. Only limited discussion of stress corrosion and static fatigue of OWG fibers has been presented in the literature to date. Some theoretical considerations of fiber fatigue have been discussed by Olshansky and Maurer,[8] and estimates of the effects of cladding glass composition on susceptibility to stress corrosion were reported by Snowden and Justice[9] based on strength measurements in water and liquid nitrogen environments. Actual static fatigue experiments on OWG fibers have not been described.

A number of aspects of OWG fiber strength and static fatigue will be discussed in this paper. These include the nature and origin of the several types of surface flaws responsible for fracture, the suitability of the Weibull model for describing flaw populations, strength - length considerations of fibers, and the effects of flaw size on static fatigue behavior. Certain difficulties encountered in the analysis of sets of fibers characterized by a wide range of critical flaw sizes will be noted.

EXPERIMENTAL PROCEDURE

High-silica optical waveguide fibers for strength and static fatigue testing were furnace drawn from several different blanks formed by vapor phase oxidation processes. Fiber diameters for different sets of specimens ranged from 104 μm to about 125 μm; standard deviation for any given set was less than 2% of the corresponding mean diameter. A coating of polyvinylidene fluoride (KYNAR®, Pennwalt Corp.) or ethylene vinyl acetate (EVA, DuPont Corp., ALATHON® - 3170) was applied on the draw. Typical coating thicknesses were approximately 5 μm for KYNAR® and 60 μm for EVA.

Specimens for strength testing were generally cut from reeled fibers greater than k km in length. For strength - length studies, specimens ranging in length from ∿0.04 m to 2.4 m (KYNAR® -coated fibers), and from 0.15 m to 20 m (EVA-coated fibers), were obtained by a random selection and cutting process. For other strength tests, samples were obtained by cutting sequentially. Strength measurements on shorter specimens were made using an Instron test machine, and for longer specimens up to 20 m in length a horizontal testing apparatus was used. For most tests, the strain rate was approximately 4×10^{-2}/min.

Static fatigue tests were carried out using sets of 40 EVA-coated fibers (104 μm diameter) 0.6 m and 2.4 m in length. Fibers comprising the various fatigue and corresponding strength-control sets were selected and grouped following a two step randomization process. Constant stresses ranging from 260 MN/m^2 to 1400 MN/m^2 were applied for periods of up to six months. Failure times were established by frequent observations of the current number of survivors in each set, and uncertainties in recorded failure times were small. Relative humidity in the test room ranged from 40% to about 70% during the six month test period.

FIBER STRENGTH AND DISTRIBUTIONS OF SURFACE FLAWS

Representative strength data for relatively short lengths of KYNAR® -coated OWG fibers are presented in Fig. 1. The distribution of strength values for brittle materials have often been found to be described by the Weibull distribution, and the distribution has previously been used by other investigators[1-4] to characterize such data for OWG fibers. Accordingly, the data in Fig. 1 are plotted on Weibull probability paper. For application of the Weibull distribution to strength data for glass fibers of various lengths L and approximately constant diameter, the cumulative probability of failure at a stress level σ can be expressed as

$$F(\sigma,L) = 1 - \exp\left[-(\sigma/\sigma_o)^m (L/L_o)\right] \quad , \qquad (1)$$

where m is the distribution shape parameter (characteristic of the distribution of flaw sizes[10]), σ_o is the scale parameter, and L_o is an arbitrary normalizing constant with units of length. Values for $F(\sigma,L)$ in Fig. 1 were calculated using median ranks.[11] An attractive feature of the Weibull distribution for fiber strength analysis is the size dependence, which, in principle, permits estimates of the strength of very long fibers to be made. At any given failure probability, it follows from Eq. (1) that the strength σ_1 of specimens of length L_1 is related to the strength σ_2 of specimens of length L_2 by

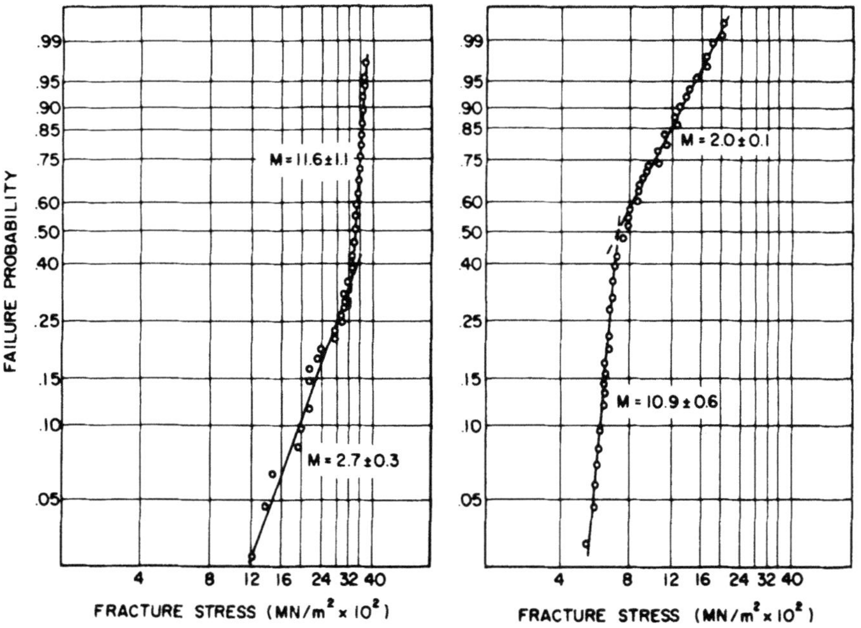

Fig. 1. Typical strength data for KYNAR® -coated fibers. Left, for specimens 0.6 m in length; right, for specimens 0.04 m to 2.4 m in length.

$$\sigma_1 = \sigma_2 \, (L_2/L_1)^{1/m} \quad . \qquad (2)$$

The critical assumption made in deriving Eq. (2) is that both m and σ_0 are independent of length, i.e., that flaws which determine the measured strengths of short specimens are of equal importance for long specimens. This assumption will be discussed in some detail below.

For both of the sets of data included in Fig. 1, a bimodal distribution of strength values (and, consequently, of surface flaws) is indicated. Values for m were determined from least-squares analyses of data for $\ln \ln (1/1\text{-}F)$ vs. $\ln \sigma$, and the indicated bounds represent 95% confidence limits. For specimens 0.6 m in length, a low strength region of higher variability (m = 2.7, corresponding to a coefficient of variation of about 37%) and a higher strength region of lower variability (m = 11.6, for a coefficient of variation of about 9%) are apparent. In contrast, for the set of fibers ranging in length from 0.04 m to 2.4 m , a low strength region of low variability (m = 10.9) and a higher strength region of high variability (m = 2.0, corresponding to a coefficient of variation of about 50%) were noted. As plotted, these data actually correspond to an identified set of critical surface flaws for the several lengths represented. Both types of bimodal behavior shown in Fig. 1 have previously been reported for OWG fibers, but the distribution characterized by an increase in slope at higher stress levels is more common and of greater physical significance. Olshansky and Maurer[8] have noted that a bimodal distribution of strength data characterized by a decrease in slope at higher stresses cannot be interpreted in terms of two Weibull distributions unless it is assumed that neither distribution extends over the entire experimental range of stresses. A low strength region of comparatively low variability must dominate measured strength values for sets of short specimens. Thus, such a bimodal character will only occur as a peculiar result of the combined effects of the particular distribution of surface flaws and the fiber test length. In the present work, distributions of strength data exhibiting three distinct regions when plotted on Weibull probability paper were occasionally observed. When longer specimens of the same type were tested, the intermediate region disappeared to give the normal bimodal character shown in Fig. 1 for 0.6 m specimens.

Although the Weibull distribution was originally established empirically, its use to represent breaking strengths of solids has been justified by Freudenthal and Gumbel[12] using extreme-value theory. However, strict application of the distribution presumes a single mode of failure. Some statistical justification for using a bimodal Weibull distribution to describe fiber strength data was sought by calculating the moments of various distributions of such data. From the calculated standardized coefficients of skewness (α_3) and kurtosis (α_4), two parameters $\beta_1 = (\alpha_3)^2$ and $\beta_2 = \alpha_4$ useful for selecting a representative distribution were determined.[13] A chart illustrating characteristic regions of the (β_1, β_2) plane for a number of different distributions is presented in Fig. 2. The figure is based on one given by Hahn and Shapiro,[13] and the dashed line represents the Weibull distribution.[14,15] When β_1 and β_2 are known, the chart can be used to identify the particular distribution most likely to describe the data being considered. For numerous sets of bimodal fiber strength data, the calculated values for β_1 and β_2 consistently fell within the beta distribution regions of Fig. 2, and far from the dashed Weibull line. For example, for a set of 126 KYNAR® -coated fibers 0.6 m in length obtained by interrupting the draw process, and which includes the 0.6 m specimens represented in Fig. 1, the calculated values $\beta_1 = 2.42$ and $\beta_2 = 4.85$ suggest a J-shaped beta distribution. A histogram of strength values for these fibers shown in Fig 3 confirms the J-shape. For similar fibers tested after being reeled, the calculated values $\beta_1 = 0.09$ and $\beta_2 = 1.64$ suggest a U-shaped beta distribution. The corresponding histogram, also shown in Fig. 3, confirms the U-shape.

The beta distribution, discussed in detail by Hahn and

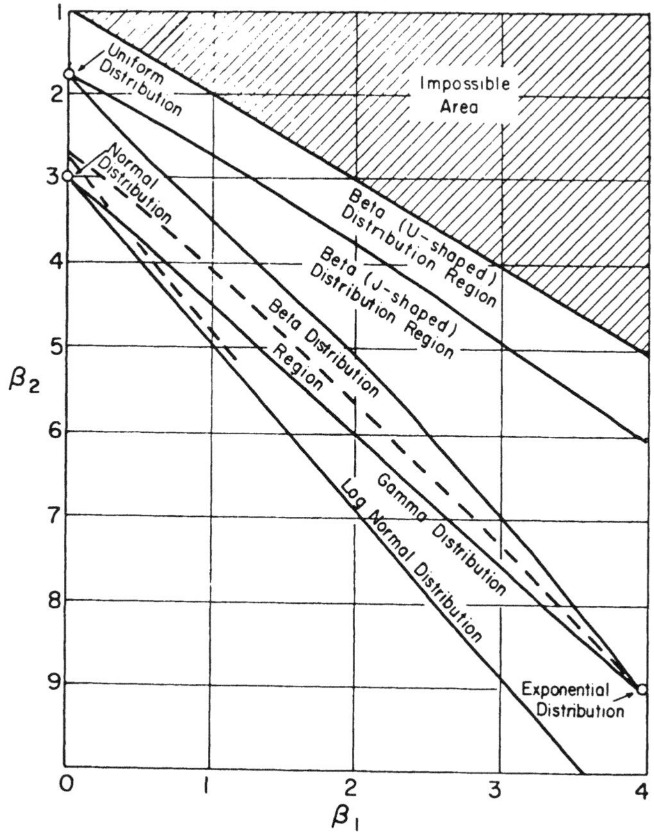

Fig. 2. Regions in the (β_1,β_2) plane for various distributions. The dashed line corresponds to the Weibull distribution.

Shapiro,[13] is a distribution used to model variates whose values are limited to a finite interval. Strength data for brittle materials is of this type, with a lower bound of zero being reasonable, and an upper bound equal to some realistic theoretical maximum value. The distribution has two shape parameters, and although these parameters are easily determined from the mean and variance of a set of data, subsequent calculation of failure probabilities for various fiber stress levels is difficult. The most important feature (two shape parameters) of the beta distribution, however, can be maintained in the analysis of fiber strength data by allowing the Weibull distribution to assume different values for m and σ_0 over two distinct intervals. The usefulness of more than one shape parameter in certain instances was recognized by Weibull,[7] who used a bimodal fit to identify specimens of different types in a mixed population. For the set of OWG fibers not reeled

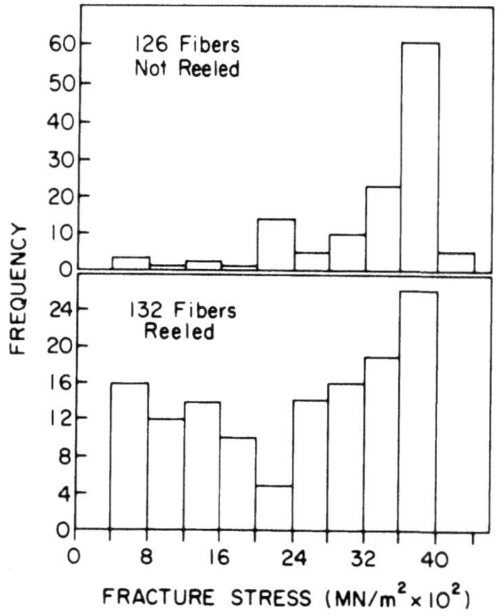

Fig. 3. Histograms of fiber strength data for KYNAR® -coated specimens 0.6 m in length.

shown in Fig. 3, a fit of the strength data made using Eq. (1) with appropriate different values for m and σ_0 in each of two regions identified by Weibull analysis gave the results shown in Fig. 4. Calculated and observed failure probabilities were found to be in good agreement except in the region of transition of failure modes for F ≃ 20%. The statistical significance of the bimodal representation was established using the χ^2 goodness-of-fit test.[16]

Several types of surface flaws which can give rise to the bimodal Weibull character for fiber strength data described above have been identified. Examination of numerous fracture origins of very low strength fibers using a scanning electron microscope has revealed the presence of foreign particles of up to 10 μm in diameter embedded in the surface. Using an x-ray energy dispersion unit attached to the microscope, these particles have generally been identified as various refractory oxides. These flaws must result from furnace contamination during drawing, and a wide range of particle sizes can be expected.[3] Surface damage of this type undoubtedly contributes to the low strength and high variability region of the Weibull plot shown in Fig. 1 for fibers 0.6 m long, as well as to low strength and low variability data obtained for longer specimens. The lower strength portion of the Weibull plot for specimens ranging in length from ∿0.04 m to 2.4 m, also shown in Fig. 1, is dominated by critical flaws associated with the

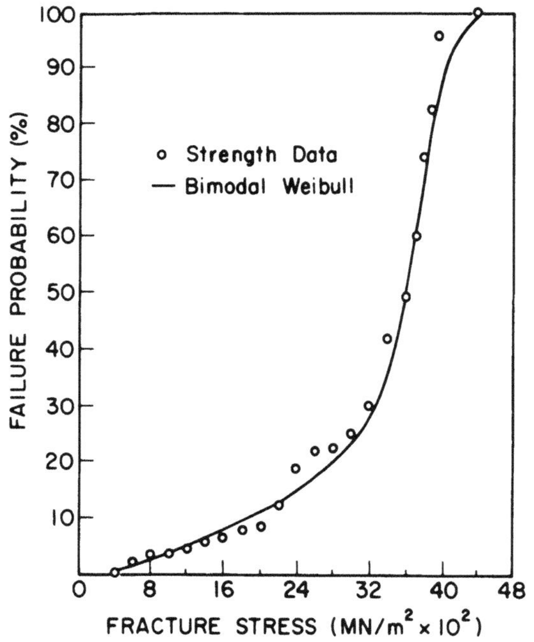

Fig. 4. Failure probability - fracture stress relations for KYNAR® coated fibers 0.6 m in length not reeled. The calculated Weibull curve is for a bimodal distribution of flaws.

longer specimens in this group. Surface contamination of OWG fibers by foreigh particles can be eliminated by laser drawing.[3] A second type of surface flaw which also contributes to low strength Weibull regions of either low or high variability depending on specimen length results from fiber abrasion during drawing and reeling. Evidence of surface abrasion of KYNAR® -coated fibers is apparent in Fig. 3, where the distribution of fiber strength values was found to be altered significantly after reeling; statistical strength degradation on reeling was about 25%. Abrasion damage was much less for EVA-coated fibers due to the greater thickness and improved uniformity of the coating. It is expected that abrasion damage of OWG fibers can also be eliminated by process improvements. Finally, flaws responsible for high strength and low variability fiber data (either for short specimens such as those of 0.6 m in Fig. 1, or for much longer specimens formed by laser drawing[3]) are believed to be predominantly intrinsic in nature, characteristic of the glass surface structure. These smallest flaws determine the upper strength limit for OWG fibers at any length.

The depth δ of a critical surface flaw corresponding to the fracture stress σ_f can be estimated using the relation

$$\delta \simeq 0.64\ (K_{Ic}/\sigma_f)^2 \quad , \qquad (3)$$

which was derived by assuming a semi-circular flaw geometry and brittle fracture.[17,18] A value for the stress intensity factor K_{Ic} of 0.714 $MN/m^{3/2}$, measured by Wiederhorn et al.[19] for high-silica glass, can reasonably be applied to the OWG fibers studied in this work. It can be shown that estimates for δ obtained using Eq. (3) are consistent with those obtained from the well-known relation between fracture stress and mirror radius r when $r/\delta \simeq 10$ as determined for a variety of materials.[20,21] The relation between σ_f and mirror radius has previously been found to hold for OWG fibers.[22] Calculated values for δ for the very low strength fibers examined using the scanning electron microscope mentioned above were found to be in very good agreement with actual measurements made from photographs of the fracture origins; the flaw size was taken to be the embedded depth of the foreign particle responsible for fracture. It is believed that an alternative relation for determining critical flaw sizes for OWG fibers reported previously[1] underestimates flaw depths by a factor of about three.

For the range of fracture stresses observed in this work, calculated flaw depths were estimated to range from greater than 2 μm to about 0.02 μm. Argon[23] has suggested that a decrease in flaw depth for glass of only two orders of magnitude over a comparable size range may correspond to an increase in flaw density of eleven orders of magnitude. For many of the fibers tested in this work, about one flaw as large as 2 μm in depth was expected per kilometer. Thus, the spacing of the smallest flaws detected is of the order of 100 Å. This revelation supports the stated belief that fracture of very high strength OWG fibers is associated with flaws characteristic of the glass surface structure. For bimodal strength data similar to that shown in Fig. 1 for specimens 0.6 m in length, the flaw size corresponding to the break in the curve between the two identified flaw regions was consistently estimated to be 0.03 - 0.04 μm (300 - 400 Å). It seems noteworthy that this critical flaw depth range is close to the maximum value of 500 Å observed for characteristic flaw lines along the boundaries of different microphase regions in borate glasses.[24] For various OWG fibers studied in this work, the cladding glass contained up to 13% B_2O_3. Similar flaw lines of up to 150 Å have been reported by Warshaw.[25]

STRENGTH - LENGTH ASPECTS OF GLASS FIBERS

Mean strengths and minimum values for sets of 20 specimens of KYNAR® -coated fibers ranging in length from ~0.04 m to 2.4 m are shown plotted as a function of gauge length in Fig. 5. The bars drawn through the mean values represent one standard deviation

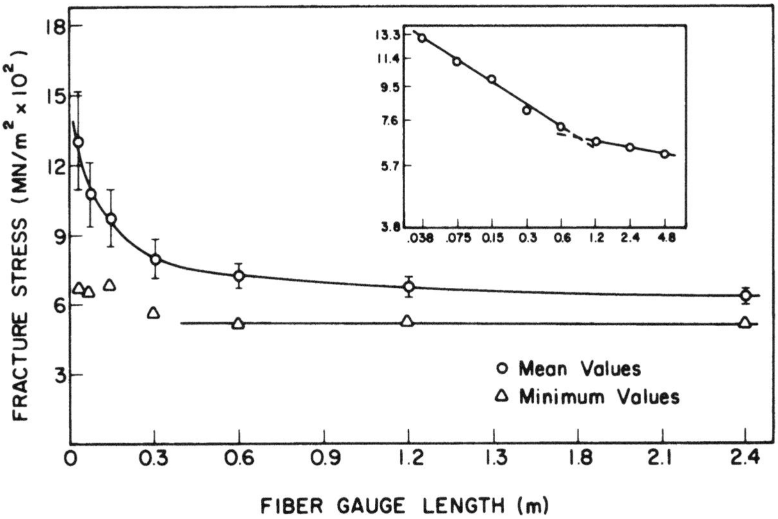

Fig. 5. Strength data for KYNAR® -coated optical fibers of different test lengths.

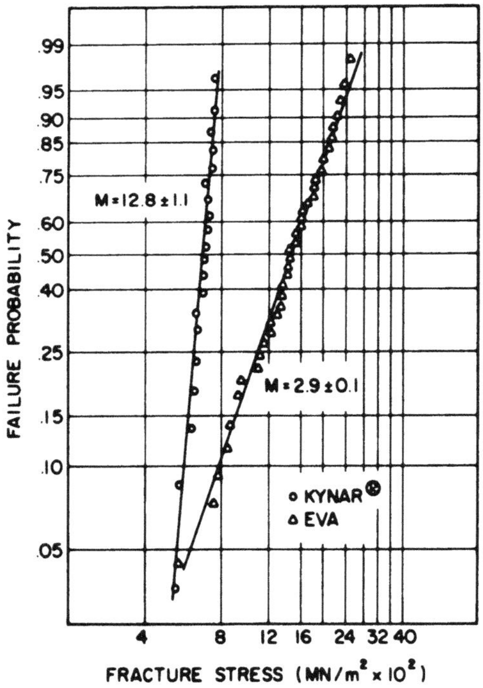

Fig. 6. Strength data for optical fibers 2.4 m in length.

calculated from the data at each length. Also included in Fig. 5 (inset) is a similar log - log plot of mean strengths for specimens up to 4.8 m in length. Data for the 4.8 m fibers was obtained by randomly pairing measured strengths for the 2.4 m specimens and taking the lower strength in each pair to be the strength of the longer fiber. A similar procedure has been used by Schonhorn et al.[3] to obtain data for much longer lengths. The decrease in both mean strength and variability shown in the figure was an expected result, because the probability of encountering larger surface flaws must increase with longer fibers. The low strengths and comparatively low standard deviations for all of the sets of specimens represented indicate the presence of a large number of surface flaws in the range 0.1 - 1.0 μm.

For the log - log plot of fiber strength vs. length data in Fig. 5, two lines were required to describe the data. As in the Weibull failure probability plots, the change in slope suggests a transition from one dominant type of defect to another.[26] A greater size effect is apparent for specimens up to about 0.9 m in length, reflecting a wide range of critical flaw sizes in fiber test populations of short lengths. The lower slope for specimens longer than about 0.9 m represents a lessening of the size effect on strength and suggests that the severity of relatively large surface flaws may be limited depending on the source of damage. Because these fibers were subjected to a proof stress during drawing, large flaws were indeed eliminated from the distribution. The fact that minimum strength values obtained for fibers $\geq$ 0.6 m in length were essentially independent of gauge length is consistent with these observations.

Each set of strength data represented in Fig. 5 was analyzed using the Weibull Distribution. No real bimodal character was observed, possibly due to the limited numbers of specimens tested for each length or to the high density of surface flaws of relatively large size. Strength data for specimens (KYNAR® - coated) 2.4 m long from this strength - length study are shown plotted on Weibull probability paper in Fig. 6. The data are adequately described by a single straight line, indicating that a single population of flaws was responsible for failure. From the Weibull analyses it was determined that the distribution parameters m and σ_0 were not constant for all lengths as required for extrapolation of strength data using Eq. (2). Rather, m increased and σ_0 decreased as the gauge length was increased, and both parameters tended to approach constant values for the longest specimens tested. The variation of m with gauge length for these fibers is shown in Fig. 7. For the longest specimens tested, the value $m = 12.8 \pm 1.1$ was obtained.

A graph of predicted strengths for 1 km lengths of KYNAR® - coated fibers based on data noted above is presented in Fig. 8.

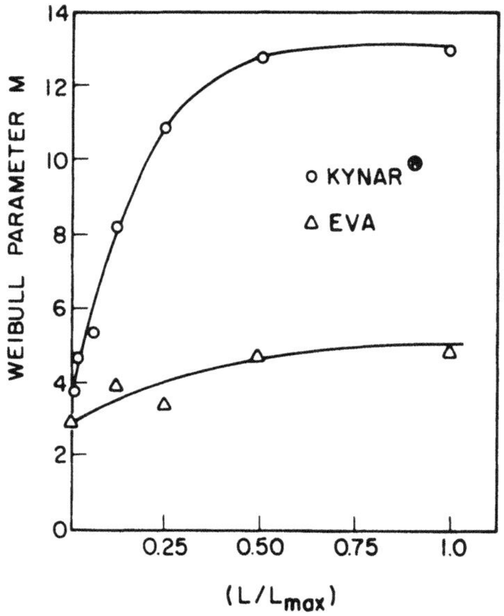

Fig. 7. Variation in the Weibull shape parameter m with gauge length for two types of fibers. For KYNAR® -coated fibers, L_{max} = 2.4 m; for EVA-coated fibers, L_{max} = 20 m.

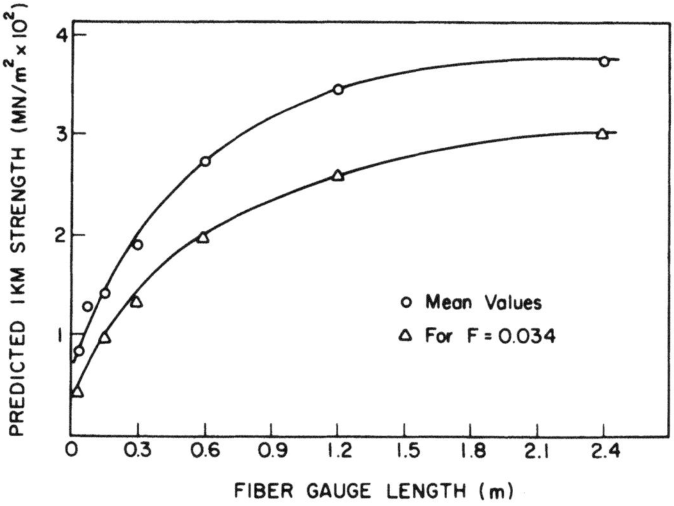

Fig. 8. Predicted strengths of 1 km long KYNAR® -coated fibers based on the Weibull shape parameter m determined for different test lengths.

The various strength values were calculated using Eq. (2) and the appropriate Weibull slope m for each gauge length shown in Fig. 7. Two different strength values from each set of data for short fibers were used to generate the curves included in Fig 8, the mean strength for each set, and the minimum value corresponding to a failure probability of 0.034. It is apparent that although the predicted strengths for 1 km long fibers is very low when the prediction is based on data for very short gauge lengths, the more realistic values based on data for longer test specimens are significantly greater. These results suggest that the strength of KYNAR®-coated fibers 1 km long should be between 300 MN/m^2 and 400 MN/m^2. The minimum measured strength for a number of sets of 50 such specimens each 20 m in length has consistently been found to fall within this range at about 350 MN/m^2. For very strong laser-drawn OWG fibers tested by Schonhorn et al.,[3] the strength distribution was characterized by a value $m \approx 25$. The predicted median strength of 1 km fibers was 4200 MN/m^2, and the minimum strength obtained on testing 84 specimens each 20 m in length was 4350 MN/m^2. Thus, extrapolation of fiber strength data for relatively short gauge lengths using Eq. (2) seems to be valid when the basis length is sufficient to give a value for the Weibull parameter m which can reasonably be expected to be similarly representative of very long lengths.

For the KYNAR® -coated fibers tested in this work, and a basis length of 2.4 m, an increase in m from 12.8 to 25 as reported for high strength fibers corresponds to an increase in the predicted strength of 1 km lengths of only about 25%. For a basis length of 20 m, the effect of such a change in m is only 16%. Routine strength - length evaluations of OWG fibers using specimens of up to only 20 m in length could thus serve as a very useful method for realistic determinations of the strength of very long fibers. This approach is believed to be more satisfactory than previously suggested procedures.[1,4]

Additional measurements of fiber strength as a function of gauge length were made using EVA-coated specimens up to 20 m long. Results were similar to those described above, and the observed variation in the Weibull parameter m for these fibers is also shown in Fig. 7. The much lower values for m for these specimens indicates a lower density of relatively large surface flaws. For the longest specimens tested, a value $m = 4.5 \pm 0.3$ was obtained. The predicted strength of EVA-coated fibers 1 km in length corresponding to this value and a basis length of 20 m was approximately 400 MN/m^2. Minimum strengths observed for sets of EVA-coated fibers containing 50 specimens each 20 m long, cut from a single fiber longer than 1 km, have also been close to 400 MN/m^2.

STATIC FATIGUE OF OPTICAL FIBERS

Static fatigue data for the EVA-coated fibers tested were analyzed using the Weibull distribution as a time-to-failure model.[27] For fibers of length L subjected to a constant applied stress σ, the cumulative probability of failure at time t as a result of stress corrosion and slow crack growth (fatigue) processes can be expressed as

$$F(\sigma,t,L) = 1 - \exp\left[-(\sigma/\sigma^*)^m (t/t^*)^b (L/L_0) \right] , \quad (4)$$

where σ^*, t^*, and L_0 are constants. The Weibull shape parameter m for use in Eq. (4) is obtained from an analysis of strength data for fibers of the same type being considered in static fatigue tests. The parameter b is similarly determined from a Weibull analysis of time-to-failure data. For use in Eq. (4), b should be a constant independent of both stress and fiber length L, and a measure of the susceptibility of the glass to stress corrosion. For fibers tested in this work, values for b were obtained from least-squares analyses of ln ln (1/1-F) vs. ln t.

Static fatique results for specimens 2.4 m in length subjected to applied stresses of 783 MN/m^2 and 1100 MN/m^2 are shown plotted on Weibull probability paper in Fig. 9. Similar results were obtained for the 0.6 m specimens tested. Failure probabilities were estimated using median ranks as noted previously in considering strength data. The distribution of strength values for a control set of 40 specimens 2.4 m in length is shown in Fig. 6; the characteristic Weibull shape parameter was m = 2.9 ± 0.1. For the static fatigue data included in Fig. 9, b was found to be 0.08 ± .02 for σ = 783 MN/m^2, and 0.13 ± .01 for σ = 1100 MN/m^2. For the 0.6 m fatigue specimens, b ranged from 0.18 ± .03 for σ = 1100 MN/m^2 to 0.28 ± .02 for σ = 1400 MN/m^2.

In all cases noted above, b tended to be higher for higher stresses. Higher values for b may indicate a greater susceptibility to stress corrosion at higher stresses. A fully satisfactory explanation for this behavior has not been found. Variations in b could arise as a consequence of the particular distributions of flaw sizes contributing to fatigue test failures. Alternatively, it seems possible that the range of values obtained for b could reflect the relative importance of fatigue and aging (crack tip blunting) phenomena at different stress levels. Fundamental studies of aging have not been reported in the literature. However, the existence of a fatigue limit for glass (approximately equal to 0.2 K_{IC}) has been recognized.[28,29]

For any given failure probability $F(\sigma,t,L)$, it follows from Eq. (4) that

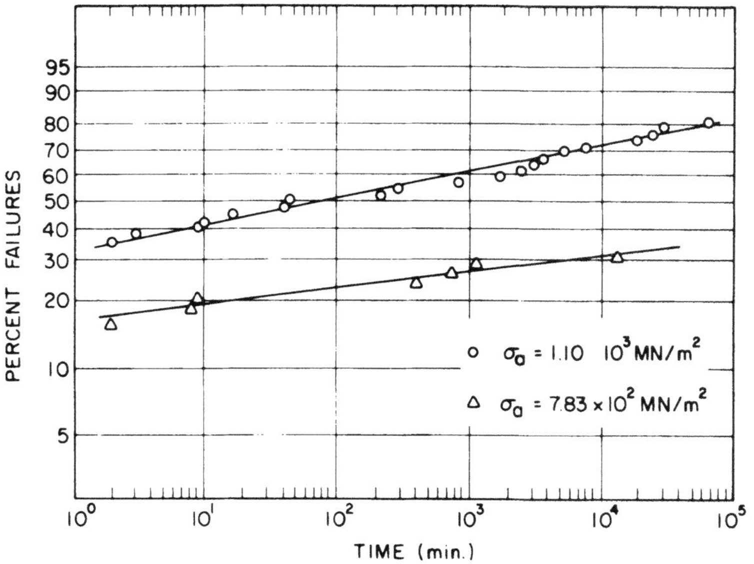

Fig. 9. Static fatigue results for EVA-coated optical fibers 2.4 m in length.

$$\sigma_1 = \sigma_2 \, (t_2/t_1)^{b/m} \, (L_2/L_1)^{1/m} \quad , \qquad (5)$$

where t_1 is the time-to-failure for a fiber of length L_1 at an applied stress σ_1 , and t_2 is the time-to-failure for a fiber of length L_2 at an applied stress σ_2. With the variation in b noted above, Eq. (5) cannot be expected to describe present fatigue test results. However, for $L_1 = L_2$, Evans and Wiederhorn[29] have derived a similar relation from fracture mechanics theory, i.e.,

$$\log \sigma_1 = \log \sigma_2 + 1/n \log (t_2/t_1) \quad , \qquad (6)$$

which indicates that the characteristic stress corrosion constant n should be equal to m/b. Estimates for n based on strength and static fatigue data were made based on this relationship. Both the absolute values for n and the possible dependence on the applied stress were of interest.

Values for m were obtained from the appropriate control set strength data. Strength values corresponding to the probability range of observed fatigue test failures were used to estimate m. For 0.6 m long fatigue test fibers, estimated values for n were approximately 26 for both of the applied stress levels indicated above. For the 2.4 m specimens, values for n of 29 and 24 were obtained at stresses of 783 MN/m^2 and 1100 MN/m^2, respectively.

An alternative method of estimating the stress corrosion constant n was considered. From the work of Evans and Wiederhorn,[29] it can be shown that for a distribution of fatigue test specimens characterized by a range of critical flaw sizes and initial stress intensity factors K_i, a plot of log K_i vs. log t should give a straight line of slope 1/(2-n). For the static fatigue tests in this work, failure times t were known for a range of failure probabilities. Values for K_i for the various applied stress levels were calculated using a form of Eq. (3), i.e.,

$$K_i = 1.25\,\sigma\,\delta^{1/2} \quad . \qquad (7)$$

The distribution of flaw depths over the required range of failure probabilities was determined from strength control data using Eq. (3).

For both the 0.6 m fibers and the 2.4 m fibers, estimates of n by the method of least-squares using this approach gave an average value of about 26, in expected agreement with previous (equivalent) estimates. However, the plots of log K_i vs. log t generally showed a change in slope corresponding to an apparent increase in n for lower values of K_i. These results could arise due to differences in the critical flaw distributions between the strength-control and static fatigue test sets. The results may also indicate a change in the effective rate of stress corrosion for low values of K due to changes in the relative importance of aging and fatigue phenomena, as suggested previously. Both explanations indicate the difficulty of analyzing fatigue test data when the range of critical flaw sizes is large.

REFERENCES

1. R. D. Maurer, Appl. Phys. Lett., 27 [4], 220 (1975)
2. C.R. Kurkjian, R.V. Albarino, J.T. Krause, H.N. Vazarani, F.V. DiMarcello, S. Torza, and H. Schonhorn, Appl. Phys. Lett., 28 [10], 588 (1976)
3. H. Schonhorn, C.R. Kurkjian, R.E. Jaeger, H.N. Vazarani, R.V. Albarino, and F.V. DiMarcello, Appl. Phys. Lett., 29 [11], 712 (1976)
4. B.K. Tariyal and D. Kalish, Mater. Sci. and Engr., 27, 69 (1977)
5. R.E. Jaeger, Am. Ceram. Soc. Bull., 55 [3], 270 (1976)
6. W. Weibull, Ing. Vetenskaps Akad. Handl., No. 151 (1939)
7. W. Weibull, J. Appl. Mech., 18, 293 (1951)
8. R. Olshansky and R.D. Maurer, J. Appl. Phys., 47 [10], 4497 (1976)
9. W.E. Snowden and B. Justice, J. Am. Ceram. Soc., 59 [9-10], 452 (1976)
10. A.M. Freudenthal, in Fracture, an Advanced Treatise, Vol. II, H. Liebowitz, ed., Academic Press, New York, 1968, Chap. 6

11. L.G. Johnson, Theory and Technique of Variation Research, Elsevier Publ. Co., New York, 1964
12. A.M. Freudenthal and E.J. Gumbel, Proc. Roy. Soc., A216, 309 (1953)
13. G.J. Hahn and S.S. Shapiro, Statistical Models in Engineering, John Wiley and Sons, Inc., New York, 1967
14. D.N. Rousu, Technometrics, 15 [4], 927 (1973)
15. A.C. Cohen, Technometrics, 15 [4], 867 (1973)
16. I. Guttman and S.S. Wilks, Introductory Engineering Statistics, John Wiley and Sons, Inc., New York, 1965
17. K.R. McKinney, J. Am. Ceram. Soc., 56 [4], 225 (1973)
18. C.F. Tiffany and J.N. Masters, in Fracture Toughness Testing and Its Applications, ASTM Special Tech. Publ. 381, Baltimore, Md., 1965
19. S.M. Wiederhorn, A.G. Evans, and D.E. Roberts, in Fracture Mechanics of Ceramics, Vol. 2. Plenum Press, New York, 1974
20. J.J. Mecholsky, R.W. Rice, and S.W. Freiman, J. Am. Ceram. Soc., 57 [10], 440 (1974)
21. D.A. Krohn and D.P.H. Hasselman, J. Am. Ceram. Soc., 54 [8], 411 (1971)
22. R.D. Maurer, private communication
23. A.S. Argon, Proc. Roy. Soc., A250, 482 (1959)
24. E.W. Sucov, J. Am. Ceram. Soc., 45 [5], 214 (1962)
25. I. Warshaw, J. Am. Ceram. Soc., 43 [1], 4 (1960)
26. G.K. Schmitz and A.G. Metcalfe, Mater. Res. Std., 7 [4], 146 (1967)
27. W. Weibull, Fatigue Testing and Analysis of Results, Pergamon Press, Oxford, 1961
28. B.J.S. Wilkins and R. Dutton, J. Am. Ceram. Soc., 59 [3-4], 108 (1976)
29. A.G. Evans and S.M. Wiederhorn, Int. Journ. of Fracture, 10, 379 (1974)

MECHANICAL BEHAVIOR OF OPTICAL FIBERS

B. K. Tariyal and David Kalish

Bell Laboratories

Norcross, Georgia 30071

ABSTRACT

The strength and static fatigue behavior of polymer coated, fused silica optical fibers is described. Experiments on glass quality and fiber processing conditions illustrate that the strength limiting flaws arise from several sources and occur with different frequencies. In turn, it is difficult to make strength extrapolations to long lengths (kilometers) based upon strengths measured on short gage lengths (meters). The probability of a failure in long lengths is analyzed from results on prooftesting of kilometer length fibers and compared to predictions from short length tests. The truncation of a strength distribution due to prooftesting is examined experimentally and the extent of dynamic fatigue that can occur during the prooftest is considered. Finally, the dynamic and static fatigue behavior of fibers with three different coatings are compared. The use of flaw growth parameters determined from dynamic fatigue experiments to predict static fatigue behavior is discussed.

INTRODUCTION

The commerical potential for lightwave communication was successfully demonstrated by the Bell System in May, 1977 when a 2.56km lightguide cable route in Chicago was put into operation.[1] This system is transmitting voice, video and data signals at a rate which corresponds to a capability of 672 telephone channels per fiber pair; each cable contains 24 fused silica low loss optical fibers 110 μm in diameter. This application obviously places a great demand on the mechanical reliability of the individual glass fibers as compared, for example, to a copper wire pair carrying a single telephone channel. Any fiber break over the cable route means a significant loss of information carrying capacity. Consequently, long length optical fibers must be manufactured, assembled in cables, installed in service and continue to operate in service with an acceptably low probability of failure due to tensile overload or to static fatigue. The major advances at Bell Laboratories and elsewhere, in this regard, have been attained through improving the surface quality of as-drawn fibers and then protecting the surface with a polymer coating.[2-5] This paper describes selected aspects of: (i), the strength characteristics of polymer coated, fused silica optical fibers; (ii), the analytical problems in relating laboratory strength measurements to the prooftest performance of large numbers of long length fibers, (iii), the use of dynamic fatigue results to establish prooftest parameters and, (iv), the relationship between dynamic fatigue and static fatigue in coated optical fibers.

SHORT LENGTH STRENGTH CHARACTERISTICS AND PROCESSING VARIABLES

The importance of glass quality, handling and fiber drawing conditions on the strength of fused silica fibers is illustrated by the strength histograms in Figures 1 to 3. In the first set of experiments, Figure 1, tensile tests were performed at a strain rate of 8.3ks^{-1} on 20m gage lengths in a series of 110 μm diameter fused silica fibers that were drawn from optical quality rods in either a graphite resistance furnace or by a 250W CO_2 laser.[6] These fibers were coated in-line with an epoxy-acrylate copolymer 45 μm in thickness and the coating was cured with ultraviolet radiation (UVC-EA). The two points of interest here are the shifts in the median strength and in the minimum strength. A significant increase in the median of the strength distribution occurs when the rod (type T08, made from fused crystalline quartz) is fire polished with an oxy-hydrogen torch prior to the furnace drawing, compare Figures 1(a) and 1(b). A slight improvement in the median strength and a pronounced increase in the minimum occurs when a Suprasil rod (synthetic silica) is fire polished and furnace drawn, compare Figures 1(b) and 1(c). A simple fracture mechanics analysis for a part-through semicircular crack normal to the fiber axis:

$$A_{cr} = (YK_{Ic})^2/\sigma_f^{\,2} \qquad (1)$$

where a_{cr} = critical flaw size, Y = geometric constant = 0.806, K_{Ic} = 0.789 $MN/m^{3/2}$ for fused silica, shows that the median critical flaw depth (ignoring dynamic fatigue) decreased from about 0.05 to 0.02 μm by switching to the Suprasil rod. Laser drawing of a fire polished Suprasil rod essentially eliminates the low strength tail of the strength distribution, Figure 1(d), giving a narrow critical flaw size distribution in the range of 0.02 μm. The laser drawing was performed in ambient laboratory atmosphere so that the rod and

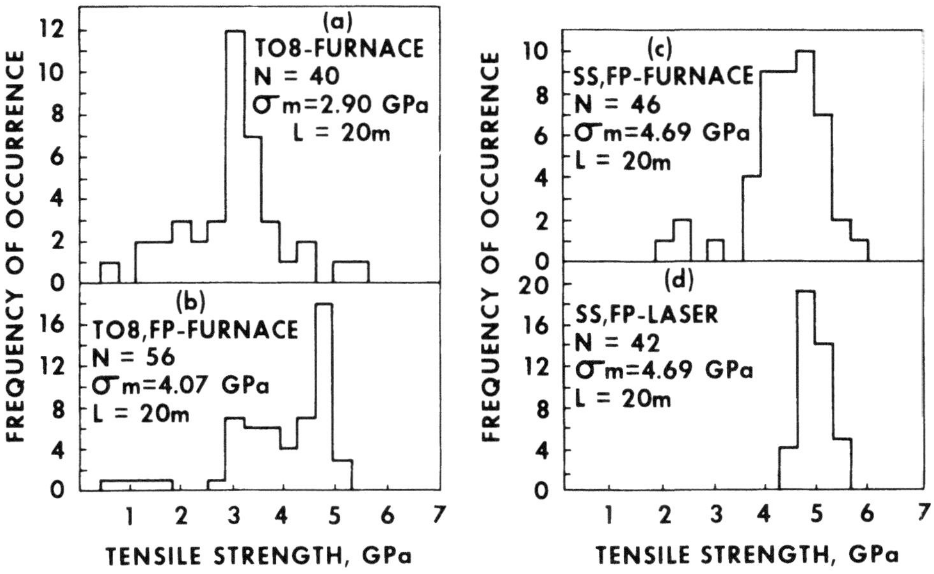

Figure 1 Fused Silica Fiber Strength Distributions for (a) TO8 Rod Drawn in Resistance Furnace (b) Fire Polished TO8 Rod Drawn in Resistance Furnace, (c) Fire Polished Suprasil Rod Drawn in Resistance Furnace and (d) Fire Polished Suprasil Rod Drawn with Laser

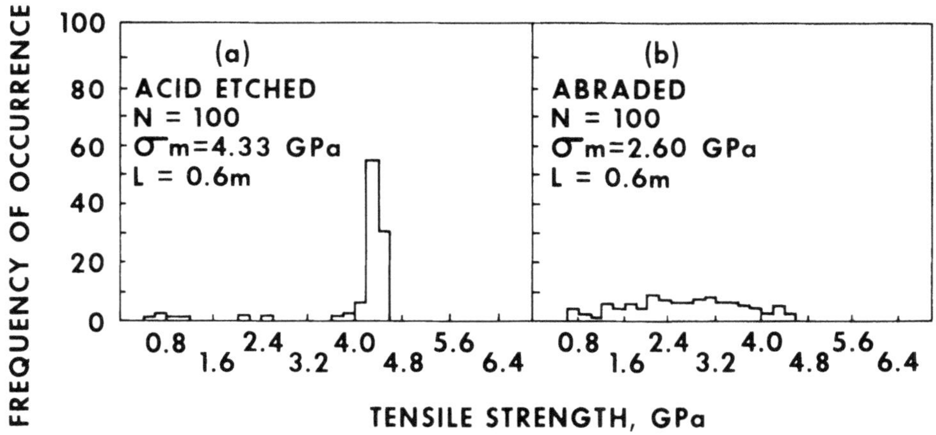

Figure 2 Strength Distributions for Furnace Drawn Optical Fibers From (a) Acid Etched Preform and (b) Paper Abraded Preform

fiber were not exposed to contamination from the hot refractory oxide and metal components of the drawing furnace.

In the second set of experiments, Figure 2, tensile tests were performed on 0.6m gage length specimens at a strain rate of $1.4ks^{-1}$. Two fibers, 1km each, were furnace drawn (graphite elements) from optical quality preforms prepared by the modified chemical vapor deposition (MCVD) process[7] starting with fused silica tubes. The fibers had a core diameter of ~55 μm and a fused silica cladding thicknesses of ~27 μm. Each was coated in-line, first with a silane compound and second with a solution of an ethylene-vinyl-acetate copolymer (sil+EVA); the coating thickness was ~50 μm. One preform was surface etched with 10% HF solution prior to drawing and the resulting strength distribution, Figure 2(a) is narrow and essentially unimodal. A standard Weibull analysis:[8]

$$\ell n \ell n(1-f)^{-1} = b \ell n \sigma_f + k \quad (2)$$

where f is the failure probability at stress σ_f and b and k are characteristics of the distribution function, gave a Weibull parameter b=21.0 and a strength median ~4.33GPa corresponding to a critical crack depth of ~0.02 μm. There are some isolated low strength values from the 100 replicate tests; the maximum critical flaw size in this group is 2.5 μm. The second preform was lightly abraded with a soft paper towel (of the type normally used to clean standard laboratory glassware). The resulting strength distribution, Figure 2(b), is broadened considerably from that of the acid etched preform. The Weibull parameter was decreased to b = 2.6, reflecting the much broader distribution, and the strength median decreased to ~2.6GPa corresponding to an increase in the median critical flaw size up to ~0.06 μm. However, the minimum strength and hence the maximum critical flaw size were not significantly altered by the abrasion.

In the third set of experiments, Figure 3, the focus of the strength question is directed at the fiber coating process. It is well-established[2] that dramatic improvements in strength can be achieved by protecting the as-drawn fiber from further damage with a coating; several polymers and application techniques are effective for this purpose. However, the process of applying a coating can itself be a source of surface damage. The fibers represented in Figure 3 were furnace drawn from preforms and coated (sil+EVA). The tests in Figures 3(a), 3(b) and 3(c) were on a 0.6m gage length and a strain rate of $1.4ks^{-1}$. The strength distributions in Figures 3(a) and 3(b) appear to be best described as bimodal with overlapping populations.[2,8] If the distribution is bimodal, then the lower population, with a mode ~1.6GPa results primarily from flaws (mode critical size ~0.16 μm) that are thought to be introduced by the coating process. The evidence for this physical explanation of

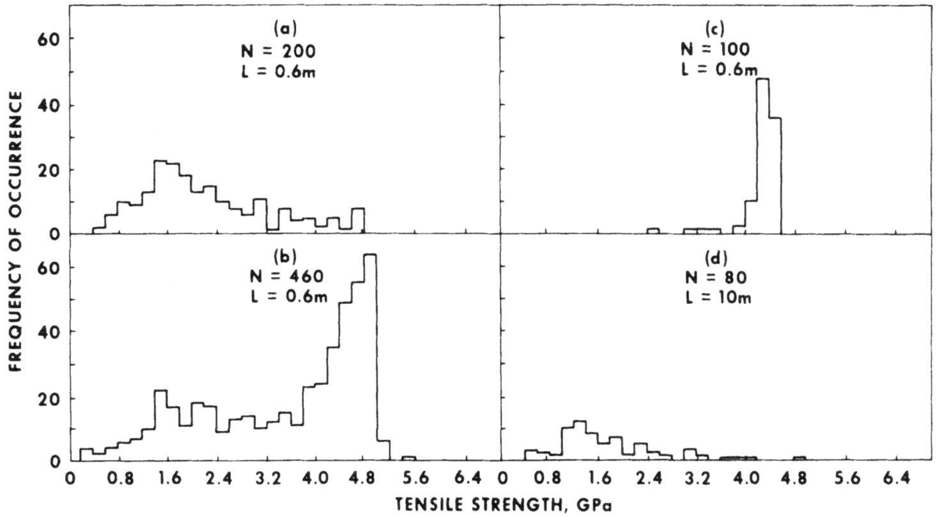

Figure 3 Strength Distributions for Silane Plus EVA Coated Fused Silica Optical Fibers Showing Improvements in Fiber Alignment and Coating Applicator Material, (a) Through (c), and Gage Length Effect for Fixed Coating Technique, (c) Versus (d).

the lower strength population is that improvements in the coating applicator materials to make them less abrasive and better alignment of the fiber through the coating reservoirs led to decreasing and then virtually eliminating the lower strength population, compare the progress from Figures 3(a) to 3(b) to 3(c), and the concurrent dominance of the upper strength population. However, the histogram in Figure 3(c) does not establish that flaws in 0.16 μm range are eliminated over kilometer lengths but merely that their frequency of occurrence has decreased to the level where the total length of fiber being tested must be greatly increased in order to find such flaws. The same type of fiber as in Figure 3(c) was tested with a 10m gage length at a strain rate of 2.1ks^{-1} to generate the histogram in Figure 3(d); the 0.61m tests sampled a total of 61m whereas the 10m tests sampled a total of 800m. The strength distribution for the 10m gage length is unimodal with a mode critical flaw size ~0.24 μm as compared to the distribution for the 0.6m length which is also unimodal but with a mode critical flaw size ~0.02 μm, an order of magnitude smaller. Of course, both flaw populations could be uncovered with a sufficiently large number of 0.6m tests since the physical nature of the fiber does not change with the test gage length, but the test program becomes unacceptably burdensome. At this time, it is not known which processing parameter(s) controls the flaw size population represented by Figure 3(d). However, the results in Figure 1 to 3 clearly show that the statistical nature of brittle fracture in fused silica optical fibers is derived from a variety of flaw origins and, as a consequence, the prediction of long length (km) strengths from short length strength measurements probably cannot be accomplished with any useful degree of reliability.

II. LONG LENGTH STRENGTH AND PROOFTESTING

A series of 329 optical fibers, each about 1.1 km in length, 110 μm in diameter, furnace drawn and coated (sil-EVA), were prooftested at 240MPa. The prooftests were performed[9] by passing a fiber from one reel to another between two double capstans which stressed and then unloaded the fibers. The spacing between the capstans was about 0.5m and the strain rate was $15ks^{-1}$. The strength data in Figures 3(c) and 3(d) for the 0.6m and 10m gage lengths were obtained on three randomly selected fibers from this lot of 329 fibers. The prooftesting resulted in 75 failures or a failure probability of 0.23 at 240MPa and a test length of 1.1km. In order to compare the short length tensile strengths to the failure rate in prooftesting, a correction must be made for the difference in strain rate[10] according to:

$$\log(\sigma_1/\sigma_2) = \frac{1}{n+1}\log(\dot{\varepsilon}_1/\dot{\varepsilon}_2) \tag{3}$$

where σ_1 and σ_2 are strengths at a constant failure probability for strain rates $\dot{\varepsilon}_1$ and $\dot{\varepsilon}_2$ and $m = 1/(n+1)$ is the slope of the log strength versus log strain rate line. Separately, in dynamic fatigue experiments (discussed in a later section) we found that $n = 30$ at ambient humidity for the dual silane plus EVA coating on fused silica fibers, so that at an order of magnitude slower prooftest strain rate, a 23% failure probability would occur at a prooftest stress of 223MPa. Probability plots (not shown here) of the strength distributions in Figure 3(c) for the 0.6m tests and in Figure 3(d) for the 10m tests gave Weibull parameters of $b = 14.2$ and $b = 2.1$, respectively. The stress for $f = 0.23$ was calculated to be 4.07GPa at 0.6m and to be 1.02GPa at 10m. The usual Weibull approach[2,8] to the strength dependence on length predicts a straight line for a plot of log strength vs. log gage length at a constant failure probability;

$$b\log(\sigma_1/\sigma_2) = \log(L_2/L_1) \tag{4}$$

where L_2 and L_1 are two different gage lengths. This extrapolation was applied at $f = 0.23$ with the data from the 0.6 and 10m tests and at a length of 1.1km failure stresses of 2.41 and 0.13GPa, respectively, are predicted. It is concluded that the gage length extrapolation based upon the 0.6m data is an order magnitude higher than the observed behavior (corrected for the strain rate difference) in prooftesting because the flaws controlling the 0.6m strength are not from the same population (physical origin) as those responsible for prooftest failures. On the other hand, the gage length extrapolation from the 10m data, which presumably represents the same flaw population as is responsible for the prooftest failures, gives a failure strength (at f = 0.23) that is lower than the prooftest result by a factor of 1.7. There are at least two reasonable explanations for this discrepancy. The 10m strength distribution Figure 3(d) while essentially unimodal is not symmetric, there are a number of high strength values that give a broad high strength tail to the histogram which may be part of a separate flaw population. Thus the distribution of low strengths (large flaws) could be artificially broadened, so that the Weibull parameter b, Equation 2, would be too low and the gage length extrapolation Equation 4, could give an unrealistically low strength as was found in the present case. An alternative explanation for the extrapolation discrepancy is that another population of even less frequent but more severe flaws than are responsible for the 10m results causes the low strength long length failures.

A separate set of 88 tensile tests were performed on 10m gage lengths of a furnace drawn and coated (sil+EVA) optical fiber; the results in the form of strength versus position along the length of the fiber are shown in Figure 4. These data on 10m lengths were used to synthesize strength distributions at 20, 30, 40 and 50m. by taking the lowest strengths from sequential groups of 2, 3, 4 and 5 fiber samples in the 10m population. Thus the 88 tests on 10m specimens were converted to groups of 44, 29, 22 and 18 tests on 20, 30, 40, and

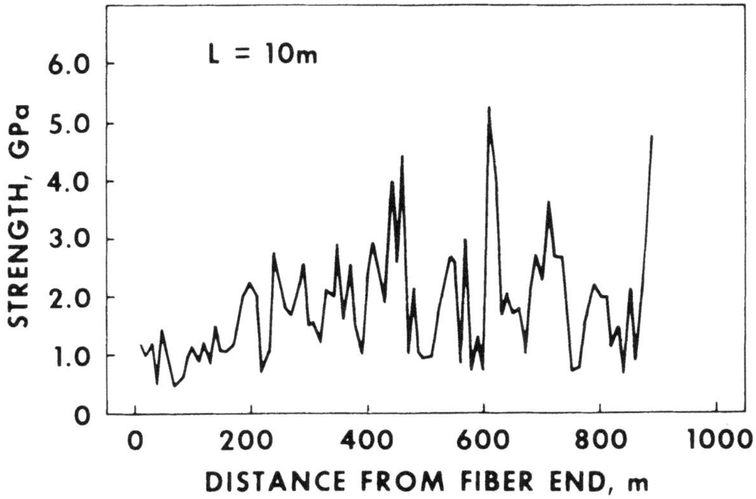

Figure 4 Strength Versus Position Along the Length of An 880m Optical Fiber

50m. gage lengths, respectively. A Weibull analysis was performed on the original 10m data as well as on each of the synthesized groups and a strength extrapolation to 1.1km at f = 0.23 was obtained, Table I. It is speculated that the continual increase in the predicted long length strength primarily reflects the rejection of data points that do not belong to the same flaw population as those responsible for the long length (prooftest) failures; the extrapolated strength approaches 240MPa as the gage length increases. One obvious deficiency with this approach is the significant decrease in data points as the synthesized gage length is increased so that the tails of the physically meaningful distribution become less well-defined and once again the accuracy of the gage length extrapolation suffers.

TABLE I

STRENGTH EXTRAPOLATIONS TO 1.1 Km LENGTH AT FAILURE PROBABILITY f=0.23 BASED UPON 10m MEASUREMENTS AND SYNTHESIZED RESULTS FOR OTHER GAGE LENGTHS

GAGE LENGTH m	AVERAGE STRENGTH MPa	STRENGTH AT f=0.23 MPa	WEIBULL PARAMETER b	EXTRAPOLATED STRENGTH f=0.23 at 1.1 Km MPa
10	1718	1005	2.09	106
20	1303	736	2.08	107
30	1111	674	2.34	145
40	968	620	2.72	183
50	857	583	3.12	217

III. PROOFTESTING AND DYNAMIC FATIGUE

The benefit of prooftesting long length optical fibers in the subsequent survival of fibers upon tensile loading of lightguide cables is documented in a separate paper.[9] Briefly, the cable tests showed that cables containing fibers which had been prooftested at 207MPa showed no fiber breaks until the strain on the cable core (containing the fibers) was approximately equal to or greater than the prior prooftest strain level. In contrast, a cable containing fibers that were not prooftested showed fiber breaks at cable core strains well below the equivalent prooftest strain level. Moreover, there was excellent agreement between the fiber failure rate in the latter cable and that in prooftesting of individual fibers. One aspect of designing a useful prooftest is to insure that no fiber damage is introduced by the prooftest itself. Two possible types of additional damage are the introduction of new flaws by the abrasion of the fiber as it moves across the loading and unloading capstan surfaces and the growth of existing flaws due to dynamic fatigue. The previous work[9] established that mechanical damage does not occur and the question of dynamic fatigue is addressed here.

A strength distribution truncation experiment was performed on 0.6m gage length tensile specimens of silane plus EVA coated optical fibers. Two groups of 100 specimens each were taken from a long fiber; the specimens were taken sequentially along the length so that consecutive specimens were distributed between the two groups. All tests were at 23°C, 40 to 50% relative humidity and a strain rate of $3.5ks^{-1}$. One group was tensile tested to failure. The second group was preloaded to 31.2N (nominally 3.28GPa) and unloaded; the fibers which failed up to this preload were discarded and the remaining fibers in this second group were retested to failure to generate a strength distribution trucated by the preload. The theoretical failure probability plots[11] and the experimental plots for the initial distribution and the trucated distributions are shown in Figure 5. The observed behavior is in excellent agreement with the theoretical behavior except for two retest data points which fall below the prestress level. It is assumed that some small amount of dynamic fatigue occurred during the prestressing cycle; samples that had flaws that were just subcritical at the onset of unloading could have grown during unloading to a size slightly larger than that to cause failure at the 3.28GPa prestress level. Obviously, any flaws that grew to a critical size

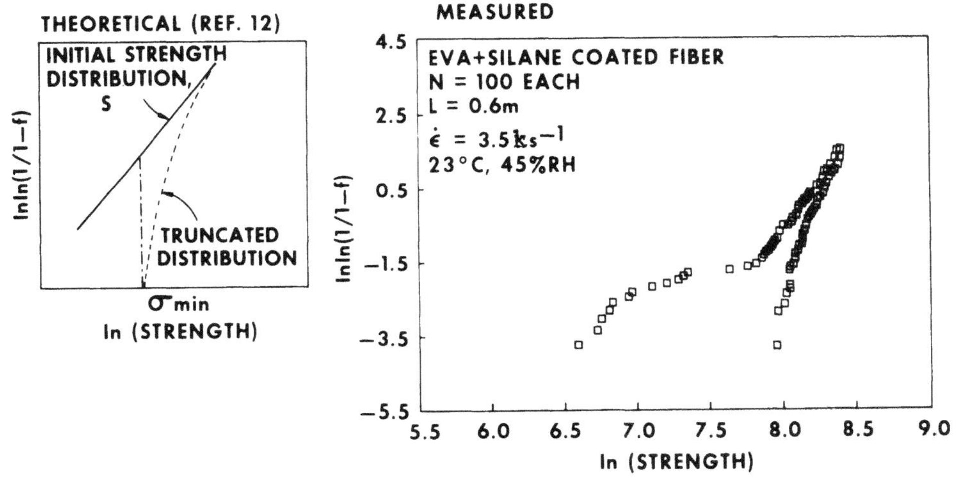

Figure 5 Initial and Trucated Strength Distributions Presented As Failure Probably Plots. The Measured Distribution Was Truncated by Preloading to 3280MPa or ℓn(Strength) = 8.096.

during the loading part of the prestressing cycle would have failed and been discarded.

These arguments apply to the long length prooftest as well. Dynamic fatigue, which occurs during the proofstressing, can shift the strength distribution slightly to lower strengths but will not affect the guaranteed minimum strength. On the other hand, dynamic fatigue which occurs during the unloading of the proofstress can reduce the guaranteed minimum strength. The maximum flaw size (and hence the guaranteed minimum strength) after unloading depends upon the competition between the crack velocity and the stress unloading rate; using the model of Evans and Fuller[11] the following is derived:

$$\sigma_g = \sigma_P \left\{1-\left[\frac{n-2}{n+1}\right]\left[\frac{AY^2K_{Ic}^{(n-2)}\sigma_p^3}{2\dot{\sigma}_u}\right]\right\}^{\frac{1}{n-2}} \qquad (5)$$

where σ_g = guaranteed minimum strength, σ_p = proofstress, $\dot{\sigma}_u$ = stress unloading rate and A, n are material crack velocity parameters[12] in $da/dt = AK_I^n$. Some examples of the effect of the unloading rate $\dot{\sigma}_u$ and of the crack velocity exponent n on σ_g are shown in Figure 6. If there is no dynamic fatigue, for example fused silica in vacuum ($n = \infty$), then $\sigma_p = \sigma_g$. Otherwise the unloading rate must be increased as σ_p is increased in order to maintain σ_g at or near σ_p. The precipitous drop in σ_g at the various combinations of $\dot{\sigma}_u$ and n in Figure 6 reflects the condition from Equation (5) when

$$\dot{\sigma}_u = \left[\frac{n-2}{n+1}\right]\left[\frac{AY^2K_{Ic}^{(n-2)}\sigma_p^3}{2}\right] \qquad (6)$$

i.e., when unloading is slow enough so that a flaw that was just subcritical at the onset of unloading can grow completely across the specimen. An n = 30 is obtained with the dual silane plus EVA coating on fused silica at 45% relative humidity. These conditions are representative of the prooftest results previously discussed where σ = 240MPa and $\dot{\sigma}_p$ = 30MPaS^{-1} so that from Equation (5) σ_g = 239.3MPa; i.e., there was negligible dynamic fatigue, the maximum critical flaw grew from 7.02 to 7.06 μm. An n = 22 is representative of a single coating of only EVA on fused silica in ambient environment. As n decreases (e.g., from 30 to 22) the material is undergoing more rapid dynamic fatigue so that a given unloading rate (e.g., 30MPaS^{-1}) is effective to a lower prooftest stress, as shown in Figure 6.

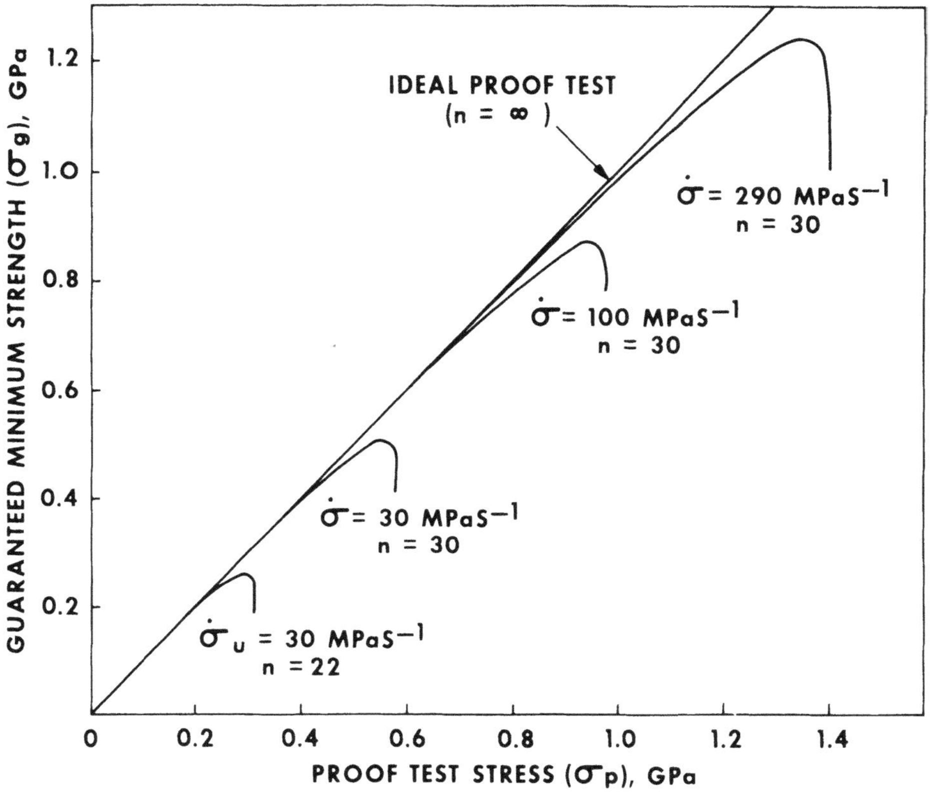

Figure 6 Effect of Unloading Rate in Prooftesting and of Crack Growth Exponent on Guaranteed Minimum Strength Versus Proof Test Stress

IV. DYNAMIC FATIGUE AND STATIC FATIGUE

The dependence of fracture strength on tensile strain rate (dynamic fatigue) was used to determine the slow crack growth parameters[13] in fused silica fibers with several coating systems. In another paper[14] it is reported that an increase in relative humidity (RH) in the range of 2 to 97 percent, promotes dynamic fatigue in an EVA coated fused silica fiber and consequently leads to a predicted acceleration of static fatigue. Measurements of the static fatigue time-to-failure at three applied stress levels in 0.6m EVA coated specimens in 97 percent RH are in good agreement with the predicted behavior based upon dynamic fatigue, Figure 7.[14]

This work has been extended to include dynamic fatigue measurements on fused silica with a silane plus EVA coating and with an UVC-EA coating at 45 and 97% RH. The experimental

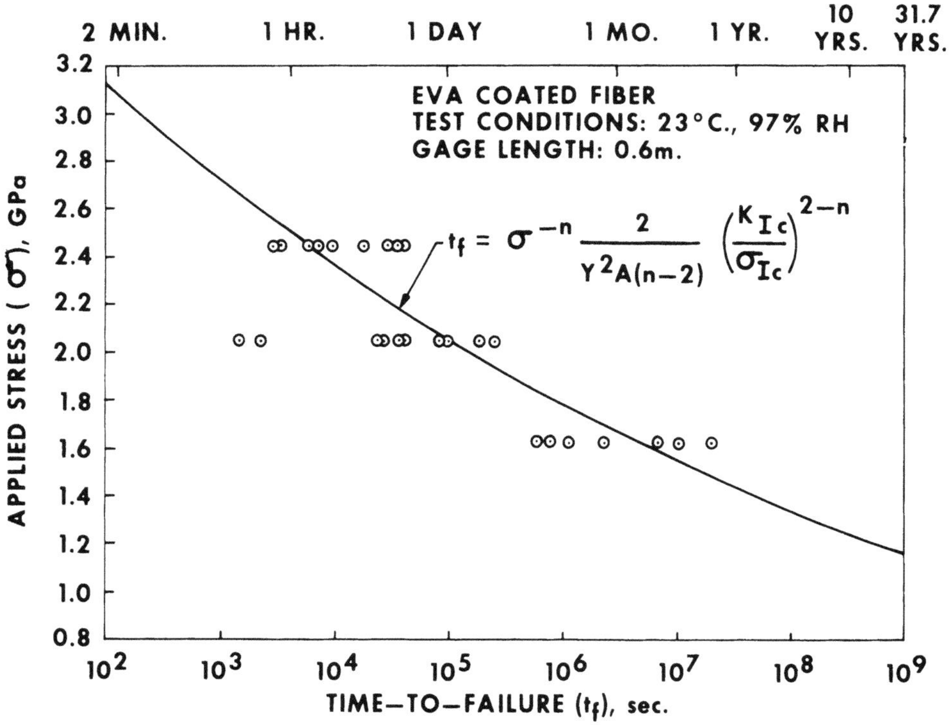

Figure 7 Comparison of Predicted Most Probable Time-to-Failure in Static Fatigue (Curve) Based Upon Dynamic Fatigue Results With Measured, Static Fatigue Failure Times, (Ref. 14), for Fused Silica Optical Fiber

procedure and analytical treatment are identical to that used with the EVA coated fiber.[14] The effect of fracture stress on strain rate ($\dot{\varepsilon}$) is given by:[12]

$$\sigma_f^{(n+1)} = \dot{\varepsilon} \left\{ 2(n+1)E\sigma_{Ic}^{(n-2)} \right\} / AY^2(n-2)K_{Ic}^{(n-2)} \quad (7)$$

Measurements of σ_f (mode value from 30 replicate tests) at $\dot{\varepsilon}$ of 0.83 to 42.2ks^{-1}, of σ_{Ic} (at -40°C in bone dry grade CO_2) and of K_{Ic}[15] for fused silica allow values of A and n to be determined for each coating. The results of these analyses, and the prior results of EVA coated fibers are summarized in Table II. In all three coating systems, the crack velocity exponent increases and ln A decreases as the relative humidity increases so that dynamic fatigue is more rapid at the higher humidity. In addition, the presence of silane compounds, as a separate layer on the fiber surface as in the dual coating with EVA or mixed in with the polymer coating as in the single coating of epoxy acrylate, apparently tends to inhibit dynamic fatigue.

Static fatigue tests were performed in a chamber that was maintained at 23°C and 97% RH. Specimens of a fiber coated with silane plus EVA and a fiber coated with the UV-cured epoxy-acrylate were tested at applied stresses of 2.46, 2.05 and 1.65GPa; 10 replicate specimens of 0.6m gage length were used. The stress levels were equivalent to those used for the fiber coated with EVA alone, Figure 6.[14] The fibers were selected so that the mode fracture strengths (on 0.6m gage length specimens at 23°C, 45%RH, and a strain rate of $1.39ks^{-1}$) were nearly equal: 4.43GPa for the EVA coated fiber, 4.35GPa for the silane plus EVA coated fiber and 4.28GPa for the epoxy-acrylate coated fiber. The static fatigue behavior of all three fibers appears to be comparable, Figure 8; data points for the EVA coated fiber from Figure 7 are reproduced in Figure 8. Thus, there is not a correlation between dynamic fatigue (short term crack growth behavior) and static fatigue (long term crack growth behavior) in the two fiber coating systems that contain silane compounds. It is postulated that the silane compounds provide an effective short term (chemical or physical) barrier in inhibiting water molecules from reaching the crack tip so that a difference among the three fibers in dynamic fatigue is measured, Table II, but that the interaction of the silane com-

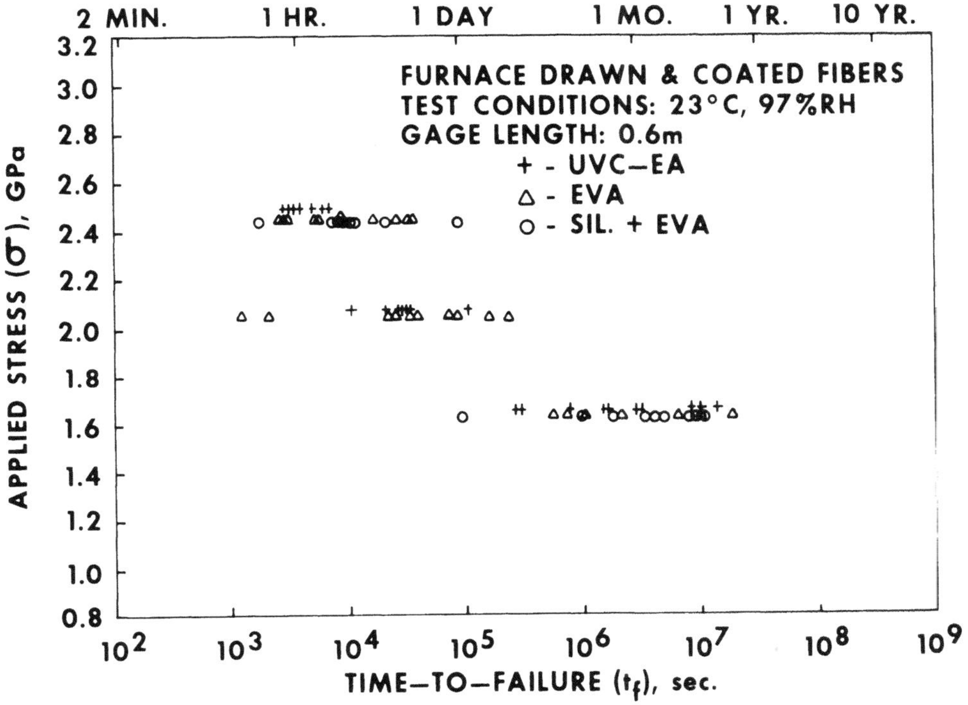

Figure 8 Time-to-Failure in Static Fatigue for Fused Silica Optical Fibers with Three Coatings: UV-cured Epoxy-Acrylate, Ethylene-Vinyl-Acetate, and a Silane Plus Ethylene-Vinyl-Acetate

pounds with the glass surface is not effective in the long term (static fatigue tests), Figure 8. Consequently, caution must be exercised in using dynamic fatigue results to predict static fatigue lifetimes.

TABLE II

SUMMARY OF CRACK VELOCITY PARAMETERS* FROM DYNAMIC FATIGUE TESTS FOR COATED FUSED SILICA OPTICAL FIBERS

COATING	% RH	n	ℓnA
EVA[14] (single coating)	2	29.2	-9.13
	45	21.8	-6.03
	71	16.2	-11.33
	97	15.3	-11.62
Silane + EVA (dual coating)	45	29.9	-9.06
	97	16.8	-12.14
UVC Epoxy Acrylate (single coating containing a silane compound)	45	27.6	-4.24
	97	25.3	-4.43

* $v = AK_I^n$ (ref. 12)

SUMMARY AND CONCLUSIONS

The statistical nature of strength in polymer coated fused silica optical fibers is derived from flaws which can originate at various stages in the glass preform fabrication, fiber drawing or coating operations. Typically, bimodal strength (flaw) distributions are observed with population modes of approximately 4.5GPa (.02 μm) and 1.4GPa (.2 μm) and minimum strengths (maximum flaws) of the order of 0.5GPa (3.3 μm). The multimode strength distribution makes it difficult to accurately predict the strength behavior in kilometer lengths. Prooftesting of long length fibers is a valuable tool for providing a guaranteed minimum strength. Experiments and analysis verify that the competition between the unloading rate in prooftesting and the crack velocity fatigue determines whether the prooftest can cause significant damage to the fiber. However, prooftest conditions have been established for coated fused silica fibers so that negligible flaw growth occurs. Dynamic fatigue experiments show that an increase in relative humidity increases the flaw growth in fused silica fibers polymer coated with ethylene-vinyl-acetate or an epoxy acrylate. Moreover, silane-type compounds on the glass surface or in the coating tend to inhibit dynamic fatigue at 45% and 97% RH but do not appear to influence long term static fatigue failure times in 97% RH.

ACKNOWLEDGEMENT

The authors wish to express their appreciation to Dr. C. R. Kurkjian of Bell Laboratories for his valuable discussions of this work and review of the manuscript.

REFERENCES

1. Bell Laboratories Record, 55 (1977) 170.

2. D. Kalish, B. K. Tariyal and R. O. Pickwick; Ceram. Bulletin, 56 (1977) 491.

3. C. R. Krukjian et al; Appl. Phys. Lett. 28 (1976) 558.

4. R. D. Mauer; Appl. Phys. Lett., 27 (1975) 220.

5. Optical Fiber Transmission II, Technical Digest of Papers Presented at Williamsburg, Va. Feb. 22-24, 1977, (publ. Optical Soc. of America, Wash. D.C., 1977) see papers in section on Fiber Materials and Fabrication Processes.

6. (a) H. Schonhorn et al, Appl. Phys. Lett. 29 (1976) 712.
   (b) R. E. Jaeger, Ceram. Bulletin 55 (1976) 270.
   (c) C. R. Kurkjian et al, in Reference 5, p. TUA2-1.

7. G. W. Tasker and W. G. French, Proc. IEEE, 62 (1974) 1281.

8. B. K. Tariyal and D. Kalish, Mat. Sci. and Eng. 27 (1977) 69.

9. B. K. Tariyal, D. Kalish and M. R. Santana, Ceram. Bulletin, 56 (1977) 204.

10. D. P. H. Hasselman et al, J. Amer. Ceram. Soc. 58 (1975) 452.

11. A. G. Evans and E. R. Fuller, Mater. Sci. and eng. 19 (1975) 69.

12. A. G. Evans, Int. J. Fract., 10 (1974) 251.

13. S. M. Wiederhorn; pp. 613-646 in Fracture Mechanics of Ceramics, Vol. 2 ed. by R. C. Bradt et al, Plenum Press, NY 1974.

14. D. Kalish and B. K. Tariyal: Paper 45-G-77 presented at 79th Annual Mtg. Amer. Ceram. Soc., April, 1977 Chicago - submitted for publication to Amer. Ceram. Soc.

15. S. M. Wiederhorn - J. Amer. Ceram. Soc. 53 (1970) 543.

SPIN TESTING OF CERAMIC MATERIALS

G. G. Trantina and C. A. Johnson

General Electric Corporate Research & Development

Schenectady, New York

INTRODUCTION

An essential step in the application of structural ceramic materials is the development of a design methodology for ceramic components. This methodology can be based on a Weibull statistical approach which allows one to predict the probability of failure of a ceramic structure [1, 2]. In addition, the effect of size and stress distribution on the fracture strength can be predicted. The size effect, where the mean strength of a component with a large volume of stressed material is much lower than the mean strength of the material as measured with small bend specimens, is due to the increased chance of encountering a larger critical flaw with increasing volume of stressed material. The validity of Weibull statistics in describing this effect must be verified with laboratory experiments. Spin testing of disks and bars provides a method of stressing a large volume of material, while bend specimens provide a convenient method of stressing small volumes. Previous spin testing of hot pressed silicon nitride disks with a hole in the center have indicated excellent agreement with predictions from four-point bend tests [2]. The objectives of this study were

1) To develop a spin test technique for ceramic bars and solid disks.

2) To demonstrate and separate the effects of size, stress distribution and biaxial stresses on the fracture strength.

3) To verify the validity of Weibull statistics in predicting these effects.

EXPERIMENTAL PROCEDURE

In order to spin ceramic materials until failure occurs due to inertia stresses, the component of interest must be physically connected to an air turbine in one of two general methods. In the first method, the two are connected with a rigid shaft to transmit the torque and define the center of rotation. In order to eliminate unacceptable vibrations, the center of mass of the ceramic component must correspond exactly with the axis of rotation of the shaft, thus requiring a precise and expensive balancing procedure. In the second method the two are connected with a flexible shaft (quill) which supports the component weight and transmits torque from the air turbine but does not define the center of rotation. The quill is attached at approximately the center of mass of the component thus allowing the component to spin about its true center of mass.

Using the second method, bars and disks were bonded to an aluminum holding fixture which was suspended on a quill below a 150,000 RPM air turbine. An alignment fixture was used to center the bars and disks on the holding fixture. A flexible epoxy adhesive was used to minimize any stresses introduced by the differential radial displacement of the holding fixture and the bars and disks. In addition, the holding fixture was designed so that the radial displacement of the holding fixture and the disk was nearly the same over the bonded area. Figure 1 shows a silicon carbide (Norton - NC435) disk bonded to the aluminum loading fixture and part of the 1/16 in. quill which attaches to the air turbine. The air turbine was mounted

Figure 1. Silicon carbide disk (2.8 in. diameter) bonded to aluminum holding fixture.

Figure 2. Spin pit showing air turbine and paraffin rings.

above the circular plate which seals the vacuum test chamber (Figure 2). Also, laying on top of the circular plate is an untested alumina disk and the broken pieces of a disk that was tested. The rings painted on the disk aid in reassembling the disk so that the location of the fracture origin can be determined. This location provides useful information in understanding the fracture process and correlating the results. A paraffin ring surrounding the specimens contains the broken pieces when fracture occurs. A paraffin ring which has contained the fracture is shown in the center of the test chamber and a new paraffin ring is shown outside the test chamber. Fracture was detected by a coil of fine wire located on the inside of the paraffin ring. When fracture occurs, the broken pieces interrupt the electrical continuity of the circuit thus providing a positive external signal that failure has occurred.

ANALYTICAL BACKGROUND

The failure mechanism of ceramic materials must be characterized so that the effects of size and stress distribution can be accounted for in the design of a ceramic structure. The strength of ceramic materials depends on the stress required to propagate small flaws which are distributed throughout the ceramic material. The fracture stress, σ_f, is related to the flaw size, a, for situations where no subcritical crack growth occurs, by the relation

$$\sigma_f = K_c/Y\sqrt{a} \tag{1}$$

where K_c is the value of the stress intensity required to propagate a crack and Y is a geometric factor which accounts for the crack shape, orientation, and location. Since K_c is a constant material property, there is a variation in strength corresponding to the square root of the flaw size distribution. This strength variation can be characterized by a weakest link model due to Weibull [3] where

$$P = 1 - e^{-R} \tag{2}$$

where P is the probability of failure and R is the risk of rupture. For the two parameter distribution (σ_o, m), R is defined by the integral of the stress, σ, over the volume, V, where

$$R = \int \left(\frac{\sigma}{\sigma_o}\right)^m dV = kV\left(\frac{\sigma_{max}}{\sigma_o}\right)^m \tag{3}$$

and σ_o is a normalizing constant, m is the Weibull modulus, k is the load factor, and σ_{max} is the maximum stress.

The Weibull modulus is a measure of the scatter in the strength distribution, a small m indicating a large amount of scatter. The

Weibull modulus can be determined from a set of strength measurements by a number of methods. One simple method involves a rearrangement of equations (2) and (3)

$$\ln \ln \frac{1}{1 - P} = m \ln \sigma + \text{constant} \tag{4}$$

and a least squares estimation of the slope, m, of the ordered strength measurements. The probability is calculated as n/(N + 1) where n is the ordering number and N is the total number of specimens.

The load factor is a measure of the uniformity of the stress distribution and is defined from equation (3) as

$$k = \int \left(\frac{\sigma}{\sigma_{max}}\right)^m \frac{dV}{V} \tag{5}$$

For uniform tension, k = 1, and for pure bending with a constant stress distribution along the specimen length

$$k_{PB} = \frac{1}{2(m + 1)} \tag{6}$$

where the (m + 1) term originates from the integral of the linear stress gradient along the beam height and the factor of 1/2 from the fact that only tensile stresses are included. For three-point bending

$$k_{3B} = \frac{1}{2(m + 1)^2} \tag{7}$$

where the additional (m + 1) term arises from the linear stress distribution along the specimen length. The magnitude of these load factors is shown in Figure 3 as a function of (m + 1). The load factors for bending emanate from 0.5 for m = 0 since only one-half of the specimen volume is loaded in tension. Since four-point bending is a combination of a linear stress gradient (three-point bending) and a constant stress along the specimen length (pure bending), the load factor is calculated by proportioning the relative amounts of linear and constant stress. The load factors for all four-point bending geometries therefore lie between the three-point bending and the pure bending lines of Figure 3.

A load factor based on surface area may be more appropriate than the volume formulation for materials that fail from surface flaws. This area treatment is the same as the volume formulation except surface area is substituted for volume in equations (3) and (5) and the load factor for area flaws in bending is

$$k_{area} = k_{volume} \left(\frac{mw}{w + h} + 1\right) \tag{8}$$

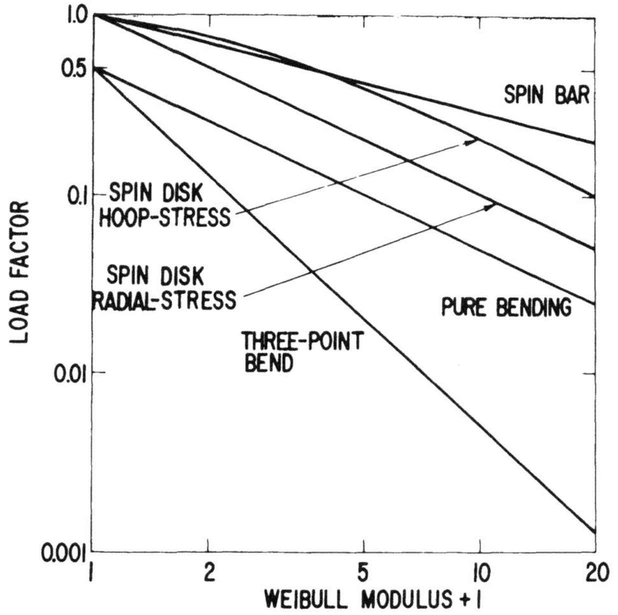

Figure 3. Volume load factors as a function of Weibull modulus + 1.

where w is the beam width and h the height.

Spin testing provides a method of applying a more uniform stress distribution than exists in bending. Bars can be spin tested to provide a uniaxial stress distribution and disks can be spin tested to provide a biaxial stress distribution. The stress distribution in a beam of length 2b spinning about its centroid is

$$\sigma = \frac{\rho\omega^2}{2}(b^2 - x^2) \tag{9}$$

where ρ is the mass per unit volume, ω is the angular velocity, and x is the distance from the centroid. Substitution of equation (9) into equation (5) results in a load factor for a spinning rectangular bar of

$$k_{SB} = \frac{2^{2m}(m!)^2}{(2m + 1)!} \tag{10}$$

where m is an integer and where σ_{max} occurs at $x = 0$. As shown in Figure 3, this load factor reduces to 1 for $m = 0$ and has a more shallow slope than that for pure bending due to the parabolic versus linear stress distributions. Since the stresses in a spinning bar vary only along the specimen length, the load factor based on an integration over the surface area has the same value. The volume integral and the area integral both reduce to an equivalent integral along the specimen length.

The radial and hoop stresses in a disk with a radius of c spinning about its centroid are given by

$$\sigma_r = \frac{3 + \nu}{8} \rho\omega^2 (c^2 - r^2) \tag{11}$$

$$\sigma_\theta = \frac{3 + \nu}{8} \rho\omega^2 c^2 - \frac{1 + 3\nu}{8} \rho\omega^2 r^2 \tag{12}$$

where ν is the Poisson's ratio and r is the radial distance from the center of the disk. These stresses are maximum and equal at the center of the disk. At the outer rim $(r = c)$ $\sigma_r = 0$ and σ_θ is a function of ν equal to 0.5 for $\nu = 0.2$. The load factor for the radial stress distribution is calculated by substitution of equation (11) into equation (5) with the result that

$$k_{RSD} = \frac{1}{m + 1} \tag{13}$$

For the hoop stress distribution in a spin disk

$$k_{HSD} = \left(\frac{1}{m + 1}\right)\left(\frac{3 + \nu}{1 + 3\nu}\right)\left[1 - \left(\frac{2 - 2\nu}{3 + \nu}\right)^{m + 1}\right] \tag{14}$$

Both of these load factors reduce to 1 at $m = 0$. For large values of m and for $\nu = 0.2$, $k_{HSD} \cong 2/(m + 1)$. In Figure 3 these load factors are plotted as a function of $(m + 1)$ where $\nu = 0.2$ was used for the hoop stress load factor. As in the spin bar, the load factors for the disk based on the integral over the surface area are the same as the volume load factors since the stress distribution varies only in the radial direction. The hoop stress over the surface at the rim contributes only an insignificant amount to the area load factor for large m. The total load factor for the spin disk can be calculated as the sum of the load factors for the radial stress and the hoop stress.

In order to compare the mean strengths of specimens with different sizes and stress distributions, equation (3) can be reduced to

$$\frac{\sigma_1}{\sigma_2} = \left(\frac{k_2 V_2}{k_1 V_1}\right)^{\frac{1}{m}} \tag{15}$$

With this expression or the corresponding expression for area, the results of bend tests and spin tests of beams and disks can be compared.

RESULTS AND DISCUSSION

Spin tests and three-point bend tests were performed at room temperature with a self-bonded silicon carbide (NC435 - Norton Co.) with a composition of about 80% silicon carbide and 20% silicon [4]. Small and large bars were tested in three-point bending to demonstrate the effect of volume or surface area, large bars were tested by spinning to demonstrate the effect of stress distribution, and disks were tested by spinning to demonstrate the effect of a biaxial stress distribution. The loading condition and specimen size are given in Table I along with the average fracture strength and the Weibull modulus. In all cases, as the effective volume (volume × volume load factor) or the effective area (area × area load factor) increases the average fracture strength decreases. Also, the Weibull modulus is nearly constant with an average value of about 7, the value used to calculate the load factors.

TABLE I - Test Conditions and Results for Silicon Carbide

LOADING CONDITION	SPECIMEN SIZE	LOAD FACTOR VOLUME	LOAD FACTOR AREA	NUMBER OF SPECIMENS	FRACTURE STRENGTH (KSI)	WEIBULL MODULUS
3-point bend	0.1×0.1×0.9* in.	.0078	.035	24	68.3	6.3
3-point bend	0.2×0.2×2.7* in.	.0078	.035	24	59.3	7.4
spin	0.2×0.2×3.0 in.	.32	.32	21	38.5	6.4
spin	0.2 in. thick × 2.8 in. dia.	.37	.37	12	28.2	7.7

*Outer span.

To correlate this strength data with a Weibull approach, equation (15) can be used. A plot of the ℓn of the fracture strength versus the ℓn of the effective volume or area should result in a straight line with a slope of $-1/m$. This data has been plotted in Figure 4 for effective volume and Figure 5 for effective area. A least squares fit of the data results in an m of 9.6 for volume and 6.3 for area. Thus, the correlation based on area results in a Weibull modulus nearly equal to the values obtained in the individual tests while the correlation based on volume results in a somewhat higher value.

As described in the previous section, the load factor for the spin disk was calculated as the sum of the load factors for the

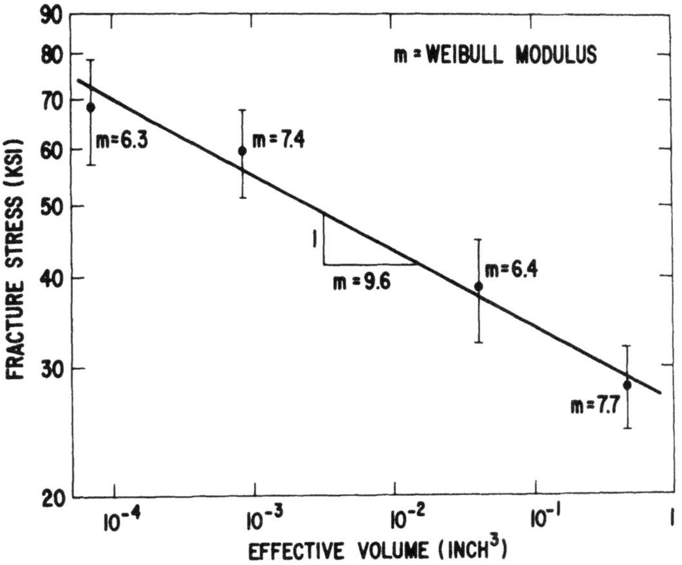

Figure 4. Average fracture stress versus effective volume (volume load factor × volume) for two sizes of bend bars, spin bars, and spin disks.

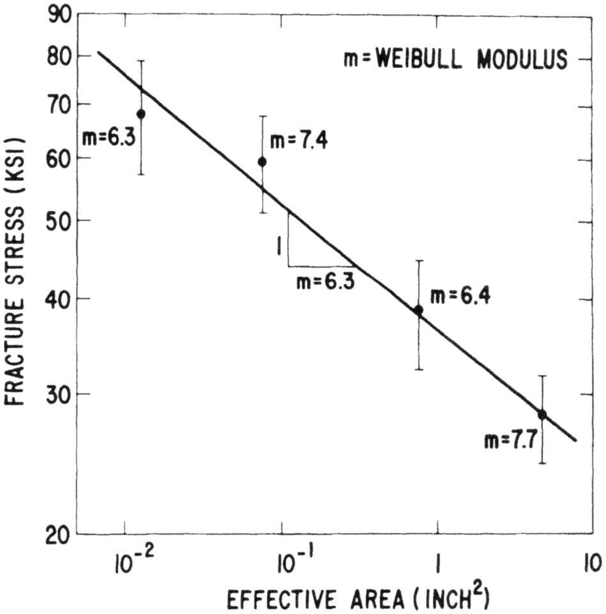

Figure 5. Average fracture stress versus effective area (area load factor × area) for two sizes of bend bars, spin bars, and spin disks.

radial stress and the hoop stress. Batdorf [5] has shown that this approach provides a lower bound to the failure probability for biaxial stresses since it is assumed that the principal stresses are applied individually to the crack. By considering the interaction of the principal stresses on the crack, an upper bound to the failure probability can be calculated [5]. For equibiaxial tension, this provides effectively an upper bound on the load factor which is $m^{0.45}$ times the lower bound on the load factor, the lower bound being the sum of the load factors for the two equibiaxial stresses. However, for the spin disk, equibiaxial stress only occurs at the center of the disk while the radial stress to hoop stress ratio reduces to zero at the surface. In any case, the upper bound on the load factor for $m = 7$ is 2.4 times the lower bound load factor used in Figures 4 and 5. A least squares fit of the data with the upper bound load factor results in an m of 10.4 for volume and 7.2 for area and an increased correlation coefficient in both cases. The Weibull modulus based on area is now more nearly equal to the values obtained in the individual tests. This good agreement of the spin test data with the correlation line indicates the necessity to account for biaxial stresses when they are of nearly the same magnitude.

Most of the fracture surfaces were observed with low-power optical microscopy to determine fracture origins. In some cases no indication of a fracture origin could be found. In many cases it was possible to distinguish between surface and volume flaws, however the fracture origin generally could not be detected. Of those specimens where it was possible to distinguish between surface and volume flaws, almost all were surface flaws. This observation is consistent with the fact that the Weibull correlation based on surface area was the more satisfactory. An SEM photograph of one of the fracture surfaces in a SiC disk is shown in Figure 6. The fracture mirror indicates a surface flaw but the fracture origin could not be detected even at higher magnification.

Additional bend tests of bars and spin tests of disks were performed with alumina. Magnesia-doped alumina was chosen, thus allowing specimens to be sintered directly to shape with both fine grain size and high density. Meller 0.3 μm Al_2O_3 powder was ball milled in toluene with 0.08 w/o of MgO (as Mg CO_3). Specimens were cold pressed, prefired at 1175°C in air, ground to the proper oversized dimensions, and final fired at 1750°C for 5 hours in wet hydrogen. The final density was greater than 99 percent of theoretical and the average grain size was approximately 8 μm. Thirty-five bars (0.15 in. × 0.15 in.) were tested in three-point bending over a span of 1.25 in. The average fracture strength was 37.1 ksi and the Weibull modulus was 12.6. Based on this data, the predicted strength of alumina spin disks (0.4 in. thick and 2.6 in. dia.) was 18.6 ksi for volume flaws and 21.8 ksi for surface flaws. The average strength of 10 spin disks was 27.4 ksi with a Weibull modulus of 6.8.

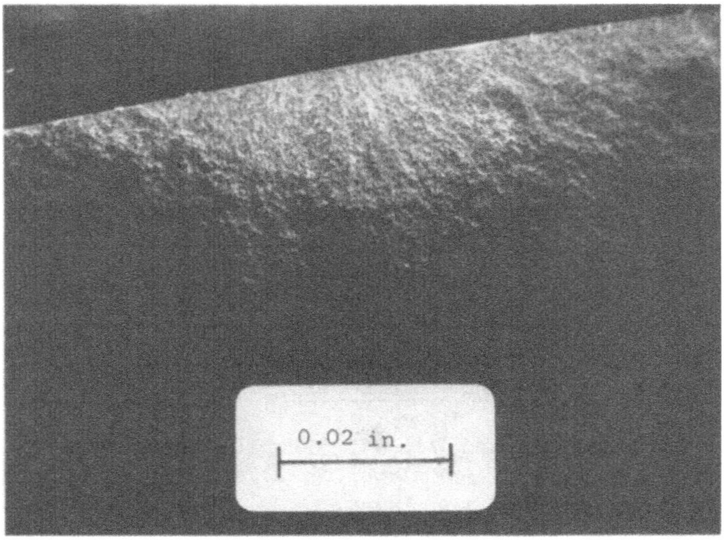

Figure 6. SEM photograph of fracture mirror indicating probable surface flaw.

Both predictions were extremely conservative and the Weibull modulus values did not agree. A possible reason for this discrepancy is that the two specimen configurations were pressed and sintered in markedly different sized bodies which may influence the type and severity of their flaw distributions.

CONCLUSIONS

A spin test technique has been developed for testing ceramic bars and disks. Results from spin tests and bend tests of disks and bars have led to the following conclusions:

1) For both the silicon carbide and the alumina that was tested, the fracture strength decreased with either an increase in volume or with an increase in the uniformity of the stress distribution (increase in load factor). For the alumina the two effects were not demonstrated independently but the fracture strength decreased with the combination of an increase in volume and an increase in the uniformity of the stress distribution.

2) For the silicon carbide, a Weibull statistical correlation based on surface area provides a very good interpretation of the effects of size, stress distribution, and biaxial stresses on the fracture strength. For the

alumina, a Weibull statistical correlation did not provide an acceptable interpretation of the results of bend tests and spin tests of disks.

ACKNOWLEDGMENT

The authors would like to thank Jack Craig for performing the spin tests and Ken Lay for fabricating the alumina bars and disks.

REFERENCES

[1] Trantina, G. G. and deLorenzi, H. G., "Design Methodology for Ceramic Structures," ASME Paper No. 77-GT-40, to be published in ASME Trans. - J. of Eng. for Power.

[2] Paluszny, A. and Wu, W., "Probabilistic Aspects of Designing with Ceramics," ASME Paper No. 77-GT-41, to be published in ASME Trans. - J. of Eng. for Power.

[3] Weibull, W., "A Statistical Distribution Function of Wide Applicability," J. of Applied Mechanics, Vol. 18, Sept. 1951, pp. 293-297.

[4] Wills, R. R. and Wimmer, J. M., "Controlled Surface Flaw-Initiated Fracture in Reaction-Densified Silicon Carbide," J. of American Ceramic Society, Vol. 59 [9-10], 1976, pp. 437-440.

[5] Batdorf, S. B., "Some Approximate Treatments of Fracture Statistics for Polyaxial Tension," International J. of Fracture, Vol. 13, No. 1, Feb. 1977, pp. 5-11.

EFFECTS OF SPECIMEN SIZE ON CERAMIC STRENGTHS

G. K. Bansal and W. H. Duckworth

Battelle
Columbus Laboratories
505 King Avenue, Columbus, Ohio 43201

ABSTRACT

Fracture stresses in specimens of four commercial polycrystalline ceramics differing in each linear dimension by a factor of four or five were measured at room temperature under controlled conditions. Data obtained were analyzed with the aid of fractographic examinations for applicability of Weibull statistics.

INTRODUCTION

Brittle fracture is triggered in a ceramic by tension acting at the site of a small discontinuity or flaw which intensifies the stress locally. Fracture occurs according to Griffith's criterion, as follows[1]:

$$\sigma_f = \frac{K_{IC}}{s} \qquad (1)$$

where σ_f is tensile stress at the flaw site, K_{IC}, is critical stress intensity factor of the material, and s is flaw severity*. If K_{IC} is considered bulk property and the ceramic contained a homogeneous population of identical worst flaws, it would be expected to fail at a unique tensile stress. However, strength values of nominally identical specimens when tested alike are usually dispersed because of specimen-to-specimen variability in fracture-initiating flaws. This variability precludes assigning

* $s = Y\sqrt{a}/Z$, where Y and Z are dimensionless parameters and a is flaw depth[2].

a unique strength value to specimens of a given size and causes a size dependence of strength. Large specimens tend to fail at lower mean strengths than small ones simply because there is apt to be a more severe flaw among the greater number of flaws in the large specimen. For specimens of the same size, the effective size is smaller when failure is by bending than by direct tension[3], because only part of the specimen is subjected to tension in bending.

A central problem in structural designing with brittle materials, results from this size dependence of fracture stress. With a size dependence, strength obviously cannot be described for purposes of structural analyses in terms of stress alone.

The object of the present research was to define and interpret the effects of size on strengths of four commercial polycrystalline ceramics. Emphasis was placed on precision in determining strengths, and fractography was used in interpreting strength-size data.

MATERIALS AND METHODS

A glass-ceramic, a conventionally sintered alumina, a hot-pressed alumina, and a hot-pressed silicon nitride were studied. Physical properties of each material are given in Table 1. Grain-sizes (G) and densities (ρ) are as reported by the manufacturers. Young's moduli (E) were determined in direct compression from measurements of load as a function of average strain. Critical stress-intensity factors, K_{IC}, were determined by the double-torsion technique[4].

TABLE I. MATERIAL PROPERTIES

	G, μm	ρ, g/cm^3	E, GNm^{-2}	K_{IC}, MNm$^{-3/2}$
Glass-Ceramic[(a)]	1-2	2.60	114	2.38 ± 0.08
Sintered Alumina[(b)]	5	3.75	318	3.84 ± 0.05
Hot-Pressed Alumina[(c)]	1-2	3.90	413	4.19 ± 0.15
Hot-Pressed Si_3N_4[(d)]	1-2	3.20	310	4.24 ± 0.30

(a) Corning's Pyroceram 9606 (c) AVCO's 99.8% Al_2O_3
(b) 3M Co.'s Alsimag 614 (d) Norton's NC-132

Bend-test specimens for strength determinations were cut from billets with a diamond saw, and were finish-ground parallel to the tensile-stress direction with a 320-grit diamond wheel. Edges were rounded slightly by polishing with a 1-μm diamond paste to prevent edge-initiated fractures.

Room-temperature bend tests were conducted on specimens of two sizes of each material. Small specimens of all four materials were 0.1 by 0.2 by 1.5 in. Each linear dimension of the large glass-ceramic and sintered-alumina specimens was five times that of the small specimen; the large hot-pressed alumina and silicon nitride specimens were larger by a factor of four.

Specimens were tested in 3- or 4-point bending using the bend-test fixture designed by Hoagland, et. al.[5]. Only small specimens were tested in 3-point bending over a span of 1.25 in. In the 4-point bend tests on small specimens, outer and inner spans were 1.25 and 0.75 in., respectively, and on large specimens they were four or five times those on small specimens, corresponding to the ratio of linear dimensions between the large and small specimens.

Strength tests on materials except silicon nitride were conducted under conditions that either restricted or enhanced subcritical crack growth. Specimens were tested in dry nitrogen at a stress rate of 100 MNm^{-2}/sec to restrict subcritical crack growth, and in distilled water at a stress rate of 4 MNm^{-2}/sec to enhance such growth. Specimens of Si_3N_4 were tested in laboratory air (relative humidity 45%) at a stress rate of 100 MNm^{-2}/sec.

Strengths, σ, were calculated from the expression, $\sigma = Mc/I$, where M is the applied moment, c is one-half the specimen thickness, and I is the cross-sectional moment of inertia.

After strength testing, the site (whether surface or subsurface) and the type of the fracture-initiating flaw in each specimen were identified by examination of the fracture surface using both optical[6] and scanning-electron microscopy[6,7,8].

RESULTS

Strengths of the glass-ceramic specimens, excluding those which failed from edge-initiated fractures, are given in Table II. Strength specimens were cut from three billets (each 1 x 5 x 10 in.). Because of a different microstructure, strength levels of specimens from Billet B material differed significantly from the others, necessitating separate strength analyses of specimens

TABLE II. BEND STRENGTH OF THE GLASS-CERAMIC

Environment	Specimen Size	Type of Loading	Fracture Origin	No. of Observations	Average Fracture Stress (MNm^{-2})	Coefficient of Variation (%)
				Billet A		
Dry N_2	Small	4-point	Surface	7	308	2.5
			Pore	1	243	
Dry N_2	Large	4-point	Surface	3	317	4.4
			Pore	2	265,245	
				Billet B		
Dry N_2	Small	4-point	Surface	8	395	4.2
			Pore	1	326	
Dry N_2	Large	4-point	Surface	2	378,396	
			Pore	3	296	1.1
			Surface-pore	1	383	
Dry N_2	Small	3-point	Surface	11	383	3.5
				Billet C		
Dry N_2	Small	4-point	Surface	5	317	2.2
Water	Small	4-point	Surface	4	204	1.5
Water	Large	4-point	Surface	4	199	1.0

from each billet. This seriously limited the number of replicate specimens available for statistical treatment of strength data. Strengths of specimens of the other three materials are given on 2-parameter Weibull plots (i.e., as log σ versus log log [1/(1-P)], where P is the probability of fracture).

DISCUSSION

Strength of Glass-Ceramic Specimens

As shown in Table II, fracture stress of the glass-ceramic was dependent on whether failure initiated at a machine flaw on the tensile surface or at a subsurface pore; higher fracture stress values were associated with surface origins. It will be noted that subsurface origins were observed only infrequently in tests of small specimens in dry nitrogen, and they were not observed in any specimens tested in water. The increased frequency of subsurface pore origins in large specimens tested in dry nitrogen is attributed to a sparse pore population in the material. The absence of subsurface origins in specimens tested in water is, of course, explained by the larger size of surface flaws in this environment coupled with the expected lack of an atmospheric effect on fracture initiation at internal sites.

An important finding is the lack of an observed strength dependence on specimen size among specimens fracturing from surface origins. In these instances, large and small specimens from each billet exhibited essentially the same average fracture-stress values and small (<5%) coefficients of variation. This finding applies to tests conducted in both water and the dry environment, and to comparable data from 3- and 4-point bend tests.

Conventionally strength values from surface and subsurface flaw origins are averaged together in assigning a strength value to a given ceramic. For purposes of academic interest, strengths from tests in the dry environment have been calculated in this way. As shown in Table III, these values exhibit a size effect due to the greater frequency of pore origins in the large specimens.

TABLE III. CONVENTIONAL TREATMENT OF BEND STRENGTH DATA OF THE GLASS CERAMIC

Specimen Size	Average Strength (MNm^{-2})	Coefficient of Variation (%)	No. of Specimens
	Billet A		
Small	299	6.7	8
Large	292	15.4	5
	Billet B		
Small	387	5.6	9
Large	341	15.2	6

Strength of Sintered-Alumina Specimens

Strengths of all specimens, excluding those exhibiting edge fractures, given in Fig. 1 indicate that in each environment the stress for any failure probability within the experimental limits of 0.1 and 0.9 decreased with increasing specimen size. The Weibull plots are reasonably linear, indicating that each set of strength data can be described by an equation of the form[3,9]:

$$P = 1 - \exp\left[- \int_{S \text{ or } V} (\sigma/\sigma_0)^m \, dS \text{ or } dV \right] \qquad (2)$$

where m is Weibull modulus, σ_0 is a normalizing constant, and the integral is taken over the volume or surface under tension depending on whether subsurface or surface flaws controlled fracture. m, as determined by the slopes of the plots, is constant (i.e., 34) only for specimens of differing size tested in water. For tests in the dry environment, m decreases with increased specimen size; values are 34, 26, and 11 for the small 3-point, small 4-point, and large 4-point bend specimens, respectively.

The equivalance of Weibull moduli from tests of large and small specimens in water (m = 34) and small specimens in dry N_2 (m = 34 or 26) suggests that fractures in these specimens were a consequence of flaws from the same statistical population.

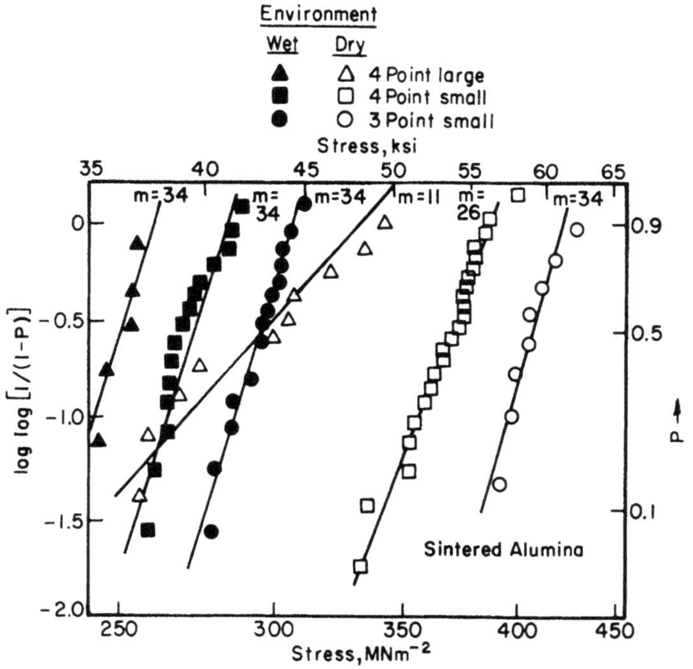

Fig. 1. Weibull Plots for Strength-Size Data of Sintered Alumina

This matter was investigated further by calculating the strengths in water from independent measurements of crack velocity as a function of stress-intensity factor[8]. Table IV shows good agreements between the calculated and measured strengths indicating that the same surface flaws initiated subcritical crack growth of specimens in water and catastrophic fracture of small specimens in dry nitrogen. Ratios of mean strengths were also calculated using[3]:

$$\bar{\sigma}_1/\bar{\sigma}_2 = (S_2/S_1)^{1/m} \tag{3}$$

and were compared with experimental ratios (Table V). Eq. (3) requires constancy of m; the significant differences between m's for data from large and small specimens tested in dry nitrogen precludes similar analysis for tests in the dry environment*.

* Differences in stress distributions in the 3- and 4-point bend specimens of the same actual sizes made the effective size of the 4-point specimen larger than that of the 3-point specimen by $\simeq(3m+5)/5$. This factor was calculated by determining the size of a direct-tension member that has the same P-σ relation as the specimen subjected to nonuniform tension[3]. For the two sizes of 4-point bend specimens, similarity allows use of actual surface-area ratio in Eq. (3); i.e., $S_2/S_1 \simeq 25$.

TABLE IV. CALCULATED AND MEASURED STRENGTHS OF SINTERED ALUMINA SPECIMENS IN WATER

Type of Loading	Specimen Size	σ_f, MNm^{-2} Calculated	Measured Avg.	Measured Std. Dev.
3-point bend	Small	288	295	9.4
4-point bend	Small	263	271	9.5
4-point bend	Large	243	250	5.5

TABLE V. OBSERVED AND CALCULATED MEAN STRENGTH RATIOS FOR THE SINTERED ALUMINA

Specimens Considered	Environment	Observed	Calculated*
Small 3-point: small 4-point	Water	1.09	1.09
Small 3-point: large 4-point	Water	1.18	1.20
Small 4-point: large 4-point	Water	1.08	1.10
Small 3-point: small 4-point	Dry N_2	1.11	1.09

* Using m = 34.

In tests in the dry N_2, the larger m exhibited by small specimens (m = 34 or 26) than by large specimens (m = 11) indicates less dispersion in the size of strength-controlling flaws in the small specimens. Fractography supported this indication. The microscopic inhomogeneities at fracture origins in the small specimens ranged from 40 to 70 μm in size whereas those in the large specimens ranged from 150 to 400 μm. This also suggests a sparser population of the flaws that initiated fracture in large specimens in dry N_2 than of the smaller surface inhomogeneities that initiated fracture in the small specimens. Presumably, if a sufficiently greater number of small specimens had been tested in the dry environment, some would have exhibited "large-flaw" failures. In this event, one would not expect mean strengths of the small specimens to be much different from those reported here, but the low-probability region of the probability-strength curve should be affected significantly. The Weibull plot in this case would be complex (e.g., two straight lines), reflecting a fundamental change in the nature of fracture at some stress level. Fig. 2 demonstrates that is indeed the situation[10].

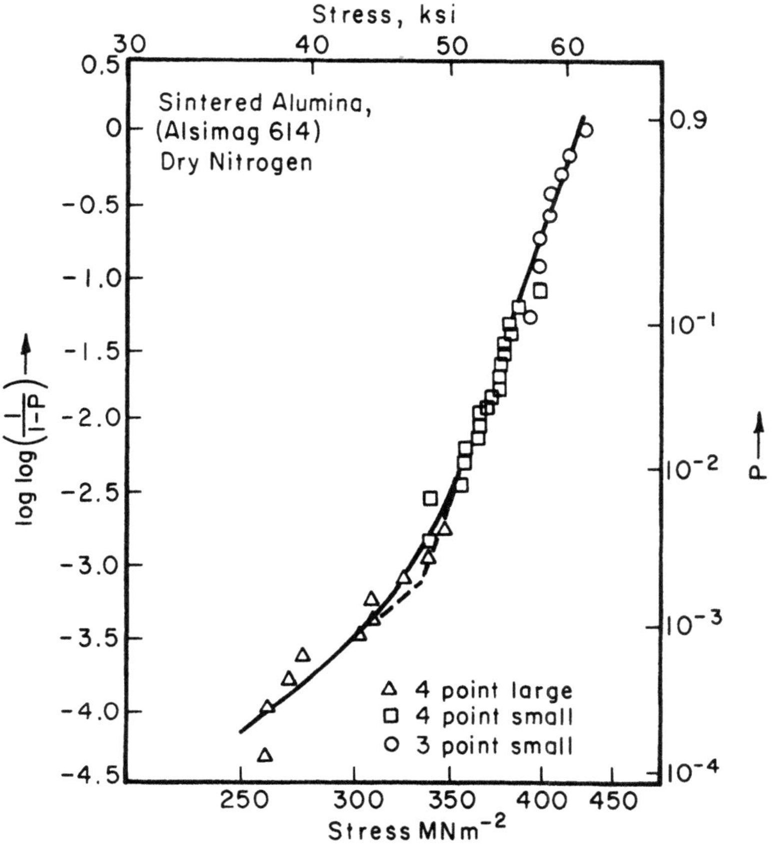

Fig. 2. Weibull Plot of the Data Obtained on Three Different Sizes of Sintered Alumina Tested in Dry N_2 (see Fig. 1). Failure probabilities for the small and large 4-point bend data have been normalized to those of the small 3-point bend specimen (Ref. 10).

Strength of Hot-Pressed Alumina Specimens

The data (Fig. 3) show a qualitative trend of decreasing strength with increasing specimen size in each environment. With an exception of strength data from 3-point bend tests, the Weibull plots do not exhibit the linear relationship required for applicability of Weibull's 2-parameter function. Large specimens both in dry N_2 and water failed from several different types of intrinsic flaws in each environment. However, apparent m's for large specimens were similar in the two environments which indicates that similar flaw populations controlled fracture in both environments. The lower strengths in water of course resulted because subcritical crack growth preceded fracture.

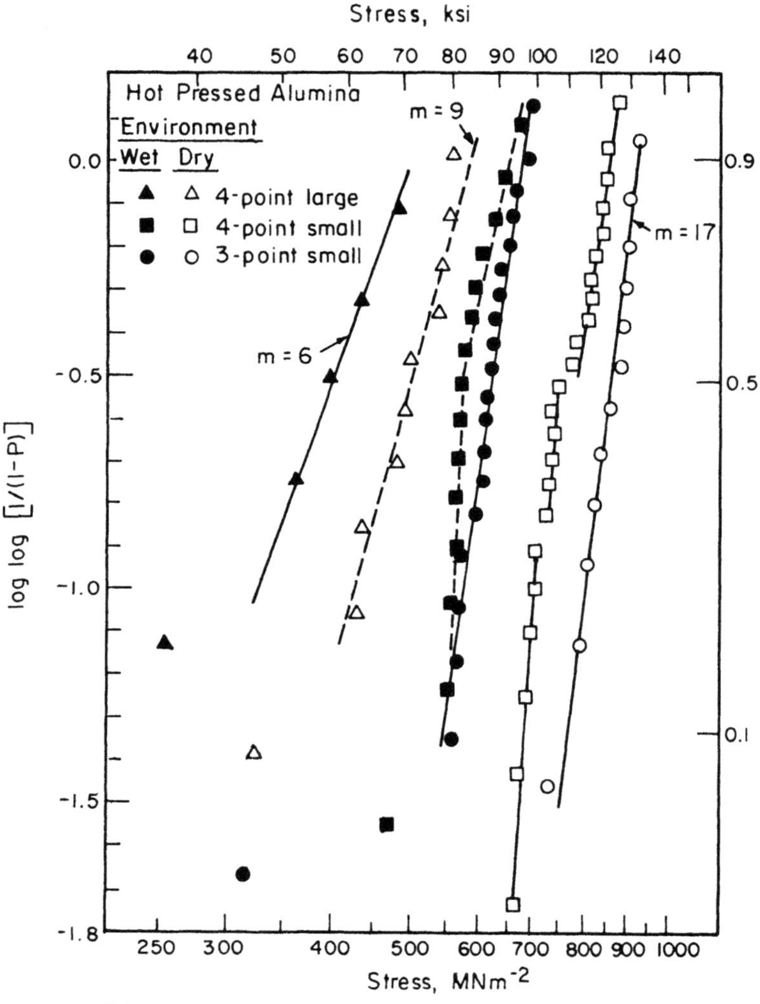

Fig. 3. Weibull Plots of Strength-Size Data of Hot-Pressed Alumina

m for 3-point bend specimens tested both in dry N_2 and water was calculated to be ~17. Although, the fit of the 4-point bend data in dry N_2 to the 2-parameter function is poor, similarity of the slope to that of 3-point bend data suggests similar flaw populations controlled fractures in each.

Strength data obtained in water on 4-point small specimens show two distinct regions in Fig. 3. Three different explanations are possible, as follows: (1) two flaw populations controlled fracture, (2) Weibull's 2-parameter function does not apply, and the 3-parameter function is more appropriate in describing this set of data, and (3) some variable(s) in addition to flaw size is affecting strength. No significant differences were observed in the type of fracture-initiating flaws in these specimens. Also, the parameter, σ_u, which is a measure of stress for zero probability of fracture, should not vary with strength levels and environments[3]. Therefore, the third possibility was explored.

K_{IC} values which controlled strengths of individual specimens were calculated from fracture-mirror analysis using the following equation[6]:

$$K_{IC} = 0.425\ \sigma_f\ r^{1/2} \tag{4}$$

where σ_f is the fracture stress at the origin, and r is the radius of the mirror which surrounds the flaw. K_{IC} values so calculated for specimens of different sizes tested in water and dry N_2 gas are given in Table VI.

TABLE VI. K_{IC} VALUES CALCULATED FROM STRENGTH-TESTED SPECIMENS OF HOT-PRESSED ALUMINA

Specimen Size	Type of Loading	Environment	K_{IC}, $MNm^{-3/2}$
Small	3-point	Dry N_2	4.15 ± 0.20
Small	3-point	Water	4.98 ± 0.38
Small	4-point	Dry N_2	4.35 ± 0.27
Small	4-point	Water	5.40 ± 0.52
Large	4-point	Dry N_2	4.02 ± 0.19
Large	4-point	Water	4.23 ± 0.26

Table VI shows no significant differences in the calculated K_{IC} values between large and small specimens tested in dry N_2. Also, tests of large specimens in water gave K_{IC} values similar to those obtained in dry N_2. However, K_{IC} from tests in water on small specimens exhibited higher values and larger dispersions; lower K_{IC} values were generally associated with specimens which failed at higher fracture stresses. Recently, Hübner and Jillek[11] have observed similar effects on K_{IC} of an alumina ceramic. They have attributed the increase in K_{IC} to a microcracked zone forming a three-dimensional network of cracks ahead of a natural crack tip in the presence of moisture-related crack growth. The similarity of K_{IC} values for large specimens tested in water and dry N_2 indicates the absence of the microcracked zone in these specimens. Although, some subcritical crack growth may have occurred in large specimens in water, the extended crack invariably linked with an intrinsic hetrogeneity, i.e., large pore, inclusion, or large-grain cluster, which apparently stopped the microcracking in the immediate vicinity of the critical flaw boundary. If K_{IC} varies among individual small specimens tested in water, as indicated, it is surprising that the Weibull modulus for these specimens is nearly the same as that of 3-point bend specimens tested in dry N_2.

Ratios of mean strengths ($\bar{\sigma}$) of small specimens tested in 3- and 4-point bending were calculated using Weibull's surface area formulation (Eq. (3)), and m = 17. Table VII gives the calculated and observed ratios.

TABLE VII. OBSERVED AND CALCULATED MEAN STRENGTH RATIOS FOR HOT-PRESSED ALUMINA SPECIMENS

Specimens Considered	Environment	Observed	Calculated
Small 3-point: small 4-point	Dry N_2	1.12	1.15
Small 3-point: small 4-point	Water	1.04	1.15

* Using m = 17.

Table VII shows a good agreement between the observed and calculated ratio for tests in dry N_2. This indicates that variability in the same flaw population was the primary cause of the size effect on strength. However, the observed and calculated ratios differed by ~10% for tests in water; the calculations indicating a larger effect of specimen size than that observed.

This can be explained as follows. In both the 3- and 4-point bend specimens, fractures are believed to be initiated from flaws from a single population since fractography revealed no differences in flaw types among individual specimens, and the average size of the flaw prior to subcritical crack growth is larger in 4-point bend specimens on the basis of dry N_2 test results. Because a larger initial flaw size gives a larger crack extension[12], the average critical flaw depth in 4-point bend specimens tested in water is expected to be larger than that in 3-point bend specimens both by virtue of initial distribution of flaw sizes and of subcritical crack extension. Because K_{IC} increases with the amount of slow crack growth, 4-point bend specimens are therefore expected to exhibit higher strengths than those predicted.

Correlation of strengths of the large and small specimens is precluded by the fact that flaws from different populations were responsible for failure in the two cases, both in water and dry N_2.

Strength of Silicon Nitride Specimens

Fig. 4 shows a qualitative trend of decreasing strength with increasing specimen size for this material. Data obtained on

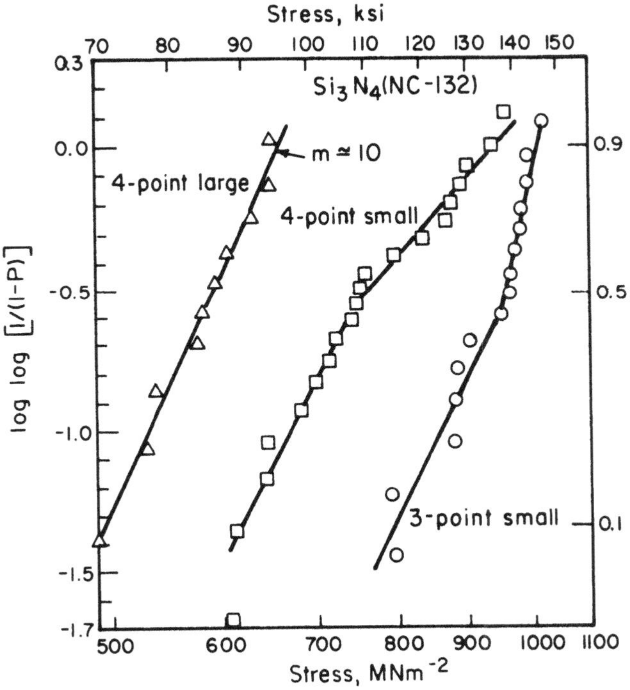

Fig. 4. Weibull Plots of Strength-Size Data of Silicon Nitride

large specimens give a reasonable fit to Weibull's 2-parameter function, but plots of strength data from small specimens tested in 3- and 4-point bending are not linear. Three different explanations for the nonlinearity, as stated earlier for the case of the hot-pressed alumina, are possible. No significant differences were observed in the type of fracture-initiating flaws in this material. Regardless of specimen size and type of testing, fractures invariably occurred on the tensile surface, and flaws were of same kind. Thus, the explanation based on multimodal distribution of flaws is contradicted. Also, Weibull's 3-parameter function can only describe the strength data represented by squares because the plot in Fig. 4 is concave downward. It can not describe the strength data represented by circles which yield a concave upward plot[13]. The inability of Weibull's 3-parameter function in describing both sets of data tends to preclude its usefulness in explaining the nonlinearlity. Therefore, the third possibility of some other variable affecting the strength distribution was explored.

Freiman and coworkers[14] have observed variable critical stress-intensity factors, K_{IC}, associated with individual small specimens of this particular Si_3N_4 material, suggesting that K_{IC} is a local property. We have also observed a variation of K_{IC} among individual small specimens, with no consistent trend of variable K_{IC} with strength levels. K_{IC} values calculated from fracture-mirror radii measurements[6] ranged from 3.2 to 4.8 $MNm^{-3/2}$ for small specimens. However, calculated K_{IC} values from large 4-point bend specimens were $4.20 + 0.15$ $MNm^{-3/2}$, which agree well with the value of 4.24 ± 0.30 $MNm^{-3/2}$ independently measured by the double-torsion technique (Table I). It should be pointed out that the mirror boundaries in small specimens were difficult to define precisely; the variability in K_{IC} could be associated with this experimental difficulty and not be real. In this case, some other explanation for the nonlinearity of strength data is required.

It is interesting to note, however, that if Weibull modulus from large specimens, i.e., m = 10, is used to calculate the average strengths of small specimens from the average strength of large specimens, a reasonable agreement with measured strengths is obtained (Table VIII).

TABLE VIII. MEAN STRENGTH RATIOS FOR HOT-PRESSED SI_3N_4

Specimens Considered	Observed	Calculated
Small 3-point: small 4-point	1.204	1.215
Small 3-point: large 4-point	1.585	1.603
Small 4-point: large 4-point	1.316	1.319

CONCLUSIONS

Observed size dependencies of strength resulted from specimen-to-specimen variations in "worst" flaws, such that a severe worst flaw was associated with larger effective sizes subjected to tension; strength, therefore, decreased with increased specimen size. For each ceramic, a single Weibull function was inadequate to describe the strength dispersion over the entire range of observed strength values. This situation resulted in three of the four ceramics studied because more than one population of worst flaws was present in the material, and the population which dominated depended on specimen size, testing environment, and strain rate.

In one of the ceramics, hot-pressed alumina, K_{IC} also varied among specimens. The variation occurred in a consistent manner, increasing with extent of subcritical crack growth. In this case, the effect of variable K_{IC} on strength must be determined independently and strength values adjusted accordingly prior to any statistical treatment of the data to define the size dependence of its strength. In the one ceramic where no evidence was found of more than one population of worst flaws, hot-pressed silicon nitride, the observed strength dispersion for specimens of all sizes studied could not be described clearly by a single Weibull function. The reason was not firmly established, but the existence of a variable K_{IC} in the material is indicated.

Generally, in cases where two worst flaw populations were present, one population was wholly or partially associated with surface finishing and the other with microstructural features. Strength of glass-ceramic specimens surprisingly did not exhibit a size dependence when failure resulted from flaws associated with surface finishing.

ACKNOWLEDGMENT

The research was supported by the Office of Naval Research under Contract No. N00014-73-C-0408, monitored by Dr. A. M. Diness.

REFERENCES

1. Griffith, A. A., Phil. Trans. R. Soc., A221, 163-198 (1920).
2. Bansal, G. K., J. Am. Ceram. Soc., 59 [1-2] 87-88 (1976).
3. Davies, D.G.S., Proc. Br. Ceram. Soc., 22, 429-52 (1973).
4. Williams, D. P. and Evans, A. G., J. Test. Eval., 1 [4] 264-70 (1973).
5. Hoagland, R. G., Marschall, C. W., and Duckworth, W. H., J. Am. Ceram. Soc., 59 [5-6] 189-92 (1976).

6. Bansal, G. K., Phil. Mag., 35 [4] 935-44 (1977).
7. Bansal, G. K., Duckworth, W. H., and Niesz, D. E., Am. Ceram. Soc. Bull., 55 [3] 289-92, 307 (1976).
8. Bansal, G. K., Duckworth, W. H., and Niesz, D. E., J. Am. Ceram. Soc., 59 [11-12] 472-78 (1976).
9. Weibull, W., J. Appl. Mech., 18 [3] 293-97 (1951).
10. Johnson, C. A. and Prochazka, S., Quarterly Progress Report #3, prepared for Naval Air Development Center under Contract No. N62269-76-C-0243 (January, 1977).
11. Hübner, H. and Jillek, W., J. Mat. Sc., 12, 117-25 (1977).
12. Evans, A. G., Int. J. Fract. Mech., 10 [2] 251-59 (1974).
13. Pears, C. D. and Starrett, H. S., Technical Report No. AFML-TR-66-228 (March, 1967).
14. Freiman, S. W., Mecholsky, J. J., and Rice, R. W., U. S. Naval Research Laboratory, Private Communication (1977).

INDENTATION FRACTURE AND STRENGTH DEGRADATION IN CERAMICS

B.R. Lawn and D.B. Marshall

Department of Applied Physics

The University of New South Wales, Australia

ABSTRACT

Brittle ceramic components may suffer severe degradation as a result of localised cracking in contact (notably impact) situations. Such situations, although complicated by a variety of factors, e.g. nature and extent of damage modes operating during the contact, prior state of the exposed surface and the environment, indenter geometry, are most conveniently analysed in terms of "indentation fracture mechanics". This paper accordingly sets out to review the basic principles of indentation fracture and to indicate how these principles may be applied to practical contact problems, with special emphasis on the problem of contact-induced strength loss.

1. INTRODUCTION

Ceramic components in systems exposed to certain unfavorable in-service environments can suffer considerable surface damage from ostensibly minute contact events. Basically, the contacts generate intense stress concentrations, which give rise to characteristic indentation fracture patterns. With the increasing call for ultra-strong materials in high technology it has become more and more necessary to pay attention to the degrading effects that small-scale indenting particles (including liquid drops) may have on mechanical properties. The cumulative effect of a large number of particle contacts can cause rapid erosion and wear of a brittle surface; in structural applications a single contact can reduce the strength of a component by more than an order of magnitude.

"Real" contact situations are very complex. Intuitively, one might expect a large number of variables to enter a general description. Material properties (e.g. toughness, hardness, stiffness), prior state of the brittle surface (e.g. flaw density, level of surface stress), indentation parameters (e.g. indenter geometry, load rate): these are some of the factors which need to be considered. Despite this apparent complexity, a sound basis for classifying and analysing indentation fracture patterns has now been established.[1] Based on whether the contact is essentially elastic ("blunt" indenters) or plastic ("sharp" indenters), the scheme provides a rationale for investigating a wide range of practical problems in ceramics engineering.

In this presentation we survey the basic principles of "indentation fracture mechanics". Our aim is to bring together those results most pertinent to practical applications: details of the formulations have been reviewed elsewhere,[1] and are not included here. In describing geometrical features of the indentation patterns we shall see that certain types of crack tend to penetrate the specimen, and accordingly bear strongly on strength properties, while others tend more to turn back toward the indented surface, and thereby relate more closely to surface-removal properties. Particular attention will be paid here to the question of strength degradation, this lending itself most readily to straightforward analysis in terms of a well-defined indentation crack system.

2. INDENTATION FRACTURE MECHANICS

The experimentation involved in indentation fracture testing is simple. A standard indenter is loaded onto the surface of a test material at a controlled rate, and the attendant crack growth monitored. For general purposes the commercial hardness or crosshead testing machines available in most materials science laboratories are perfectly adequate for delivering the load, and optical techniques applied either during or after indentation usually provide sufficient information on the crack behaviour. More detailed studies may in some cases demand more elaborate equipment,[2] e.g. environmental chamber, impact apparatus, acoustic sensors (notably for opaque specimens).

What has emerged from comprehensive investigations of indentation fracture patterns on brittle materials, particularly from "model" studies on silicate glasses, is a convenient classification system based on the essential nature of the contact deformation. Indenters are considered "blunt" or "sharp" according to whether the contact is governed by equations of elasticity or plasticity.[1] In their applications to practical contact fracture situations these two categories may be regarded as limiting cases.

In establishing a theoretical framework for analysing the different types of indentation fracture pattern, one seeks to incorporate the accessible contact parameters into a Griffith-Irwin fracture mechanics formulation.[3] First it is necessary to specify the stress field through which the cracks evolve; in this context it is important to appreciate that the general indentation field contains a significant tensile component, even though the loading is ostensibly compressive.[4] It is the distribution of tensile stresses which ultimately determines the crack driving force and the crack path. A characteristic contact dimension and contact pressure uniquely determine the spatial extent and intensity (i.e. the scale) of the field.

Given the indentation field, one must then ask where and how the cracks initiate. This raises the issue of nucleation centers for the overall fracture process, so the availability of pre-present flaws, either on the surface or in the bulk, becomes an important factor. In a controlled experiment one may choose to regulate the flaw population by a suitable surface abrasion (flaw introduction) or etch (flaw removal) treatment. A simplistic view of the initiation process in the indentation problem is one of providing a precursor energy barrier to full-scale development of the cracks, this being manifested experimentally as a fracture threshold in the applied loading.[5]

Once initiated, the crack propagates in accordance with some extension criterion. It is usual to formulate any such criterion in terms of a convenient fracture mechanics parameter representing the driving force on the crack, e.g. the "crack extension force" G or "stress intensity factor" K.[3] Generally, the cracks extend under either equilibrium or kinetic conditions. In the first of these, the critical requirement for extension is determined by a Griffith energy balance expressible as $G_c=K_c^2(1-\nu^2)/E=2\Gamma$, with Γ the fracture surface energy, K_c the toughness, E Young's modulus and ν Poisson's ratio; equilibrium conditions usually obtain in high vacuum or inert environments, or at low temperatures. In the second case the crack growth is rate controlled, so that a crack velocity requirement $v_c=v_c(K)$ or $v_c(G)$ is applicable; such conditions are met when the fracture takes place in a hostile chemical environment. For a well-defined indentation crack system one then computes an appropriate fracture mechanics function in terms of indentation load P and crack length c, i.e. $K(P,c)$ or $G(P,c)$. This provides sufficient information, in principle, to predict the scale of fracture under any specified loading conditions.

Let us now investigate the different categories of indentation fracture pattern, model and real, in turn. We focus particular attention on crack growth under equilibrium conditions: kinetic effects are regarded as secondary (but by no means unimportant) in the fracture mechanics analysis.

2.1 "Blunt" Indenters

The archetype of the fracture pattern produced under conditions of essentially elastic loading is the classical Hertzian "cone crack". This fracture configuration is most easily obtained with a spherical indenter on a flat specimen surface. Considerable effort has been directed to fracture mechanics studies of cone crack evolution.[6-14]

Figure 1 shows schematically the sequence of events for one complete indentation cycle. The form of the tensile field, with degree of shading representing level of stress, is indicated in diag. (a). While all stress components directly below the elastic contact area are compressive, as one might expect, elsewhere the major principal stress becomes tensile. The tension is particularly strong in a thin, "surface skin" region outside the contact circle. Cone fracture then develops as follows:

(a) the spherical indenter subjects pre-present surface flaws (shown as short dashed lines in diagram) to increasing tensile stress outisde the expanding contact circle; (b) upon attaining a critical "Griffith configuration" a favourably located flaw runs around the contact to form a surface "ring crack",[6] its simultaneous downward growth attenuating to a greater or lesser degree as it extends out of the surface skin region; (c) on further loading, the contact continues to expand and the ring crack is driven downward in controlled fashion, either in equilibrium[6] or kinetically,[10] deviating outward to avoid the compressive zone; (d) the tension over the area of the crack, although now weak, cumulates with growth in the net driving force until a major instability attains, whence the ring spontaneously develops into a full cone crack (this corresponding to the usually observed critical event in the conventional Hertzian test); (e) on still further loading the crack continues in controlled extension (until the contact circle engulfs the surface ring crack and closes the fracture interface, in which case secondary, tertiary etc. ring cracking may follow); (f) finally, on unloading, the cone crack tries to close and heal to recover the stored elastic energy and surface energy, but is prevented from doing so completely because of mechanical obstruction ("debris", fracture steps, chips, etc.) at the interface[15] (it is for this reason, of course, that the remnant cracks remain visible) - if the unloading is rapid, the

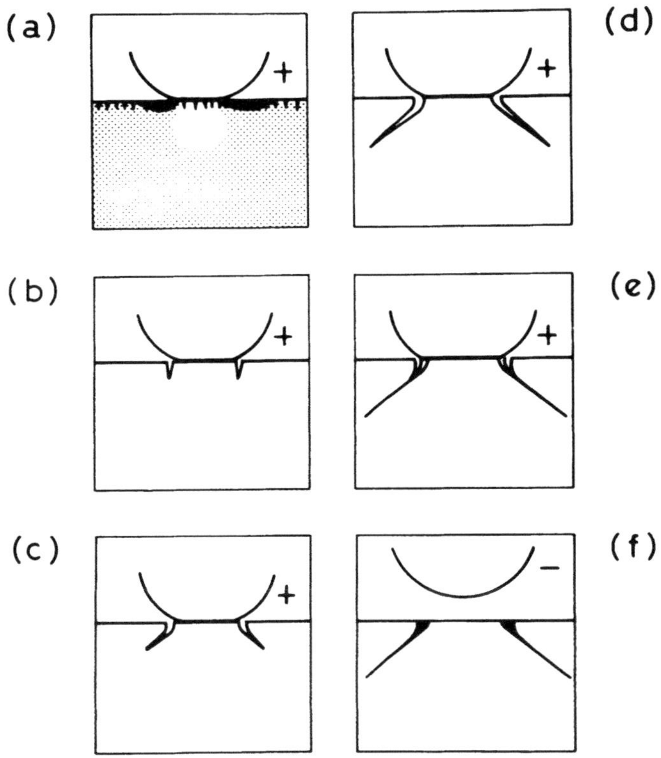

Figure 1: Evolution of cone crack pattern during one complete loading (+) and unloading (-) cycle. Stress field indicated in (a).

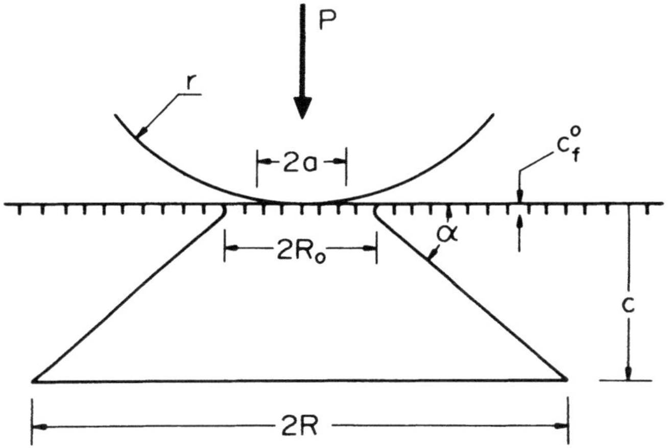

Figure 2: Parameters of Hertzian cone crack system.

base of the cone may turn up in "hat brim" fashion.[16]

The indentation parameters used in quantifying the Hertzian fracture process are shown in Figure 2. To establish the scale of the elastic field one resorts to the equations of Hertzian contact.[17] The radius of the contact circle establishes the spatial extent of the field:

$$a = (4kr/3E)^{1/3} P^{1/3} \tag{1}$$

where r is the sphere radius, and $k=(9/16)[(1-\nu^2)+(1-\nu'^2)E/E']$ is a dimensionless constant involving the Young's moduli E, E', and Poisson's ratios ν, ν', of specimen and sphere respectively. Likewise, the mean indentation pressure establishes a convenient unit of stress:

$$p_o = P/\pi a^2 = (3E/4\pi^{3/2}kr)^{2/3}P^{1/3} . \tag{2}$$

Note that for a given load and sphere size the scale of the field is uniquely determined by elastic constants. In a low-energy impact situation the impulsive load $P(v)$ delivered by a sphere is readily determined from the Hertzian theory, in which case the quantities a and p_o may equally well be written in terms of impact velocities.[8,17]

With the application of equilibrium fracture mechanics to the Hertzian configuration one seeks first to specify the threshold conditions under which the cone crack is made to "pop in". On the assumption that the indented surface contains an abundance of large initiating flaws the critical load is found to be[6]

$$P_c = \alpha_E r K_c^2/E \tag{3}$$

where $\alpha_E=\alpha_E(E/E',\nu)$ is a dimensionless constant of the specimen/indenter materials system. An important conclusion to be drawn here concerns the independence of P_c on the effective initial flaw size c_f^o. This is a consequence of the existence of the precursor energy barrier stage (c) in the evolutionary sequence of Figure 1. Equation (3) has been confirmed in a number of experimental studies[7,8,12,18-21]. The formulation actually breaks down in the limit of small flaws or large spheres,[6,8] whence the energy barrier diminishes to zero and the cone crack accordingly develops direct from stage (b) to stage (d) in the evolution. It also tends to break down as the flaw density becomes sparse, such that the expanding contact circle has to "search" for a suitable flaw; statistical considerations then enter the problem.[1] Nevertheless, Eqn. (3) remains a useful basis for conservative design in that predictions beyond its range of validity will tend to underestimates in P_c. One drawback is the inability of the theory to predetermine

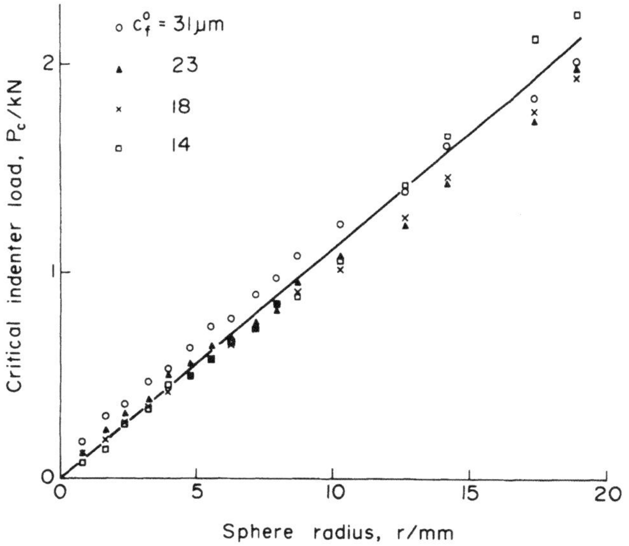

Figure 3: Critical load to cone crack formation as function of sphere indenter radius, steel on soda-lime glass. Glass surfaces pre-abraded to produce controlled flaw population, sizes as indicated. Each point is mean of at least ten tests (standard deviation typically < 10%). Note $P_c \propto r$, but independent of c_f^o. After Ref. 8.

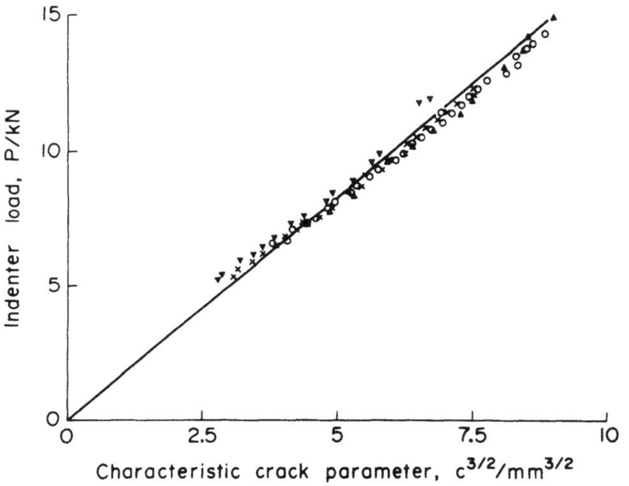

Figure 4: $P(c)$ data for fully-developed cone cracks in soda-lime glass. Truncated sphere indenter. (Measurements made from observations of base radius of cone, using $c = R\tan\alpha$, Fig. 2.) Each symbol represents a separate crack. After Ref. 1.

the constant α_E to much better than an order of magnitude; for this purpose an experimental calibration is desirable. Data in Figure 3 for tests with steel spheres on abraded soda-lime glass surfaces,[8] in conjunction with independent data on K_c and E for the glass, provide suitable information to effect the calibration, and serve also to highlight the main features of the threshold equation.

Once the cone crack is fully developed, further growth becomes less sensitive to events at the contact zone. The configuration most amenable to fracture mechanics analysis is the limiting one of an ideal cone ($R>>R_0$), for which one obtains

$$P/c^{3/2} = \beta_E K_c \qquad (P>P_c) \qquad (4)$$

where $\beta_E = \beta_E(\nu)$ is a dimensionless constant of the specimen material. This result is seen to be totally insensitive to variations in c_f° and a, as one would anticipate from an essentially point-contact configuration. Any errors incurred by neglecting the non-zero surface diameter of the truncated cone in the derivation of Eqn.(4) reflect as overestimates in the predicted crack size, once again consistent with conservative design. The constant β_E is readily calibrated from equilibrium cone crack growth data, such as those for soda-lime glass in Figure 4.

2.2 "Sharp" Indenters

When the contact is essentially plastic* the indenter is deemed to be sharp. Such indenters are typified by the diamond pyramids or cones used in routine hardness testing, although even a sphere is capable of generating precursor plastic flow in less brittle materials.[22-24] Evaluation of sharp-indenter fracture mechanics is relatively recent.[1,4,5,25]

The sequence of events for one complete indentation cycle is shown schematically in Figure 5. In this case the evolution is more complex than for blunt indenters: the elastic/plastic field of the applied loading becomes significantly modified by residual stress effects.[26] During the loading half-cycle the tensile field, depicted by shading in diag. (a), peaks directly below the indenter point, where the greatest concentration of deformation and elastic/plastic constraint occurs. This phase of the evolution proceeds in the following way:[4]

---

* We use the term "plastic" loosely here to include any mode of irreversible deformation, including densification.

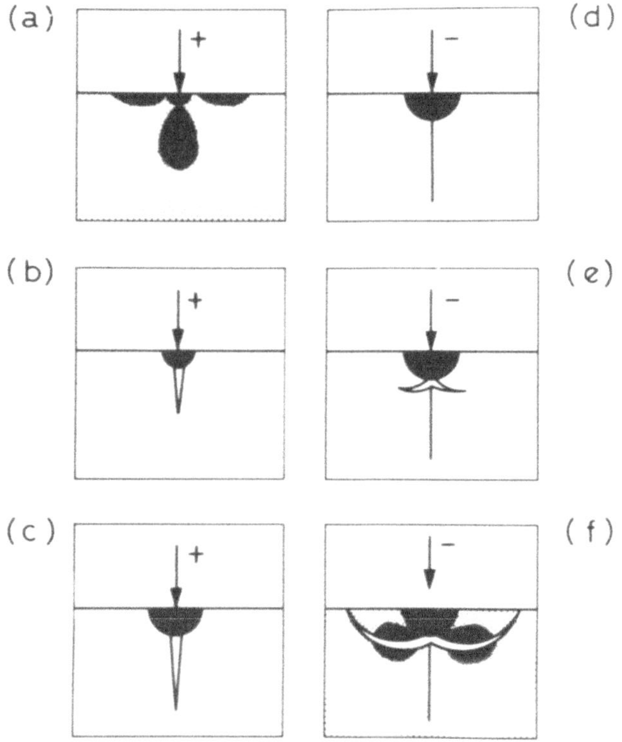

Figure 5: Evolution of median/lateral crack pattern during one loading (+) and unloading (-) cycle. Dark region represents plastic zone. Point-loading and residual stress fields indicated in (a) and (f), respectively.

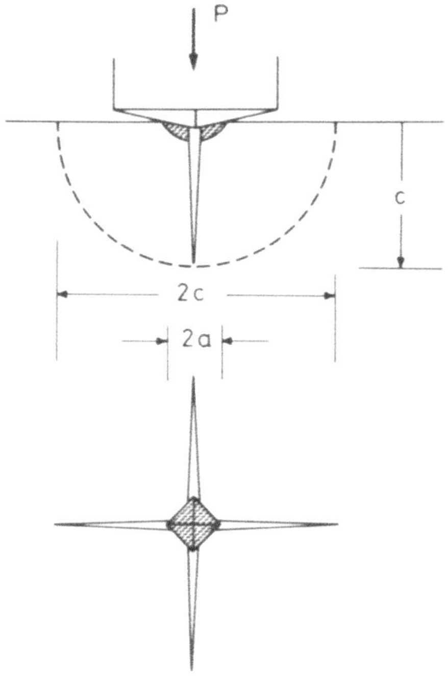

Figure 6: Parameters of median crack system (lateral cracks not shown).

(a) the plasticity processes subject sub-surface flaws to a "yield-controlled" level of tensile stress (stresses on pre-present surface flaws being relieved somewhat by the absence of elastic/plastic constraints at the free surface); (b) on the spatial extent of the field reaching some critical level[27] one or more of the flaws grow suddenly into sub-surface penny-like cracks, so-called "median cracks", on symmetry planes containing the load axis and major impression diagonals; (c) increased loading expands the contact and drives the median cracks stably downward beneath the indenter and simultaneously upward to intersection with the free surface at the sides, until the geometry tends ultimately to a semicircular profile with linear radial traces extending from the impression corners on the specimen surface (hence an alternative terminology, "radial cracks", although this designation is reserved by some for near-surface, radially directed fractures which are sometimes observed to initiate from particularly large surface flaws[22,23]).

Throughout the indentation cycle the stress field is augmented by incompatibility effects at the elastic/plastic interface.[26] Thus the field acquires a residual component, and it is the residual stresses which begin to dominate as the indenter is unloaded from the specimen. Diag. (f) in Figure 5 indicates the approximate distribution of residual stresses: most notable is the distortion of the tensile maxima upward toward the free surface. Hence the sequence in the unloading half-cycle:[4]

(d) as unloading starts the walls of the median cracks begin to move together, but fracture debris and residual tension once more prevent total closure (indeed, the residual stresses may actually drive any partially formed median cracks to full development[5]); (e) just prior to full removal of the indenter the residual stresses become sufficient to initiate a completely new system of sideways spreading, saucer-like "lateral cracks"; (f) the lateral cracks continue their spread as the indenter is finally removed, and may actually intersect the free surface to produce a chip[28] (growth may persist long after completion of the indentation cycle if the environment is reactive).

Characteristic indentation parameters, such as those shown in Figure 6 for the most widely used Vickers pyramid arrangement, provide the necessary specifications for a fracture mechanics analysis. An idealised sharp-indenter situation invokes the concept of geometrical similarity[29] such that the mean contact pressure is invariant with load:

$$p_o = \text{const.} = H \qquad (\text{all } P) \qquad (5)$$

where H defines the hardness of the specimen. Then the impression half-diagonal is given by

$$a = (1/\Lambda\pi H)^{1/2} P^{1/2} \quad : \tag{6}$$

for the Vickers geometry of Figure 6 the dimensionless indenter constant Λ is $2/\pi$. It is seen that at a specified load the scale of the field is uniquely determined by plastic properties.

Since the indentation stress field generated during the loading half-cycle is primarily determined by the contact forces, one may reasonably neglect residual stresses in deriving working fracture relations for the median cracks: effects due to the residual component of the field may then be introduced as simple modification factors,[30] which may in turn be absorbed into convenient calibration constants in the equations. Starting with the threshold conditions for the formation of the median cracks, again assuming an abundance of large initiating, sub-surface flaws, the critical load is[27]

$$P_c = \alpha_P K_c^4/H^3 \tag{7}$$

where $\alpha_P=\alpha_P(\nu$, elastic/plastic zone parameters) is a dimensionless constant of the specimen/indenter system. As with its counterpart Eqn. (3) for blunt indenters, Eqn. (7) does not involve the initial flaw size c_f°. In this case the computed critical load represents an absolute minimum value for initiation, requiring the presence of an optimal flaw favourably located within the elastic/plastic field; if the density of flaws is sparse such that this requirement is not automatically met, the contact must continue its expansion until a suitable alternative crack source is encountered, whence the value of P_c increases and becomes subject to statistical variations. While Eqn. (7) has not been put to rigorous experimental scrutiny it does bear out empirical findings of higher thresholds in tougher ceramics.[22] It would appear that with an appropriate calibration of α_P the equation should serve as an adequate basis for conservative design.

At full development of the median cracks, growth continues with increasing load according to[5]

$$P/c^{3/2} = \beta_P K_c \qquad (P>P_c) \tag{8}$$

where $\beta_P=\beta_P$(effective indenter angle) is a dimensionless constant of the specimen/indenter system. This result is of precisely the same form as Eqn. (4) for well-developed cone cracks: both median and cone cracks extend ultimately on a near-circular front, and accordingly comply with the basic fracture mechanics relations for "penny-like" cracks.[5] We note again the independence of the result on c_f° and a, consistent with the mouth-loading geometry of

Figure 6. Figure 7 shows some median crack data for Vickers pyramid tests on soda-lime glass, from which a calibration value of β_P may be readily determined.

An interesting extension of Eqn. (8) for well-developed median cracks is the case of a biaxially pre-stressed surface, such as in tempered glasses or other ceramics where a residual surface compressive stress σ_R is introduced to oppose fracture. The fracture mechanics relation then modifies to[31,32]

$$P/c^{3/2} = \beta_P[K_c + \sigma_R(\pi\Omega_c c)^{1/2}] \tag{9}$$

where Ω_c is a dimensionless constant $\approx 4/\pi^2$ for penny cracks of characteristic dimension c. [Comparative measurements of $P(c)$ for the material in its tempered and untempered (control) states accordingly permits an evaluation of σ_R.[32]]

Unfortunately, the fracture mechanics of lateral cracking is not so well understood. It will be recalled that lateral cracks develop toward the end of the unloading half-cycle, and thereby associate directly with the residual, elastic/plastic mismatch stress field. The true nature of this field is only beginning to be investigated in a proper analytical way. Several workers have proposed oversimplistic models in which the indentation is viewed in terms of an expanding spherical cavity in an infinite elastic/plastic medium. But such models completely ignore one of the most important features of the lateral crack system, its clear tendency to depart from spherical symmetry in its growth and interact with the free surface. Recent stress field analyses by Perrott[26] represent an important first step in incorporating essential surface effects. Empirical observations of lateral crack growth in several ceramics[22,33] do show the same qualitative trends as for the other crack types in their well-developed form, namely an increase in crack size with peak indentation load.

2.3 "Real" Indenters

How does our analysis of blunt and sharp indenters above help us to understand complex contact events which occur at the surfaces of ceramic structural components during in-service operation? Indenting particles in such situations will generally have a complicated topography of rounded and angular corners and edges, so that the nature of the contact may lie anywhere between the extremes of ideally elastic and ideally plastic. What is more, an indenter of the simplest geometry may act as though blunt in early phases of an indentation cycle, and as sharp in advanced phases of the same cycle.[1] However, none of these complications detract from the present, idealised indentation fracture scheme which takes

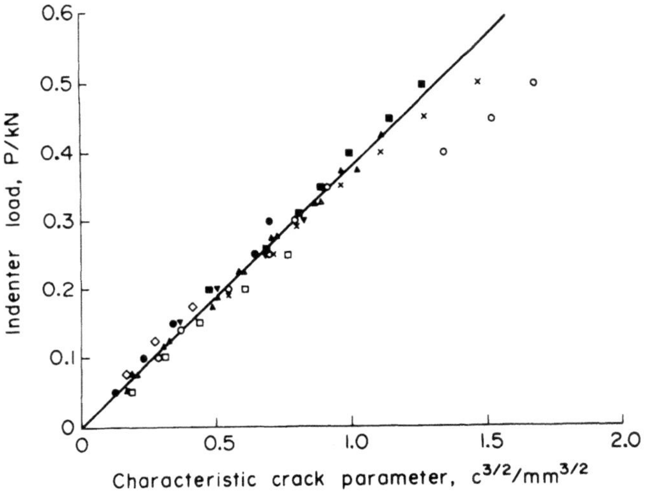

Figure 7: *P*(*c*) data for fully-developed median cracks in soda-lime glass. Vickers indenter. Each symbol represents a separate crack. After Ref. 4.

TABLE 1: Essential features of indentation fracture evolution

Blunt Indenters	Sharp Indenters
Initial contact elastic	Initial contact plastic
Tensile component of contact field initiates crack from surface flaw outside contact circle	Tensile component of contact field initiates cracks from subsurface flaw at elastic/ plastic interface
Well-developed cone crack extends on circular front	Well-developed median cracks (loading) and lateral cracks (unloading) extend on circular fronts

Figures 1 and 5 as outer bounds to real behaviour.

It is accordingly of some interest to compare and contrast the main features of the fracture mechanics for the blunt and sharp indenter cases. This is done in Table 1. We note that the greatest differences in behaviour occur in the initial stages of contact, where threshold loading is highly sensitive to flaw response in the widely variable near-surface stress field. Generally speaking, the threshold for crack initiation with sharp indenters is some orders of magnitude less than that with blunt indenters. The greatest similarities occur once the indentation cracks have become well developed, since by this stage the fracture geometry is universally penny-like about an effective point-contact center. Thus in applying our blunt/sharp scheme to real contact situations we must expect the indentation fracture predictions to be more reliable at higher loads; of course, it is the severe contact which is of most concern in preventitive design.

A complete account of any real contact event may involve any number of variants in fracture behaviour. Our choice of glass as a model brittle material conveniently avoids the factors of microstructure and anisotropy which play such an important role in determining the mechanical properties of most ceramics. These factors, while perhaps not altering the basic mechanisms of indentation fracture, may have an important bearing on the realm of validity of any specific fracture mechanics relation:[22,23] thus, for instance, striking differences arise in the Hertzian fracture relations for SiO_2 in going from the amorphous form (silica) to the crystalline form (quartz).[2] Additional variables may enter via velocity effects, either as a result of a reactive environment, whereby subcritical growth can accentuate the fracture,[1] or in high velocity impacts, where dynamic interactions can change the entire nature of the contact field.[23] Friction forces, both in static[34] and sliding[35] contact, constitute yet a further consideration. In applying the equilibrium equations to any practical contact problem it is important to be aware of the effects of such complicating factors in the prospective damage pattern.

Turning now to practical applications of indentation fracture, we may note that certain of the crack types are likely to assume dominant roles in certain properties. Medians are the most penetrative of all cracks, and are therefore the most dangerous as far as strength is concerned. Lateral cracks conversely interact most strongly with the indented surface, hence relate most closely to erosion and wear. Cone cracks represent an intermediate configuration.

3. CONTACT-INDUCED STRENGTH DEGRADATION

The question of strength degradation lends itself more readily than any other practical problem to indentation fracture analysis. Evans[36] was the first to propose a strength theory in terms of formal indentation mechanics, at a time when empirical evidence for the highly deleterious effect of localised contacts on specimen integrity was beginning to mount.[37-41] Since the initial work of Evans a comprehensive research program has been initiated to refine and simplify the theory and to determine suitable parameters for design.[42-45]

Figure 8 depicts the nature of the problem. In (a) a brittle surface is subjected to an in-service contact event which introduces a penetrant crack of characteristic length c. If the contact is sufficiently severe the indentation crack will constitute the "dominant flaw" in the system. In (b) the surface becomes stressed in tension, most usually by flexure in the case of beam or plate components. The susceptibility to failure from the contact site is then relatively high, and the material is thereby degraded.

The indentation fracture approach to the problem takes as its starting point a standard equation for the strength,

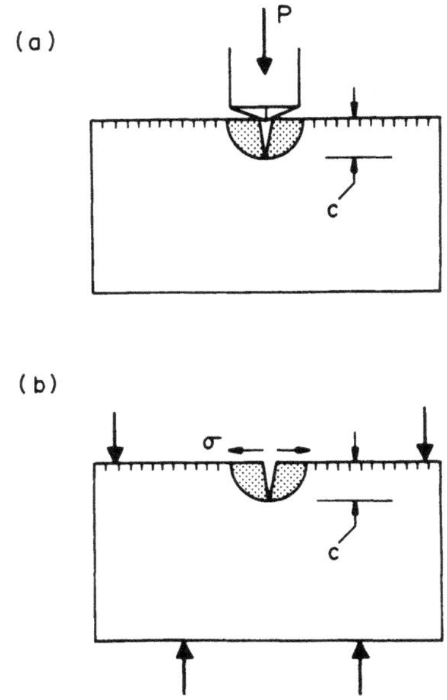

Figure 8: Contact-induced strength degradation: (a) indentation event introduces crack (dominant flaw); (b) subsequent tensile loading (flexure) leads to failure.

$$\sigma = \sigma(c_f), \tag{10}$$

in terms of the effective length c_f of the dominant flaw. In the extreme of light contact loading the size of pre-present flaws will govern the strength, in which case the prior history of the component remains the determining factor. In the opposite extreme of heavy loading c_f identifies with the size of the indentation cracks, whence one may resort to appropriate fracture relations to generate a strength degradation function $\sigma(P)$, or, in an impact situation, $\sigma(v)$, for a given specimen/indenter system. Material fracture parameters then enter the formulation in an explicit and simple way. In this manner one obtains suitable expressions for a priori prediction of strength under any specified loading conditions without the need to resort to extensive strength testing; in principle, the only experimentation that is necessary are "calibration" tests to determine indentation constants (α and β terms, e.g. per Figures 3, 4 and 7) and the material parameters.

In practice, a limited strength testing program is considered desirable to verify essential predictions of the theory. This involves indenting individual specimens, each at a prescribed load, either statically or in impact, and then testing the specimens to failure in a four-point bend test (indented surface on tension side).

To illustrate these ideas we survey the results of studies on two classes of material, those without and with residual surface stresses. Again attention is focussed on equilibrium fracture conditions.

3.1 Non-tempered Materials

Consider an ordinary ceramic material free of any form of surface stress prior to indentation/strength testing. Then the standard strength equation assumes the form

$$\sigma = K_c/(\pi c_f)^{1/2} \tag{11}$$

where c_f is now to be regarded as the effective length of an "equivalent through (straight-edged) crack". The appropriate value of c_f now depends on the relative severity of the indentation-induced crack and pre-present flaws: below some "cutoff load" P^* we have $c_f = \mathring{c}_f$, the characteristic size of the pre-present flaws; above P^* we have $c_f = \Omega_c c$, with the dimensionless constant Ω_c a calculable geometrical factor conjugate to the indentation crack depth c. The cutoff load is determined either by the threshold load P_c at which full-scale indentation fracture initiates, or by the crossover load P' at which the condition $\Omega_c c = \mathring{c}_f$ is satisfied, whichever is the greater. Then by recourse to Eqns. (3) and (4) for blunt indenters, or Eqns. (7) and (8) for sharp indenters, one obtains the following strength degradation expressions for general static loading:

$$\sigma = K_c/(\pi c_f^\circ)^{1/2} \qquad (P \leq P^*) \qquad (12a)$$

$$\sigma = [(\beta K_c^{\,4})^{1/3}/(\pi\Omega_c)^{1/2}]/P^{1/3} \qquad (P > P^*) \qquad (12b)$$

where we have now dropped subscripts in β. We may expect the degradation function $\sigma(P)$ to exhibit an abrupt discontinuity at P^* when $P_c > P'$, corresponding to the sudden development of a dominant indentation crack. Again, if the impulsive load $P(v)$ in a low-energy impact situation can be evaluated for a given specimen/indenter system, Eqn. (12) can be simply modified to an equivalent degradation function $\sigma(v)$.

Indentation/strength data for soda-lime glass serve to confirm the main features of the theory. In Figure 9 for blunt indenters,[42,44] and Figure 10 for sharp indenters,[43] the curves represent a priori prediction from Eqn. (12) (or, in the impact results, from the corresponding, velocity-modified equation[44]), and the data points represent experiment. The following regions of behaviour may be distinguished:

(i) $P < P^*$; At low loads (or velocities) there is negligible degradation, regardless of the nature of the contact, and prior surface damage controls the strength. In this particular region there is some advantage to be gained by careful surface preparation.

(ii) $P = P^*$; At intermediate loads the strength suddenly declines. The results indicate a very much lower threshold for crack initiation in glass for sharp indenters than for blunt indenters, as one might expect for a geometry of relatively high stress-concentrating power: in Fig. 9 the decline is abrupt, at $P^* = P_c > P'$, whereas in Figure 10 it is smooth, at $P^* = P' > P_c$. In the case of blunt indenters, where the effects of the threshold are manifest, the cutoff is seen to be independent of c_f°, in accordance with Eqn. (3). There is little to be gained by refined surface preparation in this region.

(iii) $P > P^*$; At high loads all strength data asymptotically approach the limiting curve for ideal penny-like indentation cracks. The relatively slight, $P^{-1/3}$ falloff demonstrates a remarkable capacity for structural components to withstand severe contacts (provided the component is not so thin that spontaneous failure conditions are approached during the contact event itself). The remnant strength is now totally independent of c_f°, and indenter geometry enters in an insensitive way through the geometrical terms

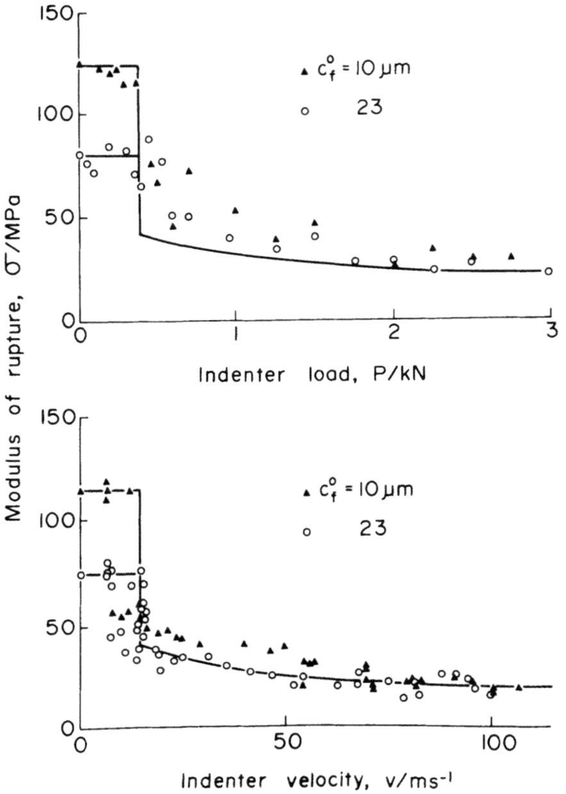

Figure 9: Strength degradation of soda-lime glass, spherical indenters, under conditions of static (top diag., r=1.6 mm) and impact (bottom diag., r=0.8 mm) loading. Note diminishing effect of abrasion flaw size as severity of loading increases. After Refs. 42 and 44.

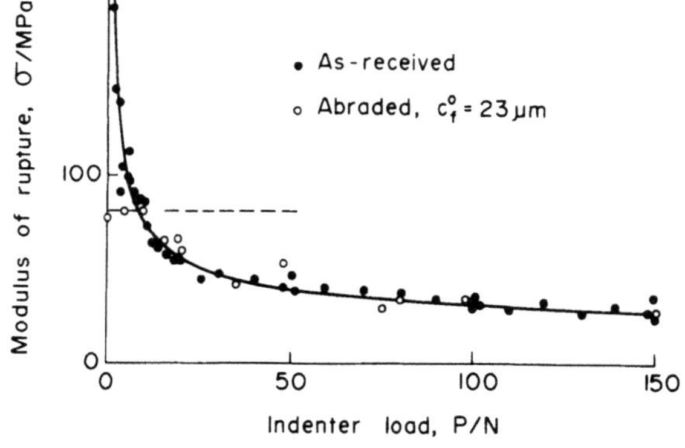

Figure 10: Strength degradation of soda-lime glass, Vickers pyramid indenter, for abraded and as-received surfaces. After Ref. 43.

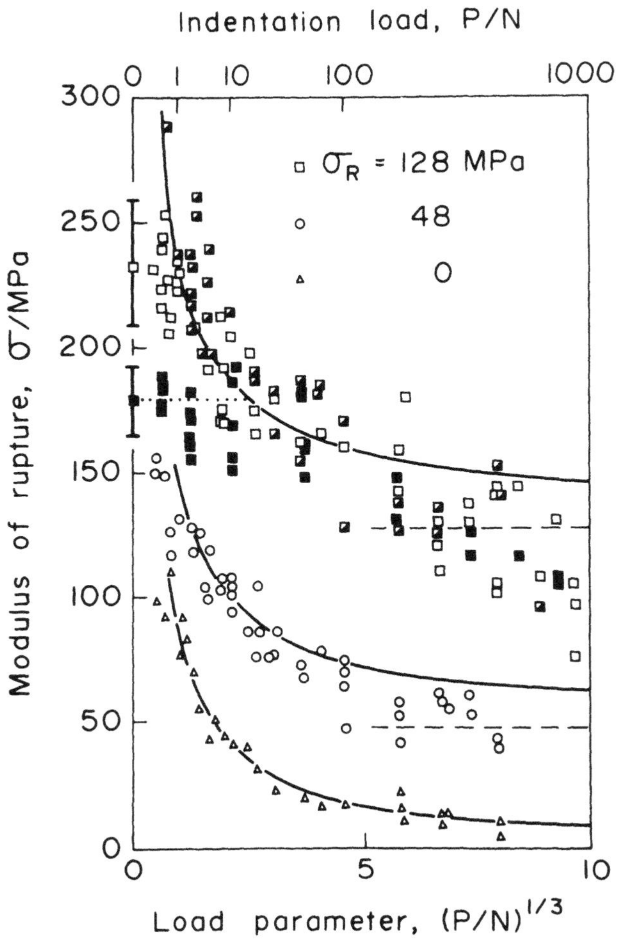

Figure 11: Strength degradation of thermally tempered soda-lime glass, Vickers pyramid indenter. Data for as-received (open symbols), pre-abraded (closed symbols) and pre-etched (half-closed symbols) surfaces. Note significant strengthening with increasing degree of tempering. After Ref. 45.

β and Ω_c in Eqn. (12). An initial state of surface perfection is of no consequence in a general severe contact situation: in this context it is worth recalling (Sect. 2.3) that sharp indenters tend to initiate indentation cracks from sub-surface flaws, and that even ostensibly blunt indenters begin to penetrate and act as though sharp at high load.

3.2 Tempered Materials

It is one thing to be able to predict the resistance to strength degradation for a given material, and quite another to improve it. Surface etching has been generally seen as a means of raising strength, but, as the previous subsection has amply demonstrated, any such gains are more than negated by a single, subsequent contact event. The most practical method available to ceramic materials, especially to glasses, is to put the surfaces in residual compression.[46] This may be done by a variety of physical or chemical "tempering" processes.

Suppose then that a surface subjected to some such tempering process contains a residual compressive stress σ_R. This stress will act biaxially in the surface, but will change and gradually become tensile below the surface toward the inner regions of the plate (in order to satisfy the requirements of zero net force across any section of the specimen). Provided the depth of the dominant flaw is small compared with the characteristic depth of the compression zone, the strength in Eqn. (11) modifies simply to[45]

$$\sigma = K_c/(\pi c_f)^{1/2} + \sigma_R \quad . \qquad (13)$$

The elimination of c_f from this equation is carried out as before, but with the appropriate indentation fracture relation itself incorporating a residual stress term: thus for sharp indenters, for instance, Eqn. (9) replaces Eqn. (8) in the analysis. Surface compression therefore increases the resistance to degradation on two counts: first, the size of the indentation crack introduced is smaller, for a specified load, Eqn. (9); second, the compression represents a closure stress which has to be overcome before any surface flaw can begin to experience tension in subsequent flexure, Eqn. (13). The strength degradation characteristic $\sigma(P)$ becomes an increasing function of σ_R.

Figure 11 accordingly shows indentation/strength data for thermally tempered soda-lime glass using a sharp indenter.[45] At low loads, $P<P^*$, there is the usual dependence of the strength on prior surface condition. At high loads, $P>P^*$, the curves once again show a steady decline, but now tending asymptotically to the strength limit σ_R for large flaws in Eqn. (13). (Some "overshoot" in the data fit to the predicted curves occurs at high loads, indicative of departure from the assumption of zero stress gradients in the fracture mechanics.[45]) At all loads the strength of the glass is dramatically improved by the tempering process.

3.3 Implications in Design

The above analysis has certain implications in the design of ceramic components for structural applications. We may summarise these as follows:

(i) Strength predictions for a given specimen/indenter (target/projectile) system can be made *a priori*; routine, static indentation tests provide the necessary contact coefficients. (In principle, a limited indentation/strength experiment can determine the necessary material constants, and even the residual stress term, as well.[45]) Moreover, the evaluated degradation curves usually (not always) fall below the data points, so the theory tends to be conservative in strength prediction.

(ii) The classification into blunt and sharp indenters provides upper and lower bounds to the strength characteristics of real contact systems.

(iii) The formulation is explicit in material parameters, and accordingly provides a sound basis for materials selection. In particular, toughness K_c is the controlling material parameter in the degradation function Eqn. (12), and should be maximised. Stiffness E and hardness H enter only as secondary parameters, through the threshold relations Eqns. (3) and (7): these parameters should be minimised.

(iv) Contact parameters are important in determining threshold conditions, but diminish in their influence on strength properties at high contact loads. Flaw parameters play an even less significant role in the degradation mechanics.

(v) Resistance to strength degradation can be substantially improved by introducing compressive stresses into the surface.

4. CONCLUDING REMARKS

This paper has sought to review the theoretical framework of indentation fracture mechanics, and to demonstrate the pertinence of this framework to the strength of materials. We have not dwelt on indentation fracture as a mechanical test in its own right, with its capacity for providing fundamental information on fracture energies, crack velocities, flaw statistics, etc. in a uniquely simple and economic way. Nor have we given explicit attention to the relevance of contact-induced cracking, notably lateral cracking, to the important surface-removal properties erosion and wear, grinding and abrasion, and comminution. Some of these additional facets of indentation theory have been considered in other review papers,[1,47] some are addressed elsewhere in this volume, and others remain subjects for future study.

ACKNOWLEDGEMENT

The authors acknowledge the Australian Research Grants Committee for funding of the indentation fracture program at U.N.S.W.

REFERENCES

1. B.R. Lawn and T.R. Wilshaw, "Indentation Fracture: Principles and Applications", J.Mat.Sci. 10, 1049 (1975).

2. M.V. Swain, J.S. Williams, B.R. Lawn and J.J.H. Beek, "A Comparative Study of the Fracture of Various Silica Modifications Using the Hertzian Test", J.Mat.Sci. 8, 1153 (1973).

3. B.R. Lawn and T.R. Wilshaw, "Fracture of Brittle Solids", Cambridge University Press, Cambridge, 1975. Chs. 3,4,8.

4. B.R. Lawn and M.V. Swain, "Microfracture Beneath Pointed Indentations in Brittle Solids", J.Mat.Sci. 10, 113 (1975).

5. B.R. Lawn and E.R. Fuller, "Equilibrium Penny-Like Cracks in Indentation Fracture", J.Mat.Sci. 10, 2016 (1975).

6. F.C. Frank and B.R. Lawn, "On the Theory of Hertzian Fracture", Proc.Roy.Soc.Lond. A299, 291 (1967).

7. B.R. Lawn, "Hertzian Fracture in Single Crystals with the Diamond Structure", J.Appl.Phys. 39, 4828 (1968).

8. F.B. Langitan and B.R. Lawn, "Hertzian Fracture Experiments on Abraded Glass as Definitive Evidence for an Energy Balance Explanation of Auerbach's Law", J.Appl.Phys. 40, 4009 (1969).

9. F.B. Langitan and B.R. Lawn, "Effect of a Reactive Environment on the Hertzian Strength of Brittle Solids", J.Appl.Phys. 41, 3357 (1970).

10. A.G. Mikosza and B.R. Lawn, "Section-and-Etch Study of Hertzian Fracture Mechanics", J.Appl.Phys. 42, 5540 (1971).

11. T.R. Wilshaw, "The Hertzian Fracture Test", J.Phys.D.: Appl. Phys. 4, 1567 (1971).

12. J.S. Nadeau, "Hertzian Fracture of Vitreous Carbon", J.Am. Ceram. Soc. 56, 467 (1973).

13. F.C. Roesler, "Brittle Fractures Near Equilibrium", Proc. Phys.Soc.Lond. B69, 981 (1956).

14. M.V. Swain and B.R. Lawn, "A Microprobe Technique for Measuring Slow Crack Velocities in Brittle Solids", Internat.J.Fract. 9, 481 (1973).

15. J.S. Williams, B.R. Lawn and M.V. Swain, "Cone Crack Closure in Brittle Solids", Phys.Stat.Sol.(a) 2, 7 (1970).

16. C.J. Culf, "Fracture of Glass Under Various Liquids and Gases", J.Soc.Glass Tech. 41, 157 (1957).

17. S.P. Timoshenko and J.N. Goodier, "Theory of Elasticity", McGraw-Hill, New York, 1970. Ch. 13.

18. F. Auerbach, "Measurement of Hardness", Ann. Phys.Chem. 43, 61 (1891).

19. F.C. Roesler, "Indentation Hardness of Glass as an Energy Scaling Law", Proc.Phys.Soc.Lond. B69, 55 (1956).

20. J.P.A. Tillet, "Fracture of Glass by Spherical Indenters", Proc.Phys.Soc.Lond. B69, 47 (1956).

21. B.D. Powell and D. Tabor, "The Fracture of Titanium Carbide Under Static and Sliding Contact", J.Phys.D: Appl.Phys. 3, 783 (1970).

22. A.G. Evans and T.R. Wilshaw, "Quasi-Plastic Solid Particle Damage in Brittle Materials: I. Observations and Analysis", Acta Met. 24, 939 (1976).

23. A.G. Evans and T.R. Wilshaw, "Dynamic Solid Particle Damage in Brittle Materials: An Appraisal", J.Mat.Sci. 12, 97 (1977).

24. M.V. Swain, "Indentation Plasticity and the Ensuing Fracture of Glass", J.Phys.D: Appl.Phys. 9, 2201 (1976).

25. B.R. Lawn, T. Jensen and A. Arora, "Brittleness as an Indentation Size Effect", J.Mat.Sci. 11, 573 (1976).

26. C.M. Perrott, "Elastic/Plastic Indentation: Hardness and Fracture", Wear, in press.

27. B.R. Lawn and A.G. Evans, "A Model for Crack Initiation in Elastic/Plastic Indentation Fields", J.Mat.Sci., in press.

28. B.R. Lawn, M.V. Swain and K. Phillips, "On the Mode of Chipping Fracture in Brittle Solids", J.Mat.Sci. 10, 1236 (1975).

29. D. Tabor, "Hardness of Metals", Clarendon Press, Oxford, 1951.

30. J.J. Petrovic, R.A. Dirks, L.A. Jacobson and M.G. Mendiratta, "Effects of Residual Stresses on Fracture from Controlled Surface Flaws", J.Am.Ceram.Soc. 59, 177 (1976).

31. B.R. Lawn and D.B. Marshall, "Contact Fracture Resistance of Physically and Chemically Tempered Glass Plates: A Theoretical Model", Phys.Chem.Glasses 18, 7 (1977).

32. D.B. Marshall and B.R. Lawn, "An Indentation Technique for Measuring Stresses in Tempered Glass Surfaces", J.Am. Ceram.Soc. 60, 86 (1977).

33. M.V. Swain, "Microcracking Associated with the Scratching of Brittle Solids", this volume.

34. K.L. Johnson, J.J. O'Connor and A.C. Woodward, "The Effect of Indenter Elasticity on the Hertzian Fracture of Brittle Materials", Proc.Roy.Soc.Lond. A334, 95 (1973).

35. B.R. Lawn, "Partial Cone Crack Formation in a Brittle Material Loaded with a Sliding Spherical Indenter", Proc.Roy.Soc.Lond. A299, 307 (1967).

36. A.G. Evans, "Strength Degradation by Projectile Impacts", J.Am.Ceram.Soc. 56, 405 (1973).

37. H.P. Kirchner, R.M. Gruver and R.E. Walker, "Strength Effects Resulting from Simple Surface Treatments", in The Science of Ceramic Machining and Surface Finishing, N.B.S. Spec. Publ. 348, 1972. p. 353.

38. R.M. Gruver and H.P. Kirchner, "Effect of Surface Damage on the Strength of Al_2O_3 Ceramics with Compressive Surface Stresses", J.Am.Ceram.Soc. 56, 21 (1973).

39. R.M. Gruver and H.P. Kirchner, "Effect of Environment on Penetration of Surface Damage and Remaining Strength of Al_2O_3", J.Am.Ceram.Soc. 57, 220 (1974).

40. R.M. Gruver and H.P. Kirchner, "Effect of Leached Surface Layers on Impact Damage and Remaining Strength of Silicon Nitride", J.Am.Ceram.Soc. 59, 85 (1976).

41. J.J. Petrovic, L.A. Jacobson, P.K. Talty and A.K. Vasudevan, "Controlled Surface Flaws in Hot-Pressed Si_3N_4", J.Am.Ceram. Soc. 58, 113 (1975).

42. B.R. Lawn, S.M. Wiederhorn and H.H. Johnson, "Strength Degradation of Brittle Surfaces: Blunt Indenters", J.Am.Ceram.Soc. 58, 428 (1975).

43. B.R. Lawn, E.R. Fuller and S.M. Wiederhorn, "Strength Degradation of Brittle Surfaces: Sharp Indenters", J.Am.Ceram.Soc. 59, 193 (1976).

44. S.M. Wiederhorn and B.R. Lawn, "Strength Degradation of Glass Resulting from Impact with Spheres", J.Am.Ceram. Soc., in press.

45. D.B. Marshall and B.R. Lawn, "Strength Degradation of Thermally Tempered Glass Plates", J.Am.Ceram.Soc., in press.

46. F.M. Ernsberger, "Strength of Brittle Ceramic Materials", Bull.Am.Ceram.Soc. 52, 240 (1973).

47. B.R. Lawn and D.B. Marshall, "Mechanisms of Micro-Contact Fracture in Brittle Solids", in proceedings of Conference on Lithic Use-Wear, Simon Fraser University, Canada, to be published.

INDENTATION INDUCED STRENGTH DEGRADATION AND STRESS CORROSION OF TEMPERED GLASSES

M.V. Swain, J.T. Hagan and J.E. Field

Physics and Chemistry of Solids, Cavendish Laboratory

Madingley Road, Cambridge, U.K.

ABSTRACT

Flaws introduced into physically tempered laths of glass with a Vickers indenter have been used in controlled fracture experiments in flexure. Provided the flaw length extended beyond the dimension of the indentation fracture always initiated from this site on bending. After making allowance for plastically induced residual stress fields about the indentation site it was possible to use this technique to determine the tempering level of the glass. It was also possible by varying the stress rate to determine the stress corrosion behaviour of these tempered glasses. The results were analysed in terms of a recent fracture mechanics treatment by Hagan et al[1] of stress corrosion cracking in tempered glasses. This approach enabled the stress corrosion index n, in $V = AK^n$ to be determined. It was also observed that the extent of the fracture mirrors in these indented and subsequently fractured laths was a sensitive function of loading rate and environment. A simple relationship between the ratio of the surface to bulk dimensions of the fracture mirror in both materials is derived.

INTRODUCTION

The increasing use of tempered glasses and ceramics necessitates a better understanding of the behaviour of these materials under sustained load and in environments that cause stress corrosion cracking. Unlike annealed glasses and ceramics it is not possible to follow the growth of planar through-going cracks because of the varying stress field introduced by the toughening process. Thus it is necessary to infer from slightly more indirect methods the stress corrosion behaviour of these materials. Two such methods which are suitable for tempered materials are the variation of time-to-failure with applied stress, and fracture stress with stress rate. To justify these approaches a fracture mechanics analysis of the stress intensity factor of a crack in a tempered glass plate in bending was carried out[1]. Here we consider the application of this analysis to three specific aspects of fracture of tempered materials; (i) the determination of the original compressive stress induced by tempering, (ii) an estimate of the stress corrosion characteristics of tempered glasses, and (iii) a discussion of fracture mirrors formed on tempered and annealed laths in bending.

THEORETICAL CONSIDERATIONS

For thermally tempered glasses the stress distribution through the thickness of a plate is parabolic[2] and may be written

$$\sigma_t(x) = -\sigma_c\left(1 - \frac{6x}{d} - 6\left(\frac{x}{d}\right)^2\right) \quad (1)$$

where σ_c is the maximum surface compressive stress and d the plate thickness. The superimposed stress distribution arising from flexure is

$$\sigma_\beta(x) = \sigma_\beta^* \left(1 - \frac{2x}{d}\right) \quad (2)$$

where σ_β^* is the outer fibre tensile stress. The resultant stress distribution through a thermally tempered plate in bending is the sum of equations (1) and (2). With this prior distribution of stresses the stress intensity factor for a planar through-going crack of length c is given by[1],

$$K = Y(\sigma_\beta^* - \sigma_c)(\pi c)^{\frac{1}{2}}\left[1 + \alpha\left(\frac{c}{d}\right) + \beta\left(\frac{c}{d}\right)^2\right] \quad (3)$$

where

$$\alpha = \frac{2}{\pi}\,\frac{6\sigma_c - 2\sigma_\beta^*}{\sigma_\beta^* - \sigma_c} \quad \text{and} \quad \beta = \frac{-3\sigma_c}{(\sigma_\beta^* - \sigma_c)}$$

and Y is a modifying factor $\sim 1.04 \times 2/\pi$.

The above analysis makes no allowance for the change in bending moment during crack propagation but should still provide a good approximation for $c \ll d$. The variation of stress intensity factor with crack length according to equation (3) is shown in figure 1a for a thermally tempered beam, σ_c = 60 MPa, and σ_β* = 0, 1.5 and 2 times σ_c. Plotted in figure 1b is the term in square brackets in equation (3) with σ_c = 0. This term might be considered as an additional modifying factor, it also enables comparison with similar modifying terms for a through-going planar crack[3] and a semi-circular surface crack[4], both in pure bending. The present analysis with σ_c = 0 is very close to the analysis of Shah and Kobayashi[4] and follows that of a through-going crack[3] provided c < 0.10 d. As shown in figure 1a, 0.1 d is well beyond the K_{1c} value for even modestly tempered materials.

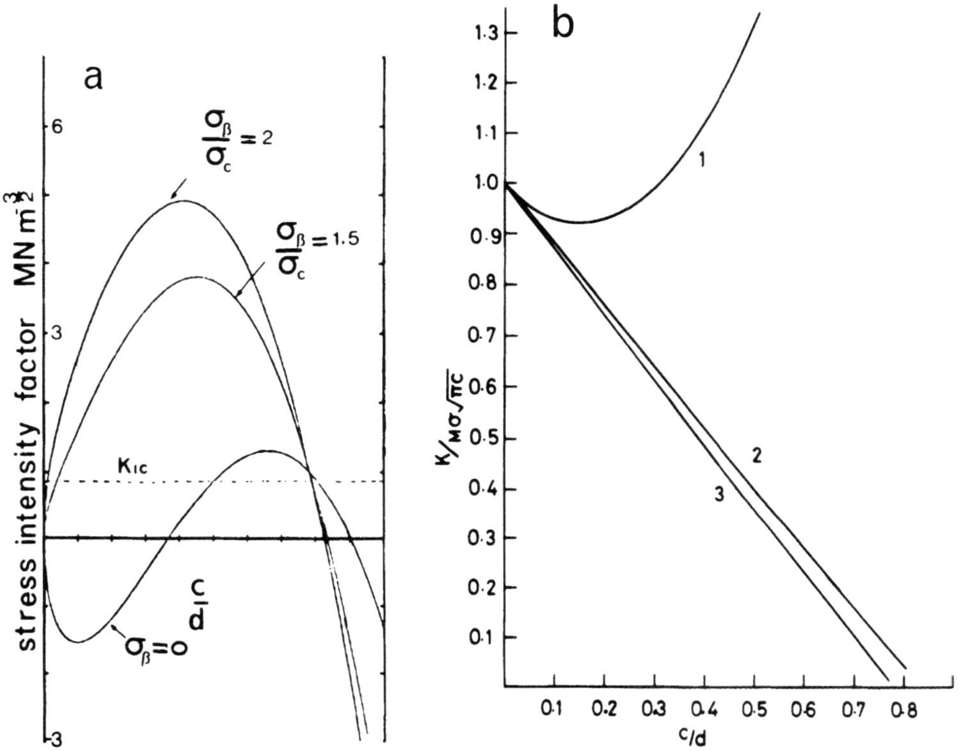

Figure 1. a) Variation of stress intensity factor with normalised crack length for three conditions of σ_β/σ_c; K_{1c} = 0.8 MPa$^{\frac{1}{2}}$; b) Variation of modifying factor with normalised crack length: 1, through-going planar crack; 2, a semi-circular surface crack; and 3, equation (3) with σ_c = 0.

EVALUATION OF SURFACE COMPRESSIVE STRESSES

The usual method of determining the compressive stresses in the surface of a tempered glass is by measuring the tensile stress level at the centre of the plate, and from equation (1) the surface compressive stress is double this value. Unfortunately it is not always possible to do this, for instance in curved plates and opaque materials, and this necessitates other methods. One such method is the use of indentation fracture mechanics[5,6]. Another approach is by the introduction of a well-defined crack in these materials and then loading to fracture. Rather than rely on fortuitous random flaws in the tensile surface of the glass from which to initiate fracture, it is more appropriate to introduce flaws of a well-defined character. The advantages of this approach are that the fracture stress is highly reproducible and the initiation site well defined. Flaws of this type may be introduced with a pointed indenter, typically a Vickers pyramid which produces well-defined and measurable flaws emanating from the corners of the indentation. This approach has enjoyed considerable popularity recently[8], but appears to have one minor problem associated with it, namely, residual stresses. The resulting residual crack-tip stress intensity factor is a consequence of plastic deformation about the permanent impression formed during indentation. The magnitude of the residual stresses has been estimated[8,9], and more recently measured directly by photoelastic observations[10]. The typical reduction in K_{1c} resulting from these residual stresses is $\sim$30% for a number of brittle materials.

Vickers pyramid indentations were made on soda-lime glass tempered to three different levels and a tempered alumino-silicate glass. Flaws of very nearly semi-circular shape were introduced. The surface flaw size was measured and the depth was taken as half the surface trace. The soda-lime specimens were plates 200 x 35 x 6 mm and the alumino-silicate beams 130 x 10 x 5 mm. All specimens were loaded to fracture in a four-point bending jig at a cross-head speed of 20 mm/min in a silicone oil environment to minimise slow crack growth.

The magnitude of the tempering stress level of the soda-lime glasses was determined from equation (3) using only the first term within the square brackets (because $c/d \ll 1$) and upon making the proper reduction in K_{1c} as mentioned above. The results along with the manufacturers' quoted values of nominal tempering stress are shown in Table (1).

For the alumino-silicate glass the tempering stress level was known to be 80 MPa and hence it was possible to determine the critical stress intensity factor from equation (3). The variation of the uncorrected stress intensity factor with initial flaw size is shown in figure 2. Applying the same correction as found for

Table 1: Flaw size, fracture stress and surface compressive stress of three tempered soda-lime glasses

Glass	Flaw size (μm)	Fracture Stress (MPa)	Compressive Stress (MPa)	
			From eq.(3)	Manufacturers' value
A	99.6 ± 1.5	87.7 ± 0.7	54.6 ± 1.2	52
B	84.4 ± 1	127.1 ± 2.5	95.5 ± 3.6	140
C	57.3 ± 2.2	246.5 ± 5.2	207 ± 5	220

other materials a mean value of the critical stress intensity factor is 0.86 ± 0.09 $MN\,m^{-3/2}$ compared with a value of 0.91 ± 0.02 $MN\,m^{-3/2}$ obtained by Wiederhorn[11] for an aluminosilicate glass of similar composition.

STRESS CORROSION OF THERMALLY TEMPERED MATERIALS

As mentioned above the stress corrosion behaviour of tempered materials can only be determined indirectly. One such method is the variation of the lifetime or time to failure of a material with an

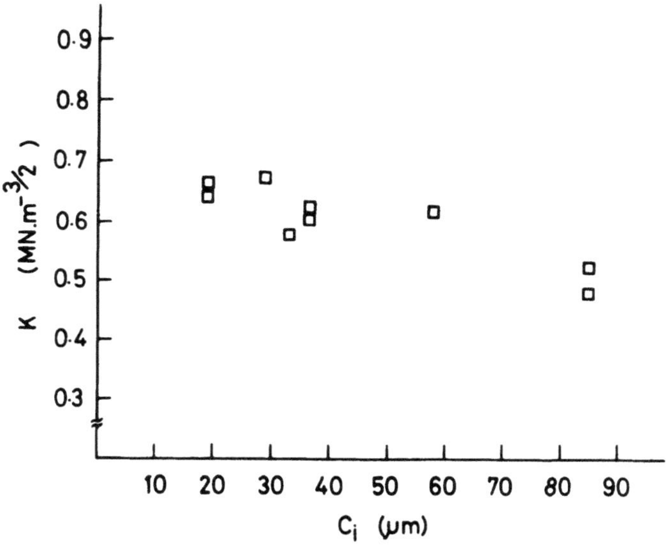

Figure 2. Variation of uncorrected critical stress intensity factor of aluminosilicate glasses as a function of initial flaw size. All tests done in silicone oil.

initial flaw c_i under a stress σ_β*. The time to failure is then given by

$$\tau = \int_{c_i}^{c_f} \frac{dc}{V} \tag{4}$$

where V is the crack velocity. It is known that crack velocity in glass and ceramics is a sensitive function of stress intensity factor and may be conveniently expressed as

$$V = AK^n \tag{5}$$

where A and n are material constants that vary with environment. Equation (4) may now be written

$$\tau = \int_{c_i}^{c_f} \frac{dc}{AK^n} \quad . \tag{6}$$

Upon substituting for K from equation (3), taking only the first two terms in the square brackets, and for large n it may be shown[1] that

$$\tau = \frac{2\, M(c_i/d)\, c_i^{\frac{2-n}{2}}}{A(\sigma_\beta{*}-\sigma_c)^n (\pi)^{n/2}(n-2)} \tag{7}$$

where $M(c_i/d)$ is a dimensionless factor, that accounts for the crack tip growing into a more tensile region, and is given by

$$M(c_i/d) = 1 - \frac{n(n-2)c_i\alpha}{d(n-4)} \quad . \tag{8}$$

For σ_β* only slightly greater than σ_c this function decreases rapidly with c_i/d, for $\sigma_\beta{*} \simeq 3\sigma_c$ it is almost a constant. Equation (7) then simplifies to

$$\tau(\sigma_\beta{*}-\sigma_c)^n = \text{constant} \ . \tag{9}$$

Similarly it can be shown that the modified expression relating fracture stress $\sigma_{\beta f}$ to the stress rate $\dot{\sigma}$ is given by

$$(\sigma_{\beta f}-\sigma_c)^{n+1} = \gamma\, \dot{\sigma} \tag{10}$$

where γ is a constant that incorporates the crack propagation constants A and K_{1c}.

Observations were made on annealed soda-lime glass and thermally tempered aluminosilicate glass in aqueous environments. The cross-head speed was varied from 0.01 to 20 mm/min corresponding to a stress rate variation from 0.025 MN s^{-1} to 25.6 MN s^{-1}. Flaws were introduced in a similar manner to that described earlier. The indenter load was 200 gm and initial flaw size, 2c, was 110 μm and 39 μm for the annealed and tempered glass respectively. The fracture stress as a function of stress rate is shown in figure 3. A stress corrosion index, n, was obtained from a least squares fit of the data in figure 3 and values of 27 and 18 were obtained for the aluminosilicate and annealed soda-lime glass respectively. These values compare with that of 25 and 16 obtained by Wiederhorn and Bolz[12] on these glasses. Similar results were obtained by analysing the results of time to failure data of Kirchener and Walker[13] on thermally tempered polycrystalline alumina and other data[1] on tempered glass.

FRACTURE MIRROR OBSERVATIONS ON ANNEALED AND TEMPERED GLASSES

The fracture surfaces of beams broken in the previous studies were subsequently examined. The dimensions of the surface mirror, that is to the onset of severe misting, were measured along and perpendicular to the surface. From these measurements and a knowledge of the fracture stress the "branching" stress intensity factor, K_{BB}, was determined. A number of studies[14,15] have shown the branching stress intensity factor to be a constant independent of the loading rate or environment. More recently it has been used to estimate the initial flaw size[16]. It should be pointed out that in computing K_{BB}, no account is taken of the crack velocity, and hence it will be

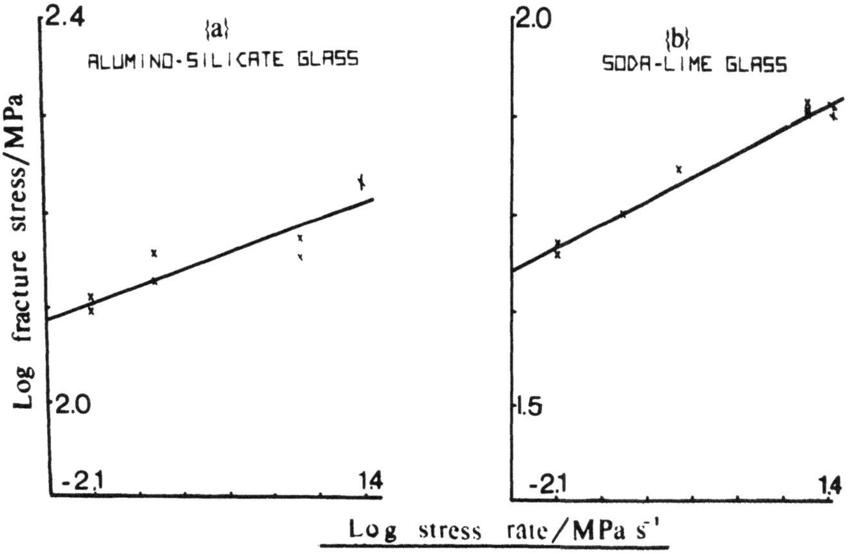

Figure 3. Dependence of the fracture stress on the stressing rate for thermally tempered aluminosilicate (a) and annealed soda-lime glass (b).

an overestimate. However K_{BB} does provide a useful and convenient method of post-failure analysis of fractured brittle materials, especially when subjected to complex stress fields.

Some typical observations of fracture mirrors in annealed soda-lime and tempered alumino-silicate glass are shown in figures 4 and 5. From these and other observations K_{BB} was computed and in figure 6 it is plotted for both glasses in both environments and all loading rates. The values obtained for annealed soda-lime glass are in reasonable agreement with other workers. As seen in figure 4 (initial flaw size a constant, $2C_o \sim 55 \pm 4$ μm) the fracture surfaces vary considerably between fast and slow loading rates, particularly for annealed soda-lime glass. For small flaws and fast loading rates

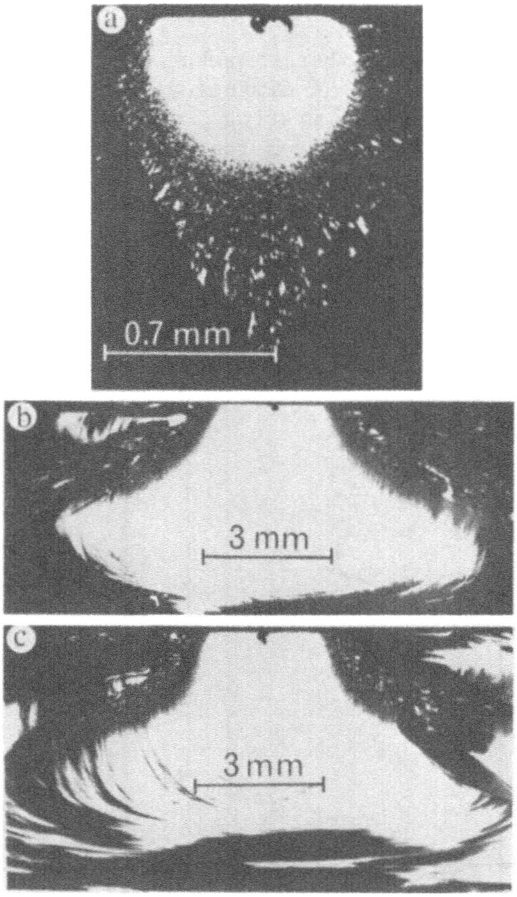

Figure 4. Fracture mirrors of annealed soda-lime glass in water at three different loading rates: a) 10 mm/min; b) 0.1 mm/min; and c) 0.01 mm/min.

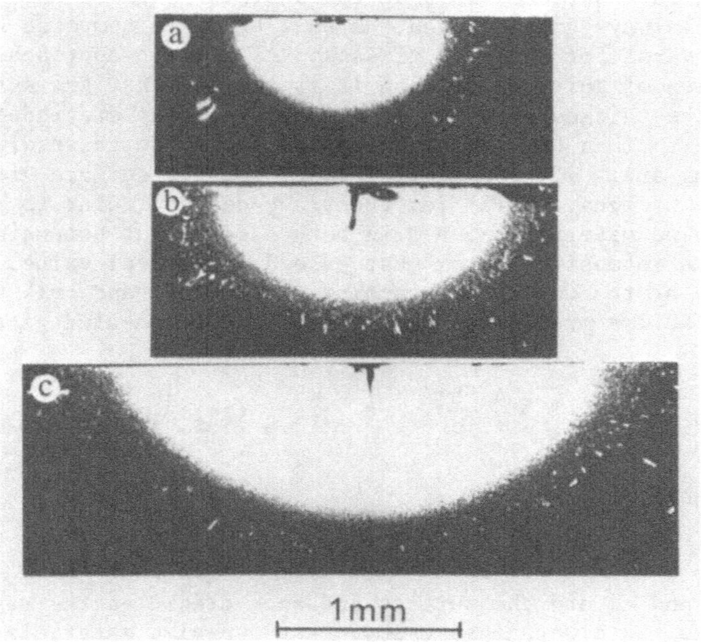

Figure 5. Fracture mirrors of tempered alumino-silicate glass in water at three different loading rates: a) 10 mm/min; b) 0.1 mm/min; and c) 0.01 mm/min. Note the orthogonal initial flaw at the origin of the mirror in b) and c).

the mirror is almost semi-circular about the original flaw, figure 4a; whereas for slow loading rates mist and hackle regions only form in a narrow region adjacent to the tensile surface, figure 4c. The fracture mirrors of the tempered alumino-silicate glass were invariably surrounded by mist and hackle. Again for small flaws (2Co 73.5 ± 2.4 μm) at fast loading rates the mirror was almost semi-circular, figure 5a; whereas for slower loading rates the mirrors were more elliptical with the surface the longer axis, figure 5c. It was observed that by varying the loading rate from 0.01 to 10.00 mm/min the ratio of the final mirror size to initial flaw size for soda-lime glass in water ranged from 22 to 52. Over the same loading rates and flaw size in silicone oil the ratio varies from 12 to 26. For the tempered alumino-silicate glass in water under the same conditions it varied from 28 to 54, and in silicone oil from 14 to 25. These results suggest that contrary to previous observations[16] about the insensitivity of the final mirror size to the initial flaw size, this ratio varied from 12 to 54 and was strongly dependent on the initial flaw size, loading rate and environment.

One significant difference between the tempered and annealed glass fracture surfaces is that the surface trace of the fractured mirror is always greater than the bulk trace in tempered glasses and vice versa for annealed glasses. As a first approach to an explanation of this observation it is assumed that the surface and through-the-thickness traces of the growing flaw are independent of each other. Then the stress intensity factor in these directions in bending are: $K_s = Y\sqrt{\pi c_s}\,(\sigma_\beta{*} - \sigma_c)$ for the surface crack, and K_B, given by equation (3) for the trace perpendicular to the surface. For annealed materials $\sigma_c = 0$ in both cases. For branching to occur the stress intensity factor must exceed a critical value, K_{BB}, and therefore at the onset of branching the surface and bulk stress intensity factors must equal K_{BB}. That is for annealed glasses

$$Y\,\sigma_\beta\,(\pi c_b)^{\frac{1}{2}}\,(1 - \frac{4}{\pi}\frac{c_b}{d}) \quad = \quad Y\,\sigma_\beta\,(\pi c_s)^{\frac{1}{2}} \tag{11}$$

which reduces to

$$c_s \quad = \quad c_b(1 - \frac{8c_b}{\pi d}) \tag{12}$$

where c_s and c_b are the surface and bulk crack lengths respectively. That is, c_s is always less than c_b for annealed materials. Alternatively one can express c_b in terms of c_s in which case one finds that branching into the bulk will only occur provided the surface mirror is less than d/10. For surface mirrors approaching d/10 the

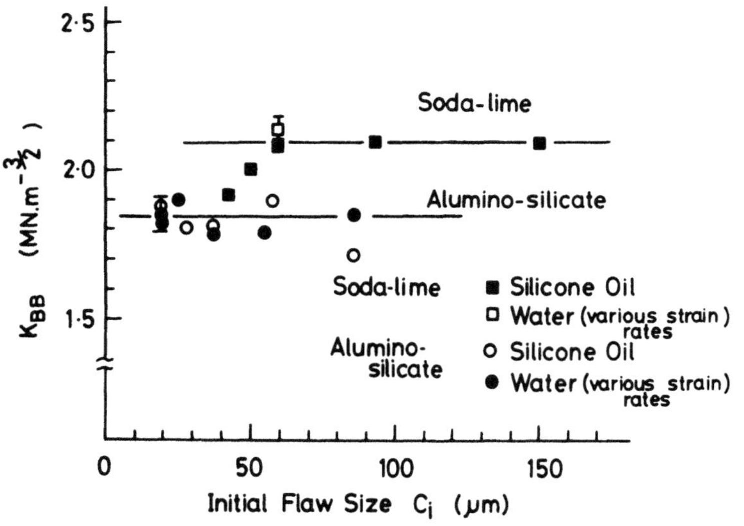

Figure 6. Dependence of the stress intensity factor at branching on the initial flaw size in various environments for annealed soda-lime glass and tempered alumino-silicate.

ratio of the bulk to surface traces of the fracture mirror are 2 to 1 whereas for small mirrors $c_b \simeq c_s$.

One can consider another approach to branching where more stringent conditions are imposed, namely, that the stress intensity factor not only attains a critical branching threshold but is also an increasing function of crack length at branching, that is dK/dc > 0. For annealed materials this condition implies, for the surface trace,

$$\frac{dK}{dc} = \frac{Y}{2}\left(\frac{\pi}{c_s}\right)^{\frac{1}{2}} \sigma_{\beta f} \qquad (13)$$

which is greater than 0 for all c_s. For the trace into bulk material,

$$\frac{dK}{dc} = Y\sigma_{\beta f}\,\frac{(\pi d - 12c_b)}{2\pi^{\frac{1}{2}} d\, c_b^{\frac{1}{2}}} \qquad (14)$$

which is >0 for $c_b \leqslant \frac{\pi d}{12}$.

These two expressions suggest that branching on the surface will occur provided K_{BB} is attained, whilst into the bulk branching will only take place for crack lengths less than $\pi d/12$. The above expression also means that if branching does occur before $\pi d/12$ then beyond this point it is possible for branching to cease when K falls below K_{BB}.

For tempered materials following the same approaches it may be shown[17] that the first condition leads to

$$c_s = c_b(1 + 2\alpha\cdot\frac{c_b}{d}) \qquad (15)$$

where α is given by equation (3). Thus c_s is always greater than c_b provided $\sigma_\beta \leqslant 3\sigma_c$. Applying the second criteria leads to an involved dependence upon α and β. However a plot of K versus crack length through the bulk, figure 1a, does show a maximum within the thickness of the plate. For tempered glass plates these two criteria suggest that c_s will become exceedingly greater than c_b as the fracture stress drops, provided branching into the bulk occurs.

Measurements of the ratio of c_b to c_s in annealed soda-lime glass varied from 1.7 to 1.2 over the strain rate range studied. For alumino-silicate the ratio of c_s to c_b was from 1.5 to 1.0. These results are in reasonable agreement with the simplified explanation given above.

Finally, in a few cases it was observed that where misting and

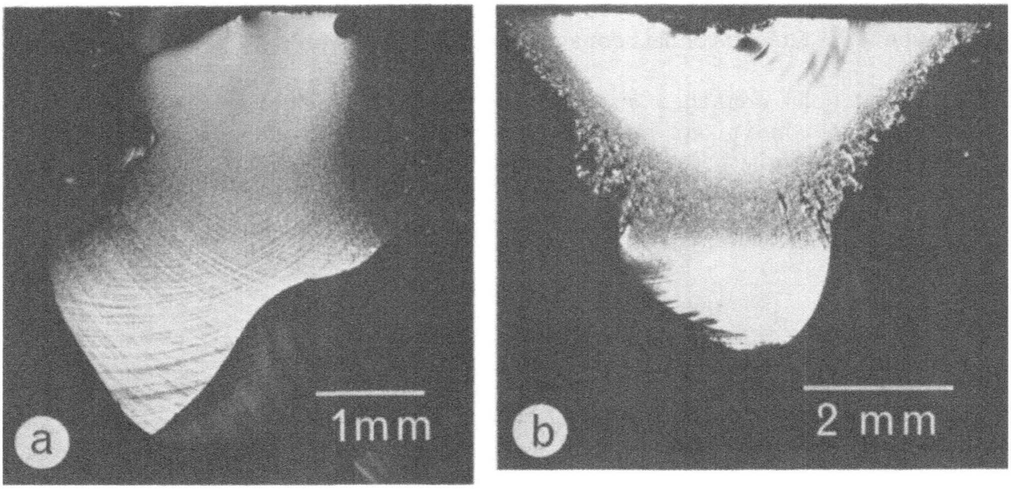

Figure 7. Fracture mirrors in (a) annealed and (b) thermally-tempered ($\sigma_c \sim 50$ MPa) soda-lime glass showing the transition from mirror to misting (hackle) back to mirror again.

sometimes hackle completely surrounded the initial flaw on continued crack growth smooth fracture mirror reappeared. Two examples of this are shown in figure 7. In figure 7a misting has completely surrounded the mirror in soda-lime glass but on continued crack growth it has completely disappeared. In figure 7b, a mildly tempered soda-lime glass ($\sigma_c = 50$ MPa) misting as well as hackle is apparent surrounding the mirror zone, but on further growth into the bulk it has disappeared. The tempering level and applied bending stress in figure 7b correspond quite closely to the situation in figure 1a with $\sigma_\beta = 1.5\ \sigma_c$. These observations suggest that crack velocity of a flaw growing through the thickness of a plate in bending attains a maximum velocity and then slows down as it approaches the rear surface. This view has recently been confirmed by subjecting the growing crack in bending to ultrasonic irradiation[18].

CONCLUSION

The use of indentation induced surface flaws together with an analysis of the stress intensity factor for through-going cracks in thermally tempered plates has enabled the determination of the surface compressive stress, the stress corrosion behaviour, and a rationalisation of the fracture mirror morphology of thermally tempered glasses. Upon making allowance for residual stress effects remaining after indentation it was possible to determine the surface compressive stress or if this were known the fracture toughness of the material. A simple extension of this approach, by varying the loading rate and environment, enabled the stress corrosion behaviour of thermally tempered alumino-silicate and annealed soda-lime glass to be determined. The stress corrosion index, n, in both cases was in good agreement with that found for these materials using conventional methods. It was also found that the fracture mirror dimensions of annealed and tempered glasses were very sensitive to loading rate, initial flaw size and environment. The ratio of the initial flaw size to the final mirror dimensions varied over the range 55 to 12 depending upon the environment and loading conditions. This observation is somewhat different from that reported by Mecholsky et al[16]. On the other hand the quasi-static branching stress intensity factor was a constant independent of environment, loading rate and initial flaw size. Finally, some criteria are presented for the occurrence of branching, and the ratio between the surface and through the thickness distances to branching for annealed and tempered materials.

ACKNOWLEDGEMENTS

This work was supported by the S.R.C. and M.O.D. (PE).

REFERENCES

1. Hagan J.T., Swain M.V. and Field J.E. (1977) 4th Int. Fract. Conf. Proc. Waterloo, Canada (in press).
2. Zijlstra A.L. and Burggraaf A.J., J. Non-Cryst. Sol. 1 (1968) 49.
3. Brown W.F. and Srawley J.E., A.S.T.M. S.T.P. No. 410 (1967) 13.
4. Shah R.C. and Kobayashi A.S., A.S.T.M. S.T.P. No. 513 (1972) 3.
5. Marshall D.B. and Lawn B.R., J. Am. Ceram. Soc. 60 (1977) 86.
6. Swain M.V., Hagan J.T. and Field J.T., J. Mater. Sci. (1977) in press.
7. Hagan J.T., Swain M.V. and Field J.T., Glass Techn. (1977) in press.
8. Petrovic J.J., Jacobson L.A., Talty P.K. and Vasudevan A.K., J. Am. Ceram. Soc. 58 (1975) 113.
9. Swain M.V., J. Mater. Sci. 11 (1976) 2345.
10. Marshall D.B. and Lawn B.R., J. Am. Ceram. Soc. (1977) in press.
11. Wiederhorn S.M., J. Am. Ceram. Soc. 52 (1969) 99.
12. Wiederhorn S.M. and Bolz L.H., J. Am. Ceram. Soc. 53 (1970) 543.
13. Kirchener H.P. and Walker R.E., Mat. Sci. and Eng. 8 (1971) 301.
14. Johnson J.W. and Holloway D.G., Phil. Mag. 14 (1966) 731.
15. Kerper M.J. and Scuderi T.G., Bull. Am. Ceram. Soc. 43 (1964) 622.
16. Mecholsky J.J., Freiman S.W. and Rice R.W., J. Am. Ceram. Soc. 57 (1974) 55.
17. Hagan J.T. and Swain M.V., to be published.
18. Swain M.V. and Hagan J.T., to be published.

COMPRESSIVE MICROFRACTURE AND INDENTATION DAMAGE IN Al_2O_3

J. Lankford

Southwest Research Institute

San Antonio, Texas 78284

INTRODUCTION

During the impact of ceramics by small hard particles, indentation damage is caused, which can lead to erosion and/or strength degradation of the ceramic. It is suspected that certain types of indentation-induced flaws may be generated by the compressive fields known to be associated with such loading. Accordingly, a research program was carried out to assess the mechanisms of damage in a strong ceramic subject to compressive loading, and to determine, if possible, the relevancy of such mechanisms to indentation damage.

Initial results, indicating that the early stages of compressive damage in Al_2O_3 consist of twinning and twin-induced microcracking, have been reported.[1] In addition, it was shown that the compressive strength was strain rate sensitive. In the present paper, the earlier results are extended over wider ranges in temperature and strain rate, and interpreted through acoustic emission experiments designed to detect the onset of microplasticity. The results of an SEM study of indentation in the same material are also reported, and it is shown that compressive damage in the form of twinning is important in nucleating the lateral microcracks thought[2,3] to be responsible for erosion. Based on the compressive test-acoustic emission data, the particle velocity regime over which the twin/microcrack erosion mechanism seems relevant is estimated.

EXPERIMENTAL PROCEDURES

The compression test set-up and procedures have been described in detail elsewhere,[1] and are discussed here only briefly. Compression specimens were fabricated in the form of right circular cylindrical specimens cut from as-received rods of polycrystalline Lucalox* alumina, having a 25μm grain size. The ends of the 1.25 cm long x 0.625 cm dia. specimens were ground parallel and polished to a mirror finish. Compression tests were carried out using platens of Coors AD 999 alumina.† Strain rates ranged from $7x10^{-6}$ sec^{-1} to $2x10^{3}$ sec^{-1}, with the latter rates being achieved by means of the Hopkinson pressure bar; temperatures were varied from -200° to 900°C. Not all tests were carried out to failure; in particular, certain specimens were loaded to predetermined levels, during which loading period acoustic emission was monitored, and then unloaded and coated with palladium for study in the SEM to detect any damage which might have been incurred.

Acoustic emission (AE) was monitored during certain compression tests within the frequency domain 100 kHz to 1 MHz, with a transducer resonance of 160 kHz. The signal was passed through an 80 dB gain amplifier, a filter, and an adjustable discriminator, whose level, once set, remained constant for all tests. Pulses from the discriminator entered a 5-decade counter and a digital waveform recorder together with the compressive stress/time signal. Following the test, the recorder was played back through a strip-chart recorder to yield as function of time both stress and accumulated counts. The best fit curve to the latter was differentiated and plotted versus the former to provide the count rate as a function of stress. This method of treating the AE information is similar to that used by others[4,5] studying the tensile strength of ceramics.

In addition to the compressive strength tests, some of the as-fired rods were indented statically with a diamond pyramid microhardness indenter. Individual grains were selected for multiple indentation at sequentially higher loads, with indenter orientation maintained constant within a given grain. For comparison, the polished surface of a large (0110) oriented, sapphire single crystal was indented in the same way. Following indentation, the specimens were coated and examined in the scanning electron microscope.

RESULTS

Compressive Strength

The compressive strength σ_f of Lucalox as a function of tempera-

---

*Type GW, General Electric Lamp Glass Division, Cleveland, Ohio.
†Coors Porcelain Company, Golden, Colorado.

ture T behaves in a manner reminiscent of its tensile strength. As shown in Figure 1, the strength, for two different strain rates ($\dot{\varepsilon}$), decreases with increasing temperature, then rather suddenly begins to increase, before finally settling back into an inverse relationship with temperature. Bend tests performed on the same material[6] in air and vacuum exhibit qualitatively the same behavior, except that the average strength is lower by an order of magnitude, and the strength minimum occurs at around 350°C rather than 150° to 250°C. Extrapolating the lower strain rate data to absolute zero yields a projected ultimate compressive strength of about 4200 MN/m^2.

The thermally activated nature of the compressive failure process is further exemplified by the dependence of strength upon strain rate, plotted in Figure 2. Over nine orders of magnitude in strain rate, the strength experiences approximately a 45% increase. However, it would appear, based on the ultimate strength of 4200 MN/m^2 derived from the σ_f-T data, that at room temperature the σ_f-$\dot{\varepsilon}$ effect should persist to strain rates in excess of 10^6 sec^{-1}.

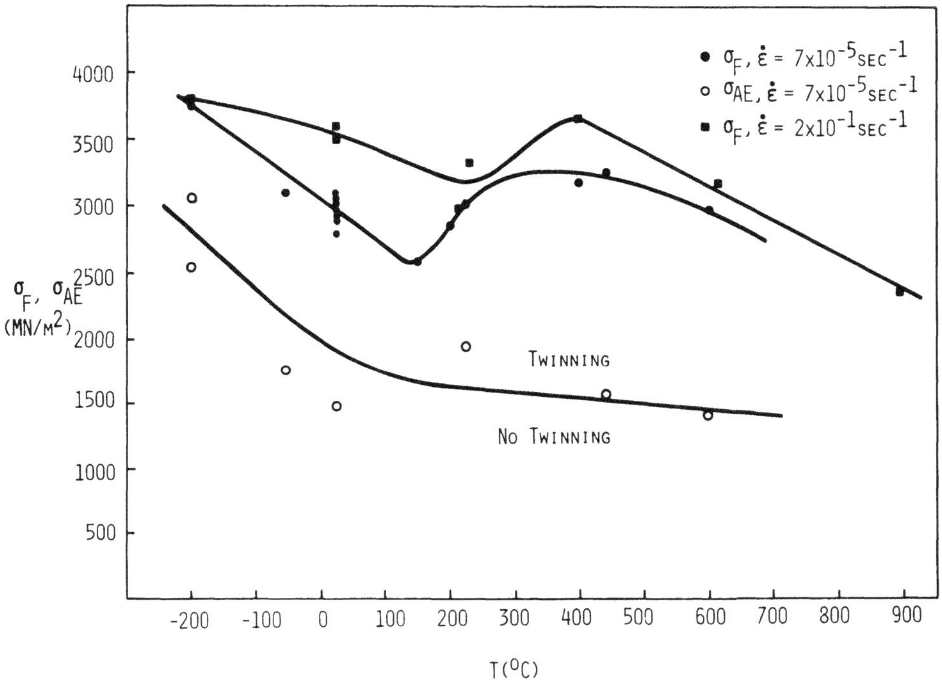

Figure 1. Compressive Strength and Threshold Stress Level for Acoustic Emission Versus Temperature.

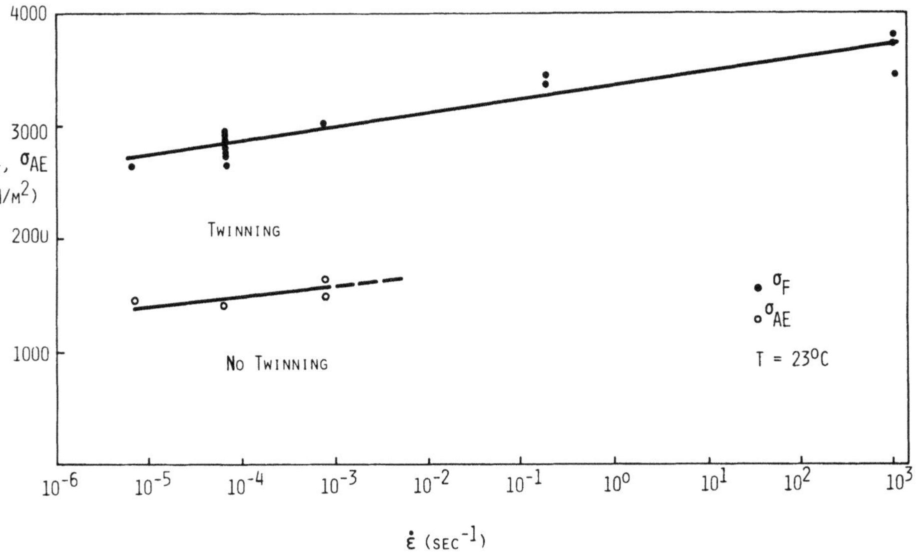

Figure 2. Compressive Strength and Threshold Stress Level for Acoustic Emission Versus Strain Rate

Acoustic Emission

The onset of acoustic emission due to damage (σ_{AE}) is defined as the stress level at which the acoustic emission count rate first begins to exceed the background rate, usually a few counts per second, determined by the discriminator level. In Figures 1 and 2, σ_{AE} is plotted as a function of temperature and strain rate, respectively. It is apparent that damage begins at stress levels well below that required to cause fracture. The curve for acoustic emission stress level parallels that for failure strength from -200 to around 200°C in Figure 1, but does not reflect the subsequent increase in strength. Similarly, the σ_{AE}-$\dot{\varepsilon}$ data plotted in Figure 2 follow the σ_f-$\dot{\varepsilon}$ reasonably well over the limited range in $\dot{\varepsilon}$ within which AE was obtained. This would be expected from the σ_{AE}-T results, i.e., as was the case with decreasing T, the strength increases with $\dot{\varepsilon}$ from that at $\dot{\varepsilon} \sim 10^{-4}$ sec^{-1}, T = 23°C. The strain rate range for AE was limited to that shown in Figure 2 by the fact that at higher strain rates, the transducer was forced to count at rates in excess of its capability to respond.

As in the case of σ_f, an extrapolation to absolute zero may be carried out for σ_{AE}, yielding (Figure 1) an ultimate stress for twinning of $\sim$ 3100 MN/m^2. Extrapolation of the σ_{AE}-$\dot{\varepsilon}$ data in Figure 2 to this stress level would imply the requirement of a strain rate in excess of 10^8 sec^{-1} in order to suppress twinning at room temperature.

SEM study of specimens loaded to various stress levels showed that below the threshold stress level for acoustic emission, no damage was discernible. Above this level, twinning was visible in many grains, and as stress levels were increased, still more twins were produced, and microcracks began to appear, associated with twin-grain boundary intersections and with twin plane decohesion. This aspect of compressive strength, i.e., the role of twinning in causing microcracking in Lucalox, has recently been documented extensively by the writer,[1] and no further description is probably required here. In the present context, the main point is that the thermally activated mechanism detected by the AE is deformation twinning. In addition, as documented in an earlier study[7] of tensile failure, twins formed above 400°C were generally thicker than those mucleated at lower temperatures, and there was a greater tendency for twins to themselves nucleate additional twins in adjacent grains.

Indentation Testing

In this section, the sequence of events corresponding to the early stages of indentation damage in Al_2O_3 are examined in greater detail than heretofore reported.[8] As shown in Figure 3a, the first evidence of damage in the polycrystalline material loaded to 5 gm is the indent, plus faint surface deformation markings (arrow) near one edge. At higher loads within the same grain, precisely the same pattern occurs, but the larger indentation now exhibits radial cracking, and the surface markings are more distinct. In Figure 3b, these features, caused by a 50 gm load, are clearly evident, as are additional faint markings along the walls of the indent (arrow); the intense surface markings are in the same location as those first appearing at 5 gm. It is interesting to note that the lower radial crack (R) is standing open, and that it seems to end near the apex of the indent.

In another grain (Figure 3c), indentation at 100 gm has produced deformation bands which appear to be twins, based on their resemblance to the obvious twins previously observed in compression tests.[1] Here the twins are continuous from the interior of the indent out onto the specimen surface. At the locations indicated by the arrows, the twins have begun to crack along their habit planes. Such sites appear to be the nuclei for lateral crack formation, as shown in Figure 3d; note the twins (arrow) running along the base of the lateral crack (L) in the lower portion of the photomicrograph.

At 400 gm, lateral cracking produces chipping, as in Figure 4.

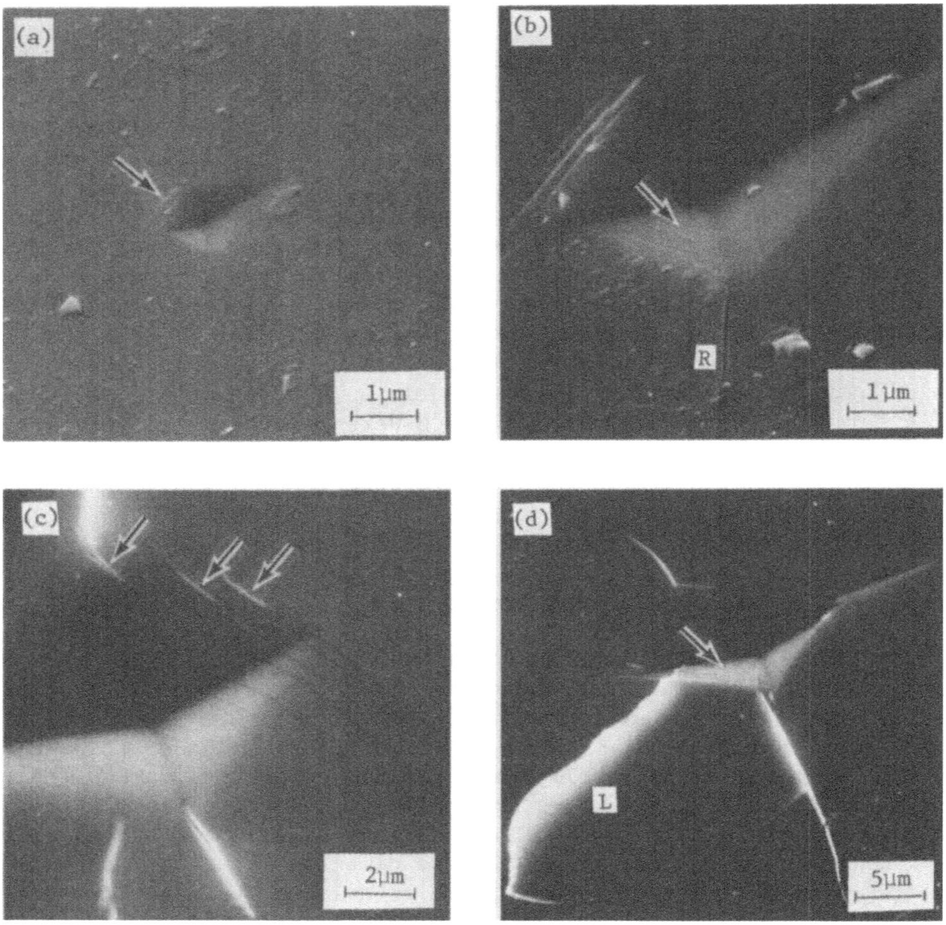

Figure 3. Indentation in Lucalox (a) 5 gm load, showing surface deformation bands (arrow) (b) 50 gm load, showing deformation bands on indent wall (arrow), radial crack at R. (c) 100 gm load, showing twin band cracking (arrows) along indent wall(d) 100 gm load, showing lateral crack (L) nucleated within indentation twins (arrow).

The origin of this crack is shown in Figure 4b, lying just below the rim of the indent. The flat plane shown (arrow) probably is a twin habit plane; note the twins on the surface to the left.

Behavior similar to that described for polycrystalline alumina is found in sapphire as well. In Figures 5a and 5b, loads of 200 gm and

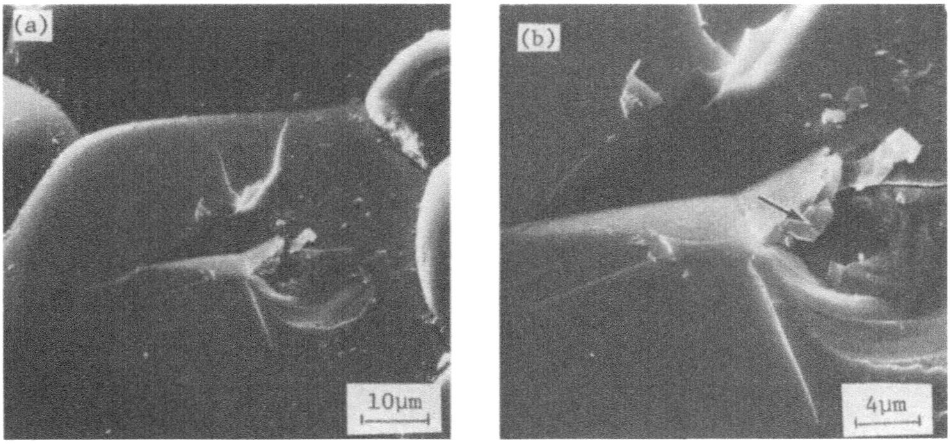

Figure 4. Indentation under 700 gm load in Lucalox (a). Chipping caused by lateral crack formation (b). Higher magnification view, showing nucleation site (arrow).

400 gm, respectively, produce lateral cracking (L) in identical locations. Because of the angle of the specimen relative to the beam, the deformation bands responsible for nucleation of the cracks do not show up, but they resemble those indicated by the arrow in Figure 5b. In Figure 5c, a 700 gm load has caused lateral crack chipping in two quadrants. A higher magnification view (Figure 5d) shows twins (arrow) and the flat twin facet (F) which is the apparent origin of the right hand spall. Twin band cracking (C) is also evident along the lower left corner of the indent.

DISCUSSION

Compressive Strength

In the temperature range from -200° to 150°C, the effect of enhanced twinning with increasing temperature is to lower the failure stress. This is easily rationalized on the basis of the microcracks nucleated by the more plentiful twins at higher temperatures. Similarly, increased strain rates inhibit twinning, hence microcrack nucleation, and thereby raise the strength.

The σ_f-T minimum is more difficult to explain. It is possible that the thickening of twins may play a role, by allowing twinned

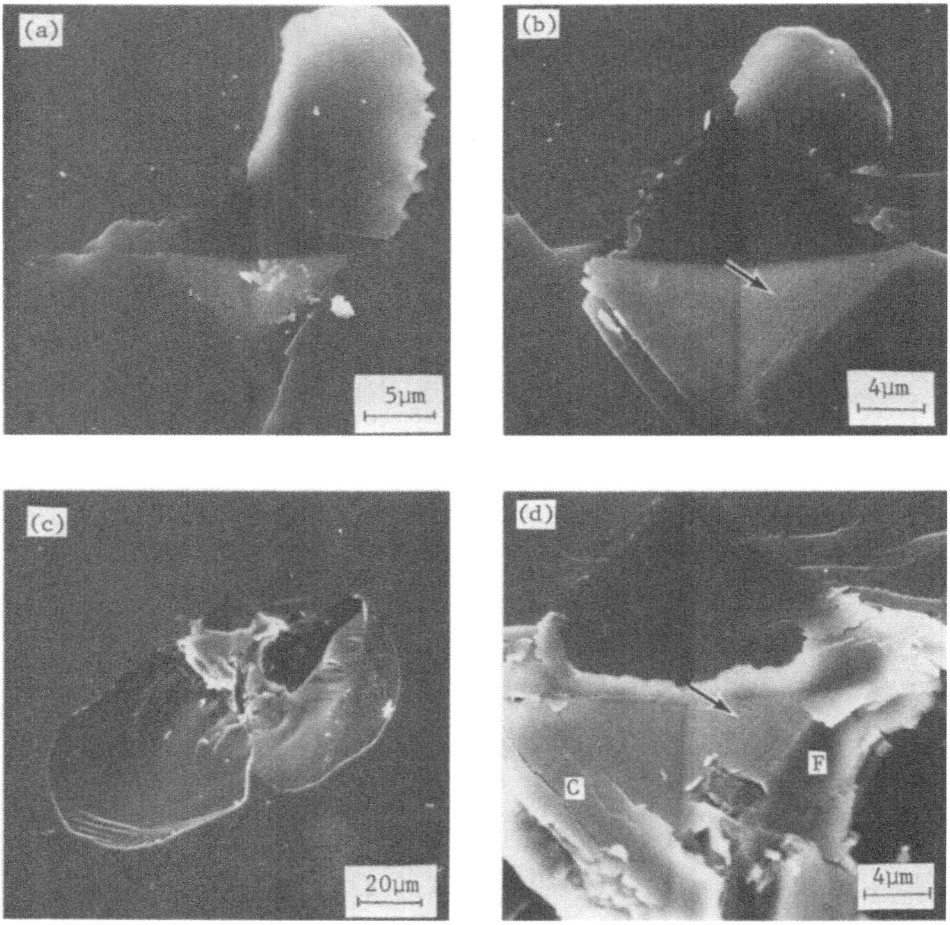

Figure 5. Indentation in sapphire (a) 200 gm load, showing lateral cracking (b) 400 gm load, showing lateral cracking as in (a), twins on indent wall (arrow) (c) 700 gm load, showing lateral crack chipping (d) Higher magnification view of (c), showing twins (arrows), twin facet (F) at lateral crack nucleation site, and twin plane cracking (c).

grains to accommodate greater strain without cracking. The process of thickening, a thermally activated mechanism distinct from that of initiation, would not be expected to cause detectable AE, and would agree with the observation of no change in the σ_{AE}-T curve in the region of the minimum. Alternatively, the minimum might be accounted for by the observed tendency for twins nucleated at and above[7] the σ_f-T minimum to in turn initiate twins in adjacent grains, rather than microcracks as at lower temperatures. The enhanced thermal activity in this case would serve to promote twinning on otherwise less highly active twin systems.

Based on conventional thermal activation theory, it is possible to calculate an activation energy for the twinning process, using the σ_{AE}-T and σ_{AE}-$\dot{\varepsilon}$ data. In its simplest form, the Arrhenius rate equation may be written as

$$\dot{\varepsilon} = \dot{\varepsilon}_0 \exp\left[\frac{-U(\sigma)}{RT}\right] \qquad (1)$$

where the activation energy $U(\sigma)$ is assumed to be a function of the effective stress σ only. Here $\dot{\varepsilon}_0$ is the strain rate corresponding to the ultimate stress achieved, R is the gas constant, and T the absolute temperature. In our case, $\sigma=\sigma_{AE}$. Assuming that the stress dependence of the activation energy is linear, we can write

$$U(\sigma) = U_0 - V(\sigma_{AE} - \sigma_0) \qquad (2)$$

where U_0 is the total activation energy of the process, V is the activation volume, and σ_0 the minimum stress required to produce fracture. Substituting Equation (2) into Equation (1) yields

$$\sigma_{AE} = \frac{U_0}{V} + \sigma_0 - \frac{RT}{V} \ln\left(\frac{\dot{\varepsilon}_0}{\dot{\varepsilon}}\right)$$

From the data obtained in the AE experiments, $\sigma_0 \simeq 1500$ MN/m^2, and $\dot{\varepsilon}_0 \cong 10^8$ sec^{-1}. On this basis, the total activation energy for twinning is estimated to be approximately 18 kcal/mole. This value is considerably lower than that determined from tests carried out at higher homologous temperatures, where dislocation mechanisms predominate. In the latter case, activation energies are on the order of 85 kcal/mole[9]. It should be noted that a valid activation energy cannot be calculated using simply the compressive strength data, since compressive failure is a composite process involving crack nucleation, growth and coalescence.[1]

Indentation

It seems clear that twins play a significant role in the nucleation of lateral cracks during the indentation of Lucalox. This is in qualitative agreement with speculation by several workers[2,3] that lateral crack formation, possibly as opposed to radial cracking, probably will be found to be controlled by the plastic damage zone beneath the indentation. In the present case, additional support for this idea is found in the extreme repeatability of the lateral microcracking sequence for a given indentation orientation, as exemplified by Figures 5a and 5b. Only crystallographically related crack initiators, such as twins, could reproduce the same crack morphology so consistently.

The suggestion that the specific lateral crack nucleation defects are twins in the case of Lucalox rests on several factors, primarily the observed twin/parent decohesion along indent walls, as typified in Figure 3c. In addition, this decohesion begins to occur in the same load range as does lateral cracking. Hockey[10] has pointed out that TEM studies show that whereas the bulk of the indentation plasticity in sapphire is confined below the indent, microtwins can extend up from this region to the surface of the indentation, where the lateral cracks seem to form.

It is important to consider the operational range (temperature, particle velocity) over which particulate erosion of alumina might be affected by a lateral microcrack mechanism based on twin-induced cracking. From the experimental measurements, the ultimate stress for twinning apparently corresponds to a strain rate on the order of 10^8 sec^{-1}. Assuming an indentation depth of about 5 μm, and an indentation strain of ~ 0.08, [11] an ultimate indenter velocity of around $5x10^3$ m/sec is calculated. Considering that twins propagate at sonic velocity[12], i.e., $9x10^3$ m/sec in alumina, the estimated ultimate velocity is reasonable.

Most of our information on the relationship between impact and twinning comes from studies of metals. In many of these systems, the number of twins increases with strain rate, although the twinning stress also increases.[12] Consequently, twin-nucleated microcracking in alumina might well be expected at high particle impact velocities. On the other hand, the speed of twin growth may limit the size of twins, which might in turn limit lateral microcrack extension. The latter is expected to be related to the initial flaw (twin) maximum dimension, and to the volume of stressed material. That is, extension of lateral cracks is controlled by tensile stresses originating during the unloading portion of the indentation event, and acting on the crack nucleating flaw according to fracture mechanics considerations. Crack extension in a particular stress field then requires the presence of a flaw in excess of some critical size.

Most particulate impact environments of engineering interest would involve temperatures above 23°C. Within the regime studied, i.e., 23°C to 900°C, σ_{AE} was essentially athermal. Consequently, it is most likely that over this temperature range, strain rate (particle velocity) effects would dominate crack initiation, with substrate temperature a secondary variable.

CONCLUSIONS

Compressive failure of polycrystalline aluminia is thermally activated, with an apparent basis in twin-nucleated microcracking. The latter process controls lateral crack formation, hence erosion, during indentation. It is likely that twinning and associated lateral crack formation in Al_2O_3 can persist to quite high velocities, on the order of $5x10^3$m/sec., during particulate impact.

ACKNOWLEDGEMENT

This research has been sponsored by the Office of Naval Research under Contract No. N00014-75-C-0668.

REFERENCES

1. J. Lankford, Jour. Mat. Sci., 12, 1977, 791.
2. B. R. Lawn and M. V. Swain, Jour. Mat. Sci., 10, 1975, 113.
3. A. G. Evans and T. R. Wilshaw, Acta Met., 24, 1976, 939.
4. A. G. Evans and Linzer, Jour. Am. Cer. Soc., 56, 1973, 575.
5. A. G. Evans, M. Linzer, and L. R. Russell, Jour. Mat. Sci., 15, 1974, 253.
6. R. J. Charles, Studies of the Brittle Behavior of Ceramic Materials, ASD-TR-61-628, 1963, 467.
7. J. Lankford, Jour. Mat. Sci. (in press).
8. J. Lankford, Scripta Met., 11, 1977, 349.
9. M. L. Kronberg, Jour. Am. Cer. Soc., 45, 1962, 274.
10. B. J. Hockey, private communication, 1976.
11. D. Tabor, "Hardness of Metals," Clarendon, Oxford, 1951.
12. S. Mahajan and D. F. Williams, Int. Met. Rev., 18, 1973, 43.

MICROCRACKING ASSOCIATED WITH THE SCRATCHING OF BRITTLE SOLIDS

M.V. Swain

Physics and Chemistry of Solids Section, Cavendish Laboratory, Madingley Road, Cambridge, U.K.

ABSTRACT

Observations of the microcracking about scratches in a number of brittle solids including sapphire and a variety of glasses are presented. The scratching was carried out primarily with a Vickers pyramid indenter under various loads at one scratching velocity. It is observed that the nature of the cracking is very similar to that occurring about a quasi-static pointed indenter. Directly beneath the indenter a well-defined median crack is formed, behind the indenter small partial Hertzian cracks form within the track and subsurface lateral cracking initiates from the region of the plastically deformed zone. With a Vickers pyramid the cracking phenomena may be conveniently divided into three regions with increasing load. At low loads ($P < 0.05$ N) no cracking is observed, intermediate loads (0.1 N $< P < 5$ N) well-defined median and lateral cracks occur, at higher loads ($P > 5$ N) the plastically deformed track appears to shatter and the extent of lateral and median cracking is less than that occurring for the higher loads in the intermediate region.

The present observations and those of others in the literature have been interpreted in terms of indentation fracture mechanics, reasonable agreement between theory and experiment is observed. The implications of this work to abrasive wear rate and residual strength of scratched bodies is pointed out.

INTRODUCTION

The shaping of brittle solids invariably involves grinding or abrading these solids with sharp and elastically harder grits. However, despite the increasing use of brittle materials the mechanisms by which material is removed during machining are still poorly understood. Of particular interest is the nature of microcracking that occurs about the contact area between abrasive and specimen. The microcracking controls to a large extent the rate of material removal, the size of subsurface flaws and hence the residual strength of ground or abraded materials.

Previous studies of the scratching of brittle solids have been carried out by Peter[1] and more recently by Broese van Groenou et al[2] and Veldkamp and Klein Wassink[3]. The latter studies have focussed more on the force ratio and specific energy associated with scratching and used this to determine the grindability of brittle solids, whereas Peter[1] concentrated on the deformation and microfracture about a scratch. Recently there has been considerable advancement in the field of indentation fracture mechanics by Lawn and coworkers[4,5,6,7,8], and the extension of these concepts to scratching of brittle solids is now possible.

In the present study the stress fields about a point contact subjected to both normal and horizontal tractions is considered to simulate the stress field about a sliding indenter. The indentation stress field in the present case differs from that of the axially loaded point contact[4] in that it is asymmetric which may reflect itself in the microcracking phenomena. Scratching experiments primarily with a Vickers pyramid indenter were carried out on a number of model brittle materials including a variety of glasses, single crystal and polycrystalline alumina. Detailed observations of the track after scratching with a scanning electron microscope indicate microcracking in regions predicted by the elastic stress field. The extent of the subsurface microcracking is predicted by a simple fracture mechanics treatment and reasonable agreement is found with observations.

Finally, consideration of previous studies by Holland and Turner[9] on the strength degradation of scratched glass plates indicates the presence of a residual stress associated with a plastic groove about a scratch. A similar residual stress has been noted about the permanent impression formed by a pointed indenter in a brittle solid[10]. It is this residual stress field and its release with time owing to stress corrosion-aided lateral cracking that appears responsible for glaziers' experience that the strength of a scratched plate of glass is least when the scratch is "hot".

STRESS FIELD BENEATH A POINTED INDENTER UNDER COMBINED NORMAL AND HORIZONTAL FORCES

(a) Elastic Stress Field

In the present case an elastic half space is subjected to a normal as well as a horizontal point load, the resultant of which is $\underline{P}$. The problem has been previously considered by Mindlin[11]. For an isotropic material of Poisson's ratio ν, the stress components in spherical-polar co-ordinates are of the form

$$\sigma_{ij} = \frac{P}{\pi R^2} \cdot (\underline{a}\, F_{ij}\{\phi\}_N + \underline{b}\, F_{ij}\{\phi\}_T)_\nu \quad (1)$$

where $F_{ij}\{\phi\}_N$ and $F_{ij}\{\phi\}_T$ are the normal and horizontal components respectively, and $\underline{a}$ and $\underline{b}$ are constants denoting the relative ratios of these two components. For instance, in a sliding friction test $\underline{a}$ would be unity and $\underline{b}$ the coefficient of friction, whereas in a grinding experiment $\underline{a}$ and $\underline{b}$ would denote the vertical and horizontal forces respectively.

The stress distribution suggested by equation (1) is valid only for an ideally sharp point which in theory will give rise to a stress infinity and produce inelastic deformation. However, for the case of a conical indenter the maximum mean contact pressure is independent of load and is given by

$$p_m = \frac{E \cot \psi}{2(1-\nu^2)} \quad (2)$$

where E is Young's modulus and ψ the half angle of the indenter. Provided p_m exceeds the hardness of the solid, permanent deformation will occur over most of the contact area. The mean contact pressure (or hardness) for a plastic impression of characteristic dimension a produced by a load P may be written,

$$p_o = P / \alpha\pi a^2 \quad (3)$$

where α is a dimensionless constant determined by indenter geometry. For the case of a sliding pointed indenter creating plastic deformation the contact pressures given by equation (3) are likely to be too low because the rear half of the contact area will be partially released from load. This would suggest that the contact pressures may be too low by a factor of ~ 2. Substituting for P from equation (3) into equation (1) we obtain

$$\sigma_{ij}/p_o = \alpha\left(\frac{a}{R}\right)^2 \{\underline{a}\, F_{ij}\,(\phi)_N + \underline{b}\, F_{ij}\,(\phi)_T\}_\nu = \alpha\left(\frac{a}{R}\right)^2 \{R_{ij}(\phi)\}_\nu \quad (4)$$

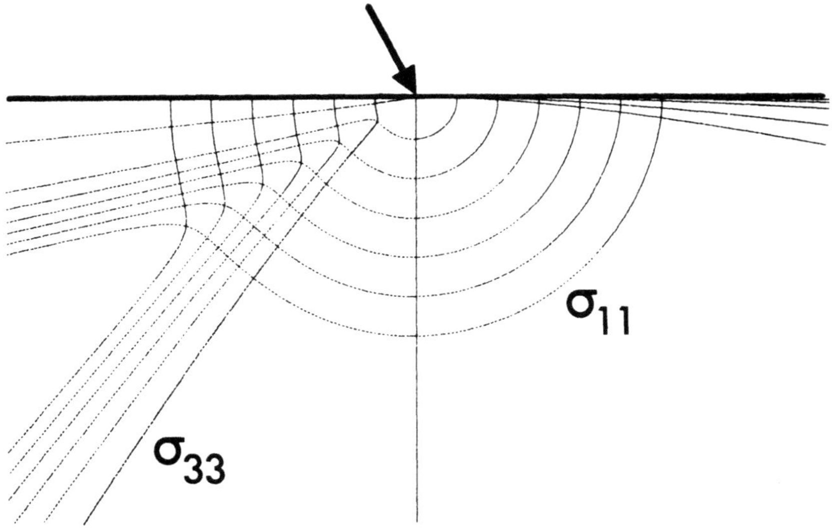

Figure 1. Side view of stress trajectories in stress field for point contact inclined 30° to the normal. Plotted for $\nu = 0.25$.

The significant features of the stress field are illustrated in figures 1 and 2 for P at 30° to the normal corresponding to a coefficient of friction of 0.5. Figure 1 is a plot of the stress trajectories of the two principal stresses σ_{11} and σ_{33}; these are the curves whose tangents indicate the directions of the principal stresses at each point and are such that $\sigma_{11} >> \sigma_{33}$ nearly everywhere in the field. In figure 2 are plotted contours of the three principal stresses.

The principal features of the stress field are that the σ_{11} and σ_{33} components are almost entirely tensile and compressive respectively, whereas the σ_{22} "hoop" stress component is tensile in a region below the indenter but compressive near the surface. The most significant difference between the axial point loading stress field and that of the present case is the asymmetry in the latter. The other important factor is the strong tension near the surface in the σ_{11} component behind the point contact, figure 2a. In this sense the stress field is somewhat similar to the sliding Hertzian stress field when friction is present[13].

(b) Inelastic Stress Field

As mentioned above plastic deformation invariably occurs about

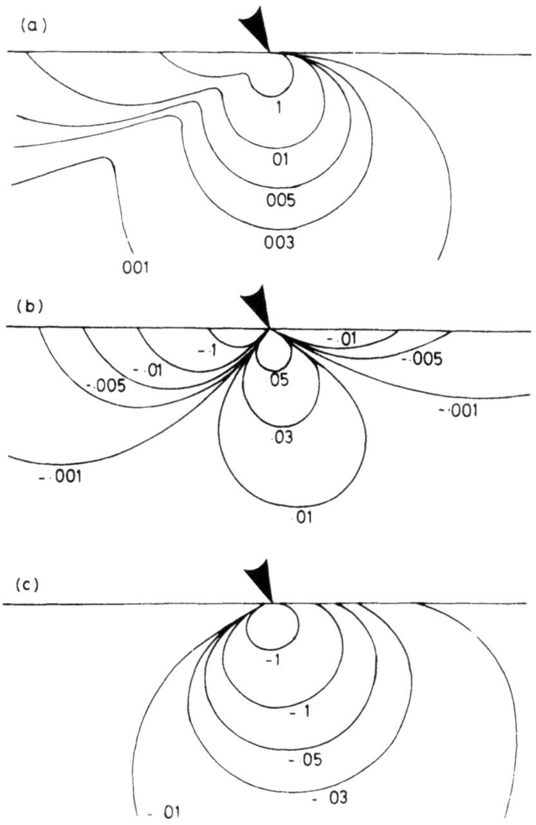

Figure 2. Contours of principal normal stresses (a) σ_{11}; (b) σ_{22}; and (c) σ_{33} under point loading as defined in figure 1. Plotted for $\nu = 0.25$ and in the plane containing the contact axis. Unit of stress is p_o and the contact diameter is shown.

a sharp indenter on loading. The stresses in the immediate vicinity of the contact site are then somewhat different from those presented above. Until recently the analysis of this problem has considered the plastic zone about a hardness impression similar to that about the expansion of a spherical cavity under internal pressure in an infinite solid[14]. Although this approach has provided some insight into the mechanics of deformation beneath a pointed indenter it neglects the existence of a free surface. It also predicts that the stresses fall off as R^{-3} from the contact site whereas the elastic solution of the previous section predicts R^{-2}.

A recent study by Perrott[15] of the elastic-plastic indentation stresses about a pointed indenter overcomes many of the above objections. The results of this analysis are such that the overall form of the elastic-plastic stress field at sufficiently large distances from the contact area differ only slightly from the elastic solution. Significant differences do occur at the elastic-plastic boundary particularly at the surface. One other interesting feature to come out of this analysis is that the hoop stress about the contact area does not become tensile until the plastic zone is sufficiently well developed, that is, it surrounds the contact site. This type of analysis has not been extended to scratching, however the elastic stress fields of the previous section should provide a good first approximation when plastic deformation is present.

(c) Unloading Stress Field

On unloading the elastically deformed hinterland about the plastically deformed zone is constrained from returning to its original equilibrium position and so gives rise to a residual stress. The analysis by Perrot[15] has also determined the stress fields upon unloading a quasi-static indenter. In this case the form of the stress field decreases in intensity slightly subsurface but increases considerably near the surface outside the contact circle.

However to date the residual stress field surrounding a scratch has not been studied analytically. As an approximation to this problem let us consider the plastically deformed material to be like

Figure 3. Stress trajectories about a hole under pressure adjacent to the surface. Trajectories emanating from the cavity are tensile and those from the surface compressive.

a cylindrical cavity under pressure adjacent to the surface in an elastic half space. Strong support for this approach comes from photoelastic observations by Dalladay and Twyman[16], of a subsurface "centre of stress" beneath the track of a scratch on glass with a diamond indenter. This problem has been analysed by Jeffery[17], from whose analysis the tensile stress trajectories may be determined, figure 3. Dalladay and Twyman[16] determined the principal stress trajectories from a photoelastic study of the strain field about scratches in glass, these are shown in figure 4. The agreement between the pressurised subsurface cavity model and Dalladay and Twyman's analysis is very encouraging.

MICROFRACTURING ABOUT SCRATCHES

Previous work[18] indicates that microcracking accompanying the scratching of brittle solids with sharp points at low loads is similar to observations of the microfracture about quasi-static pointed indenters[4]. In the case of a sliding Vickers pyramid indenter, at initial contact two median cracks form along the indentation diagonals, but only the one aligned along the prospective track ultimately survives. This favourably disposed crack simply translates along with the indenter to produce a linear fissure, while the other crack remains immobile at the starting site. The lateral cracks form continuously in the wake of the indenter from the plastically deformed zone and propagate normally outward from the median plane.

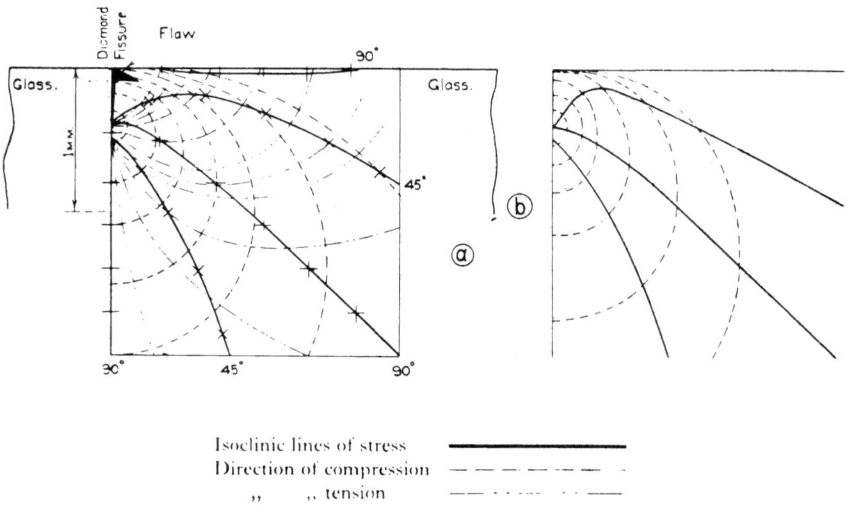

Figure 4. Principal tensile and compressive stress trajectories of the residual stress about a scratch in glass from photoelastic studies. Figure (b) is for the case when no flaw was observed beneath the scratch. After Dalladay and Twyman[16].

EXPERIMENTAL

Studies were carried out on a number of commercially available glasses and polycrystalline and monocrystalline alumina. All surfaces were highly polished. Scratching was carried out using a microhardness tester with a small motor coupled to one of the micrometer manipulators. The indenter was a Vickers pyramid and the sliding direction was parallel to the pyramid diagonal. The load was varied from 0.25 - 5 N and the scratching velocity was 1 mm/sec for the scanning electron microscope (SEM) study and 125 μm/sec for lateral crack study. Most tests were done in air and measurements of the lateral cracking were made approximately one hour after scratching. The glass samples were cross-sectioned for SEM observation by propagating a slow crack perpendicular to the scratch. The other samples were sectioned using standard metallographic techniques.

OBSERVATIONS AND DISCUSSION

(a) Scanning Electron Microscope Observations

Observations of the surface trace and cross-sectional views of a scratch formed by a Vickers pyramid drawn across the surface of soda-lime glass at 1 mm/sec under various loads are shown in figure 5. At low loads, figure 5a, the surface trace of the scratch appears completely plastic with little evidence to suggest subsurface cracking. However the cross-sectional view reveals a well-defined plastic zone beneath the track of the indenter and it is from this zone that the median and lateral cracks emanate. Other workers have occasionally observed turnings about scratches in soda-lime glass, particularly with sharper indenters[19]. At higher loads the extent of subsurface microcracking increases, figure 5b. The extent of lateral cracking has greatly increased and they often intersect the surface removing a considerable volume of material. For loads greater than 4N, figure 5c, evidence for plastic deformation about the scratch has disappeared and all that remains is shattered material about the scratch. Optical examination immediately after scratching sometimes reveals the entire plastic zone about the scratch lying intact adjacent to the path of the indenter. In figure 5c the debris was removed by cleaning prior to examination in the SEM. The well-defined median crack is clearly visible at the base of the shattered groove in figure 5c, and note how it deviates to left and right along the track. At this load it is often difficult to distinguish between diverging median cracks and lateral cracks. The other interesting feature is that the extent of lateral cracking is not as large as at lower loads, presumably because the driving force for this cracking system, the plastic zone, has been removed or greatly reduced. It is probably this last region of high indenter loads which corresponds with the model for material removal as proposed by Veldkamp

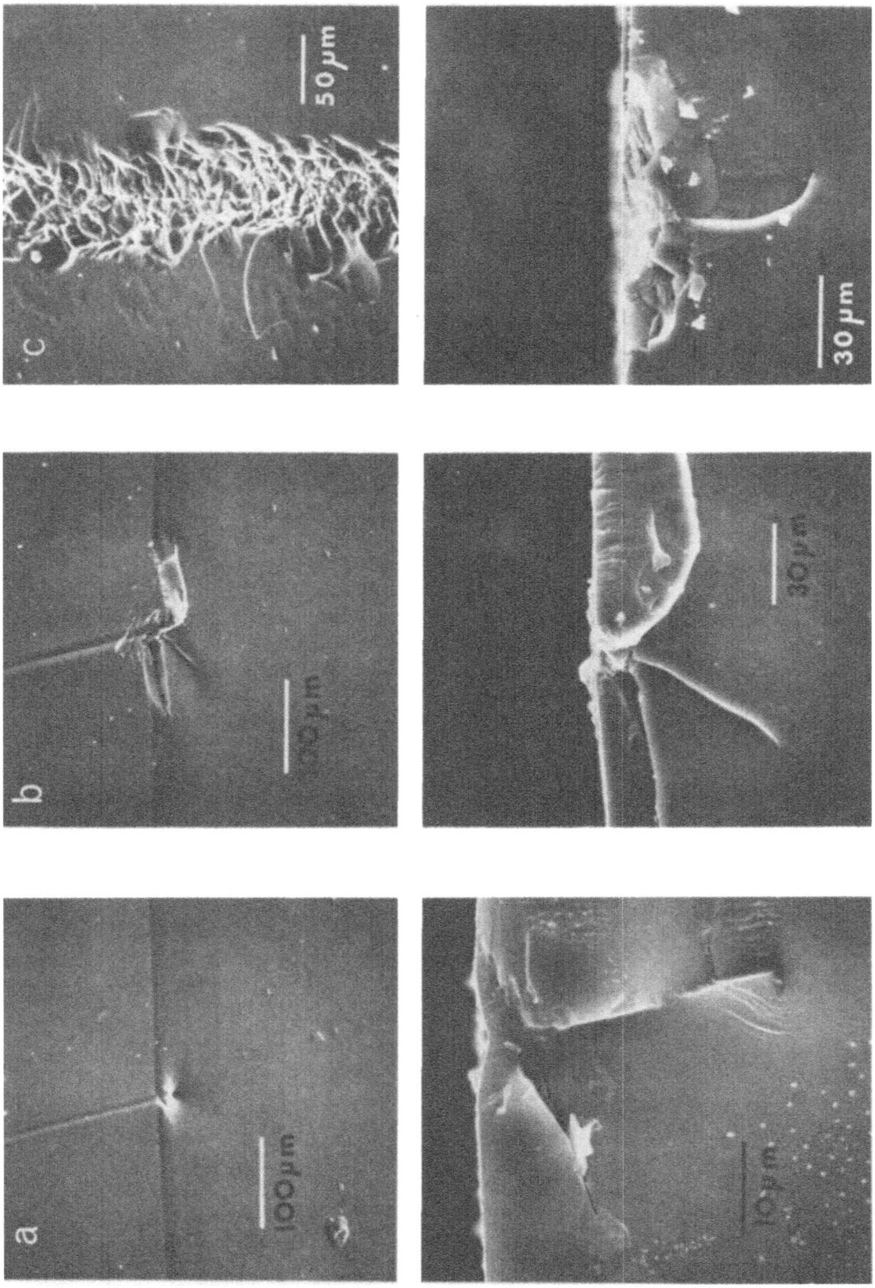

Figure 5. Surface and cross-sectional views with the scanning electron microscope of scratches on soda-lime glass made with a Vickers pyramid, velocity of indenter in all cases was 1 mm s^{-1}. Force on indenter was (a) 0.25 N; (b) 1.0 N, and (c) 4 N.

and Klein Wassink[3]. In this model, material is removed by cracks initiating ahead of the indenter owing to bending moments induced by tangential tractions. Median and lateral cracking have also been observed[20] about grooves generated with a single point grinding wheel simulator rotating at 12,000 r.p.m.

Observations were also made on hot pressed fully dense polycrystalline alumina ($\sim$ 10 μm grain size), and single crystal sapphire. Tests were only carried out under low loads (0.5 N) on polycrystalline alumina. At this load there is little evidence of either lateral or median cracking, however, there is a considerable amount of grain boundary failure about the scratch. At high magnification slip or twinning within the individual polycrystalline grains may be seen, figure 6a, and where these slip bands intersect a grain boundary, failure occurs. This observation is in agreement with transmission electron microscope studies of twin-grain boundary interaction by Hockey[21]. There is no evidence for the transmission of these slip bands from one grain to the next in figure 6. The extension of these ideas to the abrasive wear of polycrystalline alumina has been considered elsewhere[24].

Observations of the scratches in monocrystalline sapphire are somewhat similar to those in glass. All scratches were carried out on the basal plane and little evidence for anisotropy was observed. The most significant feature of the scratches however was at low loads (figure 6b) where both median cracks and bow-like Hertzian

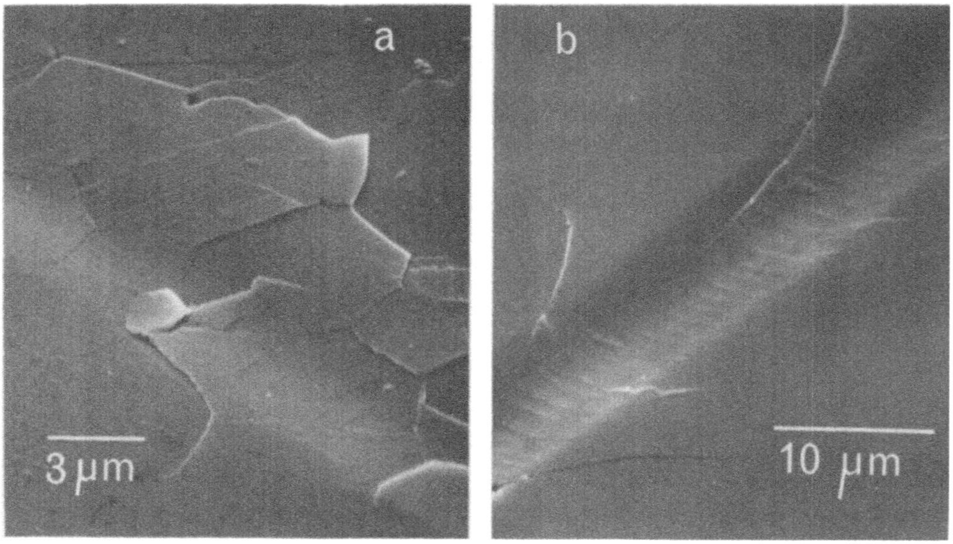

Figure 6. Surface view of microcracking about a scratch made with a Vickers pyramid on (a) fully dense hot pressed alumina, and (b) the basal plane of single crystal sapphire.

cracks within the track may be seen. Again note how the median cracks deviate to left and right from the base of the track.

(b) Median Cracking

Previous work by Peter[1] has quantified the variation of median crack depth with load and scratching velocity made with a Vickers pyramid in soda-lime glass, figure 7. The mechanics of initiation of a median crack would need to consider the subsurface flaw and stress distribution in the vicinity of the elastic-plastic boundary. The analysis would then be similar to that proposed recently by Lawn and Evans[7]. After crack initiation beneath the indenter the stress intensity factor is determined by the wedge opening forces in the surface perpendicular to the direction of travel of the indenter. However, in the present case in addition to the normal load P there is a tangential load μP, where μ is the coefficient of friction, opening up the crack the resultant of which is $P(1 + \mu^2)^{\frac{1}{2}}$. Then following Lawn and Fuller[6] the stress intensity factor at the crack-tip is given by

$$K = \frac{m_e P(1 + \mu^2)^{\frac{1}{2}}}{\tan\psi (\pi c)^{3/2}} \qquad (5)$$

where m_e is a modifying factor that takes into account slight differences in crack shape from that considered by Lawn and Fuller[6] and secondly to allow for the increased indentation pressure as mentioned in section 2. When the results in figure 7 and those of Gruver and Kirchener[19] are plotted against $c^{3/2}$ (figure 8), they very nearly fit

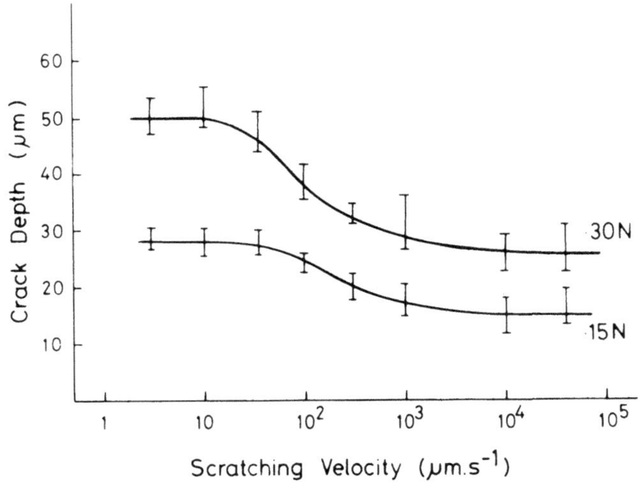

Figure 7. Extent of median cracking beneath a Vickers pyramid as a function of scratching velocity. After Peter[1].

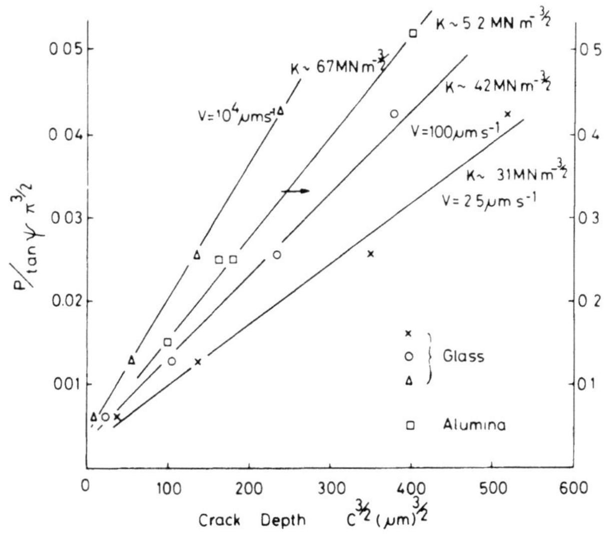

Figure 8. Data from Peter[1] and Gruver and Kirchener[19] plotted according to equation (5) with $m_e = 4$. The values of the equilibrium stress intensity factor at the different scratching velocities are labelled on the curves.

a straight line relationship except perhaps at very low loads. However, in order to obtain values of K from equation (5) that are in reasonable agreement with other workers a modifying factor of $m_e \simeq 4$ must be used.

Below a critical load on the indenter ($\sim$0.05 N) no median cracking was observed beneath the scratch[1]. A similar phenomenon was observed about quasi-static Vickers indentations on soda lime glass by Lawn et al[22]. The transition from plastic to brittle behaviour was predicted at a load of $\sim 10^{-2}$ N. The variation in the equilibrium stress intensity factor with scratching velocity as shown in figure 8 is probably a manifestation of water vapour assisted stress corrosion cracking.

(c) Residual Stress about Scratches and Residual Strength of Scratched Beams

As mentioned earlier photoelastic studies clearly indicate that residual stress exists about scratches in brittle solids. Further evidence for the presence of residual stress comes from the work of Holland and Turner[9] who measured the effect of scratch velocity on the fracture strength of scratched glass beams. These authors found that at constant load the fracture strength decreased with increasing scratch velocity. Over the range of scratching velocities studied the crack depth is independent of scratch velocity. All fracture strength determinations by Holland and Turner[9] were done at very fast loading rates in three point bending and hence the results are essentially independent of the environment. It is possible to determine the stress intensity factor K_B at fracture for Holland and Turner's results. From standard fracture mechanics handbooks

$$K_B = Y \sigma_\beta \sqrt{c} \qquad (6)$$

where σ_β is the outer fibre bend stress, Y is a geometric factor $\simeq 1.12\sqrt{\pi}$, $\sigma_\beta = 22.4$ MPa, and $c = 113$ μm from Holland and Turner's measurements. The stress intensity value from equation (6) is ~ 0.47 M N $m^{-3/2}$ which is substantially less than the K_{1c} value of ~ 0.75 M N $m^{-3/2}$ for soda-lime glass. Similar differences in the value of the measured stress intensity value versus the K_{1c} value have been noted in indentation induced flaws in glass and other brittle materials[10]. In both cases the residual stress about the scratch or indentation gives rise to an additional stress intensity factor K_R, at the crack tip which is given by

$$K_R = K_{1c} - K_B \ . \qquad (7)$$

Typically K_R is found to be $\sim 30\%$ of K_{1c}[10,20] and so contributes a considerable influence on the residual strength determination. This residual stress intensity factor may be related back to a

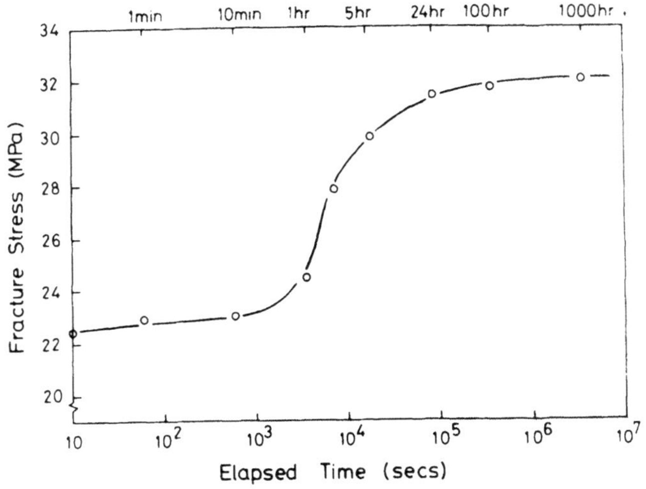

Figure 9. Fracture stress of scratched glass as a function of elapsed time after scratching. Rate of scratching was 8 cm s^{-1} and force on indenter was 5 N. After Holland and Turner .

residual stress at the crack mouth resulting from plastic deformation. The magnitude of this residual stress σ_R is $\sim H/20$ [20] in good agreement with that found about Vickers pyramid indentations in a variety of brittle solids[10].

This conclusion suggests that the residual stress is directly related to the hardness of the material. Previous work by Gunasekera and Holloway[23] has shown that the hardness of glass is strongly dependent upon the indentation time (or scratching velocity), and this hardness variation probably accounts for Holland and Turner's observations.

Holland and Turner[9] also noted that the residual strength of the scratched beams systematically increased with time elapsed after scratching as shown in figure 9. This observation confirms glaziers' experience that scratched glass is easier to break when the scratch is "hot". The difference between the instantaneous and long time fracture stress is $\sim 30\%$ agreeing with the K_R/K_{1c} ratio mentioned above, if one assumes the residual stress is zero after sufficiently long time. It is also glaziers' experience that scratching glass in the presence of paraffin makes it easier to fracture and improves the edge quality. Gunasekera and Holloway[23] have shown that paraffin does increase the hardness of glass in comparison with air. Additionally in the presence of paraffin one might expect the extent of lateral cracking to be less because of the very low water vapour pressure of this liquid, hence accounting for the better edge quality of the fractured glass. Peter[1] also noted that after scratching glass at low loads it often took some time before the lateral cracks became visible, presumably because of stress corrosion assisted slow crack growth.

From the observations of the previous section it is possible to predict the residual strength of scratched glass plates. From equations (6) and (7) the residual strength of a scratched glass plate is given by

$$\sigma_{Res} = \frac{K_B}{1.12} \left(\frac{\tan \psi}{m_e P} \right)^{1/3} \qquad (8)$$

At low loads equation (8) provides a very good estimate of the residual strength of glass plates as determined by Holland and Turner[9]. However at higher loads the above equation significantly underestimates the residual strength, presumably because of the value of m_e used.

CONCLUSION

Observations of the microfracture about scratches in brittle solids has established that it is very similar to that occurring beneath quasi-static pointed indenters, the differences in behaviour being accounted for by the assymmetric nature of the indentation stress field because of tangential tractions. There appear to be three distinct regions of material behaviour with increasing load on a pointed stylus. At very low loads (<0.05N with a Vickers pyramid) the remaining track is smooth and fully plastic with no visible cracking. This region corresponds to mechanical "polishing" of brittle solids and is characterised by a low material removal rate, and a smooth surface finish. For intermediate loads a fully plastic track and well developed median and lateral cracks occur subsurface. At higher loads (>5N) extensive crushing occurs about the region of contact. It is these latter regions that correspond to typical "abrasion" operations on brittle solids.

A simple fracture mechanics treatment is presented for the depth of median cracking beneath a pointed indenter during scratching. It was also noted that a residual crack-tip stress intensity factor exists at the crack tip in the absence of load. Finally it was suggested that the release of this residual stress by lateral cracking may account for the increased residual strength of scratched beams with time.

ACKNOWLEDGEMENTS

This work was supported by the S.R. C. and M.O.D. (PE)

REFERENCES

1. K. Peter, Glastechn. Ber. 40 (1964) 470.
2. A. Broese van Groenou, N. Maan and J.C.B. Veldkamp, Phillips Res. Repts. 30 (1975) 320.
3. J.C.B. Veldkamp and R.J. Klein Wassink, ibid. 31 (1976) 153.
4. B.R. Lawn and M.V. Swain, J. Mater. Sci. 10 (1975) 113.
5. B.R. Lawn and T.R. Wilshaw, ibid. 10 (1975) 1049.
6. B.R. Lawn and E.R. Fuller, ibid. 10 (1975) 2016.
7. B.R. Lawn and A.G. Evans, ibid. (1977) in press.
8. M.V. Swain and B.R. Lawn, Int. J. Rock. Mech. and Min. Sci. 13 (1976) 311.
9. A.J. Holland and W.E.S. Turner, J. Soc. Glass. Tech. 21 (1937) 383
10. M.V. Swain, J. Mater. Sci. 11 (1976) 2345.
11. R.D. Mindlin, Physics 7 (1936) 195.
12. I.N. Sneddon, "Fourier Transforms" (1951) McGraw Hill, N.Y.
13. G.M. Hamilton and L.E. Goodman, J. Appl. Mech. 33 (1966) 371.
14. K.L. Johnson, J. Mech. Phys. Sol. 18 (1970) 115.
15. C.M. Perrot, Wear (1977) in press.
16. A.J. Dalladay and F. Twyman, Trans. Opt. Soc. 23 (1921) 165.
17. G.B. Jeffery, Trans. Roy. Soc. A221 (1921) 265.
18. D.M. Busch and J.F. Prins, "The Science of Ceramics, Machining and Surface Finishing" NBS Spec. Publ. 348 Edts. Schneider & Rice, p.73.

19. R.M. Gruver and H.P. Kirchener, J. Am. Ceram. Soc. 57 (1974) 220.
20. M.V. Swain, to be published.
21. B.J. Hockey, Proc. Brit. Ceram. Soc. 20 (1972) 95.
22. B.R. Lawn, T. Jensen and A. Arora, J. Mater. Sci. 11 (1976) 573.
23. S.P. Gunasekera and D.G. Holloway, Phys. Chem. Glasses 14 (1973)45.
24. M.V. Swain, Wear 35 (1975) 185.

CRACK FORMATION DURING SCRATCHING OF BRITTLE MATERIALS

J.D.B. Veldkamp, N. Hattu, V.A.C. Snijders*

Philips Research Laboratories

Eindhoven, the Netherlands

ABSTRACT

In order to investigate crack formation of brittle materials during grinding, scratching experiments with diamond pyramids have been carried out in air under varying loads and scratching speeds on several materials: single crystalline MgO, polycrystalline Mn Zn- and Ni Zn ferrite and a variety of glasses. Loads between 0·05 and 5 N and speeds between 0·4 and 400 μ/s were applied. Additional experiments on one glass at about 0.5 m/s were carried out. When polycrystalline material is scratched with a leading pyramidal plane one observes with increasing speed first chips and then separate lateral cracks. With increasing load successively pure ductile grooving, lateral cracking and finally chipping are seen. On glass with increasing speed one observes successively chipping plus irregular median cracking, lateral cracking plus irregular median and sub-surface cracking and only smooth median and subsurface cracking. With increasing load the sequence of events is: ductile grooving, smooth median and subsurface cracking, irregular median and subsurface cracking plus lateral cracking and finally irregular median cracking plus chipping. This behaviour is maintained when the speed is increased to 0·5 m/s. In single crystalline MgO lateral cracks are formed at any load and at any speed. Minor median and subsurface cracks are also formed. When the experiments are carried out with a leading edge, then median cracks are also observed in polycrystalline materials together with chipping and lateral cracking.

*) College of Advanced Technology, Utrecht, Holland

For the lateral and medium cracks expressions are derived relating tangential force to stress intensity factor, crack width and groove width. For the lateral cracks plots of scratching speed versus stress intensity factor are given for Mn Zn ferrite, for one glass and for MgO. K_I values of the glass obtained at certain crack speeds in a double torsion experiment in air are shown to fall within the error bars of K_I values derived from lateral cracks obtained at the corresponding scratching speeds. The groove width at constant load is seen to decrease with increasing scratching speed.

The number of lateral cracks per unit scratch length increases with increasing load and with decreasing speed.

1. INTRODUCTION

Grinding can be considered as a multi point scratching operation. The microcracks due to scratching often determine the strength and the life time of a stressed brittle product.[1,2] Knowledge of crack formation during scratching is also essential to a better understanding of the ease of cutting and the grindability[3] of a brittle material. In the sections 1.1, 1.2, 1.3 and 1.4 first a review will be given of the litterature on the crack types which are known to develop during scratching and on the growth of these cracks.

1.1. The Stress Field, Plastic Grooving and Lateral Cracking.

In the case of a spherical indenter, sliding on the surface of a brittle material, first with increasing load purely elastic deformation takes place which is followed by ductile grooving and cracking.

The stress field due to a sphere obliquely loaded on the flat surface of a homogeneous and isotropic material was deduced by Hamilton and Goodman[4] from the stress distribution around a point indentation as derived by Boussinesq[5]. From this stress distribution the development of lateral cracks in brittle materials could be qualitatively predicted. Such cracks were experimentally found for instance by Gordon et al[6] and by Busch[7] on glass. Lawn[8] extended the work of Hamilton and Goodman and introduced the stress intensity factor for the description of the cracks.

In the case of an ideally sharp pyramid or cone, ductile grooving takes place immediately after the application of a load below a certain semi-angle of the point. (fig. 5a). This ductile grooving has been described in terms of the ploughing model[9] but also in relation with the thermally activated character of the deformation.[3]

At some distance from a sharp indenter, e.g. a pyramid, the stress field will resemble the field due to a moving sphere, hence the shape of the lateral cracks (see fig. 5b) will be rather similar in both cases.

In the films made by Busch[7] it can be seen that during scratching over a relatively short distance, rapid growth of the lateral cracks (pop-in) takes place. Over a much greater length the cracks propagate at a speed which is roughly equal to the scratching speed. This implies that scratching speed and average crack propagation speed are of the same order of magnitude.

1.2. Chipping

Ground surfaces of brittle materials often show not only grooves with seperate lateral cracks but also pieces of material broken away from the surface, the so-called chips. It can be seen in the film by Busch[7] that the majority of the chips is formed behind the scratching point (type I, see fig. 20). This is probably due to such an increase of the number of lateral cracks per unit scratch length with increasing load or with decreasing speed that new cracks can be formed in the material enclosed by the preceding crack[3]. This new crack will follow a plane of maximum tensile stress which starts at the proceeding particle and terminates perpendicular to the stress free surface of the preceeding crack[3]. Chips can also be formed due to lateral cracks proceeding in front of the scratching particle [3] (type II, see fig. 20). In that case a bending moment is exerted on the material enclosed by the crack. This means that a zone of tensile stress is present in the vicinity of the end of the crack, which enables the formation of a new crack perpendicular to the original crack surface, resulting in a chip. In both cases it is likely that also residual stresses due to the plastic grooving contribute to the crack formation. For both types of chip a mixed mode of cracking is therefore not unlikely. Chips are thus less suitable for a simple analytical description with linear elastic fracture mechanics.

1.3 Median Cracking

It has been found [10] that during scratching of glass by means of a square pyramid with a leading edge a crack (the median crack) can be formed perpendicular to the surface and through the scratch direction (see fig. 9). This crack is comparable with the median half-penny-crack obtained during application of the load in indentation experiments on glass with square pyramids [12]. For the crack the expression

$$F_n = A K r^{3/2} \tag{1}$$

where Fn = the applied load,
K = the stress intensity factor
and r = the radius of the crack,
is found.

The cracks obtained during scratching are often not initiated from the deepest point of the groove but deeper in the material, which can be explained by a nearly isostatic state of stress directly underneath the indenter[10]. Mostly the cracks propagate to the deepest point of the groove shortly after passing of the indenter[7]. The cracks are probably initiated by plastic or viscous flow[10]. Peters' results show that a minimum load is required for initiation.

It can be calculated from Peter's experiments[10] that the depth of the cracks depends on the applied force to a power of about 2/3. This is in agreement with expression 1 and with experimental results on indentation cracking[12].

The depth of the cracks is experimentally found to decrease with increasing scratching speed[10].

Maan[11] has found median cracks in single crystalline MgO during scratching with a square pyramid with a leading plane. Also indentations in brittle isotropic polycrystalline materials often show median cracks in the direction of the diagonals[13]. It can therefore be expected that median cracks also develop during scratching with a leading edge of polycrystalline materials.

1.4. Subsurface Cracking

In glass median cracks are generally accompanied by subsurface cracks, which grow roughly parallel to the surface[10] (see figs 8c and 9). A kind of subsurface cracks is also found by Maan[11] after scratching of single crystalline MgO. The subsurface cracks are probably also initiated by plastic flow at some distance from the deepest point of the groove. They initiate behind the scratching point[7]. They are comparable with the so called lateral cracks obtained in indentation experiments in glass during load relief[4]. Probably these cracks also develop under the influence of the stress field of the scratching point and of the residual stresses, which stand perpendicular to the scratch direction. Therefore the subsurface cracks may show a mixed mode of cracking. For this reason they are also less suitable for an analytical description.

From this review the question arises if different crack types develop under different conditions. Results of the appearance of cracktypes as a function of load and scratching speed are given in section 4.1.

It can be concluded that two types of crack seem to be reasonably suitable for a describtion with the aid of linear elastic fracture mechanics:the lateral and the median cracks. In contrast with the median cracks (see expression 1), such a describtion has not been given for the lateral cracks. In section 2 simple relations are derived for both types of crack between applied force, stress intensity factor, crack size and,unlike relation 1, groove width.

Also in contrast with the median cracks the growth of the lateral cracks as a function of applied load and scratching speed is, to our knowledge, not experimentally investigated. Section 4.2. gives such experimental results for single crystalline MgO, polycrystalline Mn Zn ferrite and a glass. With the aid of these results, stress intensity factors as a function of scratching speed have been calculated with the aid of the relations obtained in section 2. These results are also given in section 4.2. This section also reports about the number of lateral cracks per unit scratch length as a function of applied load and scratching speed.

The results are discussed in section 5.

2. THEORY OF THE FORMATION OF LATERAL AND MEDIAN CRACKS

Let an abrasive particle scratch the surface of a homogeneous and isotropic material in a certain direction at constant speed and at a given depth. It is assumed that the area of the groove perpendicular to the scratch direction increases proportionally with load, as observed in scratching hardness measurements with sharp cones or pyramids. Say that the cross-sectional shape of the groove is semicircular. At a certain location a crack nucleus is formed or comes within the region of sufficient tensile stress for lateral crack formation.For simplicity it is assumed that the squared length of the crack, as observed at the surface,is proportional to the surface area of the crack. We can then introduce a simple crack shape, say a cone, the curved surface of which makes an angle $90 - \delta$ with the scratch direction (see fig. 1.)

The initiation of the crack leads to a sudden decrease of the tangential force (see fig. 1). It is assumed that then the average tangential stress is no longer sufficient for ductile deformation, so that the particle must proceed by crack opening. For propagation of the crack an increasing amount of work has to be performed by the particle. At a given moment it is easier for the particle to proceed by ductile grooving, so that crack propagation stops. The work required during crack formation over a small distance dx is (see fig. 1)

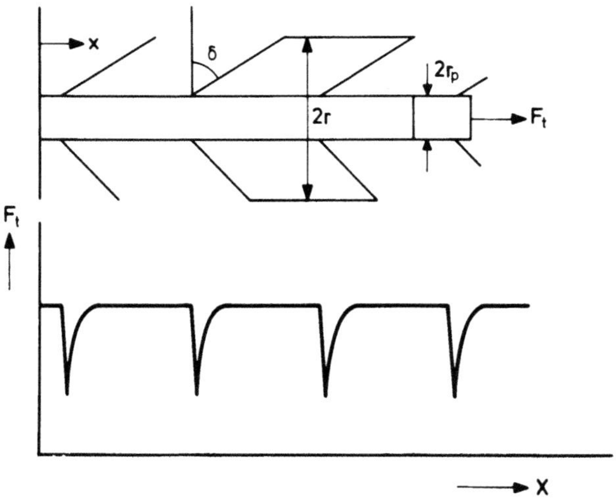

Figure 1 Schematic view from above of scratching particle and scratch. The expected force fluctuations are schematically drawn.

$$F_t\,dx = \pi r \frac{dr}{\cos\delta}\,G$$

where F_t = horizontal force on the particle,
r = half crack-width or maximum radius of the crack,
δ = angle between curved surface of the cone and cross-sectional plane of the scratch and
G = crack energy release rate.

At the moment that the transition takes place from crack expansion to ductile grooving the relation

$$F_t = \frac{1}{2}\sigma_c \pi r_p^2$$

applies (see fig. 1), where
r_p = half-width of the groove and
σ_c = average tangential stress on the particle when the deformation is plastic (σ_c is proportional to the dynamic hardness) or

$$\sigma_c \pi r_p^2\,dx = 2\pi r \frac{dr}{\cos\delta}\,G \qquad (2)$$

The original approach to crack formation given by Inglis and Griffith[15] is based a.o. on the assumption that the shape of

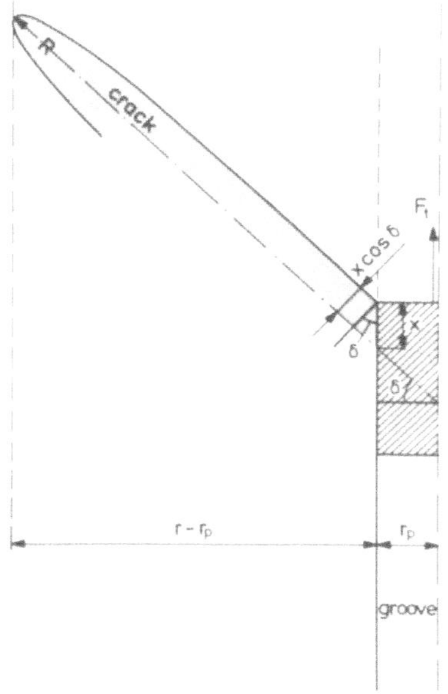

Fig. 2: Schematic view from above of scratching particle making a crack.

Fig. 3: View from above of scratch made on {100} plane in the ⟨100⟩ direction of a MgO single crystal, load 0·1 N, scratching speed 40μ/s. Incident light.

the cross-section of a crack is elliptical. When the stress intensity factor is kept constant during crack propagation it can be assumed that the degree of blunting of the crack tip is constant. For an idealised elliptical crack this means that the radius of curvature at the crack tip is constant during propagation. For the conical crack (see fig. 2) the relation between crack opening (short semi axis of the ellips), $x\cos\delta$, length of the crack (long semi axis), $(r-r_p)/\cos\delta$, and radius of curvature at the crack tip, R, can then be written as

$$x\cos\delta = A_3\sqrt{\frac{(r-r_p)R}{\cos\delta}} \tag{3}$$

where A_3 = a geometrical constant.

The relation between semi axis and radius of curvature at the tip of the expanding ellipse can also be given in such a way that the properties of the material are taken into account. It can be expected that the crack opening at a certain crack length during propagation will be proportional to the failure stress at the crack tip, σ_f, and inversely proportional to Young's modulus, E. When for simplicity a situation of plane stress is assumed, the following modification of expression (3) can be given

$$x = A_1\frac{\sigma_f}{E}\sqrt{\frac{(r-r_p)R}{\cos^3\delta}}$$

or

$$dx = \frac{1}{2}A_1\frac{\sigma_f}{E}\left[\frac{(r-r_p)R}{\cos^3\delta}\right]^{-\frac{1}{2}}\frac{R}{\cos^3\delta}\,dr \tag{4}$$

After substitution of this expression for dx in expression (2) we obtain:

$$r^2(r-r_p) = \frac{\sigma_c^2 r_p^4 A_1^2 \sigma_f^2 R}{16 G_1^2 E^2 \cos\delta} \tag{5}$$

For this situation of crack formation and for a crack length which is large with respect to the radius of curvature at the tip, a modification can be given of Neuber's[16] expression for the stress concentration

$$\sigma_f\sqrt{R} = A_2^1\sigma_c\sqrt{r-r_p}$$

where A_2^1 = a constant

In the right hand term the stress intensity factor K can be introduced

$$A_2^1\sigma_c\sqrt{r-r_p} = A_2 K$$

where A_2 = a constant

or $$R = A_2^2\frac{K^2}{\sigma_f^2}$$

This means that equation (5) can be transformed into

$$r^2(r-r_p) = \frac{\sigma_c^2 r_p^4 A_1^2 A_2^2}{16K^2 \cos \delta}$$

and because $\frac{1}{2}\pi \sigma_c r_p^2 = F_t$

$$F_t = AK\sqrt{r^2(r-r_p)}$$

where A = a constant for a particular scratching point.

An assumption in the derivation of this expression was, that groove and crack are similar in a cross-section perpendicular to the scratch direction during propagation. This is not, however, the case. Therefore expression (6) approximately has to be transformed into

$$F_t = AK\sqrt{r^2(r-Br_p)} \qquad (7)$$

where B = a constant.
A and B have to be found experimentally and are only valid for the particular particle geometry used in the experiment.

For a square pyramid with a leading plane, the tangential force F_t is given to a first approximation by[9]

$$F_t = F_n \cot \theta$$

where F_n = the applied normal force and
θ = the semi-angle of the pyramid.

The derivation of the expression describing the formation during scratching of the median crack is similar to that given above. That expression is

$$F_t = HK\sqrt{d^2(d-Cd_p)} \qquad (8)$$

where H, C = constants to be found experimentally,
d = the depth of the median crack and
d_p = the depth of the groove

Expressions 7 and 8, except for the correction factors Br_p and Cd_p, respectively, are similar to expression 1.

3. EXPERIMENTAL

The scratching experiments were carried out in an apparatus described elsewhere[4]. The apparatus was provided with a sharp square diamond pyramid (radius of curvature at the apex below 0.3 μm) with a semi-angle of 68°. The majority of the experiments was carried out with a leading pyramidal plane. Some additional experiments were carried out with a leading edge. During the experiments a constant load was present on the scratching point, which moved at constant speed. The loads

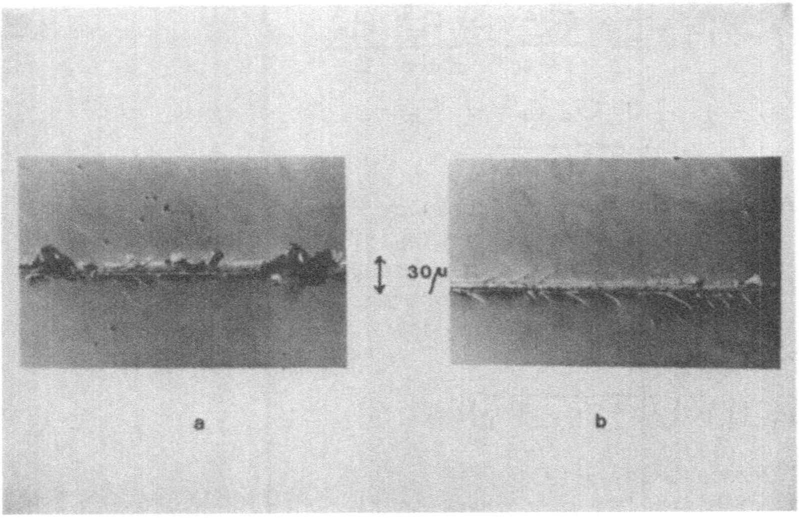

Fig.4: Influence of scratching speed on presence of crack types in crystalline materials. View from above of scratches made on Mn Zn ferrite. Load 0·5 N. Scratching speeds: 4(a) and 40 (b) μ/s. Incident light, interference contrast.

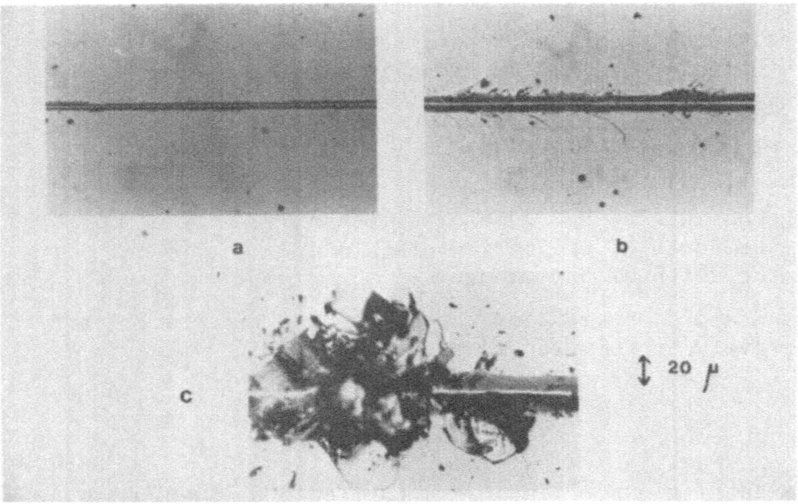

Fig. 5: Influence of load on presence of crack types in crystalline materials. View from above of scratches made on Mn Zn ferrite. Scratching speed: 400 μ/s. Loads: 0·1 (a), 0·7 (b) and 3 N (c). Incident light, bright field.

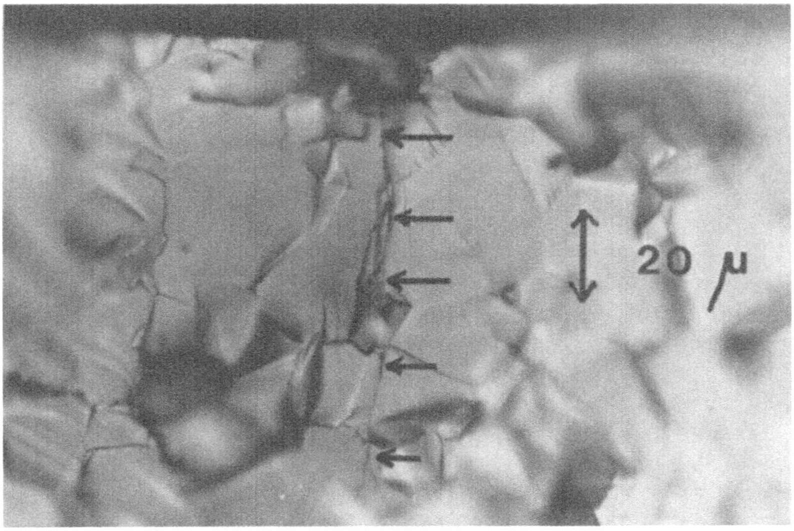

Fig. 6: Cross section of a scratch in Mn Zn ferrite. Clearly shown is the median crack. Load 1 N, speed 400 μ/s. Incident light, bright field.

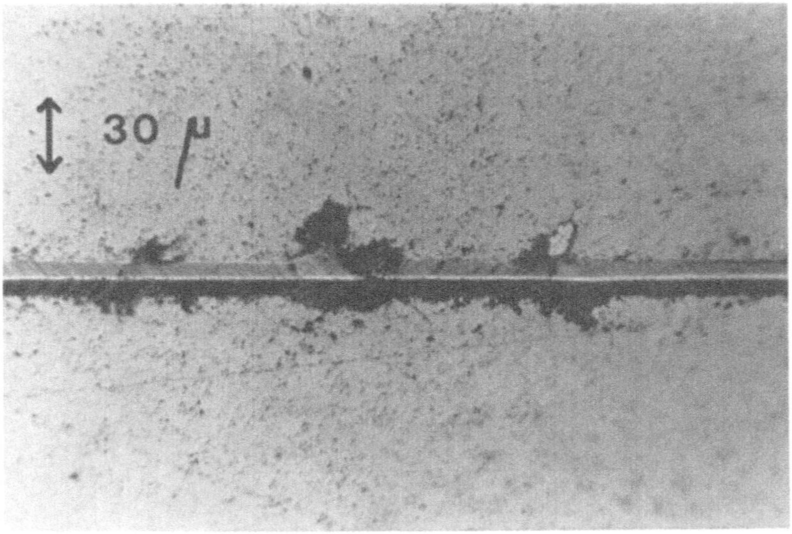

Fig. 7: View from above of scratch made on very fine grained Ni Zn ferrite. Load 1 N, speed 40 μ/s. Incident light, bright field.

used were between 0·05 and 5N. The speeds varied between 0·4 and 400μ/s. The scratch length was about 2 mm. For comparison some experiments were carried out at constant load on the glass in which speeds of roughly 0·5 m/s were obtained by hand. All the experiments were done at room temperature and at a relative humidity of approximately 55%. On one glass and on Mn Zn ferrite scratches were also made at 400μ/s in xylol.

After the test the average width of the groove ($2r_p$), the average crack width (2r) (see fig. 1) and the number of lateral cracks per unit scratch length were measured from micrographs. For the average crack width at least 20 cracks had to be present at 1·5 mm scratch length. It should be emphasized here that the measurement of r was sensitive to the optical technique used. For such a measurement, therefore, it is thus always necessary to specify the type of microscope, the objective, the optical technique used and the type of film. In our case we used a Reichert Zetopan with an Epi 28 (numeric aperture 0·55) or an Epi 11 (numeric aperture 0·25) objective in combination with the Reichert automatic camera and a Kodak plus x pan film for the quantitative measurements. For the measurements of the median cracks cross-sections of scratches were obtained by splitting the rectangular and plane parallel samples as done in the double torsion test.

The occurrence of the several types of cracks as a function of load and scratching speed was investigated by scratching a variety of glasses, coarse grained Mn Zn ferrite, very fine grained Ni Zn ferrite and single crystalline MgO. Of these materials the growth of the median cracks was only studied for one glass. The growth of the lateral cracks was studied in the same glass, in Mn Zn ferrite and in MgO. For the latter three materials we also determined static hardness and critical stress intensity factor (by means of the three-point bend test). Crack speed as a function of stress intensity factor was determined for the glass at a relative humidity of 55% with the double torsion test[17]. Data on the three materials are listed in Table I.

4. RESULTS

4.1. Development of Crack Types as a Function of Load and Scratching Speed.

For single crystalline MgO lateral cracks (fig. 3) are observed at any load and at any speed.

For polycrystalline Mn Zn ferrite one sees during scratching with a leading plane at a certain load (0·2 N) with increasing speed first chipping (0·4 μ/s) which gradually

Material	Chemical comp.	Fabr. Technique and structure	Indentation Hardness kgf/mm^2	K_{Ic} $10^6 N.m^{-\frac{3}{2}}$
MnZn ferrite	Mn 0.574 Zn 0.354 Fe 0.072 Fe_2O_4	continuous hot pressed (1350°C) and hot isostatically pressed (1280°C) grainsize 200 μm	620	1.1
Glass	in weight percent SiO_2(64.9) Na_2O_3(2.6) K_2O(6.2) CaO(1.8) SrO(0.2) BaO(12.3) MgO(1.1) Al_2O_3(3.2) Sb_2O_3(0.5) CeO_2(0.2)	pressed at 1020°C	1000	0.97
MgO	-	arc fusion technique single crystal	775 ({100} plane identation edge parallel with ⟨100⟩)	-

Table I

Data on chemical composition, fabrication, structure and mechanical properties of the materials quantitatively investigated.

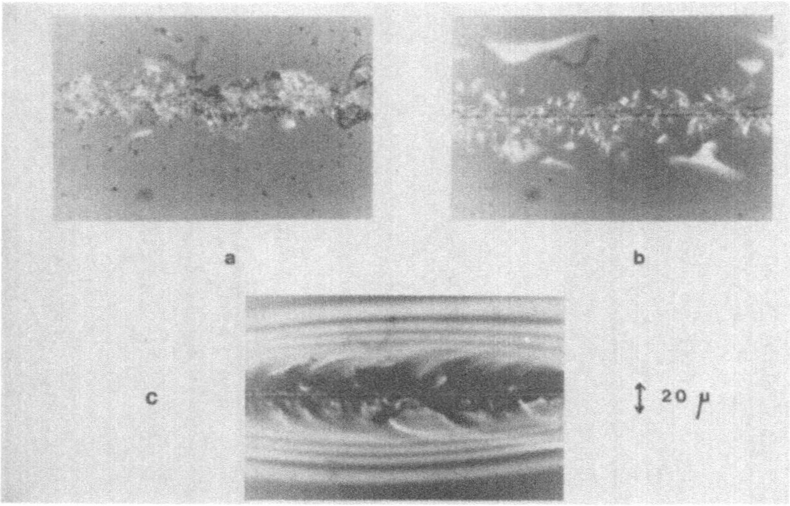

Fig. 8: Influence of scratching speed on the presence of crack-types in glass. Views from above of scratches made on BK7* glass. Load 0·2 N. Scratching speeds 0·4 (a), 40 (b) and 400 (c) μ/s. Incident light, bright field.

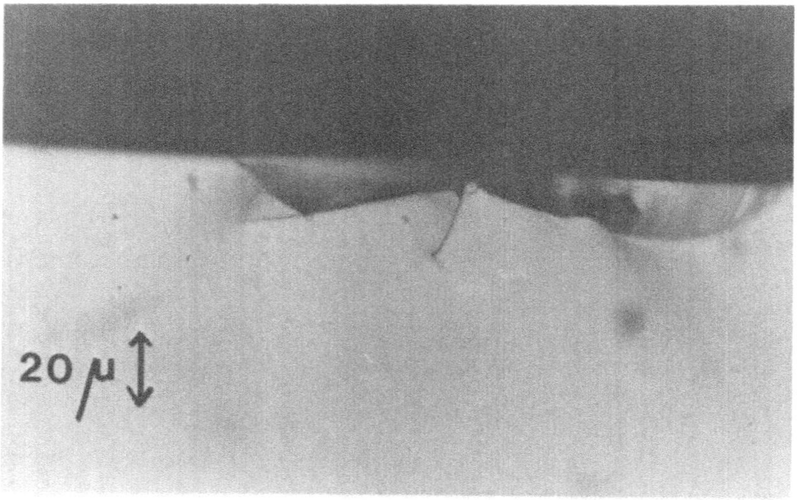

Fig. 9: Cross section of a scratch in glass. The median crack is clearly shown. The open space is bound by lateral and sub-surface cracks. Load 0·5 N scratching speed 400 μ/s. Incident light, bright field.

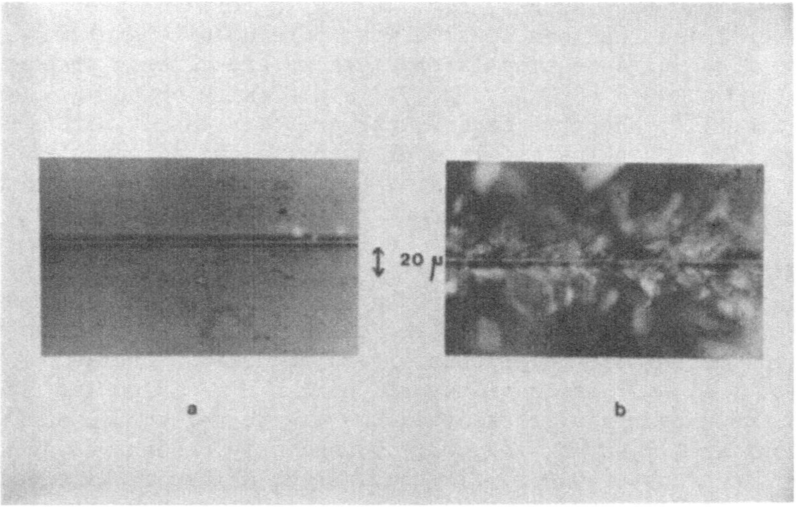

Fig. 10: Views from above of scratches made on soda lime glass, at a scratching speed of about 0·5 m/s. Loads 0·5 N (a) and 1 N (b)

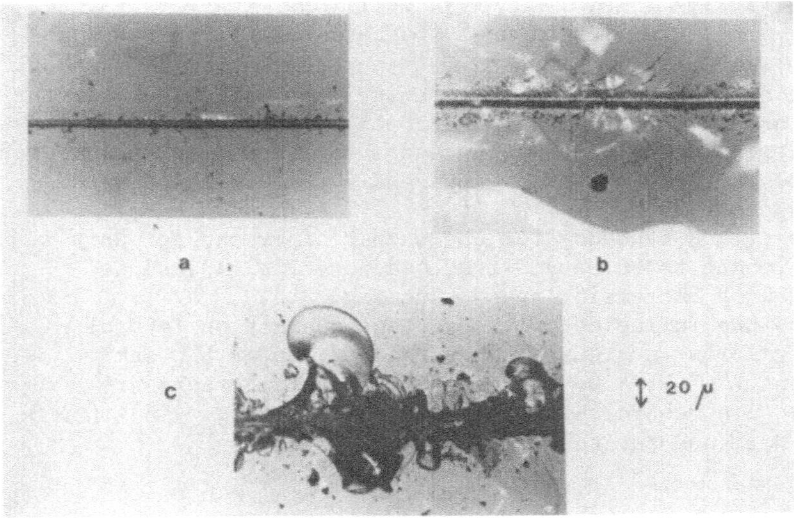

Fig. 11: Influence of load on presence of cracktypes in glass. Views from above of scratches made on SF58* glass, scratching speed 400μ/s. Loads: 0·1 (a), 0·5 (b) and 5 N (c). Incident light, bright field.

disappears (4µ/s, fig. 4a), while separate lateral cracks become visible (40, 400µ/s, fig. 4b). At a certain speed with increasing load one sees: almost pure ductile grooving (0·1 N, fig. 5a, some tiny lateral cracks are already present), lateral cracking (0·7 N, fig. 5b) and finally chipping (3 N, fig. 5c). When the experiments are carried out with a leading edge, chipping and lateral cracking are accompanied by median cracking (1 N, 400µ/S, fig. 6). For the coarse grained Mn Zn ferrite the cracks are transgranular (see fig. 4a, 4b) for a fine grained material (Ni Zn ferrite, grainsize 0·5µ, fig. 7) intergranular.

In experiments on glass with a leading plane one observes at a certain load (0·2 N) with increasing speed the successive formation of brittle chips plus irregular median cracks (0·4µ/s, fig. 8a), separate lateral cracks plus irregular median and subsurface cracks (visible due to reflection at the crack surface; 4,40µ/s, fig. 8b, 9) and only smooth median and subsurface cracks (400µ/s, fig. 8c). When, at the final speed, the force is increased, again lateral cracks are observed. This behaviour is maintained at much higher speeds (0·5 m/s). Only the subsurface cracks have almost dissappeared (0·5 N, fig. 10a). At a load of 1 N, lateral, median and subsurface cracks again appear (fig. 10b). With increasing load at a certain speed (400µ/s) the sequence is: almost only grooving (0·1 N, fig. 11a), smooth median and subsurface cracking plus lateral cracking (0·5 N, fig. 11b) and chipping almost without median and subsurface cracking (5 N, fig. 11c). In figure 12 the appearance of crack types as a function of load and scratching speed is given schematically. This behaviour appeared to hold for about half the number of glasses qualitatively investigated. For the other glasses in one scratch under the same conditions for instance both chipping and smooth median and subsurface cracking or even only grooving were observed. Apparently those glasses were inhomogeneous.

There is a tendency for the number of cracks per unit scratch length to increase with load (see fig. 13)and to decrease with increasing speed (see fig. 14).

Also the influence of the surface quality on lateral crack formation in glass is investigated. Under the same conditions (0·5 N, 400µ/s) much more lateral cracks are observed on a mechanically polished surface (fig. 15b) than on a flame polished surface (fig. 15a).

4.2. Crack Length

The length of the median cracks obtained during scratching of glass with a leading plane were about a factor 2 lower than those found by Peter[10], who scratched with a leading edge.

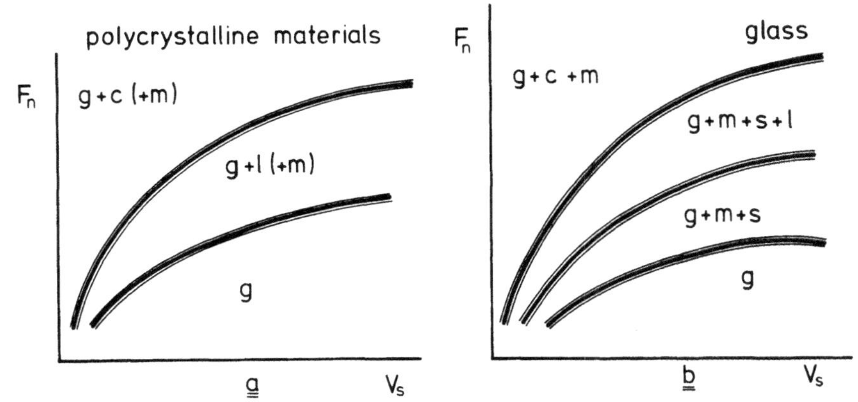

Fig. 12: Schematic representation of the influence of load and scratching speed on the appearance of cracktypes in polycrystalline materials (a) and in glass (b). g = grooving, c=chipping. l, m and s: lateral, median and subsurface cracking, respectively. In brackets: additional cracking in the case of leading edge.

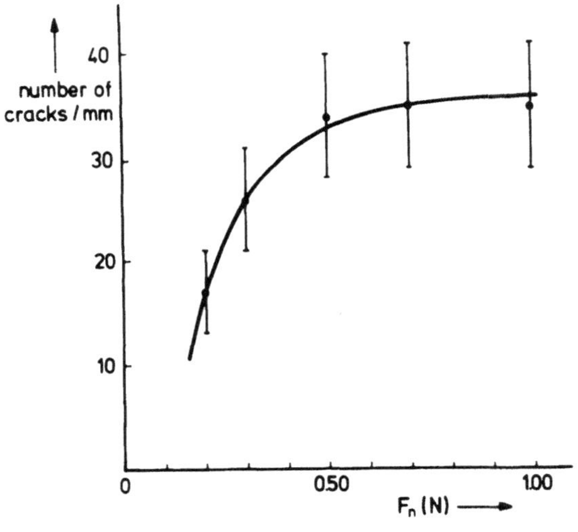

Fig. 13: number of cracks per unit scratch length as a function of load for Mn Zn ferrite. Scratching speed 400 μ/s.

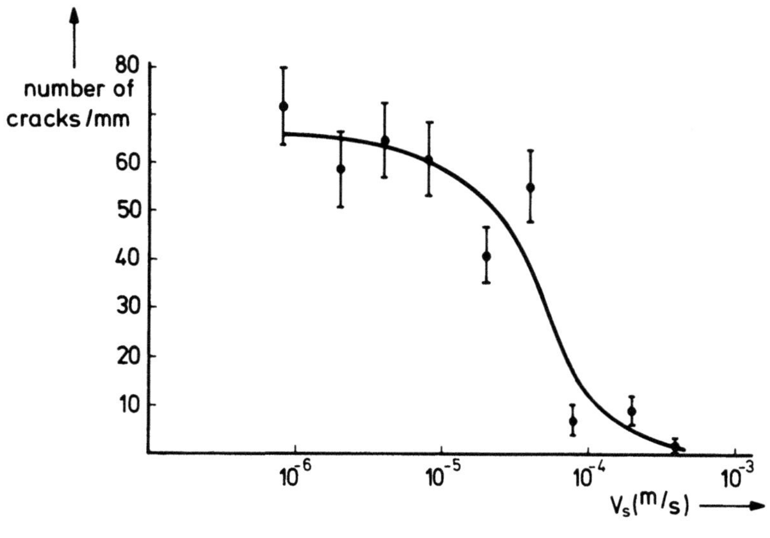

Fig. 14: Number of cracks per unit scratch length as a function of scratching speed for the glass. Load 0·2 N.

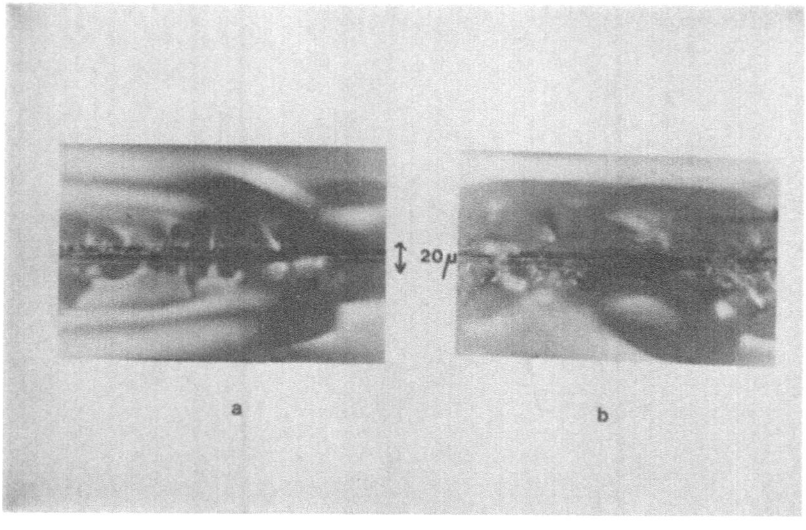

Fig. 15: Views from above of scratches made on soda lime glass with flame polished (a) and with mechanically polished (b) surface. Scratching speed 400 μ/s load 0·5 N.

The overall results were qualitatively in agreement with his findings.

For scratching with a leading plane the average width of the lateral cracks at the surface as a function of applied normal load is given for the Mn Zn ferrite at a scratching speed of 400 μ/s and at a humidity of about 55% in figure 16. At this speed too few cracks of a measurable length were present at 0·1 N. The width increases with normal force to a power of about 2/3, as expected from expression 7. From figure 16 the value of the constant B in expression 7 was calculated and was found to be - 12·4.

For glass and Mn Zn ferrite the type of cracking changes with scratching speed (see section 4.1). Thus in order to maintain one type of cracking during variation of the scratching speed the load has to be changed as well. This was not the case for single crystalline MgO. The crack width 2r as a function of scratching speed at a load of 0·1 N and at a humidity of about 55% is given for this material in figure 17.

It is well known that water plays an important role in the crack formation of glass. When water is avoided for instance by soaking the sample in water free Xylol, it can be expected that at 400μ/s the stress intensity factor of the lateral cracks for a glass, soaked in Xylol, must be near to the critical stress intensity factor K_{Ic}.

Another essential parameter for the calculation of K from the scratching result is the value of r_p which appears to decrease with increasing scratching speed. Diagrams giving hardness versus scratching speed are presented in figure 19 for the glass and for Mn Zn ferrite. For MgO only the indentation hardness at deformation speeds of 0·0012 (775 kgf/mm^2) and 0·12μ/s (970 kgf/mm^2) and the scratching hardness at 400μ/s (1600 kgf/mm^2) could be obtained. Values of the hardness at scratching speeds in between the two last speeds were estimated with the aid of the known variation of scratching hardness with speed of the other two materials.

Apart from the scratch experiment on glass in xylol K_{Ic} was determined separately for the glass with the three-point bend test. Comparison of the results of these two experiments gave the value of the constant A in expression 7, which was 2·24. With the same expression K_I as a function of scratching speed could now be calculated. These results are given in figure 18, for the glass, Mn Zn ferrite and MgO. For the Mn Zn ferrite and MgO it was indeed found that the calculated values of K_I were independent of load. For the glass, however, K_I weakly increased with load.

For comparison also crack speed versus stress intensity factor was measured for the glass in air at a humidity of 55% with the double torsion test. The results of this experiment are given in figure 18a.

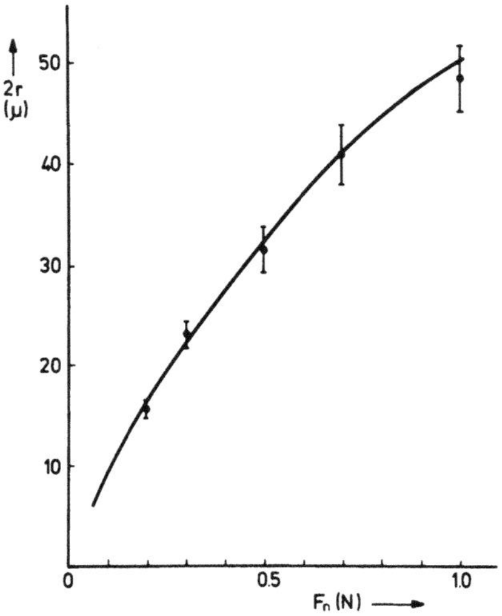

Fig. 16: Crackwidth (2r) as a function of applied load (F_n) for Mn Zn ferrite.

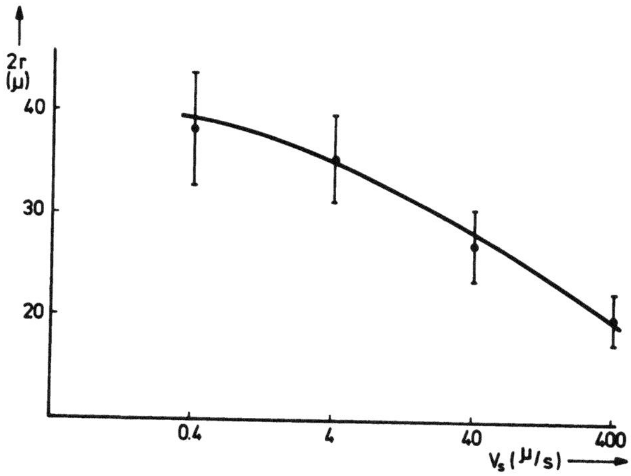

Fig. 17: Crackwidth (2r) as a function of scratching speed (V_s) for MgO.

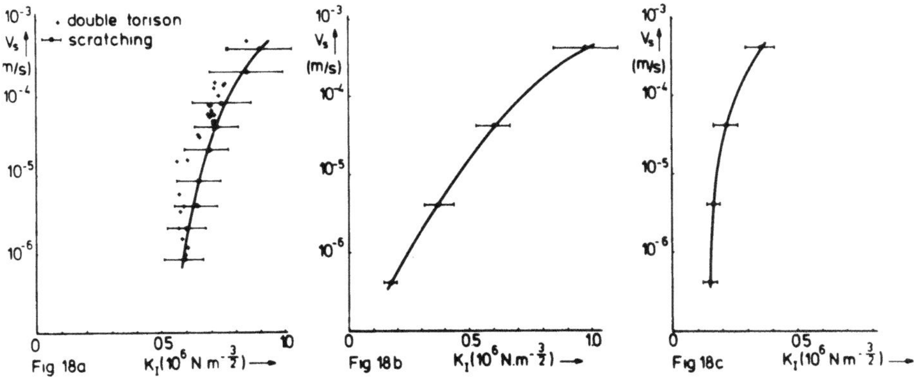

Fig. 18: Scratching speed (V_s) as a function of stress intensity factor (K_I) for the glass (a), for Mn Zn ferrite (b) and for MgO (c). Inserted in figure 18a are data of the stress intensity factor at a certain crack speed obtained from the double torsion test.

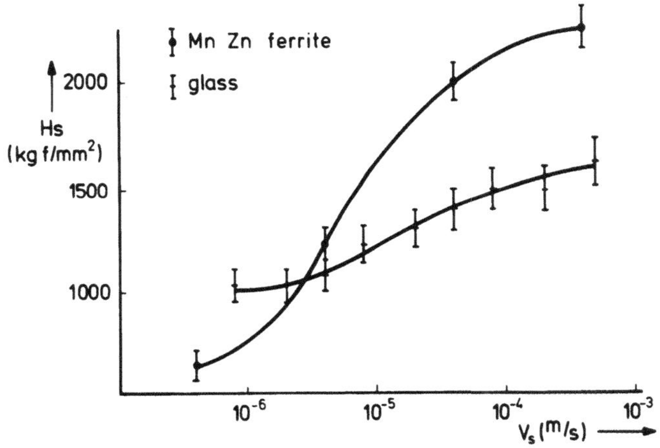

Fig. 19: Scratching hardness as a function of scratching speed (V_s) for glass and Mn Zn ferrite.

The value obtained for K_I at a scratching speed of 400μ/s for Mn Zn ferrite in air at a humidity of 55% is also clearly lower than K_{IC} which was measured with the aid of the three point bend test (table I). Therefore also on the ferrite a scratch experiment was carried out in Xylol at 400 μ/s. However, instead of the expected increase in K_I, a decrease was found.

5. DISCUSSION

5.1. Development of Crack Types as a Function of Load and Scratching Speed.

5.1.1. Crack initiation

It can be concluded from Peter's results[10] on median and from our results on lateral cracking (section 4.1) that a minimum load is required for initiation of the cracks. Such a minimum load also follows from relations 7 and 8. At the minimum load ($F_{t_{min}}$) $r=r_p$, thus expression 7 becomes

$$F_{t\,min.} = A K \sqrt{r_p^3 (1-B)} \qquad (9)$$

The scratching hardness H_s can be defined as

$$H_s = \frac{F_n}{J' \pi r_p^2},$$

where J' = a constant,

so that $$r_p = J \sqrt{\frac{F_{t\,min}}{H_s}},$$

where J = a constant.

After substitution of r_p in expression 9 the expression

$$F_{t\,min} = L \frac{K_I^4}{H_s^3}, \qquad (10)$$

where L = a constant for a particular geometry and orientation of the scratching point and for a particular crack type, is obtained.

When scratching Mn Zn ferrite with a leading plane it is found experimentally (section 4.1) that for the initiation of median cracks a higher force is required than for the lateral cracks. The reverse is found for glass. For lateral cracking it is also found experimentally that $F_{t\,min}$ is much higher for glass than for Mn Zn ferrite. For median cracking the reverse is found. With the aid of expression 9 and the results of figures 18 and 19 it can be calculated that $F_{t\,min}$ ought to be mucht higher for the glass than for the Mn Zn ferrite for median and lateral cracking as well. Apparently a basic difference is present between the deformation during scratching of a glass and of a crystalline material. Peter[10] has suggested that densification of the glass plays a role in median crack formation. The degree of densification, and probably the magnitude of the residual stresses, depends on the composition and on the thermal history of the glass[19].

When the experiments are carried out with a leading edge the measured value for $F_{t\ min}$ for median cracking in Mn Zn ferrite is much lower (section 4.1), which can be seen as a lower value for the constant L with respect to scratching with a leading plane.

5.1.2. <u>The number of lateral cracks as a function of load and scratching speed</u>.

Because the hardness is only weakly dependent on applied load[9] it can be assumed that the stress and strain fields remain roughly similar with increasing load. For cracks initiated by plastic strain it can thus be expected that the number per unit scratch length is roughly independent of applied load. However, for polycrystalline Mn Zn ferrite and also for the glass an increase of the number with increasing load (fig. 13) is found. For the greater part this increase can be explained by the increase of the number of preexisting flaws which come into a growing region of sufficient tensile stress. The importance of the preexisting flaws for crack initiation is illustrated by the difference in the number of cracks found under the same conditions on a mechanically polished surface (fig. 15b) and on a flame polished surface (fig. 15a). The fraction of the cracks initiated by plastic deformation is thus dependent on the load but also on the type of material. For single crystalline MgO probably all the cracks are initiated by plastic deformation[22].

At high loads the derivative of the number of cracks as a function of load decreases and even tends to become equal to zero (fig. 13). When at such a high load a large distance had been present between the scratching point and the last lateral crack, a preexisting flaw of a sufficient stress concentration factor should have been developed into a crack. However, when the distance is small it becomes important to consider that, in contrast with an assumption made in section 2, also during grooving the load on the scratching point still contributes considerably to the growth of the last lateral crack. As a result the magnitude of the stress required for initiation of a new lateral crack on the same flaw might be too low. This means that with increasing load, a decreasing fraction of all the preexisting flaws, developes into cracks, with as a result a decreasing derivative of the number of cracks as a function of load.

The increase of the number of lateral cracks per unit

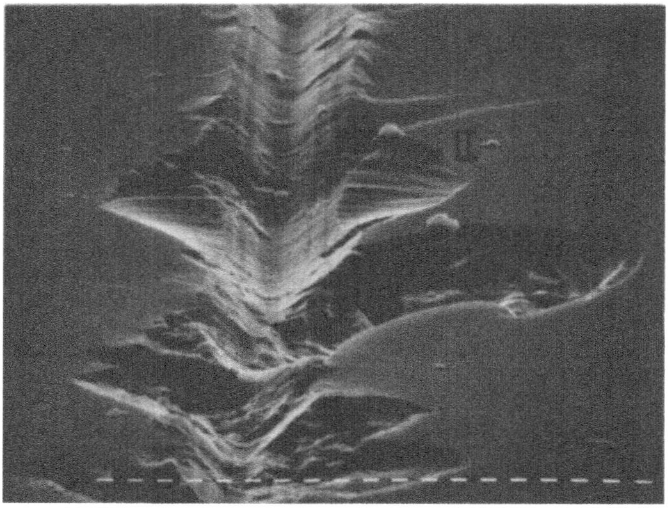

Fig. 20: Scratch made on MnZn ferrite. The two types of chip are marked with I and II. Load 0.2 N, scratching speed 4 μm/s. Scanning electron microscope.

scratch length with increasing load finally leads to the phenomenon of chipping (see fig. 20). A description of chip formation is already given in section 1.2.

The number of lateral cracks per unit scratch length and thus the chance for chipping to occur also increases with decreasing scratching speed. When the scratching speed decreases also the crack speed decreases (section 1), which means that the stress intensity factor decreases[18] (fig. 18). For this reason cracks can be initiated on preexisting flaws of a lower stress concentration factor, which increases the number of lateral cracks per unit scratch length. Above a certain scratching speed (crack speed) the stress intensity factor becomes roughly constant, which means that the number of lateral cracks per unit scratch length should become roughly independent of scratching speed.

5.1.3. The subsurface crack in glass and its interaction with other cracks.

The median cracks in glass are often accompanied by subsurface cracks, sometimes also in combination with lateral cracks. The subsurface cracks too, are probably initiated by plastic deformation at some distance from the lowest point of the groove[10]. The formation and the extent of the subsurface cracks in glass is likely to be connected with the densification which accompanies the plastic deformation underneath the scratching point[10]. This dentification probably leads to relatively high shrinkage stresses perpendicular to the scratch direction. After initiation the subsurface cracks propagate over a considerable distance parallel and perpendicular to the scratch direction. This is illustrated in figure 21, where a groove is contained by a completed subsurface crack. Their dependence on scratching speed, applied load, composition and thermal history of the glass and time after passing of the indenter is not yet clear.

The reason for the disappearance of the subsurface cracks at high scratching speeds (0.5 m/s) might be found in the relatively high temperature in the vicinity of the indenter, which leads to a reduction of the densification [19] and so of the height of the residual stresses.

In contrast with crystalline materials the number of lateral cracks does not increase gradually with decreasing scratching speed, but shows a rather sharp discontinuity (fig.14). Down to that scratching speed the stress relief due to subsurface cracking is apparently so high that almost no lateral cracks can be formed, which should mean a relatively high value for L in expression 10. With deceasing scratching (crack) speed the stress intensity factor for lateral cracking decreases, due to water that can reach the cracktip. This is,

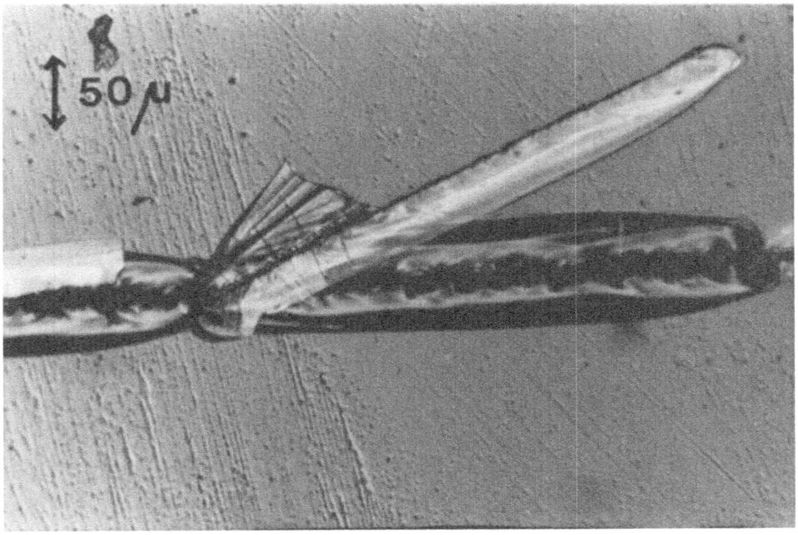

Fig. 21: View from above of scratch made on Suprasil W* glass at 400μ/s and 0.5 N. Broken out subsurface crack. Incident light.

however, not the case for the subsurface cracks. These cracks are formed behind the scratching point, where the median crack is closed, so that much less or no water can reach the crack tip. As a result the lateral cracks will be present below a certain scratching speed. Above that speed lateral cracks can be formed by increasing the load according to expression 10. As in the case of crystalline materials the increase of the number of lateral cracks with decreasing scratching speed, finally leads to chipping. During chipping almost no subsurface cracks can be observed. Apparently chips lead to sufficient relief of the shrinkage stresses due to densification, so that no subsurface cracks will develop.

5.2. Crackgrowth

Expression 8 is in agreement with Peter's[10)] experimental result that the depth of the median crack in glass depends on the applied force to a power of about 2/3. The average crack speed in the scratch direction must be equal to the scratching speed. However, during intervals much higher or lower speeds can of course be present. The depth of the median cracks clearly depends on the scratching speed and thus on the average crack speed, which means that the stress intensity factor depends on crack speed. The median cracks are thus apparently also sensitive to water, which means that the crackfront is in open contact with the surface.

The width of the lateral cracks also increases with load to a power of about 2/3 as expected from expression 6. Though the lateral cracks do not remain similar during propagation the constant B in expression 7 could be calculated within the error bands given in figure 16. The calculation of K_I as a function of scratching speed from the results on lateral cracking is only valid for homogeneous and isotropic materials, and when only lateral cracks are formed . The results obtained for single crystalline MgO, being anisotropic and having pronounced easy deavage planes, are thus not entirely comparable with the results obtained for glass and Mn Zn ferrite. In the case of the glass the lateral cracks are formed together with median and subsurface cracks. The interaction of these cracks probably leads to the weak depence of K_I on load as observed in section 4.2.

As in the case of standard fracture mechanics experiments it can be observed that cracks propagate through the grains in coarse grained materials (fig. 4) and along the grain boundaries in fine grained materials (fig. 7).

An explanation for the observed influence of water on crack formation in oxyde materials might be found in the formation of hydrous compounds (in glass: hydrous alkali silicates[20] at the freshly created surface near the crack tip and in the diffusion of ions (in glass: alkali ions[20] in the material. The stress for crack propagation at this surface will be relatively less. Moreover such a process is generally accompanied by a volume increase which leads to a certain prestress at the crack tip.

The initiation and partial propagation of the crack in plastically deformed material in the case of scratching are confusing for the comparison of these cracks with cracks obtained in standard fracture mechanics experiments. A crucial question is therefore: does this state of work hardening in brittle materials alter the crack energy release rate? In contrast with metals, highly brittle materials like sapphire do not show dislocation movement during crack propagation[21]. Therefore the crack energy release rate of cracks formed during scratching of highly brittle materials probably does not differ considerably from the release rate of cracks obtained in standard fracture mechanics experiments. As soon, however, as dislocation formation and movement play a role during crack formation, the energy release rate of cracks obtained during scratching may be expected to be different from cracks obtained in standard fracture mechanics experiments.

6. CONCLUSIONS

1. During scratching of polycrystalline materials in air with the leading plane of a square pyramid and with increasing speed, first chips are observed and then separate lateral cracks appear. With increasing load one sees successively pure ductile grooving, lateral cracking and finally chipping. In the case of a leading edge median cracks accompany chips and lateral cracks.
2. During scratching of glass in air for both orientations of a square pyramid and with increasing speed one sees successively chips plus irregular median cracks, lateral cracks plus irregular median and subsurface cracks, and finally only median and smooth subsurface cracks. With increasing load the sequence of events is: ductile grooving, smooth median and subsurface cracks, irregular median and subsurface cracks plus lateral cracks and finally chips plus irregular median cracks.
3. Lateral cracks in single crystalline MgO are observed at any speed and at any load. Minor median and subsurface cracks are also formed.
4. The lateral and median cracks will probably obey mode I of cracking. The growth of chips and of subsurface cracks will most probably show a mixed mode (I and II) of cracking.
5. The shrinkage stresses due to densification during scratching of glass probably give rise to the formation of subsurface cracks and contribute to the growth of the median cracks.
6. The extent of the lateral and median cracks depends on the applied load to a power of about 2/3. However, a correction for

the groove width has to be taken into account.
7. Scratching speed and average speed of the lateral cracks are of the same order of magnitude, provided that the pop-in length of the lateral cracks is not too large. In that case diagrams giving scratching (crack) speed versus K_I can be made.
8. The number of lateral cracks per unit scratch length increases with increasing load and with decreasing speed.
9. The median and subsurface cracks in glass are probably initiated by plastic deformation. The median cracks in crystalline materials can possibly also be initiated on structural defects as pores and grain boundaries. Depending on load and type of material, lateral cracks are initiated by preexisting flaws and by plastic deformation as well.
10. Restrictions to a general application of scratching for the determination of stress intesity factors of brittle materials can be a short length of the cracks, the work hardening that takes place around the scratching point, the concealment of the lateral cracks by phenomena as chipping and the interaction of different types of crack.

ACKNOWLEDGEMENT

The authors wish to thank Dr. A. Broese van Groenou for stimulating discussions and Dr. J.M. Robertson for providing the MgO crystal.

* BK 7 Schott, Germany
* SF 58 Schott, Germany
* Suprasil W Heraeus, Germany

REFERENCES

1. B.R. Lawn, E.R. Fuller and S.M. Wiederhorn, J. Am. Cer. Soc. 58, (1976) 193.
2. S.M. Wiederhorn, Proc. 2nd army materials techn. conf. Hyannis, Mass. 1973, 633.
3. J.D.B. Veldkamp and R.J. Klein Wassink, Philips Res. Repts. 31, (1976) 153.
4. G.M. Hamilton and L.E. Goodman, Trans. A.S.M.E. J. of applied mechanics 33, (1966) 376.
5. J. Boussinesq, See: S.P. Timoshenko and J.M. Goodier: Theory of elasticity, Mc Graw-Hill, New York 1970, 398.
6. J.E. Gordon, D.M. Marsh and M.E.M.L. Parratt, Proc. Roy. Soc. (London) A24, (1959) 65.
7. D.M. Busch, Films E1328-E1330, Inst. Wiss. Film Göttingen, 1967.
8. B.R. Lawn, Proc. Roy. Soc. (London) A299 (1967) 307.

9. A. Broese van Groenou, N. Maan, and J.D.B. Veldkamp, Philips Res. Repts. 30, (1975) 320.
10. K. Peter, Glastechn. Ber. 37, (1964) 333.
11. N. Maan, private communication.
12. B.R. Lawn and E.R. Fuller, J. of materials sc. 10 (1975) 2016.
13. B.J. Hockey and B.R. Lawn, J. of materials sc 10 (1975) 1275.
14. B.R. Lawn and M.V. Swain, J. of materials sc. 10 (1975) 113.
15. A.A. Griffith, Phil. trans. Roy. soc., 221A (1920) 163.
16. H. Neuber, "Kerbspannungenslehre", Springer, 163, Berlin 1937.
17. J.A. Kies and B.J. Clark, Proc. Conf. on Fracture, Brighton, 1969, 483.
18. S.M. Wiederhorn in "Fracture mechanics of ceramics" R.C. Bradt, D.P.H. Hasselman and F.F. Lange eds, Plenum, New York, 1974 p 613.
19. J.D. Mackenzie, Proc. ICM I IV C56, Kyoto, 1971, 347.
20. R.J. Charles, J. Appl. Phys. 29 (1958) 1549.
21. S.M. Wiederhorn, B.J. Hockey and D.E. Roberts, Phil. mag. 28, (1973) 783.
22. C.A. Brookes and P. Green, Nature Ph. Sc. 246 (1973) 119.

IMPACT DAMAGE IN CERAMICS

A. G. Evans

Rockwell International/Science Center

Thousand Oaks, California 91360

ABSTRACT

The modes of impact damage that occur in ceramic materials are described, and the present status of damage analyses that define the dependence of the damage on the primary projectile and material variables is discussed. Finally, the implications of the damage characterization for the prediction of strength degradation and erosion are assessed.

1.0 INTRODUCTION

Most materials are subject to localized damage when impacted by a high velocity projectile. This damage can be particularly deleterious in ceramics, because it is often manifested as fractures which cause substantial strength degradation. Another potentially serious consequence of the impact process is material removal, or erosion. It is the intent of this review to examine the important modes of impact damage in brittle materials, and to describe the present status of anlayses for damage characterization (in each mode). The analyses are presented with an initial emphasis on elucidating the important properties of the target and projectile that dictate the extent of the damage, followed (where possible) by a quantification of the damage extension. The damage thresholds, being key aspects of the impact process, are afforded special emphasis. Finally, the implications of the damage analyses for the prediction of strength degradation and erosion are discussed.

2.0 DAMAGE MODES

Ceramic materials impacted by projectiles exhibit several damage modes, ranging from a fully-elastic region to a plastic region. The elastic response[1,2,3] refers to a condition of undetectable plastic deformation in the target, and obtains when the projectile is relatively soft, or compressible (e.g. water or nylon). The damage that derives from these exclusively elastic stresses is determined both by the material and the topography of the target surface. For a clean single phase material and a planar surface condition, the initial damage occurs as arrays of short circumferential cracks (Fig. 1) outside a central, undamaged zone. The crack lengths tend to increase and the crack densities to decrease with distance from the impact center[1]. The outer cracks are inclined at $\sim 50^\circ$ to the surface; while the inclinations of the inner cracks are substantially more variable and can approach $\sim 0^\circ$. The subsequent damage has the same characteristics, except that the extension of the cracks formed during prior impacts often exhibits additional morphological complexity, associated with the different stress wave, crack symmetries. For example, crack extension can procede along an entirely different trajectory (Fig. 2), or secondary cracks can be initiated that then propagate according to the symmetry of the current impact. For irregular (e.g. pre-eroded) surfaces, the crack propagations are more random in character; these propagations can also be manifested on a local scale as chipping.

For homogeneous or multiphase ceramics, removal modes of damage can be observed. Inclusions, second phase particles, or grains (especially in materials with substantial elastic anisotropy) can be ejected, partially or in total. The spatial distribution of the removal sites exhibits some systematic dependence on the adjacence to the impact center, but this dependence is relatively weak.

The opposite response extreme is a plastic response, in which the target only exhibits plastic accomodation of the impact, with no detectable fracture. This regime is prevalent in metal systems[4], but also occurs in ceramics at elevated temperatures and/or small projectile sizes. In this regime, depressions develop at the impact center, and lips usually form around the periphery[5]. These lips can be removed by subsequent impacts. Evidence of material removal by localized melting has also been presented[6]. Under non-normal incidence, the depressions can be much more expansive than those observed at normal incidence. Also, material removal by a plastic ploughing process has been described[4].

An intermediate (elastic/plastic) mode of damage involves localized plastic accomodation around the impact center and fracture

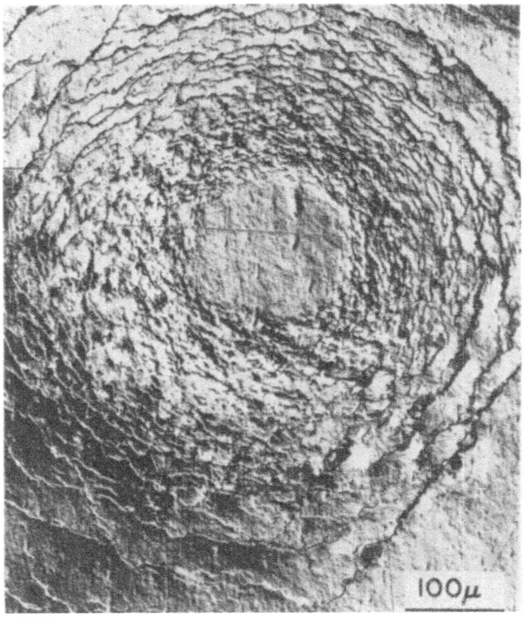

Fig. 1 Circumferential cracks produced by impact in the elastic response regime.

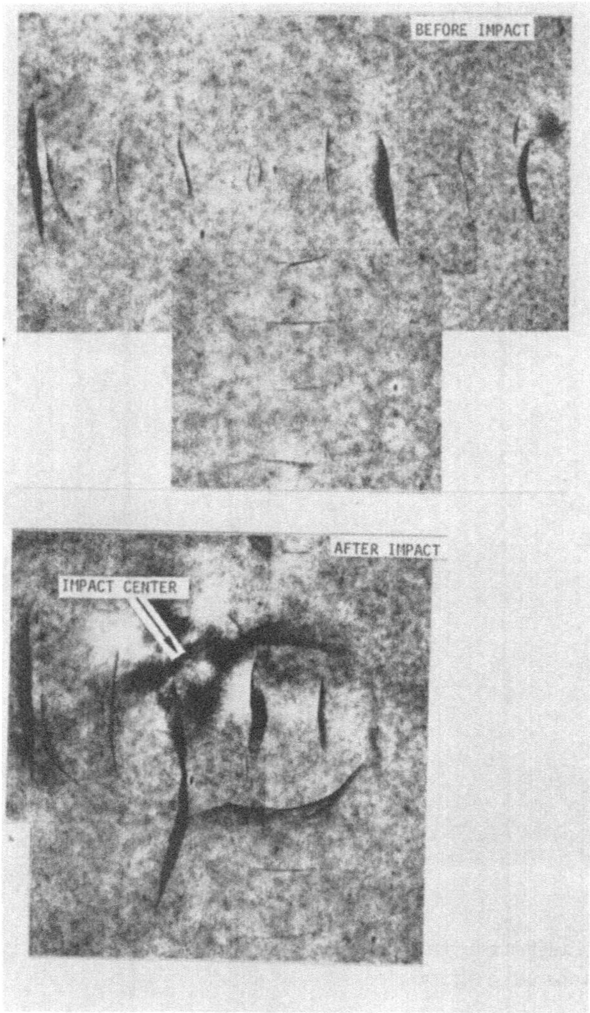

Fig. 2

The activation of pre-cracks in MgO by the stresses from a nylon sphere impact at 1100 ms^{-1}.

in both the elastic and plastic zones[1,8]. This type of damage occurs when the projectile is "relatively" hard‡ (or incompressible), and is a dominant mode of damage in ceramic systems for a wide variety of projectile conditions; it pertains to spherical or angular projectiles, planar or irregular surfaces, single phase or multiphase materials. The damage consists (in virtually all materials) of a central plastic impression and arrays of radial, conical and lateral cracks (Fig. 3a-c); although the plastic impression can be ejected at high velocity (Fig. 3d). The radial and conical cracks (the source of strength degradation) both exhibit a center of symmetry about the impact center. The crack morphologies are also strongly interactive, in that a prior radial crack (for example) can prevent the formation of a complete conical crack, and partial developments are frequently observed. The density of radial cracks tends to increase as the projectile velocity increases and as the target hardness or toughness decrease. The lateral cracks propagate roughly parallel to the surface, with a tendency for the near-surface cracks to eventually intersect the free surface to cause material removal. The lateral cracks are frequently observed to initiate at radial cracks (Fig. 3); also, their density increases as the target hardness diminishes. Additionally, there is a region below the fracture threshold, especially in very hard materials, where the formation of the plastic zone is accompanied by a small scale chipping mode of damage[8].

Additional damage modes can occur in multiphase materials, especially when the scale of the microstructure excedes the projectile diameter. Of special practical concern is the case where very hard particles are imbedded in a continuous, relatively soft matrix* (e.g. WC/Co alloys and castable refractories). In such materials, the softer phase can be susceptible to rapid material removal, causing the harder phase to project from the surface[9,10]. Thereafter, the hard phase can be sufficiently exposed that it falls-out or is ejected by a relatively mild subsequent impact.

3.0 DAMAGE ANALYSIS

3.1 Elastic Response Regime

For a fully elastic response to projectile impact most of the kinetic energy of the projectile is converted into plastic deforma-

‡The projectile hardness required for this damage regime has not been well-defined, but it is apparent that the regime still obtains when the projectile has a significantly smaller ($\lesssim 0.5$) quasi-static hardness than the target.

*The material "softness" can derive from porosity as well as from dislocation plasticity.

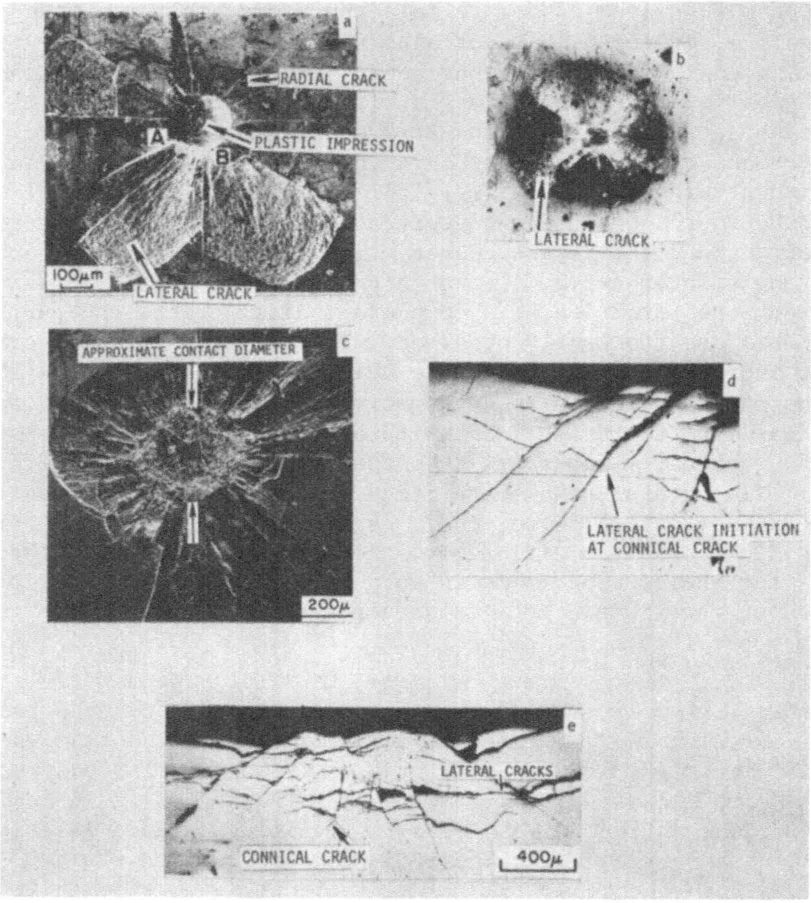

Fig. 3 Typical impact damage patterns observed in the elastic/plastic response regime: (a) is scanning electron micrograph illustrating the central plastic impression, the radial cracks and the lateral cracks that have caused material to be removed; (b) is a polarized reflected light micrograph of the same impact illustrating the symmetry of the lateral fracture; (c) is a scanning electron micrograph of a higher velocity impact showing the greater intensity of the lateral fracture and the partial removal of the plastic impression; (d) indicates the extent of the conical and lateral fracture; (e) shows the same impact but highlights the formation of lateral cracks from conical cracks.

tion energy (plastic work) in the projectile and elastic energy (which is dissipated at the sample boundaries) in the target; only a small fraction of the energy is manifested as fracture surface energy. Hence, analyses based on energy redistributions during impact have not afforded a successful description of the damage created in this region. Rather, the damage can be rationalized in terms of the dynamic tensile stresses generated by the impact and the criteria for the extension of cracks exposed to tensile stress pulses.

3.1.1 Damage Threshold. The important qualitative features of the tensile stress field generated by soft or compressible projectiles (e.g. water) on a pristine surface can be deduced from an analytic solution[11,2] for a zone of pressure impinging onto and expanding across a planar surface. This solution indicates that radial tensile stresses are generated near the impact surface behind the Rayleigh wave front, and that these stresses can be characterized by the dimensionless parameters[1]

$$\Omega = \sigma/p$$
$$R = 4v_\ell r/k^2 \qquad (1)$$
$$T = 4{v_\ell}^2 t/k^2$$

where σ is the stress amplitude, p is the impact pressure, v_ℓ is the longitudinal wave speed in the target, t is the time after the pressure is applied, r is the distance from the center of pressure application and k is a projectile dominated parameter, given approximately by;

$$k^2 \approx 2r_p v_o \qquad (2)$$

where v_o is the initial projectile velocity and r_p is the projectile radius. Solutions to this analysis[2] exhibit the characteristics depicted in Fig. 4; note the maximum in the peak tensile stress that occurs close to the pressure center. The important implication of this analysis for damage prediction are that: (i) the stress amplitude is dictated principally by the contact pressure (and hence, the projectile velocity and the Hugoniot of the projectile and target), (ii) the equivalent stress occurs at a distance from the pressure center that increases in proportion to the target wave speed, the projectile velocity and the projectile radius, (iii) the pulse duration at the equivalent position also increases in proportion to the projectile velocity and radius, but as the square of the target wave speed, a result that (as we shall see later) has important implications for damage creation. Another key aspect of the analysis is the steep gradient of stress that it predicts below the target surface (Fig. 5), and the implied depen-

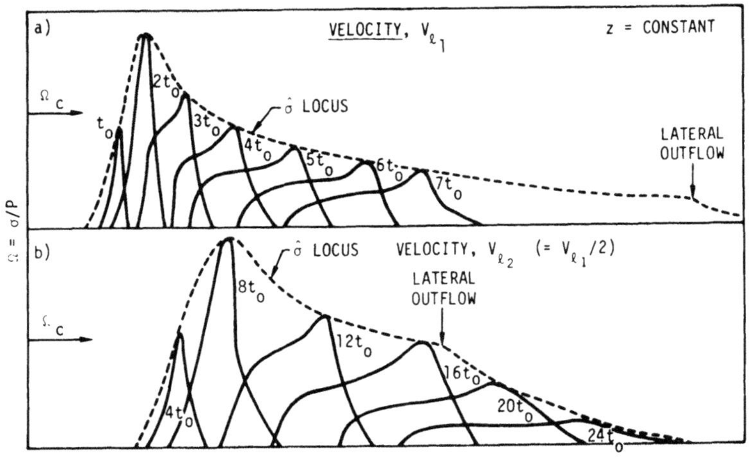

Fig. 4 The temporal and spatial development of circumferential tensile stresses generated by impact in the elastic response regime, for two values of the wave velocity.

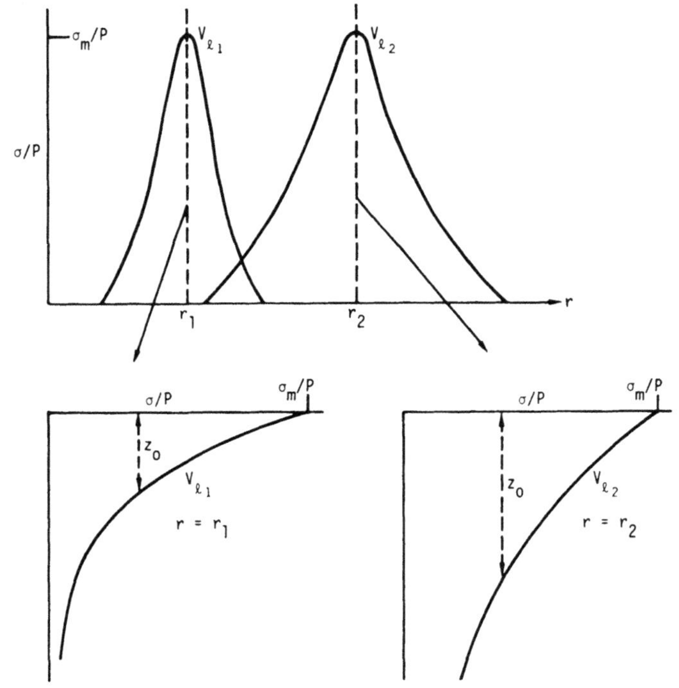

Fig. 5 A schematic indicating the locations of the maximum in the peak tensile stress and the corresponding stress gradients.

dence of the gradient on the target wave speed. Two potentially important features of the impact stress field that are not incorporated in the analytic solution are the behavior during lateral outflow of the projectile (which occurs for compressible projectiles such as water or nylon) and the details of the pressure distribution that obtain during impact. The onset of lateral outflow should coincide with a decrease in the contact pressure; thus, the peak tensile stresses generated after outflow should decline more rapidly than anticipated by Blowers' analysis (Fig. 4). The outflow process could suppress damage formation if it initiates before the maximum in the peak tensile stress can be attained. Since outflow occurs at a time after impact dictated essentially by the properties of the projectile, this possibility might obtain in materials with a low wave velocity. The impact pressure has recently been evaluated using a finite difference code[12], and has the characteristics summarized in Fig. 6. The pressure is not constant, but quickly increases to a maximum and then declines rapidly to the one-dimensional Hugoniot value. The effect of this pressure variation on the stress field has not yet been elucidated, but its short duration suggests that it could be of little consequence.

For less compressible projectiles, the target surface is elastically depressed during impact; the contact time is thus generally larger, and the stress field substantially different from the Blowers result. At very low impact velocities, the quasi-static Hertzian stresses can be approached. However, as shown by recent numerical analyses[12], the tensile stresses are generally larger than the quasi-static stresses (even though the contact pressure is similar[13]), and the tangential stresses tend to become comparable in magnitude to the radial stresses. The stresses are in fact similar to those developed, during penetration, in the elastic/plastic regime (Sec. 3.3).

The evolution of cracks within these tensile stress fields is accountable in terms of the dynamics of crack activation and growth. Recognizing that damage initiates from precursor cracks that exist in the immediate vicinity of the target surface, the growth of these cracks into observable damage is largely dictated by the crack growth curve for a semi-infinite crack[14,15], as plotted in Fig. 7; this curve can be expressed mathematically as[15];

$$\dot{a}/v_r = 1 - \frac{K_c^2\,(1-\nu^2)}{2\sigma_o^2 v_\ell\,[t^{1/2} - (t-t_o)^{1/2}\,H\,(t-t_o)]^2} \qquad (3)$$

where $\dot{a}$ is the crack growth rate, v_r and v_ℓ are the Rayleigh and longitudinal wave velocities respectively, ν is Poisson's ratio,

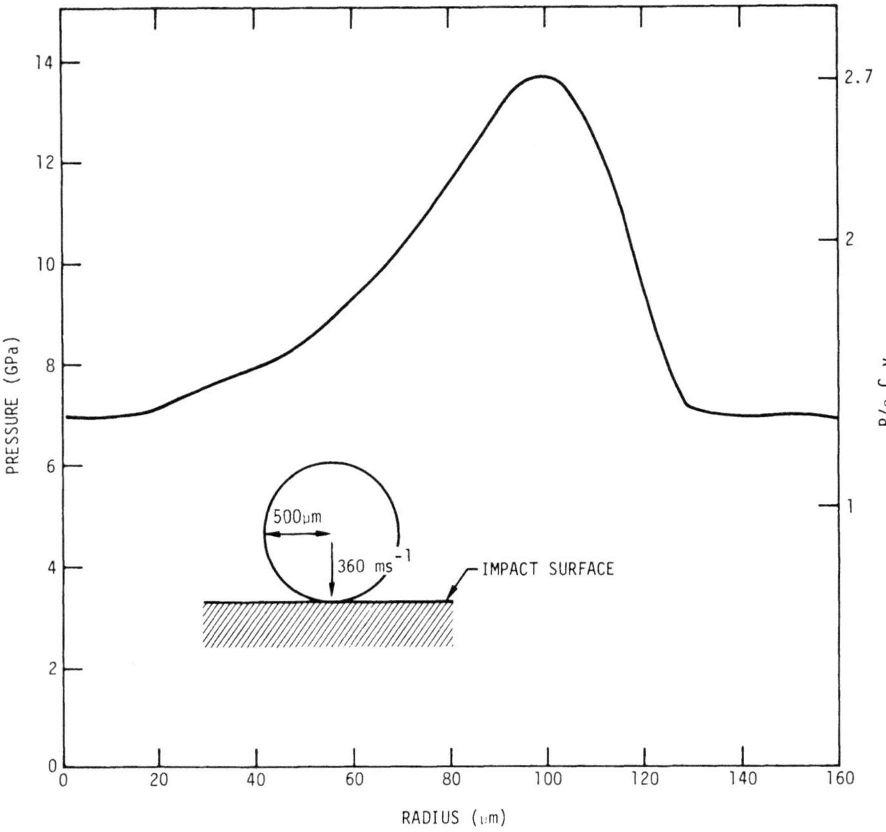

Fig. 6 A numerical computation of the impact pressure as a function of distance from the impact center.

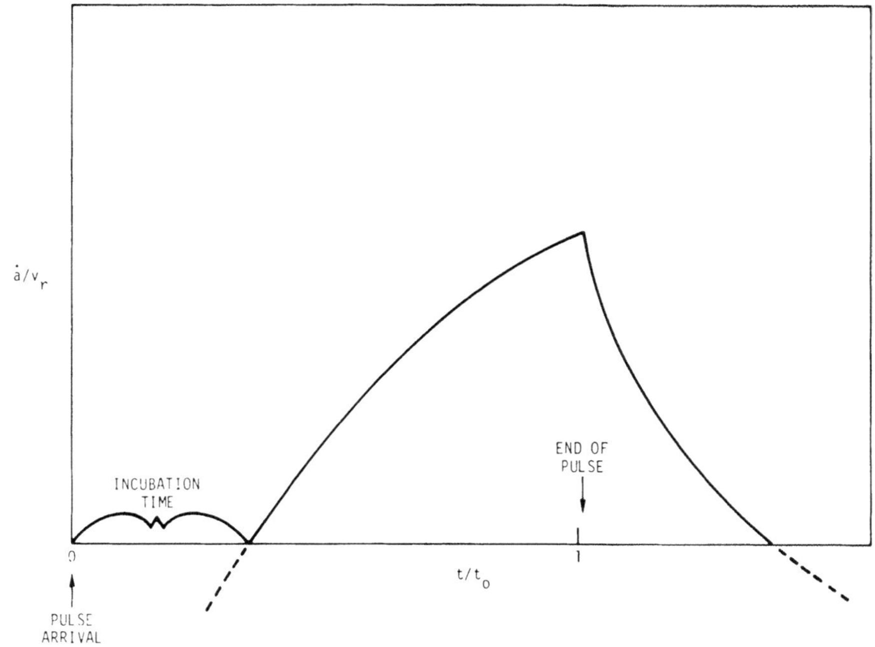

Fig. 7 A schematic showing the variation in crack growth rate for a crack subjected to a tensile stress pulse.

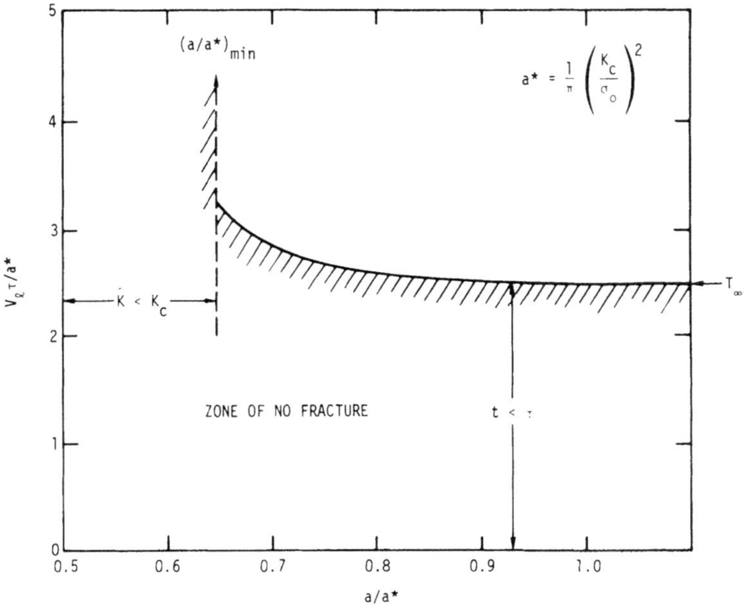

Fig. 8 The time and crack length thresholds for a finite crack subjected to a tensile stress pulse.

σ_o and t_o are the amplitude and duration of the (square) stress pulse and H is the unit step function. This relation indicates, firstly, that there is a crack growth incubation time τ given by[15];

$$\tau = \frac{\pi}{4} \frac{K_c^2}{v_\ell (1-\nu^2)\sigma_o^2} \tag{4}$$

This is similar to a result derived by Steverding[16]. Crack activation will thus be suppressed when the total pulse duration, $t_o < \tau$. This result can be expressed, using eqns. (1) and (4), in terms of a critical threshold stress σ_{oc}

$$\sigma_{oc}^2 = \frac{\pi}{T_o} \left[\frac{K_c^2 v_\ell}{k^2} \right] \tag{5}$$

where T_o is a material independent dimensionless pulse duration. The incubation time is relatively unchanged by introducing the concept of a finite crack[15,17]; but another important crack growth threshold does emerge; that of a minimum crack length a_{min}, given for a penny crack by[15,17];

$$a_{min} = \frac{\pi}{9} \left(\frac{K_c}{\sigma_o} \right)^2 \tag{6}$$

These thresholds are summarized in Fig. 8. The minor effect of the wave diffracted from the opposite crack tip on the stress intensification above the crack length threshold, indicates that the crack growth after incubation should not be very different from that anticipated by eqn. (3). There is thus a crack acceleration at $t > \tau$ given approximately by;

$$\dot{a}/v_r \approx 1 - \pi K_c^2 / 4\sigma_o^2 v_\ell t \tag{7}$$

Then, the crack decelerates at $t > t_o$,

$$\dot{a}/v_r = 1 - \pi K_c^2 / 4\sigma_o^2 v_\ell \; [t^{1/2} - (t-t_o)^{1/2}]^2, \tag{8}$$

and finally arrests. The crack increment after incubation Δa is thus;

$$\frac{\Delta a}{v_r \tau} = \frac{t_o}{\tau} - 1 - \ell n\left(\frac{t_o}{\tau}\right) + \frac{t^*}{\tau} - \frac{1}{4}\left(\frac{2\sqrt{t^*(t^*+t_o)}}{(\sqrt{(t^*+t_o)} - \sqrt{t^*})^2} \right) + \frac{1}{4} \ell n \left(\frac{\sqrt{t^*+t_o} + \sqrt{t^*}}{\sqrt{t^*+t_o} - \sqrt{t^*}} \right) = \Phi(t_o/\tau) \tag{9}$$

where t* is the crack arrest instant, given by;

$$\frac{t^*}{\tau} = \frac{1}{4}\left[\frac{t_o}{\tau} + 1\right]^2 \tag{10}$$

The dependence of the normalized crack increment $\Delta a v_r/\tau$ on the normalized pulse time, t_o/τ, is plotted in Fig. 9. By substituting for τ from eqn. (4) and t_o from eqn. (1), eqn. (9) can be conveniently expressed in terms of the material variables and the stress amplitude as;

$$\Delta a = \frac{\pi K_c^2}{8\sigma_o^2} \Phi\left[\frac{T_o\sigma_o^2}{\pi}\left(\frac{k^2}{K_c^2 v_\ell}\right)\right] \tag{11}$$

Another important aspect of crack activation concerns the presence of the steep subsurface stress gradient. This gradient can profoundly affect crack activation if a significant gradient exists over the depth of the precursors. By assuming an exponential stress gradient[15]

$$\sigma = \sigma_o e^{-z/z_o} \tag{12}$$

where z_o is a characteristic depth, it has been demonstrated that there are regions of stable and unstable crack growth (Fig. 10) - similar to the quasi-static Hertzian case[18] - and that there exists an <u>absolute</u> threshold stress for crack activation, σ_{th}, given by[15],

$$\sigma_{th}^2 \approx \lambda[K_c^2 v_\ell/k^2] \tag{13}$$

where λ is a constant.

It has not yet been possible to integrate the dynamic and stress gradient effects on crack activation, but the separate treatments indicate similar damage trends, as exemplified by the relationships for the fracture thresholds and the crack increment. A dominant damage parameter, D_1, is;

$$D_1 = \frac{K_c^2 v_\ell}{k^2} \tag{14}$$

This parameter influences both the time and stress gradient thresholds and the crack increment, such that the thresholds increase and the crack increment decreases as D_1 increases. A second damage parameter, D_2, is

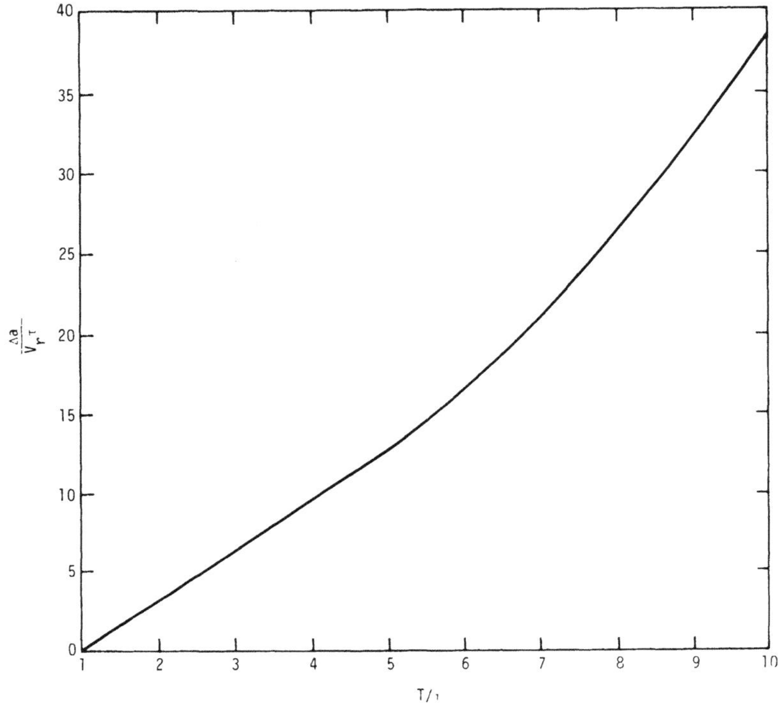

Fig. 9 The dependence of the crack length increment Δa on the duration of the stress pulse, t_o.

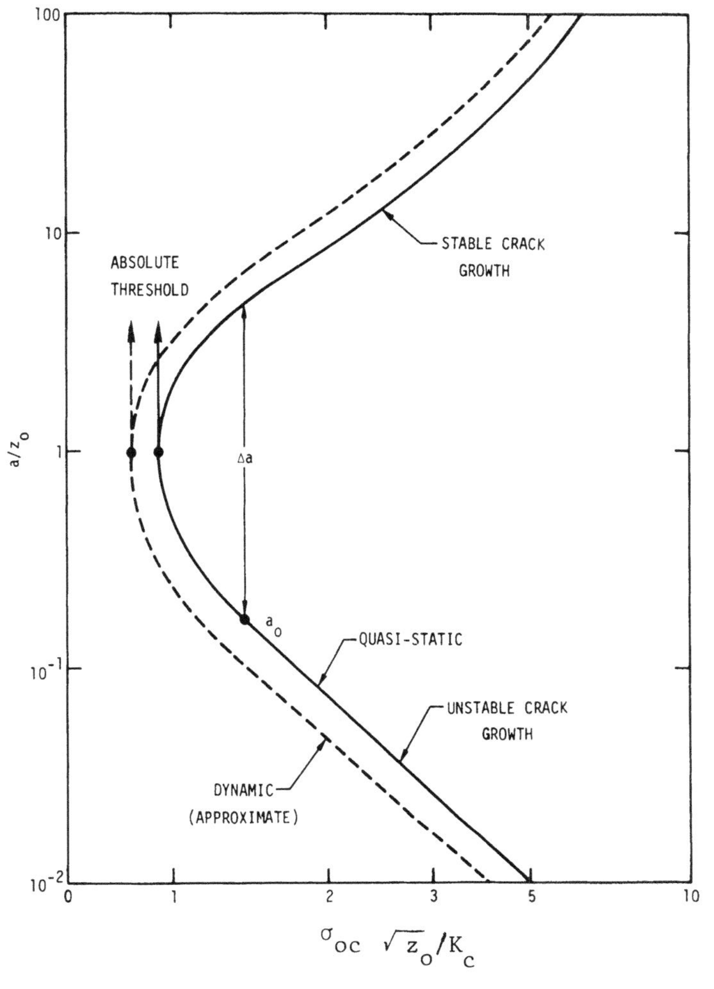

Fig. 10 Crack growth curve for a surface crack in an exponentially varying stress field.

$$D_2 = K_c/\sigma_o \tag{15}$$

which influences the crack length threshold and the crack increment. The crack threshold is enhanced by increasing D_2; but the effect on the crack increment is detremental, partially counteracting the beneficial effect of D_1 (see eqn. (11) and Fig. 9).

It is thus possible to surmise from analyses of dynamic crack activation and growth that the dominant target variables affecting impact damage are the wave speed and fracture toughness (e.g. as $K_c^2 v_\ell$), with the surface flaw depth appearing as a secondary variable in certain instances. The separate influences of the wave speed and the toughness are demonstrated by performing nylon sphere impact experiments on samples of ZnS, MgO and Si_3N_4 (Table I) containing nearly identical surface crack arrays[15] (Fig. 2). The ZnS and MgO materials have a comparable toughness, but MgO has a significantly larger wave speed (Table I): in accord with the theory, the MgO has a substantially higher threshold velocity v_c (Table I). The MgO and the Si_3N_4 have a comparable wave speed, but the Si_3N_4 has a considerably larger toughness (Table I): again, consistent with theory, the threshold velocity is much higher in the Si_3N_4 (Table I). The effect of the wave speed is larger than anticipated by the threshold expressions, which predict $v_c \propto v_\ell^{1/3}$. This could be attributed to the strong effect of v_ℓ on the crack extension (i.e. by influencing the detectability of the threshold) and/or to the relatively coarse grain structure of the ZnS which reduces the effective near-surface crack extension resistance to a value approaching the single crystal value. The effect of the toughness is similar to that anticipated by the threshold relations, i.e. $v_c \propto K_c^{2/3}$. This observation is consistent with the similar material grain sizes (MgO and Si_3N_4) in this instance, and with the crack extension being weakly dependent on the toughness (eqn. 11). The effect of the flaw depth can be discerned from Fig. 11, where it is apparent that the crack extension from relatively large pre-cracks is more extensive than that from small pre-existing cracks. However, the effect is relatively small, at least in the depth range encompassed by the present studies.

In summary, therefore, the damage threshold in this region can be substantially depressed by increasing the material's wave velocity (i.e. increasing the modulus or decreasing the density) or the toughness (especially of the near surface material) and, in certain regimes, by modifying the depth of the surface flaws. Further studies are needed to fully quantify both the effects of these variables, and the extent of the damage for a specified projectile impact.

TABLE I

The Threshold Velocities for Pre-Cracked Materials Impacted by 1mm Nylon Spheres

Material	Wave Speed (ms^{-1})	Toughness ($MPa\sqrt{m}$)	Threshold (ms^{-1})
ZnS	4.6×10^3	1	400
MgO	9.4×10^3	1	800
Si_3N_4	10^4	5	2400

3.1.2 Post Threshold Damage. Damage above the threshold can be created by mechanisms other than the generation of tensile stress pulses. These additional mechanisms are associated with the interaction of surface irregularities (prior cracks, ledges etc.) with the lateral outflow of the projectile. This process has not been analysed, but clearly has analogies in hydrodynamics which could be used to gain an appreciation for the variables that affect these damage modes.

3.2 Fully-Plastic Response

For a plastic target response the kinetic energy of the projectile is converted primarily into plastic deformation energy of the target and the projectile. This localization of the energy dissipation processes portends well for the utility of energy partitioning analyses. These analyses should recognize that the intense plastic work can lead to substantial local temperature excursions and should thus employ the plastic target and projectile properties that pertain to the impact condition (and not the quasi-static flow properties at the ambient temperature). An analysis that recognizes some of the salient features has recently been developed by Donaldson and colleagues[19]. This analysis relates the projectile penetration to the plastic and elastic properties of the target and projectile, and can be applied with good success to both ceramics and metals.

The material ejected is less well defined. The most extensively developed analysis of material removal relates to the plastic ploughing mechanism for non-normal incidence[4], in which the volume of material removed, V, is

$$V \propto \rho_p \frac{r_p^3 v_o^2 f(\phi)}{Y} \qquad (16)$$

where Y is the horizontal component of the plastic flow stress and $f(\phi)$ is the angular dependence plotted in Fig. 11. The analysis affords a reasonably accurate description of the angular dependence of the material removal (Fig. 11).

3.3 Elastic/Plastic Response

This "intermediate" regime has extensive application to ceramic systems. The damage consists primarily of fractures, which usually extend substantially beyond the deformation zones. Again, most of the projectile energy is dissipated at the boundaries by the elastic waves, and energy analyses are of minimal utility. Analyses of the dynamic elastic/plastic stresses and their influence on crack extension constitute a more definitive approach to damage prediction.

3.3.1 Post Threshold Damage. A finite difference analysis of elastic/plastic impact[8] has elucidated the following important features of the stress field (Fig. 12). The stresses within the expanding plastic zone during penetration are compressive, while the radial and tangential elastic stresses are both tensile and similar in magnitude, and exhibit a strong symmetry about the impact center. This symmetry is consistent with the observed radial and conical crack morphologies‡ (Section 2). In conjunction with both the quasi-static analogy of radial crack initiation during loading[7] and the observation of lateral crack development at radial cracks, this strongly suggests that the radial and conical cracks form during penetration. The pressure during this phase decreases from a value similar to the one-dimensional Hugoniot pressure to a value close to the quasi-static pressure.

After complete penetration, the elastic unloading wave interacts with the plastic wave and generates tensile stresses approximately normal to the surface. These tensile stresses reach a maximum near the plastic boundary at full penetration. The trajectory of the lateral crack‡, and the quasi-static analogy of lateral crack formation during unloading,[7,18] suggests that the lateral cracks develop during the projectile release phase, in the vicinity of the elastic/plastic boundary.

By analogy with quasi-static indentation[7], the amplitude of the stresses generated during penetration will be dominated by the contact pressure between the target and projectile, while the spatial extent will be determined by the contact radius, r_c. The extension of the radial or conical cracks a_r can thus be expressed in terms of these impact parameters through the functional relation[7,8];

---

‡Cracks tend to form on planes delineated by the trajectories of the maximum principal tensile stresses[18].

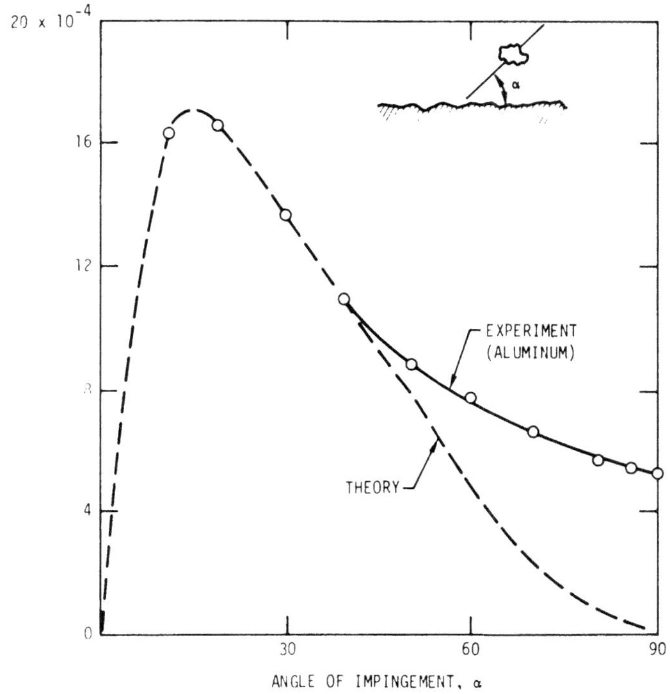

Fig. 11

The angular dependence of the material removal in the plastic response regime.

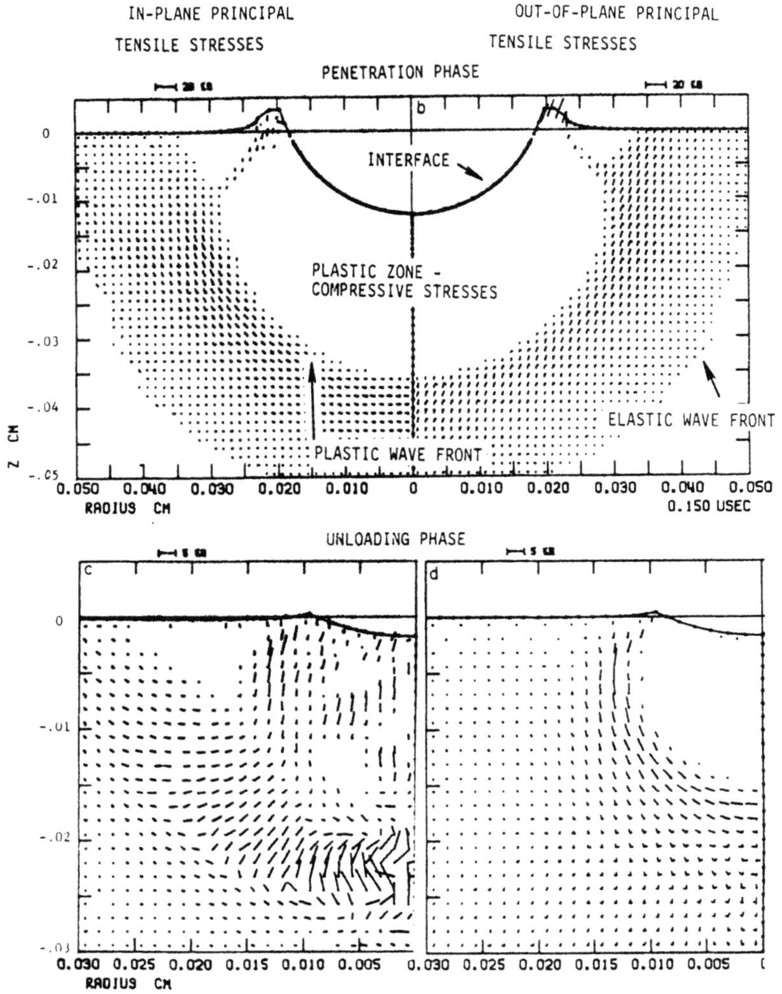

Fig. 12

A numerical analysis of elastic/plastic impact showing the principal tensile stresses that develop during penetration and release.

$$\frac{K_c}{p\sqrt{r_c}} = F\left(\frac{a_r}{r_c}\right) \tag{17}$$

where F is a function determined by the spatial and temporal details of the dynamic stress field. If it is assumed that the crack extension is dominated by the stresses that derive from the contact pressure during the earliest stages of impact, the impact pressure will be related to the one-dimensional pressure, p_1[8], by

$$p \equiv \alpha p_1 = \alpha v_o z_t z_p/(z_p + z_t) \tag{18}$$

where α is a constant (< 1, see Fig. 6) approximately independent of the target properties. For the equivalent assumption, the contact radius for small penetrations is[8];

$$r_c^{\,2} = 8\beta r_p^{\,2} v_o z_p/v_p\,(z_p + z_t) \tag{19}$$

where β is another constant approximately independent of the target properties, and v_p is the wave speed in the projectile. Further, since the spatial features of the dynamic stress field ($\sim 1/r^2$ for a spherically expanding wave front) and the quasi-static stress field ($\sim 1/r^2$ for a point force on a free surface[18]) are comparable, the functional form of F is likely to be similar to that observed quasi-statically[7];

$$F = \gamma\,(a_r/r_c)^{-3/2} \tag{20}$$

where γ is a material independent constant (the dynamic γ is likely, of course, to be larger than its quasi-static value). Combining eqns. (17 to 20), the radial or conical crack length becomes;

$$a_r = \left[\frac{(v_o r_p)^2}{K_c}\right]^{2/3} F_1(\rho_p, z_p, z_t) \tag{21}$$

where F_1 is a relatively weak function of z and ρ. A similar functional expression obtains for a quasi-static analysis of fully-elastic contact (except that the velocity exponent is smaller)[8];

$$a_r = \left[\frac{v_o^{6/5} r_p^{\,2}}{K_c}\right]^{2/3} F_2\,(\rho_p,\ \rho_t,\ z_p,\ z_t) \tag{22}$$

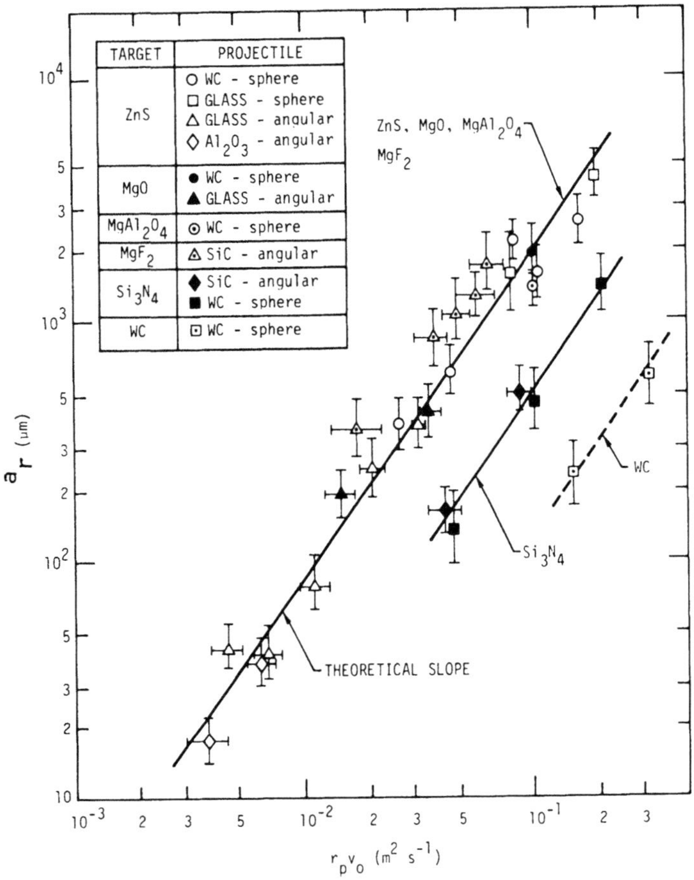

Fig. 13

The radial or conical crack length a_r as a function of the product of the projectile velocity and radius.

Experiments performed on a wide range of ceramics using several projectile types (both spherical and angular)[8] confirm the basic dependencies predicted by the theory (Fig. 13); yielding a best fit function equivalent to that derived for elastic/plastic contact.

$$a_r \approx \lambda' \left[\frac{(r_p v_o)^2}{K_c}\right]^{2/3} \tag{23}$$

where λ' is a constant.

Lateral crack formation, being dictated largely by the extent of the plastic zone at the full penetration condition, occurs (for penetrations up to the projectile center) in a zone proportional in extent to the contact radius, r_c. The contact radius is, in turn, approximately proportional to the penetration, q (especially for small penetrations); hence, the depth of lateral fracture h should exhibit a functional dependence on the target and projectile properties similar to that of the plastic penetration. The plastic penetration has been evaluated by Goodier[20] and by Donaldson[19], and can be expressed for small penetrations as;

$$\frac{q}{r_p} = \delta v_o \sqrt{\frac{\rho_p}{H}} \tag{24}$$

where δ is a constant. A general expression for the lateral crack depth might thus be;

$$\frac{h}{r_p} = F_3\left(v_o \sqrt{\frac{\rho_p}{H}}\right) \tag{25}$$

The test data confirm this primary dependence on the target and projectile properties (Fig. 14) and yield the explicit dependence[8];

$$\left(\frac{h}{r_p}\right)^2 = \lambda^* \left(v_o \sqrt{\frac{\rho_p}{H}}\right) \tag{26}$$

where λ^* is a constant.

In summary, therefore, the extension of the radial and conical fractures (which dictate the strength degradation) and the depth and extent of the lateral fractures (which determine the potential material removal zone) have been characterized in terms of the primary target and projectile variables; specifically, the toughness and hardness of the target, and the size, velocity and

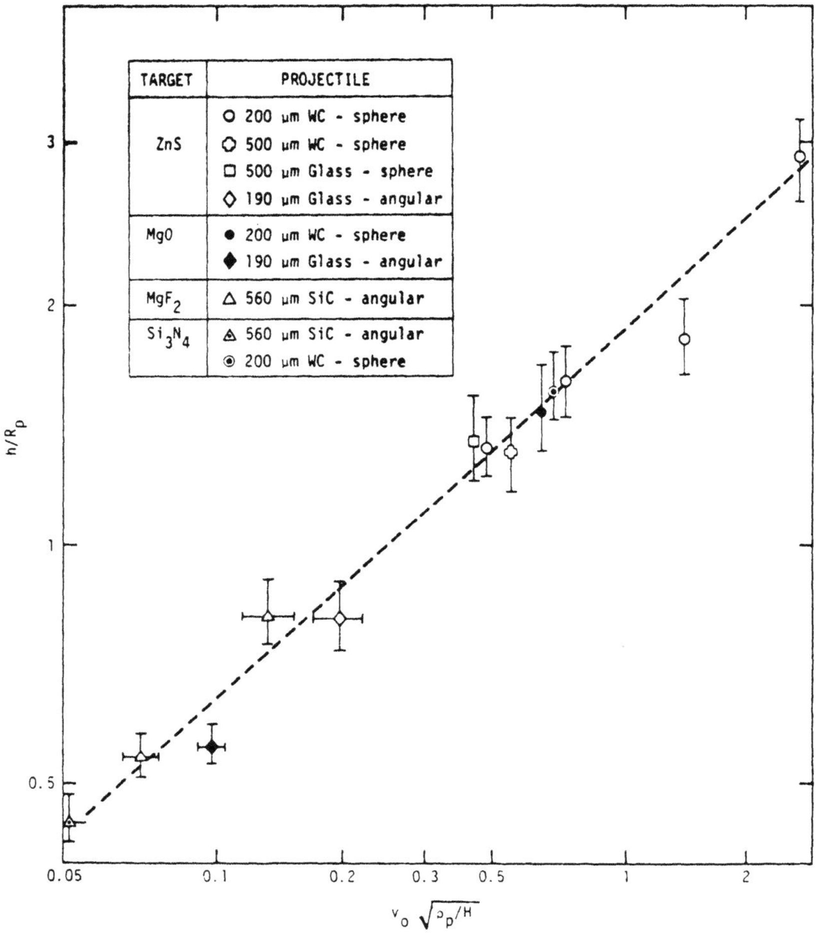

Fig. 14

The lateral crack depth as a function of the penetration parameter $v_o \sqrt{\rho_p/H}$.

density of the projectile.

3.3.2 Threshold Damage. A recent analysis of quasi-static fracture initiation[21] has shown that there is an absolute minimum initiation load, $\propto K_c^4/H^3$. The extent to which this minimum must be exceeded is dictated by the size of the largest flaw that exists in the vicinity of the elastic/plastic boundary. This, in turn, is usually related to microstructural variables such as the grain size, inclusion content etc. A dynamic equivalent of the static analyses indicates that the threshold velocity, v_c, is[8]

$$v_c \gtrsim \chi \frac{K_c^2}{r_p H^{3/2}} F_4(\rho_p, z_t, z_p) \tag{27}$$

The utility of this description of the minimum possible threshold has not been explored experimentally. Such studies are urgently needed to resolve this important facet of the impact damage problem.

3.4 Additional Damage Modes

Only cursory analyses have been performed of the damage that occurs in multiphase materials. One analysis of note[10] relates to the angular dependence, ϕ, of the damage in systems with hard and soft phases impacted by projectiles smaller than the primary microstructural entities. In this model, the soft phase erodes preferentially, and access to this phase is restricted by the presence of the harder, more erosion resistant phase. The predicted angular dependence of the erosion rate,

$$\dot{V} \propto (1-r_p \operatorname{cosec} \phi) \tag{28}$$

is consistent with experimental observations.[10]

4.0 IMPLICATIONS

4.1 Strength Degradation

The degraded strength σ_r of an impacted ceramic is dominated by the depth of the largest cracks oriented closest to the surface orthogonal. Assuming a constant coplanar strain energy release rate criterion for crack extension[11], the degraded strength for an approximately planar semi-circular impact crack (e.g. a radial crack) becomes[23]

$$\sigma_r^2 = K_c^2 \, \Theta(\theta)/4(1-\nu^2)a \tag{29}$$

where

$$\Theta(\theta) = [\cos^2\theta(\cos^2\theta+\sin^2\theta/(2-\nu)^2]^{-1}, \tag{30}$$

and θ is the angle between the normal to the crack plane and the target surface. For non-planar (e.g. conical) cracks, the equivalent relationship has not been explicitly evaluated, but the extension stress should exhibit the same inverse dependence on the crack depth.

For the elastic response regime, the expression for the crack increment (eqn. 11) induced by a stress pulse allows an approximate functional relation for the degraded strength to be derived as;

$$\sigma_r^2 = \frac{2\sigma_o^2}{\pi(1-\nu^2)} \; \frac{\Theta(\theta)}{\Phi\left[T_o\sigma_o^2 k^2/\pi K_c^2 v_\ell\right]} \tag{31}$$

The retained strength should thus be enhanced by increasing the toughness and wave speed of the material, a prediction that has not been explored experimentally.

For the elastic/plastic regime, the retained strength can be explicitly written, by combining eqns. (29) and (23), as;

$$\sigma_r^2 = \frac{\Theta(\theta) \; K_c^{8/3}}{4\lambda(1-\nu^2) \; (r_p v_o)^{4/3}} \tag{32}$$

The predicted dependence on the projectile properties, $r_p v_o$, has been verified by tests performed on glass[24]; but, the important dependence on the target toughness has not been investigated.

4.2 Erosion

The material removal caused by projectile impact is much more difficult to predict than the strength degradation. In fact, the only regime in which the material removal mechanism is sufficiently defined that a meaningful prediction can be attempted is the elastic/plastic regime. In this regime, the material removal zone, $\hat{V}$, is bounded by the zone of lateral fracture[8];

$$\hat{V} = \pi a_r^2 h = \sqrt{\lambda^*} \, \lambda^2 \frac{v_o^{19/6} \, r_p^{11/3} \, \rho_p^{1/4}}{K_c^{4/3} H^{1/4}} \tag{33}$$

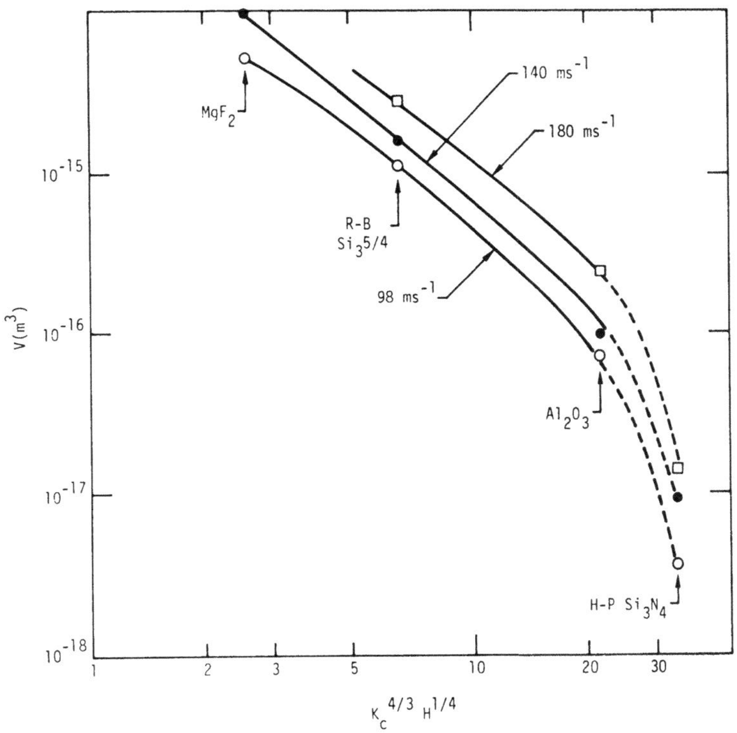

Fig. 15

The dependence of the erosion rate in five different ceramics on the toughness/hardness parameter suggested by the fracture characterization scheme.

However, only a fraction ξ of this zone is removed by a given impact, or sequence of impacts, and this fraction may exhibit important dependencies on the target or projectile variables. Nevertheless, a plot of erosion data for several ceramics (Fig. 15)[8] indicates that their erosion properties order according to the scheme suggested by eqn. (33); although the absolute values of the exponents are different (a consequence of the role of ξ).

5.0 CONCLUSION

Recent studies of impact damage in ceramics have identified several modes of damage, and have progressed toward an understanding of the physics of the damage formation processes. These are highly complex phenomena, however, and their detailed quantification will require a substantial research effort; especially if the damage characterization is to be extended into erosion prediction.

ACKNOWLEDGEMENT

The author wishes to thank the Office of Naval Research for their support under Contract No. N00014-75-C-0609.

REFERENCES

1. A. G. Evans and T. R. Wilshaw, Jnl. Mater. Sci. 12 (1977) 97.

2. W. F. Adler, Jnl. Mater. Sci., 12 (1977) 1253.

3. F. P. Bowden and J. E. Field, Proc. Roy. Soc. A282 (1965) 331.

4. I. Finnie, Wear 19 (1972) 81.

5. I. M. Hutchings, R. E. Winter and J. E. Field, Proc. Roy. Soc. A348 (1976)

6. C. E. Smeltzer, M. E. Gulden and W. A. Compton, Jnl. Basic Eng. 92D (1970) 139.

7. A. G. Evans and T. R. Wilshaw, Acta. Met. 24 (1976) 939.

8. A. G. Evans, M. E. Gulden and M. Rosenblatt, Science Center Report SC5023.9TR (Nov. 1976).

9. M. S. Crowley, Bull. Am. Ceram. Soc., 47 (1970) 481.

10. S. M. Wiederhorn, National Bureau of Standards, to be published.

11. R. M. Blowers, J. Inst. Maths. Applics. 5 (1969) 167.

12. M. Rosenblatt, California Research and Technology, unpublished results.

13. Y. M. Tsai, Int. Jnl. Solids Structures 7 (1971) 563.

14. L. B. Freund, Jnl. Mech. Phys. Solids, 21 (1973) 47.

15. A. G. Evans, Science Center, to be published.

16. B. Steverding, Fracture Mechanics of Ceramics (Ed., R.C. Bradt, D.P.H. Hasselman and F. F. Lange), Plenum, N.Y. 1974, p. 287.

17. M. Kassir and G. C. Sih, Three Dimensional Crack Problems, (Ed. G. C. Sih), Noordhoff 1974.

18. B. R. Lawn and T. R. Wilshaw, Jnl. Mater. Sci. 10 (1975) 1049.

19. C. Donaldson, ARAP Report.

20. J. N. Goodier, Proceedings of the Seventh Hypervelocity Impact Symposium, Vol. III (1965) p. 215.

21. B. R. Lawn and A. G. Evans, Jnl. Mater. Sci., in press.

22. J. J. Petrovic and M. G. Mendiratta, Jnl. Amer. Ceram. Soc. 59 (1970) 163.

23. A. G. Evans, Jnl. Amer. Ceram. Soc., in press.

24. S. M. Wiederhorn and B. R. Lawn, Jnl. Amer. Ceram. Soc., in press.

PARTICLE IMPACT REGIMES IN SINGLE CRYSTALS

Stephen V. Hooker* and William F. Adler**

*University of California, Berkeley

**Effects Technology, Inc.

INTRODUCTION

The present study illustrates the differences which occur in the particle impact response of crystalline materials which may not be anticipated from quasistatic indentation test results.

Single crystals of magnesium oxide (MgO) and calcium fluoride (CaF_2) were subjected to quasistatic bead indentations from very low to moderately high load levels as reference for the dynamic loading conditions. 1 mm diameter beads were used in all of the tests reported here. Sapphire beads were propelled at velocities up to 100 ms^{-1} using a modified air gun arrangement which allowed accurate placement of the particle on the surface, measurement of the impact velocity, and eliminated any rotational motions which would normally be imparted to the particle. Nylon beads were propelled at velocities up to 1000 ms^{-1} in the ETI exploding wire facility.(1) The upper velocity range stated for each of these experimental arrangements is not the maximum possible but is the upper limit for producing extensive damage in MgO and CaF_2.

An accurate analytical assessment of the interfacial pressure imparted to a non-isotropic target material by an impacting sphere is not available. Some idea of the relative magnitudes of the interfacial pressure for sapphire and nylon beads can be obtained from Figure 1. This pressure is calculated as a function of impact velocity according to the one-dimensional Hugoniot relations(2) and according to the Hertzian theory of impact(3) for a fused silica target material. For low velocity impacts the Hertzian calculation provides reasonable results for sapphire, tungsten carbide, and glass beads. The one-dimensional shock wave analysis would only be expected to be applicable at extremely high impact velocities. Figure 1 is intended to provide some perspective on the relative magnitude of the applied pressures for each of these loading conditions and demonstrates the significant difference in the magnitude of these pressures in discussions of the damage regimes. Although the Hertzian relation used is only applicable to isotropic materials, it is estimated (using equivalent moduli) that the pressures for MgO and CaF_2 targets can be about 100 percent higher for the Hertzian calculations. The Hugoniot relations indicate a 25 percent pressure increase for nylon on these crystals and a 60 percent increase for tungsten carbide beads.

EXPERIMENTAL RESULTS FOR MAGNESIUM OXIDE

The similarities and differences the modes of damage produced in quasistatic 1 mm sapphire bead indentations and impacts on (001) oriented MgO are described. A single nylon bead impact at 1000 ms^{-1} yielded extensive slip band formation within the underlying surface layer and a large subsurface radial fracture system which could not be produced by quasistatic load indentations.

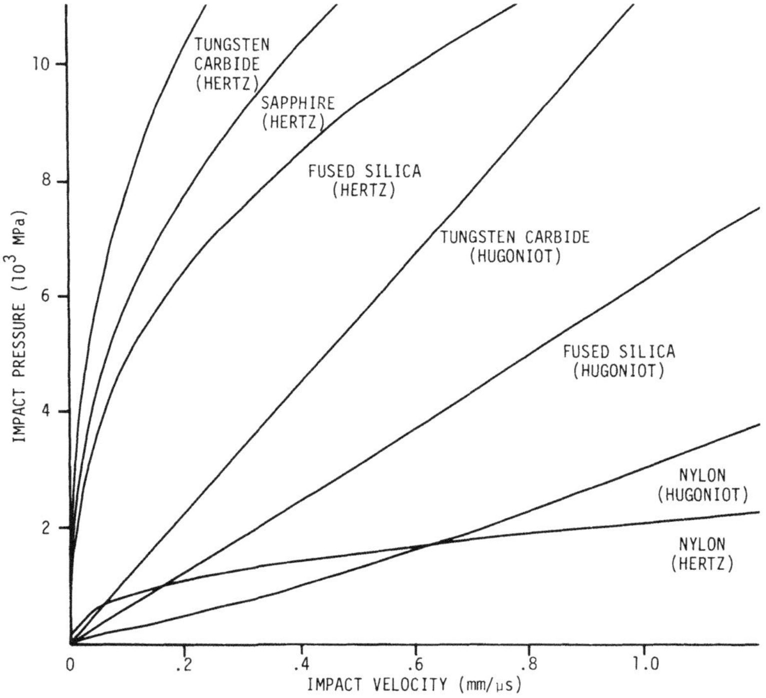

Figure 1. Comparison of Computed Interfacial Pressure Generated by Particle Impacts on Fused Silica

Quasistatic Indentations

A limited series of quasistatic indentations with 1 mm sapphire beads on cleaved and chemically polished MgO surfaces was undertaken for comparison with the bead impacts. Below the fracture threshold, rosettes of slip bands are formed with well-defined contact areas. Larger $\{110\}_{45}$ slip bands are located adjacent to the contact area with surface components extending along the <100> directions and outward into the surrounding material. A set of $\{110\}_{90}$ slip bands are also located adjacent to the contact area and extend along the <100> directions.[4] Sequential polish and etch procedures on cleaved cross-sections showed the $\{110\}_{90}$ slip bands. Previous workers have shown that surface intersects the screw components of the $\{110\}_{45}$ loops and the edge components of the $\{110\}_{90}$ loops.[5] Consequently, the $\{110\}_{45}$ slip bands form height discontinuities whereas the $\{110\}_{90}$ slip bands can only be detected by their inherent stresses.

Fracture initiation occurred at the surface with circumferential <110> slit fractures. These fractures were shallow, <15 μm in depth, and originated at the intersection of prominent slip planes as shown by Argon and Orowan for uniform compressive loading.[6] Initiation of fracture as radial <110> crack propagation in the rosette arms was observed for Vicker's indentation geometry as previously reported but is not the initiation mechanism for spherical bead indentation.[5] For spherical indentations the dislocation configuration displaces material radially outward within the <110> rosette arms so that the plastic deformations during the compression loading phase would also establish the radial stress field to initiate fractures during the pressure release phase.[5]

Increases in the indentation pressure produced more complicated surface and subsurface fracture patterns. Fracture conformed to crystallographic orientations and remained confined to the zone of plastic deformation.

Sapphire Bead Impacts

Chemically polished MgO surfaces were subjected to a series of sapphire bead impacts. 1 mm diameter beads were propelled over a sufficiently large velocity range to explore the fracture sequence. Below a fracture threshold (less than 15 ms^{-1}) the collision produced rosettes of slip bands without fracture as previously reported for rock salt crystals.[4] As for the quasistatic sapphire bead indentations glide extends further for the dislocations on the $\{110\}_{45}$ planes than the dislocations along the $\{110\}_{90}$ planes at all impact velocities.

As the impact velocity* is increased the initial fracture is a lateral cleavage fracture on the (001) surface approximately 10 to 25 μm below the specimen's surface. It is assumed this fracture forms during the pressure release phase. At higher velocities, the subsurface fracture propagates downward along $\{110\}_{45}$ fracture surfaces. The deepest penetration occurs along the <110> directions. New (001) lateral fractures tend to form instead along the <100> directions. The formation of these initial fractures does not appear to perturb the impact event. It has been observed that the initial fracture formation does not influence the rebound velocity for rigid bead impacts on glass.[7]

The impact velocities required to initiate surface fracture produce a prominant network of precursor subsurface fractures. Consistent with this, the initial surface fractures adopt several forms: radial <110> extension outward from the contact area, circumferential <110> extension adjacent to the contact area, and <100> extension above the outer edge of a lateral fracture. In all forms, the surface fracture is part of a {100} or a $\{110\}_{90}$ fracture which propagates from the subsurface network. Inspection of the surface with Nomarski interference contrast reveals the slip band discontinuities but not a central depression.

Further increases in velocity beyond the fracture threshold produce a well-defined hemispherical cavity underlain by extensive subsurface fractures. Material is displaced to form an upraised rim surrounding the crater. The $\{110\}_{45}$ slip bands which contribute to this displacement are clearly shown in Figure 2. The widest surface fractures extend radially along the center of the <110> rosette arms. These fractures result from the stresses associated with the displaced material and extend to the limit of the $\{110\}_{45}$ slip arrays. Less favorably oriented fractures tend to propagate erratically over much shorter patterns than the principal fracture. The contrast formed by these secondary radial fractures suggests the <110> rim sides are raised higher than the <110> corners so that discrete steps are formed at fracture openings. The difficulty for fracture propagation through prominant slip bands is clearly illustrated by the stepped displacements of some of the principal radial fractures.

---

*The initial pressure-velocity calibration for the air gun was found to be in error after completion of most of the sapphire bead impact studies, so precise velocities can not be stated for the damage modes described in this paper. Although the experimental arrangement is quite satisfactory, the impact velocities are inherently random; so only rough estimates of the impact velocities can be made at this time.

Further increases in the impact velocities produce bulk cleavage fractures which initiate from a point below the contact area and propagate radially outward along the {100} axes and down into the sample. The cross-sections shown in Figure 3 illustrate the extensive formation of lateral fractures and also the source of catastrophic cleavage approximately 100 μm below the surface in the vicinity of the contact zone. No defects are visible (or anticipated) so it is assumed failure initiated by a plasticity assisted mechanism. Slit fractures aligned along the $\{110\}_{45}$ traces are readily apparent in Figure 3 located lower on the cleavage surface.

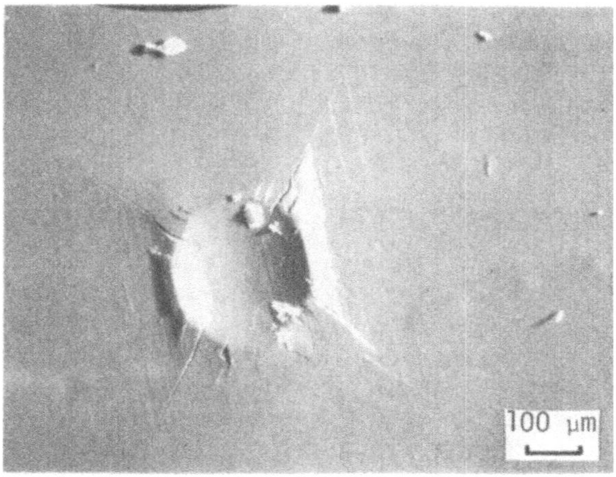

Figure 2. 1 mm Sapphire Bead Impact on (001) MgO

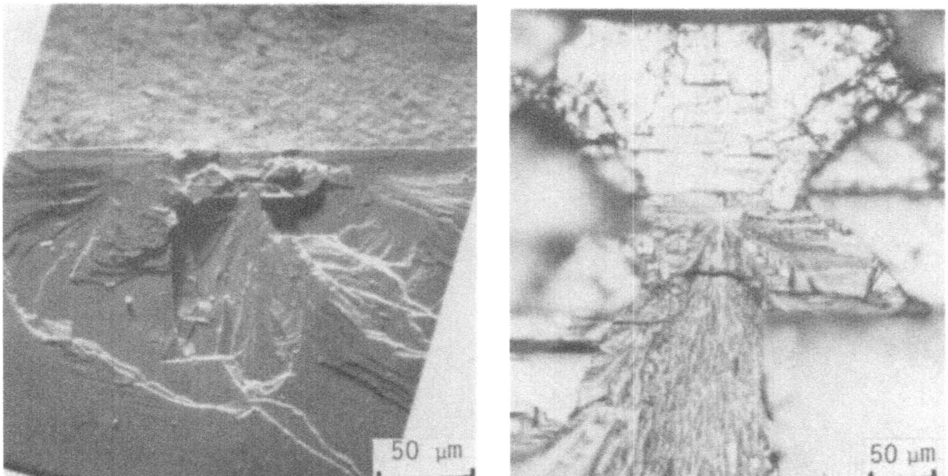

Figure 3. Catastrophic Cleavage on MgO Cross-Section due to Sapphire Bead Impact

A cross-section of the subsurface fracture and slip structure is illustrated in Figure 4. The plasticity zone clearly surrounds the fracture system. Microscopy views into the interior from both the top and cross-sectional planes indicate that the greatest strain is associated with the prominent {110} slip bands and that these bands arrest fracture propagation. Fractures tend to initiate along crystallographic planes, particularly slit fractures associated with {110} slip bands, and to a lesser extent at discontinuities such as hackle ledges. Since the fracture geometries are controlled by the slip bands, it is assumed that slip occurred during the compression phase and preceded fracture. The fracture structure does not establish when most of the fracture occurred. The fracture structure does indicate, however, that the large lateral (001) fractures which radiate outward are the last to form and occur during the pressure release phase.

Nylon Bead Impacts

A 1 mm nylon bead impacted a chemically polished MgO surface at 1000 ms^{-1}. Optical microscopy of the impact surface revealed relatively little damage as shown in Figure 5. The surface beneath the nylon contact zone underwent extensive slip along the $\{110\}_{45}$ planes to form a dense network of slip bands which penetrated 130 μm beneath the surface before interruption by the subsurface cleavage fracture. Minute fractures <20 μm in length formed along one side and corner within the contact zone. Since these fractures occurred near the deposited nylon, they probably were formed along the slip bands by the shear forces associated with the collapse of

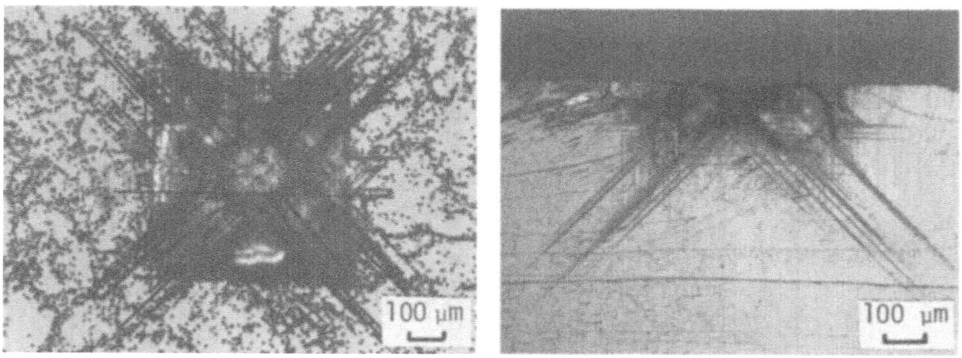

Figure 4. Surface and Cross-Section of MgO Impacted by 1 mm Sapphire Bead After Dislocation Etching

the nylon bead. Beyond the contact axis, intense slip bands formed along the $\{110\}_{90}$ plane in the vicinity of the subsurface radial fractures. These features are shown more clearly in Figure 6 after cleaving through the impact center and exposing the cleaved surface to a dislocation etchant. It may be speculated that the sequence of responses began with the extensive formation of slip bands beneath the contact zone, that this deformation generated sufficient stress to initiate the subsurface cleavage, and that propagation of these fractures generated the surrounding slip bands which occurred only near the cleavage fractures.

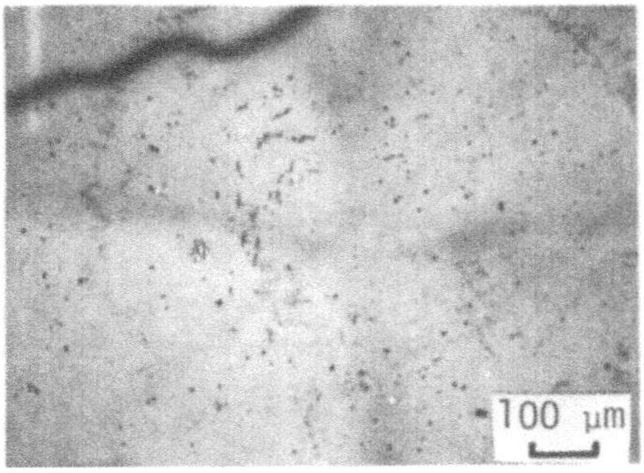

a) Reflected Illumination

b) Transmitted Illumination

Figure 5. Effect of 1 mm Nylon Bead Impact on MgO at 1000 ms^{-1}

Figure 6. Cross-Section of Fracture Surface Shown in Figure 5

The subsurface fracture is particularly important because it introduces large fractures along three orthogonal axes. The central cleavage plane shown in Figure 6 passes through the central axis of the subsurface fracture but does not include the entire fracture. The subsurface fracture is inclined approximately 4° to the sectioning face and the remainder extends beneath the matching cleavage surface. The complete fracture surfaces formed penny-shaped arcs. Markings on the fracture system indicate very rapid propagation from a common origin along the three cleavage axes.

Application of a dislocation etch to the cross-section revealed a zone of intense slip above the cleavage figure, described above, and bands of dislocations surrounding the fracture figure (some of the latter may have resulted from sectioning cleavage). Approximately 25 μm of the cross-sectioned surface was removed with the chemical polish and the specimen was etched again. This section revealed bands of dislocations surrounding the fracture figure did not penetrate very deeply into the bulk. Sequential polish and etch cycles indicated the subsurface volume adjacent to the cleavage fracture underwent extensive slip both along the $\{110\}_{90}$ and $\{110\}_{45}$ directions. The etch revealed a number of small slit fractures which initiated along $\{110\}_{45}$ slip bands and several small satellite fractures formed independently near the fracture origin.

EXPERIMENTAL RESULTS FOR CALCIUM FLUORIDE

CaF_2 displays markedly different response under all test conditions behaving in a ductile manner for quasistatic bead indentations and sapphire bead impacts but in a more brittle mode for nylon bead impacts.

Quasistatic Bead Indentations

A (001) orientation CaF_2 specimen was indented with 1 mm sapphire and tungsten carbide beads. The indentations generally produced shallow hemispherical cavities with lateral dimensions ranging from 12 to .25 μm, the smaller dimension characterized indents at the microscopic fracture threshold of 3.6 kg and the larger dimension characterized the transition to macroscopic fracture propagation. At the higher loading the circular symmetry was transformed to yield a rounded square aligned along the <110> axes consistent with the {100} <110> slip system active in the fluorspar structure.[8,9] Octagonal shapes were also observed at higher load levels.

Initial fracture formation consisted of radial fractures of nearly vertical, half-penny geometry which appeared to originate from the cavity wall beneath the surface. This radial fracture propagation is confined to the <100> and <110> axes. The zone of plastic deformation for all indents extends well beyond the fracture perimeter as determined by transmission microscopy under polarized illumination. The birefringence formed intense beams closely aligned to the <100> axes and well-defined {100} slip bands could be distinguished at least 30 μm from the impression. Analogous birefringence effects were observed about Vickers microhardness indentations. Application of an etchant to the indented surface also suggests these vertical fractures generally originate below the surface and propagate outward and upward. The second fractures to form were penny-shaped and originated from both the impression wall and the veritical fracture. Careful examination revealed no evidence of median vent formation but such fractures could have formed and essentially resealed during the indentation cycle.

Application of greater pressure to the indenter produced larger and more complex fractures as shown in Figure 7. Although no sign of crushing occurred, a complex pattern of radial and internal penny-shaped fractures formed. Gradually, the <110> fracture orientation predominated over the <100> fracture orientation (the {100} planes are secondary cleavage orientations in the fluorspar system). As the load levels increased one or two very large half-penny shaped cracks would form along the <110> axes suggesting that a median vent had extended radially from beneath the indenter to form a radial fracture during unloading. Sectioning these indents revealed inclination angles between 50 and 70° (with respect to the surface) indicating the median vent fracture approximated the {111} cleavage plane. Large radial and vent fractures were never observed on the {100} plane. However these measurements were complicated by variation in angle for individual vent fractures and activation of fracture growth by the polishing stresses.

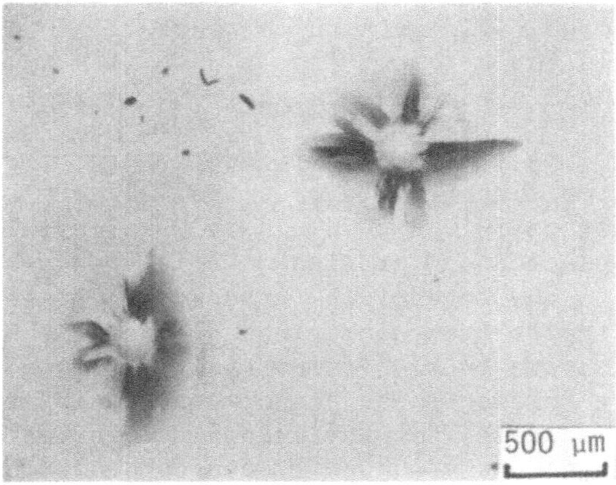

Figure 7. 1 mm Sapphire Bead Indentation on (001) CaF_2 (Transmitted Illumination)

Sapphire Bead Impacts

Both (001) and (111) surfaces of CaF_2 were impacted with 1 mm sapphire beads. At the lower velocities the impact events produced a ring fracture surrounding the impact on both orientations, shallow crushing beneath the impact site, and short radial fractures whose traces correspond to the (111) direction. The ring fractures for the (111) oriented surfaces agreed more closely with the anticipated crystallographic geometry of a hexagon (as shown in Figure 8) than the impacts on (001) CaF_2. Three lower subsurface lateral fractures radiate outward from the crushed area along <110> directions bounded by the radial fractures. In the intervening sectors, the fractures propagated steeply downward along the {111} cleavage planes. These fractures appeared to propagate more closely into the {111} geometry than the corresponding radial fractures. Crushing below the impact site appears to be a general characteristic of the sapphire bead impacts for the materials examined, although less crystallographic in nature in CaF_2 than that found in sectioned MgO.

At low impact velocities, the impact produced only a slight concavity within the ring fracture in contrast to the deep impressions under quasistatic conditions. Etching techniques reveal that the lower velocity impacts for both crystal orientations are surrounded by a field of dislocations. These dislocations conform to the anticipated geometry as shown in Figure 8. Comparison of the birefringent patterns for the impact and indentation geometries show the impact patterns are less intense and less closely aligned to the crystallographic axes; slip bands could not be discerned by this technique in the vicinity of the impact site. This implies a less coherent plastic deformation surrounds the impact sites in comparison with the quasistatic indentation sites. The fracture patterns in (111) oriented CaF_2 showed a striking symmetry in fracture lengths. Fractures in (001) oriented CaF_2 were not as symmetrical, although radial fractures tended to form along the {111} planes. This could result from the smaller angles between the surface and the cleavage planes for {001} oriented CaF_2.

Increases in impact velocity produced crater formation as described by Evans for fine-grained polycrystalline materials[10]. The intermediate stage is shown in Figure 9 in which one lateral segment has been ejected and crushed material beneath the impact axis is revealed at the edge of the crater. The lateral fracture markings suggest the fractures originated beneath the crushed zone and were bounded by previously formed radial fractures. This is clearly a pressure release phase phenomenon. Only at the higher velocities could fractures displaying the median vent structure be distinguished. These penny-shaped fractures passed beneath the indentation site and extended to form long radial fractures along the {111} cleavage planes. As the impact velocity was increased further these propagated through the central contact zone just below the surface crushing zone. Further increases in velocity produce craters and lateral cavities whose geometries were determined by the paths and intersections of these radial fractures.

Nylon Bead Impacts

Specimens of (001) oriented CaF_2 were impacted with 1 mm nylon beads in the velocity range from 675 to 885 ms^{-1}. Under these conditions the impacts produced square fracture patterns on the surface aligned along the traces of the {111} cleavage planes. These fractures tended to be short and discontinuous, frequently lying along tangentially aligned scratches. The interior of the region bounded by these square fractures appeared to be undamaged. The fractures beyond the square fracture arrays nearest the impact site were larger, were more closely aligned along the {111} orientations, and penetrated deeper. A large crystallographic truncated pyramid formed beneath the impact site which penetrated into the interior in a manner similar to the Hertzian fracture cones in glass. These characteristics are illustrated for typical impacts in Figures 10 to 12. Crystallographic fracture cones have also been observed for laser induced shock on the basal plane of quartz.[11]

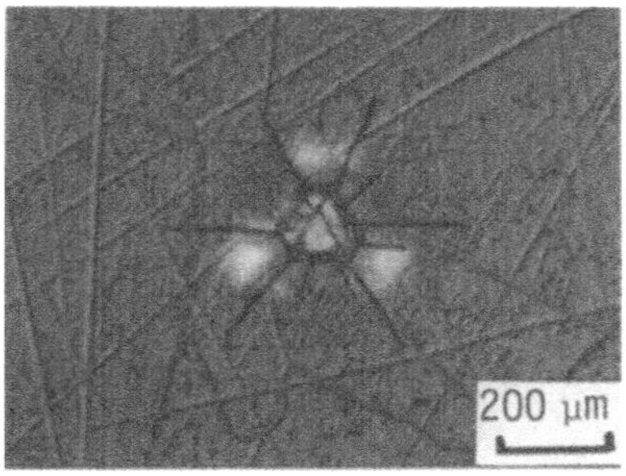

Figure 8. 1 mm Sapphire Bead Impact on (111) CaF_2 After Dislocation Etching

Figure 9. 1 mm Sapphire Bead Impact on (001) CaF_2 Showing First Stage of Material Removal

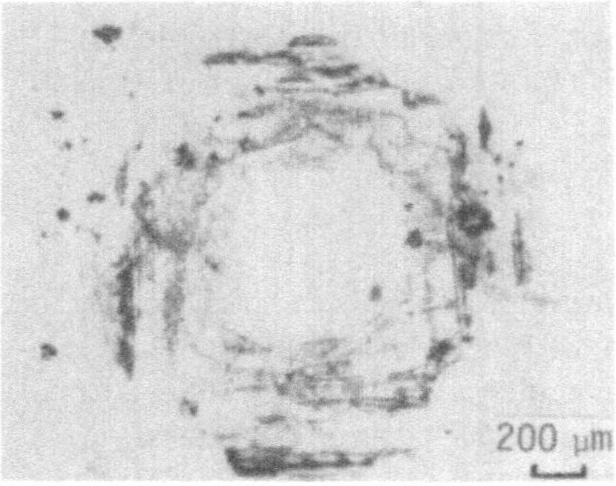

Figure 10. 1 mm Nylon Bead Impacting on (001) CaF_2 at 760 ms^{-1} (Transmitted Illumination)

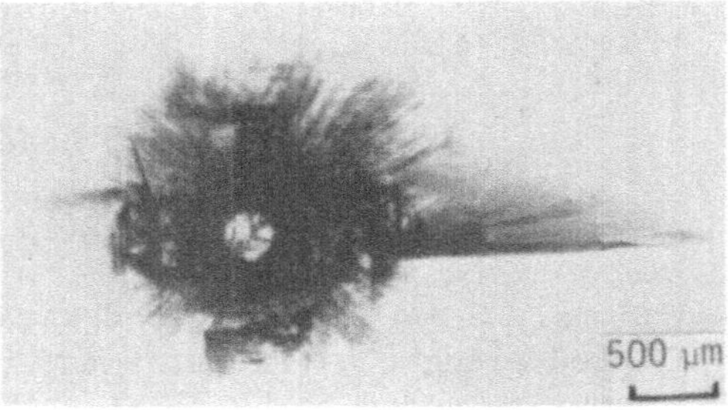

Figure 11. 1 mm Nylon Bead Impacting on (100) CaF_2 at 885 ms^{-1} (Transmitted Illumination)

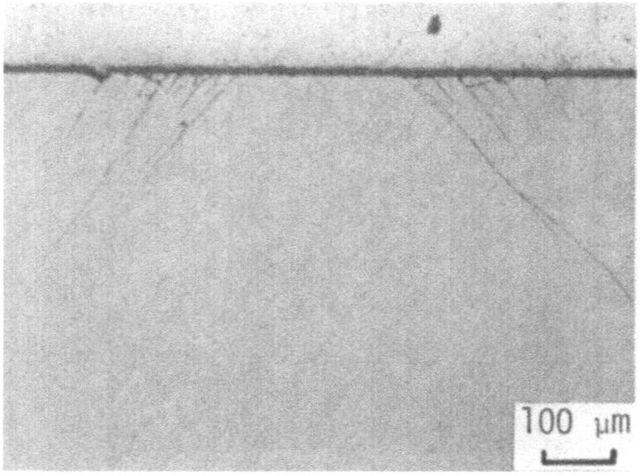

Figure 12. Lightly Etched Cross-Section of (001) CaF_2 After 1 mm Nylon Bead Impact at 760 ms^{-1}

Application of a light etchant sensitive only to sealed fracture paths revealed a more dense and circular fracture pattern existed within the square fracture pattern. These fractures were particularly evident along the <100> axes. The etchant revealed small microcleavages on the central surface. Since these are very shallow and uniform in areal density it is probable they arose from shear forces exerted by the deforming nylon bead. Application of a hydrochloric acid procedure which heavily over-etched fractures but revealed dislocations indicated the circumferential fractures were surrounded by a dense array of pits. These pits probably originated at emerging dislocations associated with fracture formation, although contributions from a frictional shearing action and formation of microfracture can't be ruled out.

The fracture pyramid is best illuminated by its own strain field as shown in Figure 13 for a partially crossed polarizer setting. Examination of the lightly etched specimen with a condensor stage at magnifications up to 300X and various working distances revealed that the pyramid originated near the innermost limit of the fracture rings as shown in Figure 12. However, both the fractographic river markings and the absence of etched surface fractures at the projected point of intersection with the impacted surface suggest many of these fractures originated 5 to 15 μm below the surface. The fracture pyramid appeared to have a coherent structure which could be traced continuously in any direction predominately along {111} cleavage surfaces.

To investigate the depth profile of the impact damage, cross-sections were prepared by grinding and polishing into several impacts. The surface of the central regions of the etched and lightly etched sections appeared to be undamaged. The central fracture pyramid penetration more deeply than the adjacent cleavage fractures. In Figure 12 the pyramid depth is 350 μm whereas the deepest adjacent fracture is 180 μm. The crystallographic angle between the (100) and (111) surfaces is 55° and the following experimental angles can be measured from Figure 12: right-51° (pyramid boundary), 48°, 58°, 61° and 60° and left-44° (pyramid boundary),41° (split in boundary), 51°, 49° and 57°. The smaller angles are due to a split in the boundary of the cross-sectioned pyramid wall on the left hand side (as seen in Figure 12), however such deviations are not representative of the cone fracture surface. It is difficult to assess the magnitude of the crystallographic influence since similar angles have been recorded for nylon bead impacts on fine grained zinc sulfide.[10] Consequently, the crystallographic cleavage requirements determine the projected geometry of the fracture pyramid but the penetration angle is also sensitive to the transient stress profile.

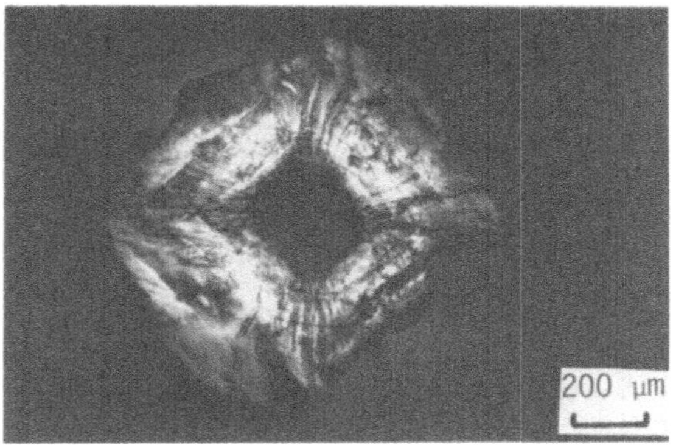

Figure 13. Pyramidal Fracture Surface in (001) CaF_2 Under Birefringent Illumination

The average fracture pyramid depths for four impacts were measured with an uncertainty of ± 0.1 mm. A plot of the cone depths as a function of impact velocity provided a straigh line extrapolation which indicated that the threshold velocity for cone formation was approximately 660 ms^{-1}. This extrapolation is consistent with highly asymmetric cone structure at the lower velocity, 675 ms^{-1}, in which only one side of the pyramid was well-developed and the opposing side was barely visible. At 885 ms^{-1} a radial fracture was produced extending several millimeters into the adjacent material (Figure 11). The velocity range over which pyramidal fracture surfaces dominate the impact damage has a lower bound of 660 ms^{-1} and an upper bound in the vicinity of 885 ms^{-1}.

CONCLUSIONS

Comparison of the sapphire quasistatic bead indentation and bead impact fracture patterns in MgO indicates the encompassment by the slip band rosette, the confinement of fracture to the {110} and {100} planes, and the fracture pattern symmetry were similar. A well-defined crater is not formed by the indentation. The lateral fractures are much smaller and less important, the volume of fractured material also appears to be much smaller, and cleavage fractures into the bulk material were not observed for spherical indentations. Nylon bead impacts produce a different damage structure. Surface fracture is minimal, confined to minute cleavage fractures resulting from interfacial shear within the contact zone. At 1000 ms^{-1} the impact stress level was sufficient to initiate two orthogonal subsurface cleavage figures containing the [001] axis passing through the center of impact and lying essentially in the (100) and (010) planes. Additional slip was evident along the periphery of three half penny-shaped arcs.

Particle damage on CaF_2 also depended on the loading rate and bead composition. Quasistatic indentation produced a hemispherical cavity surrounded by slip bands and a zone of plastic deformation. Fracture originated at the periphery of this cavity and propagated primarily along the {111} planes and secondarily along the {100} planes. Impact by sapphire beads resulted in a slight depression surrounded by a ring fracture and radial fractures. Extensive crushing occurred beneath the contact zone. Slip bands were not observed and the surrounding birefringence pattern was more diffuse than that surrounding the quasistatic indentations. Lateral, radial, and vent fractures all propagated along the {111} plane for impact on the {111} surface; fracture on the {001} plane was not as crystallographic. At high velocities, the cavity structure resembled that observed on fine-grained material.[10] Impact by nylon beads produced a damage free zone surrounded by a geometrical array of circumferential fractures aligned along the {111} cleavage planes.

A coherent pyramidal fracture surrounding the contact zone penetrated into the interior analogous to the conical fractures observed in glass for quasistatic indentations. An arry of shallower cleavage fractures extended to radial distances beyond the dominant pyramid. Ultimately all the loading conditions for (001) CaF_2 produced predominant single or multiple penny-shaped fracture radiating outward from the contact zone in the <110> directions and tending to lie along {111} planes. As described above however the events preceding the full development of this fracture system are quite distinct for each loading system.

ACKNOWLEDGEMENT

This research was supported by the Office of Naval Research under Contract No. N00014-76-C-0744, NR032-565.

REFERENCES

1. M.E. Graham, J.D. Carlyle and T.L. Menna, Rev. Sci. Instrum., 46, 1221 (1975).
2. G.E. Duvall, in Response of Metals to High Velocity Deformation (P.G. Shewmon and V.F. Zackey, ed), New York: Interscience Publishers, 1961.
3. S. Timoshenko and J.N. Goodier, Theory of Elasticity (2nd ed.), New York: McGraw-Hill Book Co., 1951.
4. K. Meyer and E. Gregert, Phys. Stat. Sol., 3, 2005 (1965).
5. A.S. Keh, J. Appl. Phys., 31, 1538 (1960).
6. A.S. Argon and E. Orowan, Phil. Mag., 9, 1003 (1964).
7. Y.M. Tsai and H. Kolsky, J. Mech. Phys. Solids, 15, 263 (1967).
8. T.S. Liu and C.H. Li, J. Appl. Phys., 35, 3324 (1964).
9. A.G. Evans et al., Proc. Brit. Ceramic Soc., 6, 173 (1966).
10. A.G. Evans and T.R. Wilshaw, J. Materials Sci., 12, 97 (1977).
11. N.E.W. Hartley and T.R. Wilshaw, J. Materials Sci., 8, 265 (1973).

A HIGH-SPEED PHOTOGRAPHIC INVESTIGATION OF THE IMPACT DAMAGE IN SODA-LIME AND BOROSILICATE GLASSES BY SMALL GLASS AND STEEL SPHERES

M. M. Chaudhri and S. M. Walley

Cavendish Laboratory

Madingley Road, Cambridge, CB3 OHE

ABSTRACT

A high speed photographic (10^6 frames per second) study has been made of the damage caused to soda-lime and borosilicate glasses by the impact of 1mm diameter glass and steel spheres travelling at speeds up to $300 ms^{-1}$. For velocities up to $180 ms^{-1}$, in both glasses the cone and radial cracks formed during loading and lateral cracks during unloading. At higher speeds the cone crack was not observed to form in the soda-lime glass whereas in the borosilicate glass the cone crack formed up to the maximum velocity studied.

The threshold velocity for the formation of lateral cracks in both glasses was found to be in the range 100 - $120 ms^{-1}$ for both glass and steel spheres. Lateral cracks are shown to form without any evidence of plastic flow in the glasses.

Contact times have been obtained from the photographic sequences and found to agree with the quasi-static theories.

Finally, the possible mechanisms for the formation of radial and lateral cracks are discussed.

INTRODUCTION

The problem of elastic contact between a solid spherical indenter and a flat solid surface was first analysed by Hertz (1) and since then it has been a subject of numerous experimental and theoretical investigations. The elastic displacements are such that on the surface the radial stress is compressive within the circle of contact and tensile outside, while the circumferential stress is compressive everywhere on the surface. Johnson, O'Conner and Woodward (2) modified Hertz theory to take into account interfacial frictional stresses and showed that if there is a mismatch in the elastic constants of the solids the maximum of the radial tensile stress occurs outside the circle of contact. This helps explain the observations that the radius of the ring crack that forms on the surface, for a critical load on the indenter, is greater than the radius of the contact circle.

If the radius of the indenter is small, plastic deformation of the substrate takes place for a sufficiently high load and then Hertz analysis no longer applies. Infact, the surface stresses change their character; the radial stress eventually becomes compressive, while the circumferential stress tensile. It is the circumferential stress that causes radial cracks, which are normal to the indenter, around plastic indentations. Experiments have shown that when the load is removed another system of cracks, almost parallel to the indented surface, is formed around the indentation. These cracks are called lateral cracks; when they intersect the indented surface, chips of the substrate are removed.

There have been only a few investigations of the behaviour of brittle solids when quasi-statically loaded with small enough spherical indenters so as to cause plastic flow (3-5). Swain and Hagan (3) and Evans and Wilshaw (4) followed the damage with an optical microscope and thus were able to distinguish between the cracks formed during loading and those formed during unloading. Evans and Wilshaw found that for a number of materials both radial and lateral cracks formed during loading, which extended during unloading. Similar observations were made from experiments on soda-lime glass indented by 0.4mm diameter tungsten carbide spheres by Swain and Hagan.

The impact of spherical particles on brittle solids can also result in the formation of Hertzian cone, radial, and lateral cracks (6-10). However, there has been very little work done on the sequence of formation of the various types of crack. It is quite clear that this knowledge will be very useful in understanding the mechanisms involved in the process of material removal and strength degradation by solid particle impact. We describe here a high speed photographic study of the damage formed in soda-lime and borosilicate glasses by the impact of 1mm glass and steel spheres. It was found that cone and radial cracks formed during the loading part of the impact, whereas lateral cracks formed during the particle rebound; most of the damage was complete within 2-3μs of the impact. Contact times were also measured and found to agree with the quasi-static theories. Preliminary accounts of these investigations have recently been described elsewhere (11,12).

THE STRESS EQUATIONS AND THE MAXIMUM IMPACT LOAD

For the purely elastic case the radius, a, of the circle of contact between the indenter of radius, R, and the flat solid surface are given by (13),

$$a^3 = \frac{4}{3} \frac{kFR}{E} \qquad 1$$

$$Z = \left(\frac{4}{3} \frac{k}{E}\right)^{2/3} F^{2/3} \quad R^{-1/3} \qquad 2$$

$$k = \frac{9}{16} \left[(1 - \nu^2) + (1 - \nu'^2) \frac{E}{E'}\right] \qquad 3$$

where F is the normal load, ν and ν' the poisson's ratios of the solid and the indenter, E and E' the Young's moduli of the solid and the indenter, and Z the distance of mutual approach of the deforming solids. Outside the circle of contact the surface radial stress, σ_r, is tensile and is given by

$$\sigma_r = \frac{(1 - 2\nu)\, F}{2\pi r^2} \qquad 4$$

where r is the distance from the axis of contact.

Using equations 1 - 3, we can also obtain a relationship between the impact velocity, and the maximum load generated. This is done by considering that the maximum elastic strain energy during the impact is equal to the kinetic energy of the impinging particle. That is,

$$\frac{1}{2} \left(\frac{4}{3} \pi R^3 \rho\right) V^2 = \int_0^Z F(Z)\,dZ \qquad 5$$

where ρ is the density of the particle, V the impact velocity and F(Z) the load corresponding to the distance of approach Z. Intergration of equation 5 gives the maximum load, F_m,

$$F_m = \left(\frac{5}{3} \pi\rho\right)^{3/5} \left(\frac{4}{3} \frac{k}{E}\right)^{-2/5} R^2\, V^{6/5} \qquad 6$$

The maximum average pressure, P_m, is then

$$P_m = \frac{1}{\pi} \left(\frac{5}{3} \pi\rho\right)^{1/5} \left(\frac{4}{3} \frac{k}{E}\right)^{-4/5} V^{2/5} \qquad 7$$

In experiments with steel spheres it was found that the spheres deformed plastically for impact velocities greater than $80ms^{-1}$; this means that equations 6 and 7 become invalid. The maximum load in this case is obtained by measuring the radius, $a_{plastic}$, of the plastically deformed flat on the sphere and using the relation,

$$F_{max.\ plastic} = \pi a^2_{plastic}\ Y_{dynamic} \qquad 8$$

where $Y_{dynamic}$ is the flow stress of the sphere. The strain rate, in these experiments is $\sim 10^5\ s^{-1}$, and therefore, the dynamic flow stress may be higher than the static value, ($\dot{\varepsilon} = 0.2a/Rt$) However, Davis (14) has shown that in hardened steels the dynamic flow stress is not significantly different from the static value. The value of the static flow stress can be obtained by using Tabor's relationship (15),

$$Y_{static} = 1/3 \times H \qquad 9$$

where H is the Vickers hardness number. As an example the maximum load due to the impact of a 1mm diameter steel sphere at $180ms^{-1}$ was 58kg.

THE IMPACT TIME

When the collision is elastic, the total time of contact of a sphere with a flat is (13),

$$t_{elastic} = 2.94\left[\frac{5}{4}\pi\rho\left\{\left(\frac{1-\nu^2}{E}\right) + \left(\frac{1-\nu'^2}{E'}\right)\right\}\right]^{2/5}\frac{R}{V^{1/5}} \qquad 10$$

Experiments showed that the glass spheres did not deform plastically on impact, and therefore, equation 10 is applicable here. On the other hand, because the steel sphere deformed plastically equation 10 will be invalid. The calculation of the contact time then becomes a complex problem. We can, however, simplify the problem by making an approximation that the total contact time consists of two parts: (i) the plastic loading time and, (ii) the elastic unloading time. It can be shown that the loading time of a sphere deforming plastically against a rigid surface is,

$$t_{plastic} = \frac{\pi}{2}\left(\frac{m}{2\pi R\ Y_{dynamic}}\right)^{\frac{1}{2}} \qquad 11$$

where m is the mass of the sphere. A similar expression has been found for the loading time of a rigid sphere on to a plastically deformed target with the dynamic hardness of the target replacing the dynamic flow stress of the sphere (16).

Note that while the elastic contact time varies slowly with impact velocity, the plastic loading time is independent of the impact velocity.

EXPERIMENTAL

The soda-lime glass and steel spheres used were 1mm in diameter: the surface roughness of the glass spheres was better than 1μm. The surface of the steel spheres was of optical quality. The Vickers hardness value of the steel sphere was 9.2 GPa. The blocks of soda-lime and borosilicate glasses used for the high speed photographic studies were of size 50mm x 50mm x 25mm and were polished to an optical quality surface finish. The impacts were made on the smaller faces, and as the total contact time of the 1mm glass or steel balls was in the range 1-2μs, the stress waves generated at the time of impact did not come back to the impact site during the contact time. A few experiments were made on the 'as-received' and fracture surfaces of soda-lime glass; the general features of the damage were similar to that on the polished surface.

We used two types of particle projection system. In one the particles were accelerated by an air gun (17) to any desired velocity in the range 10 - 350ms^{-1}. In the other, the particles were propelled by the explosion of a small detonator; velocities in the range 100 - 200ms^{-1} were obtained. First, using the air gun the nature of the damage was determined for various impact velocities, and velocity regimes of interest selected for the high speed photographic experiments. A schematic diagram of the experimental system is shown in fig. 1. The particle, P, rests on the plate, B, and when the detonator underneath the plate explodes the particle is projected perpendicular to the plate and impacts the glass block, G, normally. The event was illuminated with a Xenon filled flash tube and photographed with a rotating mirror high speed camera (Beckman and Whitley model 189) to which a microscope was attached which gave a magnification of x4 on the camera film. The detonator was fired by discharging a bank of capacitors charged to 5kV. The flash of light (140μs duration), the event, and the position of the camera mirror were electronically synchronized to within a microsecond.

The microscope of the camera was focused within the glass block on to a plane normal to the optic axis of the camera and containing the impact axis. Therefore, in the photographs the sphere is out of focus, but the cracks within the glass are well in focus. In these photographs the fractures appear dark with the background being uniformly bright. The sensitivity of the system was adequate for resolving very fine cracks.

From the high speed photographic sequences the particle impact and rebound velocities, and the time of contact were obtained.

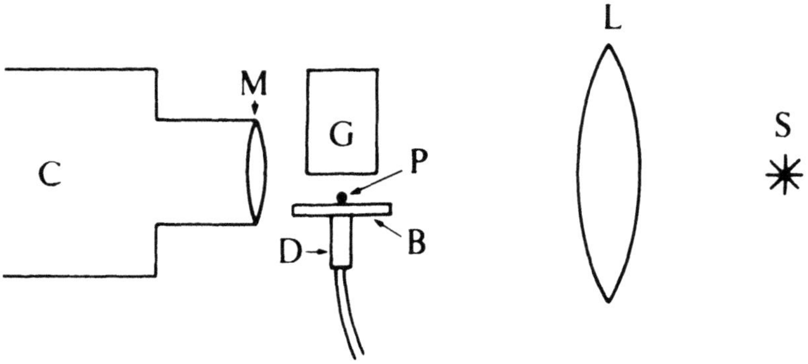

Fig. 1. Schematic of the experimental arrangement for high speed photographic studies. S, flash tube; M, microscope; C, Beckman and Whitley model 198 camera; L, condenser lens.

PHOTOGRAPHIC SEQUENCES

The high speed photography was able to show the sequences of formation of cone, radial, and lateral cracks in both soda-lime and borosilicate glasses. The sequences for the borosilicate glass are described first. Fig. 2 shows the impact of a 1mm diameter glass sphere at $125ms^{-1}$, which occurs ∿0.4μs before frame 3, resulting in the formation of a well developed cone crack of length 1.5mm. The cone crack grows at ∿$3500ms^{-1}$ during the loading part of the impact and its partial closure occurs as soon as the rebound of the sphere starts. In frame 3 radial cracks near the contact zone can also be seen. The sphere leaves the surface completely between frames 3 and 4. 3μs after frame 6 the cone crack opens up again showing the oscillatory nature of the crack (not shown in the sequence). The glass sphere was recovered and found to be undamaged. The oscillation of the cone crack and the relative time of formation of radial cracks were more clearly observed in sequences taken for higher impact velocities. In fig. 3 a steel sphere moving at $160ms^{-1}$, impacts ∿0.1μs before frame 3. This results in a cone crack (3500 ms^{-1}) which grows as the impact loading continues. Moreover, a radial crack, R, is also formed (frame 4), before the maximum impact load has been reached. The sphere starts to rebound between frames 4 and 5 and leaves the surface ∿0.1μs before frame 5. Note that the cone and radial cracks partially close during the unloading (frame 5). Furthermore, on the left of the contact area the formation and the development of a lateral crack can also be seen, the lateral

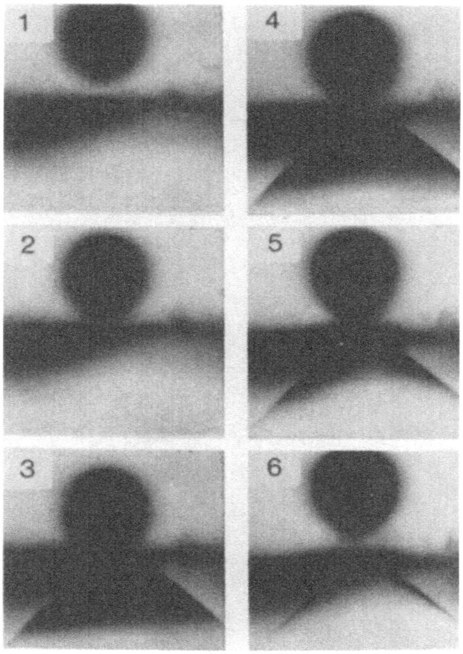

Fig. 2. The impact of a 1mm diameter glass sphere on a block of borosilicate glass at 125ms^{-1}. Interframe time 1μs.

crack becoming more distinct in the subsequent frames. An intersting observation is that after the sphere has left the surface the cone crack opens up again in frame 6, and then starts to oscillate at a frequency of 5 x 10^{5} s^{-1}. It is quite simple to show that these oscillations are not due to the stress waves reflected from the edge of the glass. Other notable observations are that the cone crack surface becomes rougher as compared with that at lower impact velocities, and as the sphere rebounds, broken up material from the contact zone is ejected normal to the specimen surface at 70ms^{-1}. Other sequences have shown that as the impact velocity is further increased more than one cone crack forms and the extent of the radial cracks increases (see ref. 12).

In the soda-lime glass also the cone crack forms very shortly after the impact. But as the sphere rebounds the end of the 'skirt' of cone crack changes its direction of motion towards the impacted surface. Fig. 4 shows the impact of a 1mm diameter glass sphere at 140ms^{-1} on a polished surface. The impact occurs ~0.6μs before frame 3 and a well developed cone crack moving at a velocity of 1550ms^{-1} forms. The sphere leaves the surface ~0.4μs before frame

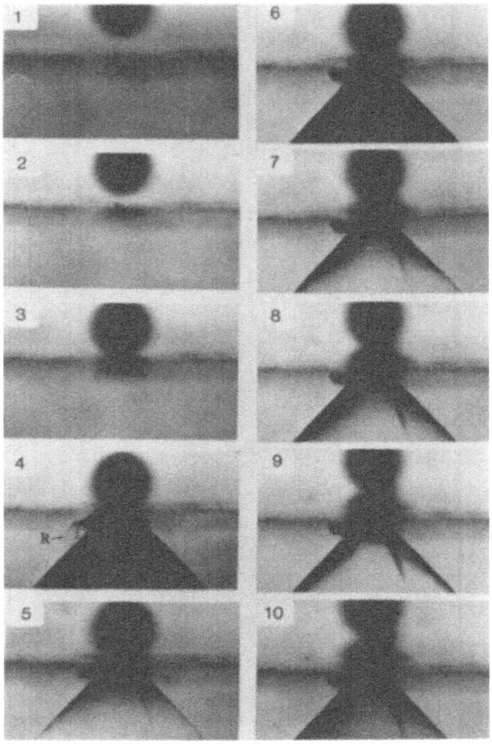

Fig. 3. The impact of a 1mm diameter hardened steel sphere on a block of borosilicate glass at $160ms^{-1}$. A cone crack and a radial crack, R, form during loading. A lateral crack forms on unloading (left of impact site, frame 5). Note the oscillation of the cone crack after the sphere has left the surface. Interframe time 1μs.

4 and this is accompanied by the end of the 'skirt' of the cone crack turning towards the impacted surface. Two radial cracks R,R can clearly be seen in this frame; one of them is approximately along the impact axis. We are not sure whether in this sequence R,R form just before the maximum load occurs or at the start of the unloading. An optical microscope examination of the impact site showed that some lateral cracks adjacent to the contact zone also formed, and that the material in the contact zone was extensively broken up. It could be argued that the crushing of the material in the contact zone plays an important role in changing the direction of motion of the cone crack during unloading. However, the crushing is not important. A sequence (not shown) of the impact of a 1mm diameter steel ball on the fracture surface of soda-lime glass at $180ms^{-1}$ showed that the end of the 'skirt' of the cone crack turned

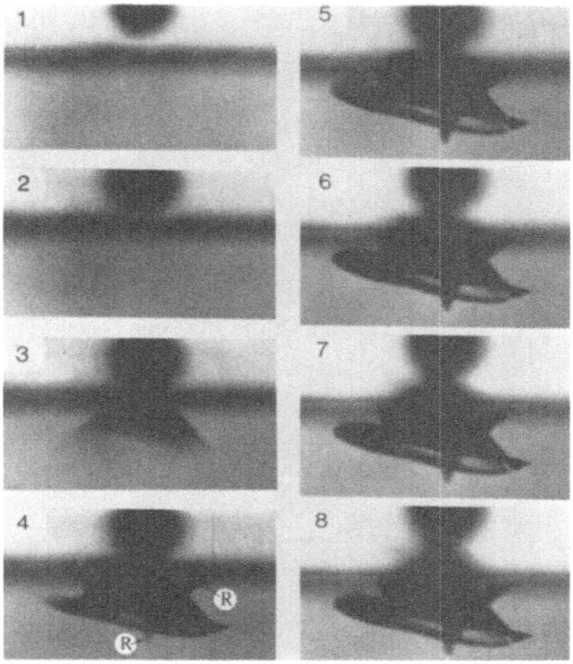

Fig. 4. The impact of a 1mm diameter glass sphere on a block of soda-lime glass at $140ms^{-1}$. Cone and radial cracks R,R form during loading; the end of the 'skirt' of the cone crack turns up as the sphere rebounds. Interframe time 1μs.

up during unloading without the material in the contact zone being broken up at all.

At higher impact velocities a considerable amount of debris of the soda-lime glass is thrown upwards at the instant of contact, and with the framing speed used the cone crack was not seen to form. A sequence of the impact of a 1mm diameter steel sphere at $200ms^{-1}$ is shown in fig. 5. The impact occurs about 0.5μs before frame 3, resulting in the formation of a large number of radial cracks; one of them is shown at R (frame 3). Some of these radial cracks travel at a velocity of $1550ms^{-1}$. The debris of the material is ejected at $\sim 380ms^{-1}$ (D in frame 4). The sphere has started to rebound in frame 4 and this results in the partial closure of R and the formation of two lateral cracks, L_1 and L_2. In the next frame a lateral crack on the left of the impact zone reaches the surface while R opens up again, indicating that, as in the borosilicate glass, the cracks in the soda-lime glass also oscillate. Examination of the

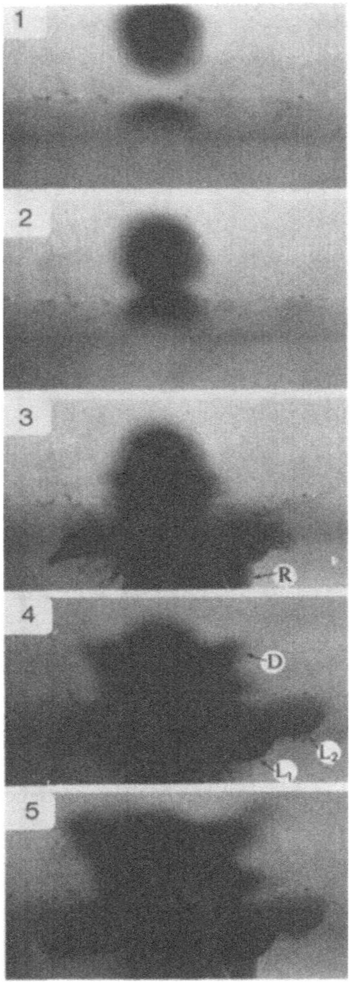

Fig. 5. The impact of a 1mm diameter hardened steel sphere on a block of soda-lime glass at $200ms^{-1}$. Debris, D, is ejected at $\sim 380ms^{-1}$ at the instant of impact. The cone crack is not observed to form. R, radial crack; L_1, L_2, lateral cracks. Interframe time 1μs.

impact site showed that at these high impact velocities most of the lateral cracks originated at the radial cracks inclined to the impacted surface.

THE CONTACT TIMES

The times of contact of the spheres with the glass surfaces were obtained from the high speed photographic sequences. At impact velocities of 200ms^{-1} or greater extensive fragmentation obscured the sphere thus making difficult the measurements. Table 1 shows the observed and calculated impact times under various conditions. It will be seen that the agreement is good.

THE NATURE OF THE FINAL DAMAGE

We found that the surface features of the damage in both glasses were similar when impacted by 1mm glass spheres at 100 - 120 ms^{-1}; these were (i) the innermost ring crack, (ii) secondary ring cracks of diameter greater than that of (i) and with the material between the rings broken up, and (iii) radial and lateral cracks. At higher impact velocities the damage in the two glasses was somewhat dis-similar. In soda-lime glass at an impact velocity 140ms^{-1} the extent of the radial and lateral cracks became larger than for impacts at 120ms^{-1}, and the region within the innermost ring crack fragmented extensively. The cone crack still formed but with the end of its 'skirt' turned up (see, for example, fig. 4). With further increase in the impact velocity the volume of the crushed zone and the size and number of radial and lateral cracks became larger. Moreover, the cone crack formed was unclear.

TABLE 1: CONTACT TIMES OF 1mm DIAMETER GLASS AND STEEL SPHERES ON SODA-LIME AND BOROSILICATE GLASSES

Target	Pro-jectile	Impact velocity ms^{-1}	Rebound velocity ms^{-1}	Calculated contact time μs	Observed contact time μs
soda-lime	glass	160	120	1.25	0.6 ± 0.3
soda-lime	glass	140	100	1.3	1.2 ± 0.3
soda-lime	steel	180	135	1.9	1.5 ± 0.3
boro-silicate	steel	160	120	1.9	2.0 ± 0.3
boro-silicate	glass	125	110	1.3	0.7 ± 0.3

In the borosilicate glass also the increase in the impact velocity resulted in the increase in the size and the number of radial and lateral cracks. However, the cone crack formed up to the highest velocity (i.e., 340ms^{-1}) used both for glass and steel spheres. The threshold velocity for the break-up of the material in the innermost ring crack for 1mm glass spheres was greater than 170ms^{-1}.

Another important observation was that for both the glasses the radius of the innermost ring crack was independent of the impact velocity, though there was some scatter in the individual cases. Similar observations have been reported by Kirchner and Gruver (8).

We have also examined the impact sites for any permanent deformation of the material within the innermost ring crack. The methods used were interferometry and tracing the surface profile with a 'Talysurf'. In the former after the impact the surface of the glass was coated with a thin metallic film and the damage zone examined with an interference microscope using the mercury green line λ = 564nm (A similar method has been used by Davies (14) to detect the very initial stage of plastic flow by indentations in metals. Swain and

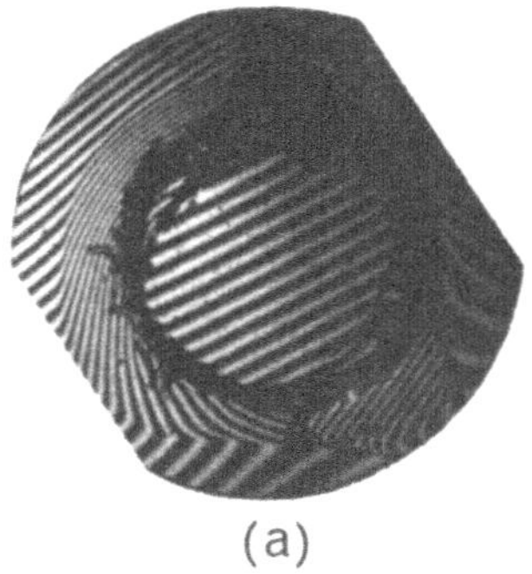

(a)

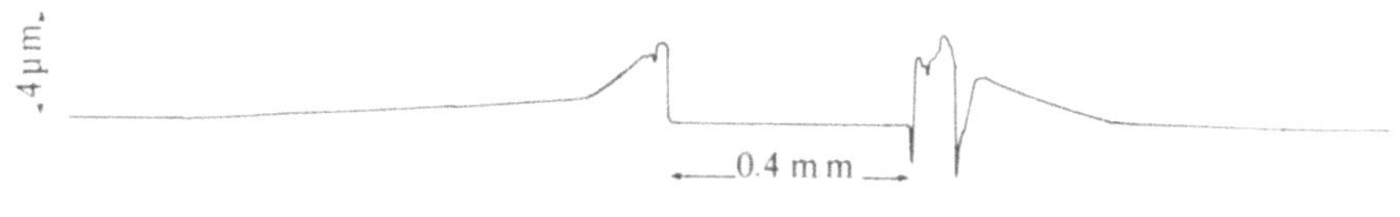

Fig. 6. The damage in soda-lime glass by the impact of a 1mm diameter glass sphere of velocity 120ms^{-1}. (a) is an interferogram, showing that the region within the innermost ring crack is flat; (b) is a trace of the surface of the damage, which confirms this. Note that the region outside the ring crack has risen above the original level of the surface.

Hagan (3) have also used this technique for measuring the depth of plastic indentations in soda-lime glass). It was found that the fringes within the innermost ring crack were straight lines, indicating that this region was flat and no plastic flow had taken place. A typical interferogram is shown in fig. 6(a). Note that though there is no evidence of any plastic flow radial and lateral cracks have formed. Outside the innermost ring crack the fringes are curved. This is due to the surface of the glass surrounding the ring crack having risen above the original level. This point is made clearer by the surface profile of the damage region, fig. 6(b). Again, the zone within the innermost ring is completely flat and is on the same level as the surface well outside the damage zone, whereas the surface around the ring crack has risen above the original level by about 4μm. No evidence of any plastic flow was found within the innermost ring crack in the borosilicate glass either.

We now consider the possibility of plastic flow in these impact experiments. The static hardness of soda-lime glass is 5.6 GPa which is measured for a loading time of 60s. However, the hardness is dependent upon the loading time as shown by Gunasekera and Holloway (18). They found that when the loading time was decreased from 10s to 10^{-3}s the hardness increased by $\sim$3 times. As in our experiments the loading time is $\sim 10^{-6}$s, further increase in the hardness is possible. Therefore in the impact experiments full plasticity may occur when the impact pressure generated is greater than $\sim$17 GPa.

Using equation 7 the maximum average pressure corresponding to the impact of a glass sphere travelling at $120 ms^{-1}$ is $\sim$5 GPa, and therefore plastic flow is not likely. Indeed full plasticity may not occur even for an impact velocity greater than $200 ms^{-1}$. In the case of the impact with a steel sphere no plastic flow was observed up to $120 ms^{-1}$, though the steel ball deformed plastically (the deformation pressure being $\sim$3.1 GPa). With the increase in the impact velocity the impact pressure will not increase significantly, though the size of the plastically deformed flat on the sphere will, as was observed.

DISCUSSION

It has been shown that in both soda-lime and borosilicate glasses cone, radial and lateral cracks formed on impact with 1mm diameter glass and steel spheres without any evidence of plastic flow in the impacted materials. The threshold velocity for the formation of radial and lateral cracks for both glasses was in the range 100 - 120 ms^{-1}; at this velocity the steel sphere deformed plastically while the glass sphere showed no deformation. The sequence of formation of the various crack systems has been revealed by the high speed photography. Up to an impact velocity of $180ms^{-1}$ a cone crack followed by radial cracks formed during the loading part of the impact and lateral cracks during unloading. In both glasses radial and lateral cracks appeared to initiate at the innermost cone crack. However, in the case of the soda-lime additional lateral cracks formed by the turning up of the end of the 'skirt' of the cone crack as shown in fig. 2.

The formation of radial cracks around plastic spherical indentations can be explained on Johnson's (19) theoretical treatment of the problem. He showed that during loading the circumferential stress is tensile in the vicinity of the elastic/plastic boundary. During unloading on account of the residual plastic strain, the circumferential stress outside the elastic/plastic boundary is also tensile; this has been thought to be responsible for the formation and growth of radial cracks during unloading (3,5).

The lateral crack formation and growth has also been considered (3) to be a consequence of the residual plastic strain during unloading, though at present a theoretical treatment of this situation does not exist.

A question relevant to our observations is what causes the formation of radial and lateral cracks in the purely elastic case. Tsai and Kolsky (7) have suggested that the radial cracks are initiated within the bulk of the specimen at a depth of 1.3a from the centre of the contact circle of radius, a, where the circumferential stress is tensile. However, they did not observe the sequence of formation and growth of the cracks. Moreover, our high speed photographic sequences have shown that the radial cracks initiate on the surface where the Hertzian circumferential stress is compressive. Therefore the origin of the radial cracks observed in these experiments is different.

We describe below some possible mechanisms for the formation of radial and lateral cracks.

As the circle of contact spreads over the ring cracks during loading the surface displacements on the sphere and the indenter will be radially towards the impact axis. This would bring the surfaces of the rings into contact, thus generating frictional forces; these forces will produce tensile circumferential stress, which could produce radial cracks.

During unloading there will be interfacial friction between the surfaces of the cone crack within the compressive zone under the sphere. The frictional force generated could lead to the formation of lateral cracks from near the contact zone during unloading.

Another important factor is the oscillation of the cone crack which is indicative of the 'overshoot' beyond the equilibrium position of the 'wings' of the cone crack and the region within the innermost ring crack. One effect of the overshoot could be to lever the 'wings' of the cone crack towards the specimen surface. For large displacements this could cause lateral cracks from the end of the 'skirt' of the cone crack (see fig. 2).

At high impact velocities (i.e. > $200 ms^{-1}$) extensive crushing and debris ejection from within the contact zone occurs during loading. This leads to penetration of the sphere into the material thus producing tensile circumferential stress and consequently, radial cracks.

We have also shown that there is good agreement between the observed contact times and those calculated using the quasi-static theories. Further, the radius of the innermost ring crack has been found to be independent of the impact velocity and equal to that of the ring crack formed with quasi-static loading. This suggests that the load required to form the innermost ring crack in the impact experiments is the same as that in the quasi-static experiments. It is therefore concluded that as far as the time of contact and the maximum impact load are concerned quasi-static theories are adequate for the velocities used. However, the use of quasi-static approach in calculating the length of the cone crack leads to values which do not agree with the experimental values (8,12).

We think that the error occurs because in the impact experiments the load is applied only for a short time (i.e. $\sim 1 \mu s$) and the cone crack grows only during this time. Therefore, a more appropriate relationship will be

$$C = V_{av} (t - t_o) \qquad 12$$

where C is the size of the cone crack, V_{av} the average crack velocity, t the loading time, and t_o the time to form the innermost ring crack. As in these experiments V_{av} in the borosilicate glass has been found to be significantly greater than in the soda-lime glass, the size of the cone crack in the borosilicate glass will be larger than that in the soda-lime glass. This has been confirmed by our experiments.

At high impact velocity V_{av} will approach the terminal velocity throughout the impact time with t_o reducing to an insignificant fraction of t. Therefore the size of the cone crack will be

$$C = V_{terminal} . t \qquad 13$$

A brief comment should be made on the high crack velocities

which were observed especially during the early stages of the crack growth. Velocities of $1550ms^{-1}$ and $3500ms^{-1}$ were found for the soda-lime glass and the borosilicate glass, respectively. At these high propagation velocities we did not observe any bifurcation of the crack. It seems possible that the particular nature of the stress field produced by the impact of a sphere helps the formation of such fast cracks.

ACKNOWLEDGMENTS

We would like to thank Drs. J.E. Field and J.T. Hagan for many helpful discussions. The experimental work reported here was done by one of us (S.M.W.) and formed a final year undergraduate project at this laboratory.

REFERENCES

1. H. Hertz, 'Miscellaneous papers' Chapt. 5 and 6 (Macmillan and Co. Ltd., New York 1896).
2. K.L. Johnson, J.J. O'Connor and A.C. Woodward, Proc. Roy. Soc. A 334, 95 (1973).
3. M.V. Swain and J.T. Hagan, J. Phys. D: Appl. Phys. 9, 2201 (1976).
4. A.G. Evans and T.R. Wilshaw, Acta Met. 24, 939 (1976).
5. C.J. Studman and J.E. Field, J. Mater. Sci. 12, 215 (1977).
6. J.P.A. Tillet, Proc. Phys. Soc. Lond. B 69, 47 (1956).
7. Y.M. Tsai and H. Kolsky, J. Mech. Phys. Solids 15, 263 (1967).
8. H.P. Kirchner and R.M. Gruver, Mat. Sci. and Eng. 28, 153 (1977).
9. A.G. Evans and T.R. Wilshaw, J. Mater. Sci. 12, 97 (1977).
10. G.P. Cherepanov and V.B. Sokolinsky, Eng. Fract. Mech. 4, 205 (197 (1972).
11. M.M. Chaudhri, C.G. Knight and M.V. Swain, 12th Int. Cong. on High speed Photography, Toronto, August 1976.
12. C.G. Knight, M.V. Swain and M.M. Chaudhri, J. Mater. Sci. 12, 1573 (1977).
13. S.P. Timoshenko and J.N. Goodier, 'Theory of elasticity' Chapt. 12 (McGraw-Hill, New York 1970).
14. R.M. Davies, Proc. Roy. Soc. Lond. A 197, 416 (1949).
15. D. Tabor, 'The hardness of metals' Chapt. 4 (Oxford University Press, London 1951).
16. ibid, Chapt. 8.
17. I.M. Hutchings and R.E. Winter, J. Phys. E. 8, 84 (1975).
18. S.P. Gunasekera and D.G. Holloway, Phys. Chem. Glasses 14, 45 (1973).
19. K.L. Johnson, 'Engineering plasticity' p.341, edited by E. Heyman and F.A. Leckie (Cambridge University Press, London 1968).

LOCALIZED IMPACT DAMAGE IN A VISCOUS MEDIUM (GLASS)

H. P. Kirchner and R. M. Gruver

Ceramic Finishing Company, State College, PA

I. INTRODUCTION

Two basic modes of localized contact damage have been observed in brittle and semi-brittle materials under "static" loading conditions (1,2,3). Blunt indenters, such as relatively large diameter spheres, form Hertzian cone cracks. Smaller spheres and sharp indenters such as abrasive grain, indent the surfaces forming radial cracks and lateral vent cracks. Recent investigations of impacts of glass plates by glass spheres have shown that Hertzian cone cracks, radial cracks and lateral vent cracks are formed (4,5). Branching of the Hertzian cracks is common and can be used to estimate the fracture stresses (6). At high impact velocities both blunt and sharp particles cause crushing and the fracture patterns become very complex. Evans and Wilshaw (3) have shown that, for "static" contact stresses, the transition from Hertzian damage to indentation damage observed with decreasing brittleness in a series of materials, results from a decrease in the radial tensile stress (σ_{11}) and a change in sign of the tangential stress (σ_{22}) from compressive to tensile.

A convenient way to vary the brittleness or deformability is to vary the temperature. In the present investigation the above-mentioned transition was investigated for impact conditions using glass plates heated to various temperatures. The mechanical properties of viscous bodies such as glass differ from those of plastic bodies mainly because of the absence of a yield point. Care must be taken in the application of concepts such as hardness to such materials.

Figure 1. Multiple Images of Incoming and Rebounding Sphere Obtained by High Speed Photography (IX).

II. PROCEDURES

PPG float glass plates (1/4 x 1-1/4 x 1-1/4 in.) were abraded on the side opposite that formed in contact with the tin by dropping -80 +150 mesh SiC grain from a height of one meter. The plates were firmly supported by a steel block and impacted by glass spheres (3 mm. diameter) at velocities ranging up to 138 ms^{-1}, as described previously (4). The incoming and rebounding velocities of the spheres were measured by high speed photography (Figure 1). An electronic stroboscope (General Radio No. 1538-A), flashing at rates up to 150,000 times per minute, was used to obtain multiple images on Polaroid film. To improve the reflectivity of the glass spheres, they were coated with a thin layer of aluminum by vacuum deposition.

Electric heaters were used to heat the glass plates to temperatures as high as 770°C. When the temperature was constant and uniform, the front heater was removed and the gas gun was fired to impact the specimen immediately, before substantial cooling occurred. The temperatures were varied from 440° to 770°C varying the viscosity of the glass from about 10^{17} to 10^{7} poise.

The glass plates were coated with a thin layer of soot. The impression, formed in the soot by the impact, was used to locate the impact site and to measure the contact radius (a_{max}). Hertzian cracks were observed at many of the impact sites. The symbols used to describe these cracks are indicated in Figure 2.

The impact damage was characterized by optical and scanning electron miscrosopy. The soot was removed and the Hertz crack radius and the radii of the radial cracks, if present, were measured in the underlying surface. Then, the specimens were cut through the impact site and polished. The cone angle (α) and depth (L) were measured on the polished surface or by focusing

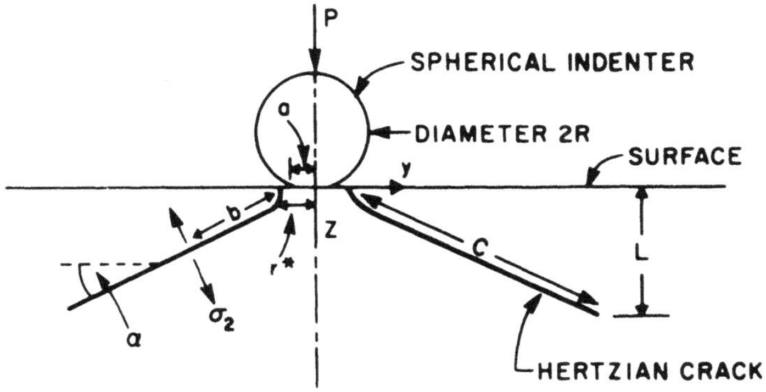

Figure 2. Hertzian Cone Crack.

into the specimen from that surface. The characteristics of the radial and lateral vent cracks were also measured.

III. RESULTS AND DISCUSSION

A. Characterization of Elevated Temperature Impact Damage

Indentations. The glass plates were impacted at velocities close to 50 ms^{-1} at various temperatures to determine the temperature at which indentation first occurred. Photographs of the impact site were taken under inclined illumination so that, when the surface was indented, light was reflected from one side of the impact site and the other side appeared dark. No evidence of indentation was observed at temperatures up to about 525°C but above this temperature indentation occurred, the degree of indentation increasing with temperature. Based on this information, impact velocities and temperatures were chosen for a more extensive investigation. Specimens were impacted at velocities close to 50, 75, 100 and 125 ms^{-1} and at temperatures from 440 to 770°C. The extent of the damage increased with impact velocity, as expected. Indentations, Hertzian cracks, radial cracks, lateral vent cracks and crushing were observed.

An impression of the observed variations in the indentations can be obtained from the scanning electron micrographs in Figure 3. At 720°C and 54 ms^{-1} there is a small indentation surrounded by a complex pattern of ring cracks with seven radial cracks. At 720°C, but with a much higher impact velocity (125 ms^{-1}), the impact site is crushed. The material between the lateral vent cracks and the surface chipped away revealing the surface of the crater and the

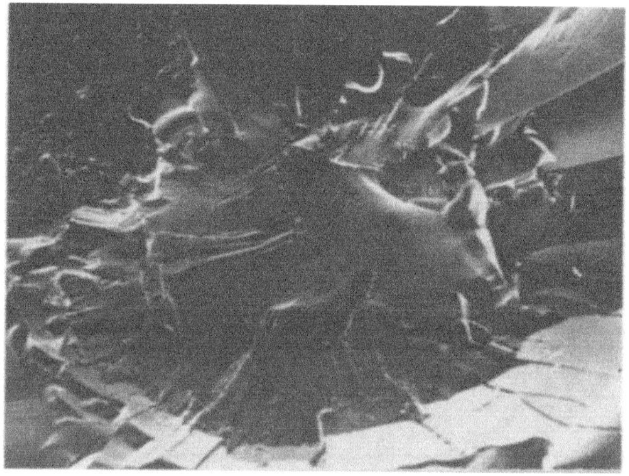

720°C 125 ms^{-1}

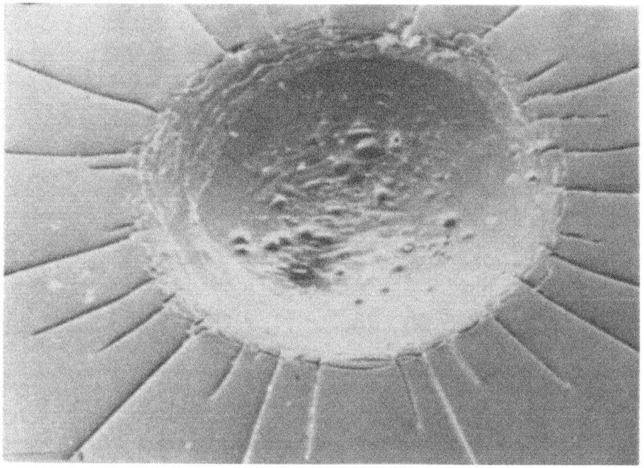

770°C 128 ms^{-1}

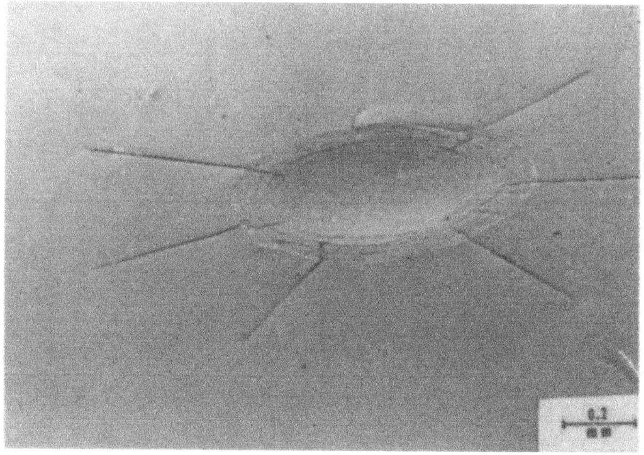

720°C 54 ms^{-1}

Figure 3. Impact Damage at Various Temperatures and Impact Velocities.

radial cracks which formed before the lateral vent cracks. At 770°C and 125 ms^{-1}, a much larger indentation formed with a profusion of radial cracks, but the crushing was absent because of the greater deformability at the higher temperature.

A profile of an indentation formed by impact at 700°C and 74 ms^{-1} is given in Figure 4. The vertical scale is 43 times the horizontal scale so that the indentation appears to be much deeper than it actually is. The profile shows that, in this case, there is at least one ring crack in the indentation. There is a rim, 5 μm above the surrounding material, pushed up by the impact. The rim contains approximately three ring cracks at the location of this particular profile. This indentation is about 20 μm deep including the rim and is about 800 μm in diameter.

Hertzian Cone Cracks. Hertzian cone cracks were observed for impacts at 50 ms^{-1} at temperatures ranging up to about 700°C. At 650 and 683°C, there is a tendency to form two concentric cone cracks. At 700°C, the cone crack depth is much less than those observed at lower temperatures. At higher temperatures the cone cracks were not observed. Except near 700°C, the depths of the cone cracks did not vary significantly with temperature. At these temperatures the "ring" cracks, which are the intersection of the Hertz cone with the surface, typically form spirals as shown in Figure 5. Branching of these spirals is common as shown in Figure 6 in which the specimen has been etched to reveal the underlying pattern of cracks. All of the branches are inclined at a small angle from the propagation direction of the spiral crack.

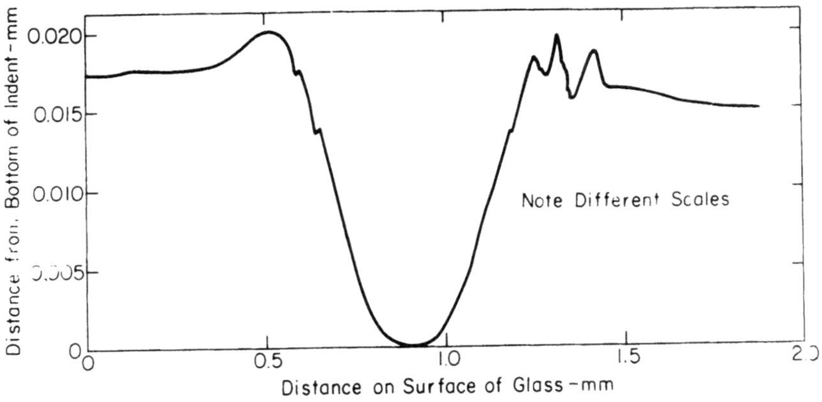

Figure 4. Profile of Indentation Formed by Impact at 700°C and 74 ms^{-1}.

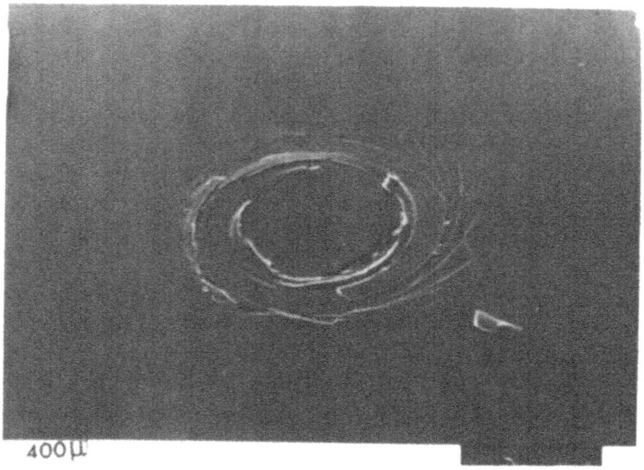

Figure 5. Spiral Crack (580°C, 52 ms^{-1}).

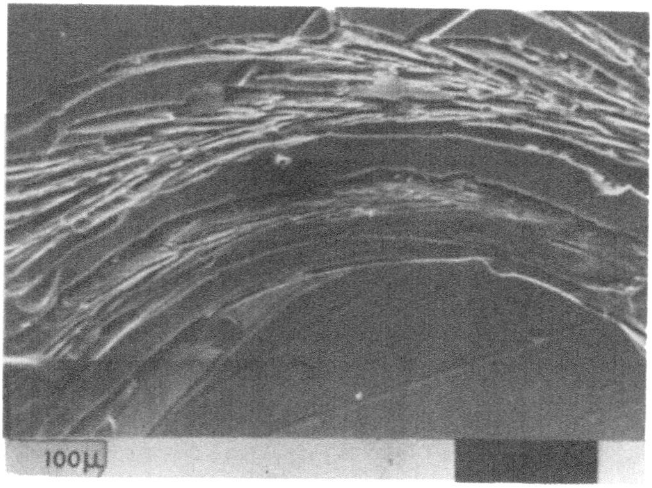

Figure 6. Branching of Spiral Crack (580°C, 52 ms^{-1}).

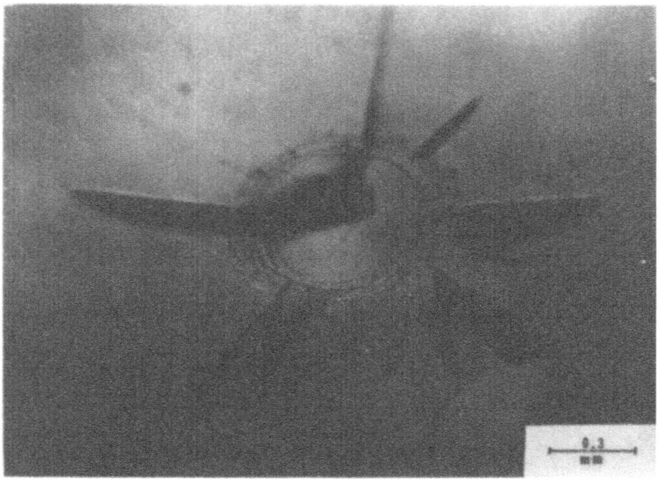

Figure 7. Radial Cracks (770°C, 57 ms^{-1}).

Radial Cracks. The shapes of radial cracks are illustrated in Figure 7 for a specimen impacted at 770°C and 57 ms^{-1}. This photograph was chosen because the damage is less complex than in many other cases and the specimen could be oriented to reflect the light from the radial cracks rather than transmitting it to the camera, thus revealing the radial cracks as dark patches with curved outlines. There is little indication of radial crack formation under the indentation. Therefore, these cracks did not form originally as median vent cracks.

The tips of the radial cracks were examined in detail and were found to be elliptical (Figure 8). The radii of crack tips, formed at various temperatures and velocities, were measured and are reported in Table I. These radii increase strongly with temperature.

There is a strong increase in the number of radial cracks with increasing impact velocity and temperature. One might expect the number of these cracks to be proportional to the total strain at the periphery of the indentation. Taking the radius of the indentation (a) as proportional to the circumference and assuming the tangential stress varies as $V^{6/5}$ as it does for elastic impacts, the number of radial cracks was plotted versus a $V^{6/5}$ in Figure 9. The essentially straight line is consistent with the above-stated expectation.

Table I. Radial Crack Tip Radii.

Velocity ms^{-1}	Temperature °C 700	720	740	770
	Crack Tip Radius - μm			
100	---	---	0.35-0.5	---
75	---	---	0.3	0.5-0.55
50	0.1	0.25	0.35	0.6

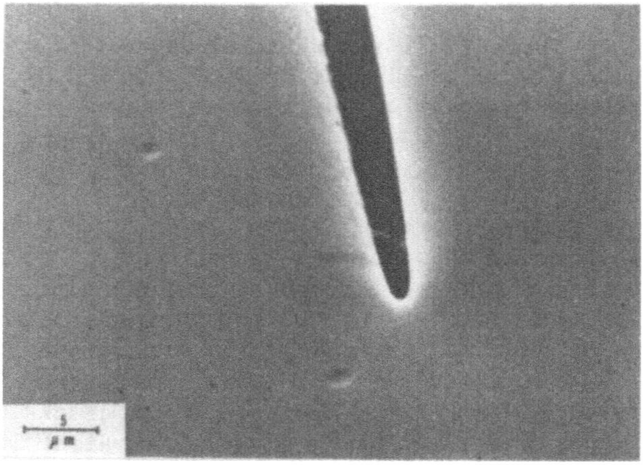

Figure 8. Radial Crack Tip (770°C, 57 ms^{-1}).

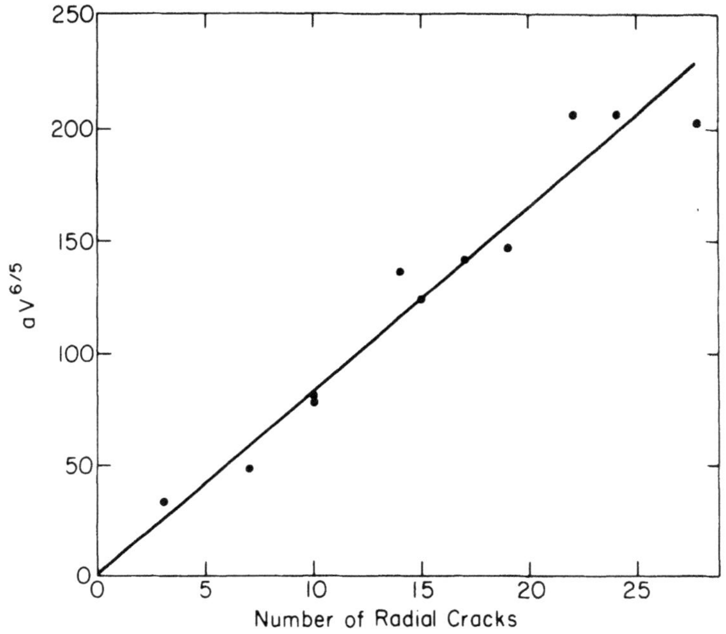

Figure 9. $aV^{6/5}$ vs. Number of Radial Cracks.

Lateral Vent Cracks. Lateral vent cracks form at various levels under the surface and propagate more or less parallel to the surface. The present results confirm Evans' suggestion that the maximum depth of the lateral vent crack formation increases with the indentation radius. Unusually long lateral vent cracks are formed at high temperatures (Figure 10). These cracks can be responsible for chipping (Figure 3) and may lead to an effective mode of material removal.

B. Energy Loss Mechanisms at Elevated Temperatures

At low temperatures, the principal energy loss mechanism for impacts on firmly supported glass plates is the stress wave energy (7). The onset of indentation at elevated temperatures introduces another important energy loss mechanism. Although, the coefficient of restitution (for impact against a stationary body this is the ratio of the rebounding velocity to the impact velocity) does not seem to have particular fundamental significance, it is a convenient way to compare the impacts at various temperatures and velocities. The variation of the coefficient of restitution was plotted for five different temperatures (Figure 11). A change in physical response occurs in the

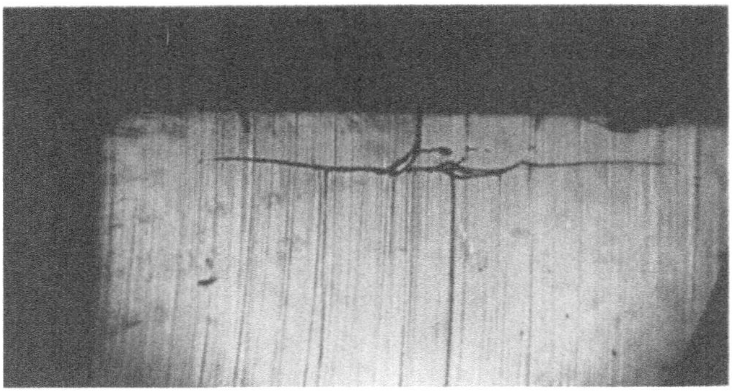

Figure 10. Lateral Vent Crack Formed at 740°C and 100 ms^{-1}(12 X).

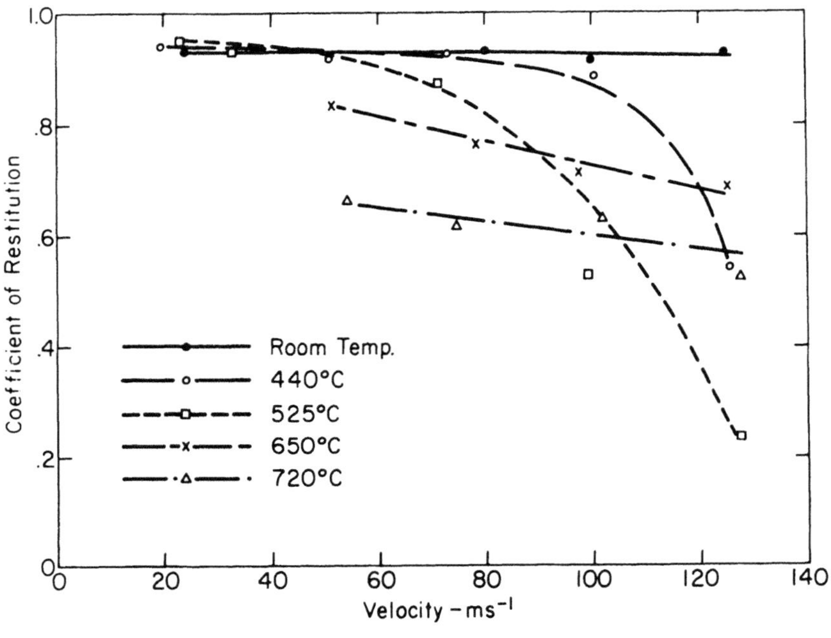

Figure 11. Coefficient of Restitution vs. Impact Velocity at Various Temperatures.

temperature range between 525 and 650°C. It will be recalled that this is the temperature range in which indentations were first observed. At lower temperatures the coefficient of restitution is very high at low velocities but decreases at high velocities, apparently because of increased stress wave energy loss and crushing. At higher temperatures the coefficient of restitution is never high because of the indentation energy losses. However, the decrease at higher velocities is less than that of the lower temperature specimens, apparently because there is substantially less crushing.

The kinetic energies of the impacting and rebounding spheres were calculated and the kinetic energy losses were determined by difference. In the low velocity-low temperature regime in which the stress wave energy is the principal loss mechanism one might expect to find that the losses increase as the 2.6 power of velocity as indicated by Hunter (8). The slopes in Figure 12 indicate these exponents which range from 2.3 - 2.7. However, at elevated temperatures the energy required to form the indentation becomes important so one cannot assume that one loss mechanism dominates. Nevertheless, the velocity exponents remain approximately the same indicating that the velocity exponents for the two mechanisms, stress wave and indentation energy, are not very different. At 440 and 525°C, the slopes increase at high velocities as a result of energy losses caused by crushing.

IV. CONCLUSIONS

The transition from Hertzian cone crack formation to indentation damage (radial and lateral vent cracks) was observed in glass plates impacted by glass spheres at velocities ranging up to 138 ms^{-1} and at temperatures ranging from 440 - 770°C. At impact velocities of about 50 ms^{-1}, the lowest temperature at which evidence of indentation was observed was 580°C. Cross sections of impacted specimens showed that Hertzian cone cracks persisted in these specimens, becoming much shorter at 700°C and finally disappearing at temperatures of 720°C and higher. Radial crack formation was common at temperatures of 700°C and higher. The radii of the crack tips of radial cracks were substantial suggesting that they may have propagated as blunt cracks. Lateral vent cracks also formed and were responsible for considerable chipping. The depths of the lateral vent cracks increased with the radii of the indentations.

The impacting and rebounding velocities were measured and were used to calculate the coefficients of restitution. Comparisons of these results showed that there was a change in the impact damage mechanism in the temperature range between 525 and 650°C. This temperature range includes 580°C, the temperature at which indentation was first observed.

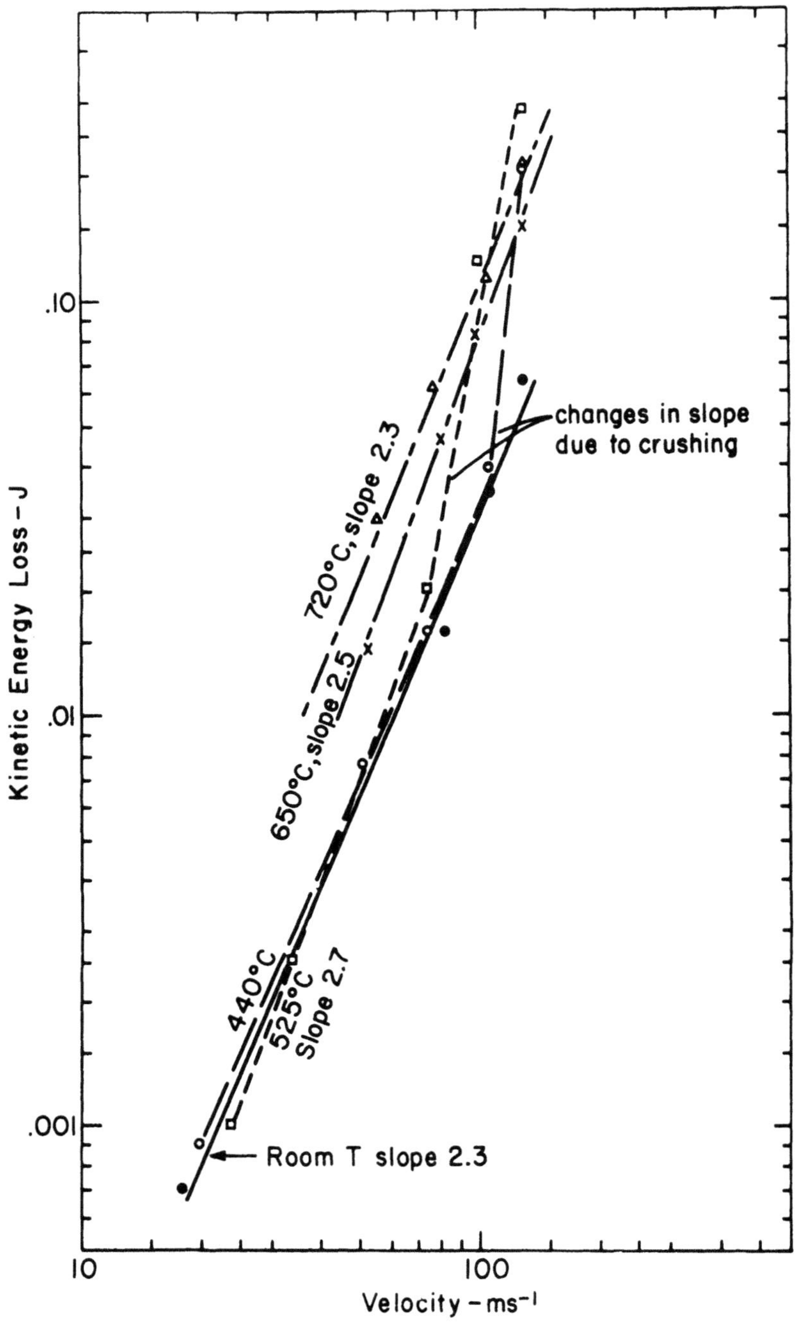

Figure 12. Kinetic Energy Loss vs. Velocity.

The kinetic energy losses were calculated and the velocity exponents were determined. At 525°C and lower temperatures, the velocity exponents ranged from 2.3 - 2.7, close to the values expected for the velocity dependence of the stress wave energy. At higher temperatures the indentation energy becomes an important loss mechanism but the slopes remain the same, indicating a similar velocity dependence for this second loss mechanism. When crushing occurs at temperatures of 440 and 525°C it introduces another loss mechanism having a greater velocity dependence.

REFERENCES

1. B. R. Lawn, S. M. Wiederhorn and H. H. Johnson, "Strength Degradation of Brittle Surfaces - Blunt Indenters," J. Amer. Ceram. Soc. 58 (9-10) 428-32 (1975).
2. B. R. Lawn, E. R. Fuller and S. M. Wiederhorn, "Strength Degradation of Brittle Surfaces: Sharp Indenters," J. Amer. Ceram. Soc. 59 (5-6) 193-197 (1976).
3. A. G. Evans and T. R. Wilshaw, "Quasi-Static Solid Particle Damage in Brittle Solids," Rockwell International Science Center First Report, Contract No. N00014-75-C-0669, (February, 1976).
4. H. P. Kirchner and R. M. Gruver, "Localized Impact Damage in Glass," Mater. Sci. Eng. 28, 153-160 (1977).
5. H. P. Kirchner and R. M. Gruver, "Localized Impact Damage in Stressed Members," Mater. Sci. Eng. 28, 249-255 (1977).
6. H. P. Kirchner and R. M. Gruver, "Branching of Hertz Cracks," Fracture 1977, Vol. 3 (1977) p. 959-964.
7. H. P. Kirchner and R. M. Gruver, "The Effect of Localized Damage on Energy Losses During Impact," Submitted for Publication.
8. S. C. Hunter, "Energy Absorbed by Elastic Waves During Impact," J. Mech. Phys. Solids 5, 162-171 (1957).

EROSION OF BRITTLE MATERIALS BY SOLID PARTICLE IMPACT

B. J. Hockey, S. M. Wiederhorn, and H. Johnson

Institute for Materials Research
National Bureau of Standards
Washington, D. C. 20234

ABSTRACT

Results are presented which show, that in addition to fracture, plastic deformation occurs and plays an important role in the erosion of brittle materials by solid particle impact. These conclusions are supported by transmission electron microscopy studies of impact damage produced in a wide variety of brittle materials and by erosive wear studies on silicon nitride and alumina. The erosive wear studies also indicate that while erosion resistance is primarily determined by fracture toughness and hardness, the relative importance of these materials parameters depends on the test conditions (e.g. temperature and angle of particle impingement).

1. INTRODUCTION

Because most ceramics are inherently brittle, they are susceptible to localized surface cracking in solid particle impact situations. Cracks formed in strong ceramics by a single impact event cause serious strength degradation for modest impact velocities (1). Moreover, continual, multiple impacts result in appreciable material loss by erosive wear. Thus, the damage resulting from solid particle impact leads to a deterioration in performance, and even failure of vital ceramic components. Despite these practical consequences, very few comprehensive erosive wear studies have been made on brittle materials. More importantly, the actual mechanisms that govern the erosive wear process in brittle materials are not well understood.

Attempts to describe the erosive wear behavior of brittle

materials have tended to be semi-empirical. Nevertheless, considerable insight into this problem has been made by a recognition of the similarity between quasi-static indentation and dynamic particle impact in the region of subsonic impact velocities. In this approach, one draws upon the similarity between the damage produced in both types of contact and uses quasi-static indentation theory to relate the extent of damage (and thus volume of material removed) to impact parameters and material properties.

This approach was first applied to an analysis of the erosive wear of brittle materials by Sheldon and Finnie (2, see also 3), who assumed that particle-target interactions were perfectly elastic. The extent of damage or volume removed was described on the basis of a classical Hertzian analysis of fracture beneath a spherical indenter. Inherently, this elastic/brittle treatment leads to the conclusion that the erosive wear of a brittle material depends on the fracture toughness (4) and on the size and distribution of pre-existing flaws (2,3).

The general assumption made by Sheldon and Finnie, that most solid particle impact events in brittle materials can be described as purely elastic contacts can be questioned. In most practical situations, surfaces are subjected to impact by angular, irregular-angular-shaped particles. Typical contact situations, therefore, resemble quasi-static indentation with indenters of small radii of curvature (i.e. sharp indenters (4)). These indenters are known to cause plastic flow (5,6) in brittle materials. Because of this, recent treatments of elastic/plastic contacts in brittle materials have suggested the importance of the flow properties (hardness) of the material as well as fracture toughness in determining the extent of fracture and erosive wear (7,8,9).

In this paper, the essential features of the quasi-static indentation approach to impact and erosive wear are examined on the basis of results from two recent experimental studies. Accordingly, the formal structure consists of two parts. In the first part, optical and transmission electron microscopy observations are used to characterize single particle impact damage in various brittle materials. In particular, emphasis is placed on illustrating the essential role of plasticity in impact situations and on demonstrating the correlation between "sharp" particle impact and quasi-static indentation. In the second part, comparative erosive wear data for alumina and silicon nitride are used to describe the dependence of erosive wear on particle velocity, angle of impingement, and temperature. These data are then discussed in the light of the single particle damage observations, and also in terms of the relative importance of fracture toughness and hardness.

2. SINGLE PARTICLE IMPACT DAMAGE

2.1 Materials

To elucidate the basic mechanisms of erosive wear of brittle materials and to provide a clearer description of the residual damage that affects material properties, solid particle impact damage studies were made on a variety of brittle materials: silicon; germanium; magnesium oxide; aluminum oxide; silicon nitride; and silicon carbide. For the most part, these materials were selected because of their practical importance and because they represent brittle materials that differ in their atomic bonding, hardness, and fracture toughness.

2.2 Experimental Procedure

All studies were conducted using an erosion tester previously described by Wiederhorn and Roberts (10). Sample surfaces were subjected to a controlled flux of abrasive grit under fixed conditions of particle velocity, angle of impingement, and temperature (up to 1000°C). To distinguish individual particle damage events - for a given set of experimental conditions - only small quantitites of abrasive grit were allowed to impinge on polished (chemically or mechanically) sections. The polished sections were then examined optically to characterize the nature and extent of surface damage. In most cases, the impacted samples were thinned and examined by transmission electron microscopy to reveal details of the subsurface damage. For these samples, the thinning procedure closely followed that used in previous indentation and mechanical abrasion studies on ceramics (5,6).

2.3 Results

2.3.1 Impact Surface Damage. Room temperature, normal-incidence impact damage studies were made using measured particle velocities (11) in the range, 30 to 100 m/sec. 100 mesh (150 micron size) silicon carbide abrasive grit was used in tests on Al_2O_3, SiC and Si_3N_4; whereas 150 mesh (∿70 micron size) alumina grit was used in tests on Ge, Si, and MgO*. Both types of abrasive grit were composed of irregularly-shaped, angular solid particles, Fig. 1, which according to the scheme advanced by Lawn and co-workers (1) fall into the "sharp" indenter/particle category.

*Because the prethinned (∿100 to 150 micron) sections of Ge, Si and MgO necessary for subsequent TEM examination tended to shatter in tests involving the larger particles, the 150 mesh alumina grit had to be used on these specimens.

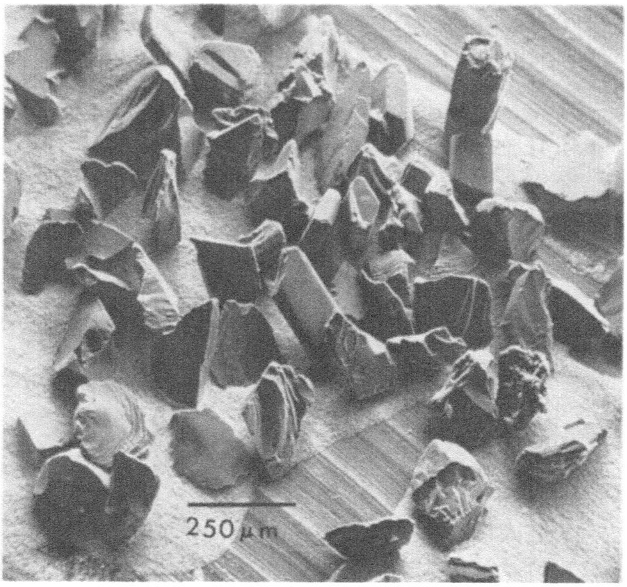

Figure 1. Scanning electron micrograph illustrating angular morphology of silicon carbide particles used in impact damage and erosive wear studies.

Because the irregular impacting particles can strike the surface in any arbitrary orientation, individual impact sites were found to differ considerably in their detailed appearance. Despite this difference, the residual impact damage was found to closely resemble that produced quasi-statically in brittle materials using "sharp" indenters. In this regard, particle impact usually resulted in the formation of a permanent surface crater (indicative of surface penetration), surrounded by a well-defined, associated crack pattern. As illustrated in Figure 2, the localized fracture patterns produced by normal-incident impact were identical in all respects to those produced by quasi-statical indentation. The fracture patterns usually contained both median and lateral cracks. In all materials studied, material loss from the surface occurred as the result of the extension of the lateral cracks to form a surface chips, Figure 3. This observation clearly indicates the importance of fracture to the process of erosion in brittle materials, and therefore suggests that the fracture toughness is one of the primary material parameters to be considered in problems of erosive wear.

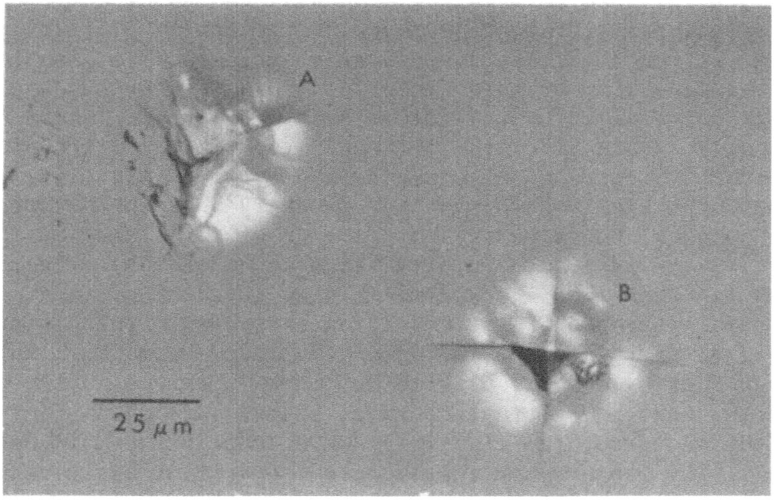

Figure 2. Comparison of surface damage produced in Sic by: (A) 90° impingement (150 micron size SiC particle at 90 m/s) and (B) quasi-static indentation (Vickers diamond pyramid, 400 gram load).

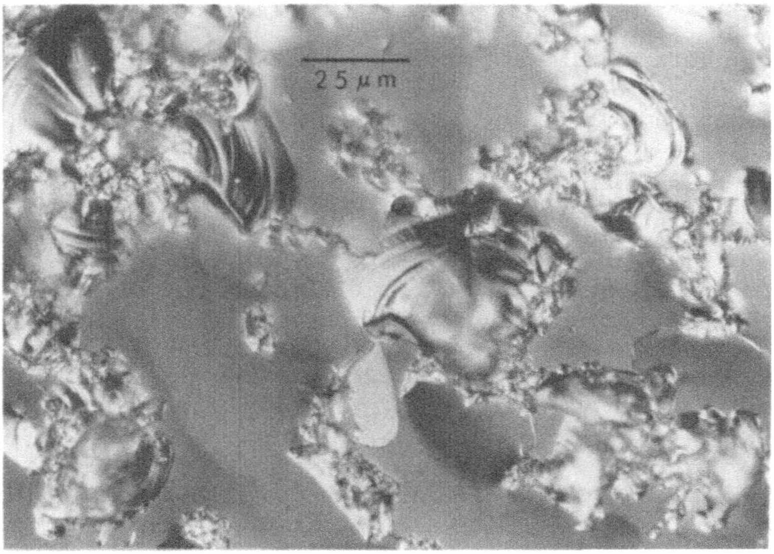

Figure 3. Surface of damage produced in Germanium by 90° inpingement impact (70 micron size alumina at 90 m/s). Note material romoval by the extension of lateral-vent type cracks.

2.3.2 Plastic Deformation: Occurrence and Extent. Although the materials studied in this paper are generally considered to be brittle, it can be demonstrated that plastic flow is also important to the erosive wear of these materials. As in previous studies of indentation sites in brittle materials (5,6), transmission electron microscopic examination of room temperature impact sites clearly revealed the occurrence of plastic flow by slip. This was evidenced not only by a highly deformed zone directly beneath the area of actual contact, but also by the presence of a residual dislocation pattern in the adjacent vicinity, Figures 4a and 4b. The transmission electron microscopy studies also revealed distinct differences in material response to the impacting particles. The differences appeared to be related to the nature of atomic bonding. In covalently-bonded Ge and Si, plastic deformation extended only slightly beyond the region of actual contact, Figure 4a. As a result, the radial dimension of the elastic-plastic zone in these materials was only slightly larger than the surface impression, which was small when compared to the associated fracture pattern, Figure 3. In contrast, particle impact of ionically bonded MgO resulted in extensive plastic flow. As seen in Fig. 4b, deformation extends far beyond the region of actual contact and often encompasses the associated fracture pattern.

Observations on Al_2O_3 and SiC, which possess partial ionic and covalent bonding, revealed a somewhat intermediate behavior during room temperature, normal-incidence impact. For these materials, the radial extent of the deformation was estimated to be about twice the lateral dimensions of the residual impact impression. However, under most conditions, the extent of deformation was considerably smaller than the length of the cracks.

2.3.3 Effect of Temperature. Surface damage produced in Al_2O_3 and Si_3N_4 by solid particle impact at temperatures up to 1000°C was investigated by optical microscopy. Impact damage in Al_2O_3 was also examined by transmission electron microscopy. Over the temperature range studied, no difference was found in the general morphology of the impact fracture patterns, Figures 5a and 5b. For normal-incidence impact, localized fracture, consisting of lateral and median vent cracks, remained evident, thus indicating that material removal still occurred predominantly by surface chipping.

Transmission electron microscopy observations on Al_2O_3, however, did reveal a considerable increase in the relative extent of the deformation associated with impact at high temperatures. In contrast to room temperature results, the radial extent of deformation by slip at 1000° was found to be comparable to the radial extent of fracture (which appeared unchanged between 25° and 1000°C). In addition, impact at 500°C and above resulted in a significant increase in deformational twinning. As a result, high

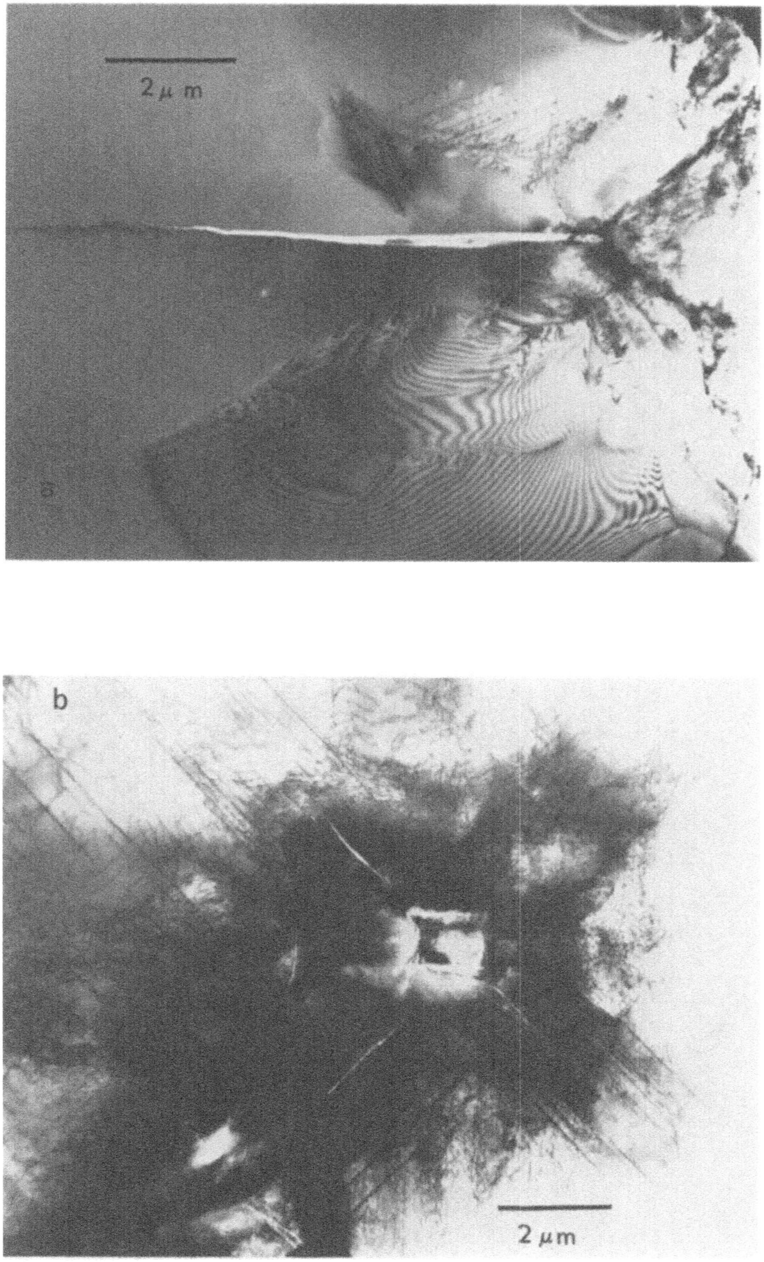

Figure 4. Transmission electron micrographs illustrating subsurface damage produced in (a) silicon and (b) MgO under similar impact conditions (90 micron size alumina particle at 90 m/s).

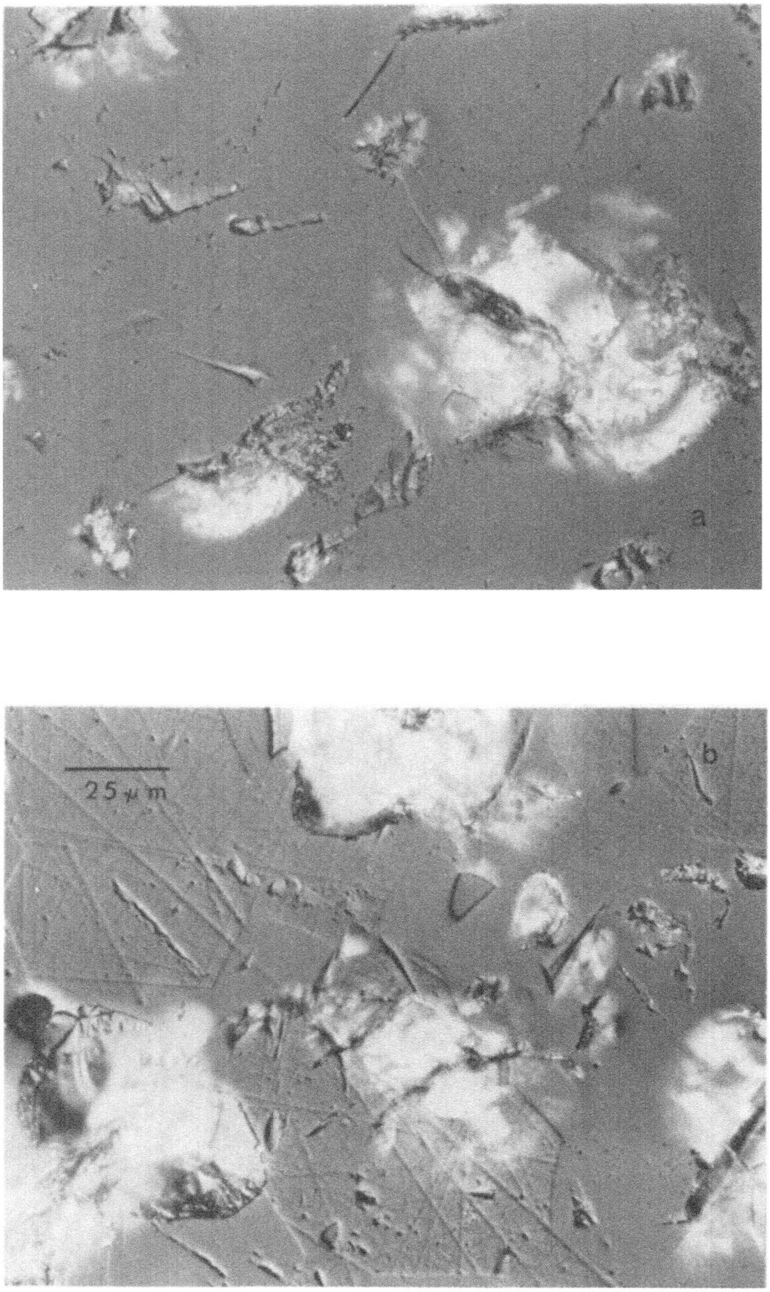

Figure 5. Normal-impingement impact damage produced in Al_2O_3 at (a) 25°C and (b) 1000°C.

temperature impact also resulted in numerous interactions between slip bands, twins, and grain-boundaries outside the region of contact.

2.3.4 Effect of Angle of Impingement. Observations on the various brittle materials included in this study showed that both the nature and extent of impact damage depends upon the angle of particle impingement. Characteristic changes that evolve as the angle of impingement is reduced to 15° are illustrated in Figures 6a and 6b. Predictably, the most obvious change is a considerable reduction in the extent of localized surface cracking. In addition, cracks produced during low angle impact were predominantly of the lateral vent type; median vent crack formation appeared to have been almost completely surpressed under these conditions.

Another characteristic of low or oblique angle impact, Figure 6a, is the formation of shallow, residual surface impressions, typically elongated along the horizontal component of particle motion. Again, transmission electron microscopy studies on SiC and Al_2O_3 confirmed that these surface impressions are the result of plastic flow, Figure 6b. Thus, oblique angle impact of the brittle materials investigated in this study is seen to result in the ploughing of material parallel to the surface. This finding suggests that at low angles of impingement plastic flow plays an important role in the erosion process and, as will be discussed, may dominate the erosion process under certain conditions (Section 4).

3.0 Erosive Wear Studies

The observations described above will now be correlated with studies of the erosive wear of hot-pressed silicon nitride and two grades of alumina (Al_2O_3) that differed microstructurally. One grade of alumina was a large-grained (∿30 μm), sintered product; the other was a fine-grained (∿5 μm), hot-pressed product. In this section, erosive wear data, describing the dependence of erosion on particle velocity, temperature, and angle of impingement are presented and compared.

3.1 Experimental Procedure

The erosive wear tests were conducted using the hot erosion tester described by Wiederhorn and Roberts (10). In this study, bulk samples (typically 1/2 x 1/2 in by 1/4 in thick) were eroded by the action of 100 mesh SiC abrasive grit (∿150 μm particle size), Fig. 1. Tests were conducted on each of the materials at temperatures ranging from 25°C to 1000°C. Angles of particle impingement were varied from 15° to 90°. In all tests at oblique angles of impact, the leading edge of the test samples were protected to avoid extraneous edgewear. Rates of erosive wear were determined by measuring sample weight loss after long-term exposure.

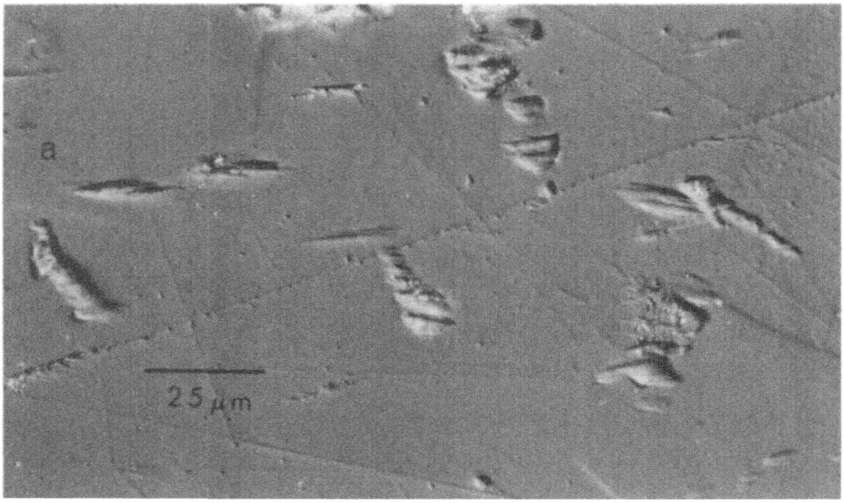

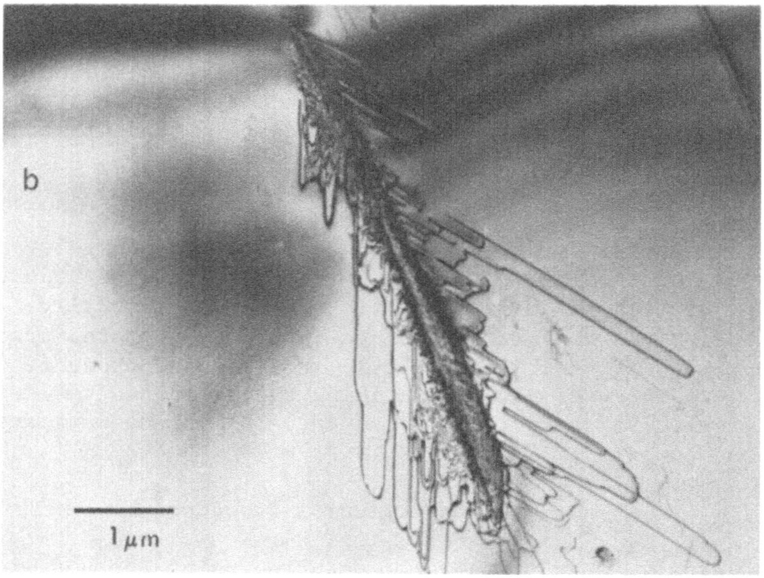

Figure 6. (a) Optical micrograph showing series of shallow surface impressions produced in Al_2O_3 by 15° impingement impact. (b) Transmission electron micrograph showing region beneath shallow surface impressions produced under similar impact conditions in SiC. Presence of dislocations emerging from contact zone and absence of cracks confirm fully plastic nature of impact event.

The erosion rates are expressed here in terms of the normalized quantity: sample weight loss per unit weight of abrasive actually striking the surface.

3.2 Results

Representative wear data, pertinent to a general description of the observed erosive wear behavior, are shown in Figures 7, 8, 9, and 10. Figures 7, 8, and 9, illustrate the dependence of erosion rate on particle velocity and temperature for the different materials. Figures 10, a and b, illustrate the dependence of erosion rate on angle of impingement at 25° and 1000°C, respectively. In each plot, the data points represent average values obtained from a minimum of three determinations; usually, however, five or more tests were made for each condition. Finally, Figures 11, a and b, illustrate the change in surface morphology that occurs when the angle of particle impingement is reduced from 90° to 15°.

3.2.1 Dependence on Particle Velocity. For all test conditions, erosive wear increases with increased particle velocity. As indicated in Figures 7 and 8, erosion rate, E, could be expressed as a simple power function of the particle velocity, V: $E \propto V^n$. On the basis of a linear least squares fit of each set of data, values of the velocity exponent, n, were found to range from 1.7 to 2.7, depending on temperature, angle of impingement, and the material tested. Although the value of n was found to increase with temperature, a consistent relationship could not be established between n and other experimental variables.

3.2.2 Temperature Dependence. A significant effect of temperature on the erosive wear of the materials studied was only apparent in the data for oblique angles of incidence (i.e. α=15° and 30°). This is illustrated in Figure 7 and in Figure 8, where 90° and 15° impingement data for each material are compared.

As seen in Figure 7, erosive wear rates at 90° impingement are, for all practical purposes, independent of temperature between 25° and 1000°C. For each material, changes in rate of wear or slope, n, with temperature were too small to be considered statistically significant. The 90° impingement data, moreover shows that hot-pressed silicon nitride is more resistant to erosion than the hot-pressed alumina (by a factor of ∿1.5 on a volume loss basis), which in turn, is more resistant than the sintered alumina (by a factor of ∿2-3).

In contrast to the results described above, data obtained using particle impingement angles of 30° and 15° both* indicated

*For brevity, the data obtained for 30° impingement are not presented in this paper.

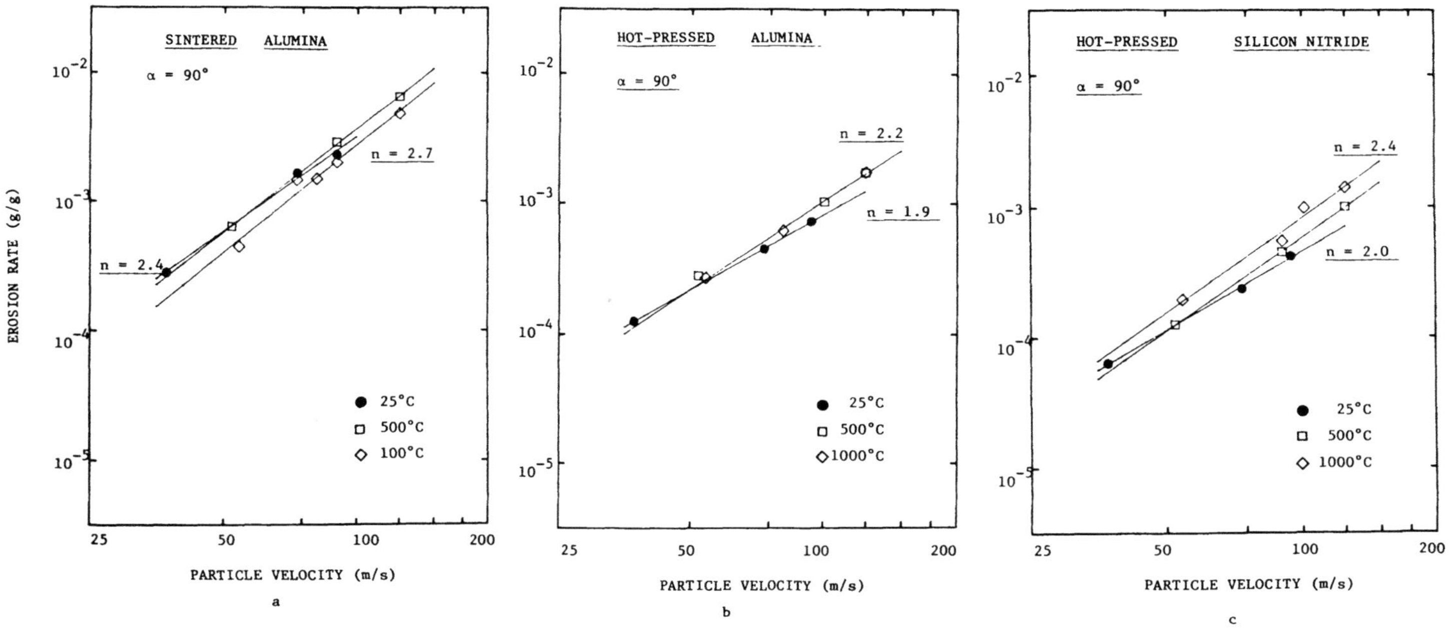

Figure 7. Erosive wear rate as a function of particle velocity and temperature for 90° impingement: (a) Sintered alumina; (b) Hot-pressed alumina; (c) Hot-pressed silicon nitride.

that erosion is affected significantly by temperature for oblique angles of impact. Under these conditions the erosive wear data for each material exhibited a substantial increase in slope, n, as the temperature was increased from 25°C to 1000°C. This increase can be seen in Figures 8a, 8b, and 8c, where 15° impingement data at 25°C and 1000°C are compared for each material. For hot-pressed silicon nitride, Figure 8a, the value of n increased from 1.7 at 25° to 2.7 at 1000°C. Although the erosion behavior of the hot-pressed alumina was similar to that of the hot-pressed silicon nitride, the increase in n from 2 to 2.6 was not as great as for the hot-pressed silicon nitride. A similar observation is made for the sintered alumina, for which n increased from 2.1 to 2.6 over the same temperature range. The erosion behavior of the sintered material, however, is seen to differ significantly from that of the hot-pressed materials, in that erosive wear rates are lower at 1000°C than at 25°C over most of the range of particle velocities used in these tests (Figure 8c).

The net effect of temperature on erosion wear at oblique angles of impingement is illustrated in Figure 9 for the 15° data. At 25°C, Figure 9a, hot-pressed silicon nitride exhibits better erosion resistance than the two types of alumina, while the hot-pressed alumina exhibits better erosion resistance than the sintered alumina. At both 15° and 30° impingement, the room temperature erosion behavior of the materials was comparable to that found at 90° impingement. In contrast to these results at 25°C, results obtained at 1000°C for both 30° and 15° impingement, Figure 9b, show little difference in the erosive wear of these materials. Thus at 1000°C, the erosion behavior at oblique angles of impingement differs significantly from that found at α=90°. This change in erosive behavior at high temperatures and oblique angles of impingement appears to be related to an increase in the role of plastic flow in the erosion process (see Sections 3.2.3 and 4.0).

3.2.3 Dependence on Angle of Impingement. Measurements of the effect of impingement angle on the erosion rate were also made to investigate mechanisms of erosion. Past studies (12,13) have shown that erosion by a brittle mode is characterized by a maximum in erosion rate at 90° impingement and by a relatively rapid decrease in erosion rate as the impingement angle is decreased. Ductile erosion, by contrast, is characterized by a peak in the erosion rate at an angle of impingement of ∿15-20° with a decrease in the erosion rate at higher and lower angles of impingement. It is generally recognized that most metals undergo combined modes of erosion. In the present investigation, the observed variation in rate of erosion with impingement angle for each material also indicates a combined mode of erosion.

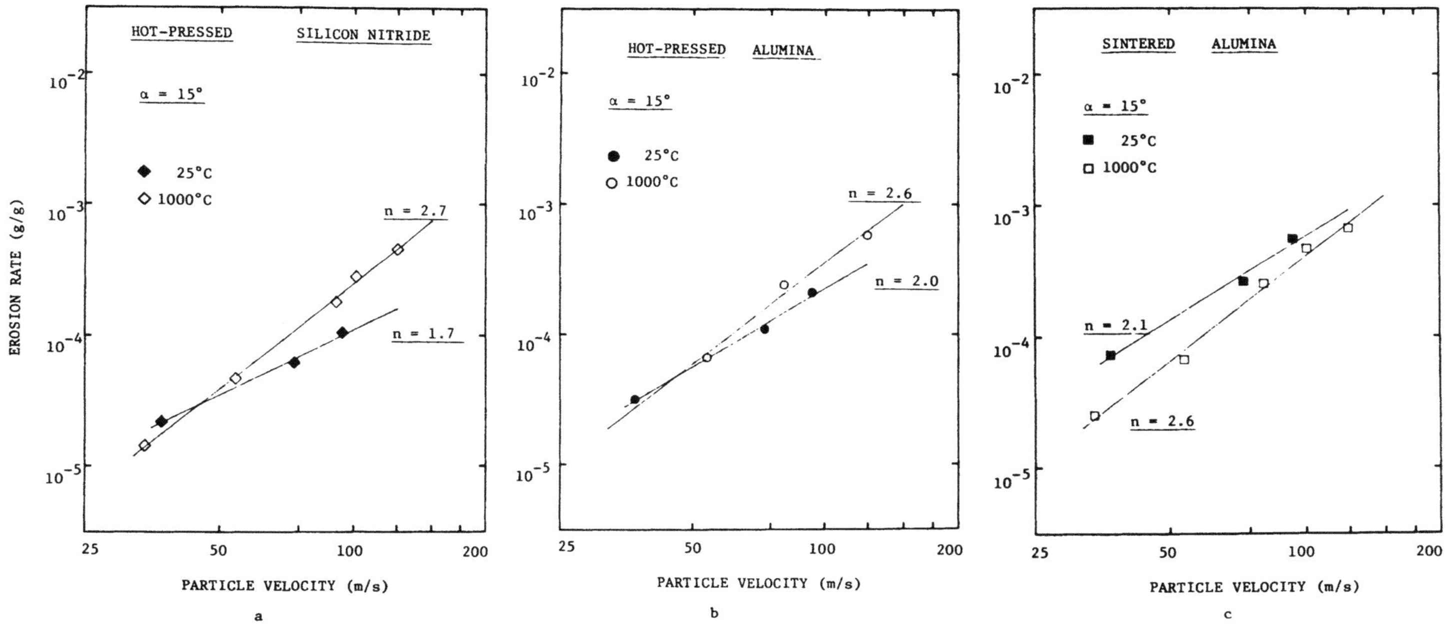

Figure 8. Erosive wear rate as a function of particle velocity for 15° impingement at 25° and at 1000°C: (a) hot-pressed silicon nitride; (b) hot-pressed alumina; (c) sintered alumina.

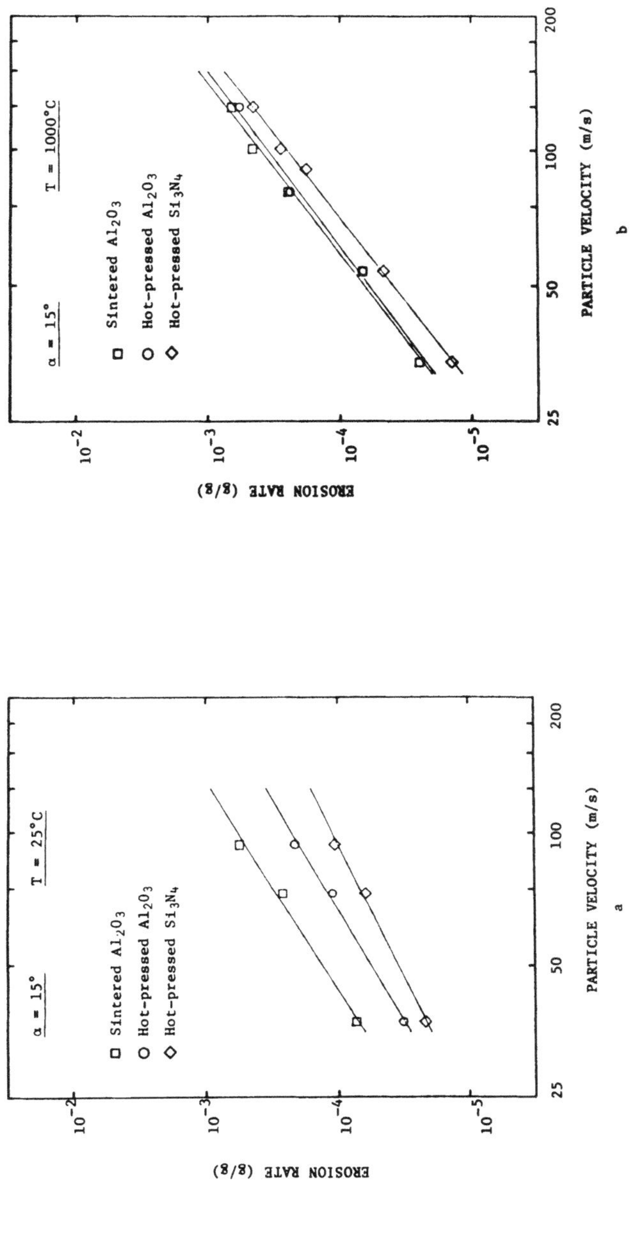

Figure 9. 15° impingement erosive wear data for hot-pressed silicon nitride, hot-pressed alumina, and sintered alumina compared at: (a) 25°C and (b) 1000°C.

In Figures 10a and b, erosive wear rates for silicon nitride are plotted as a function of impingement angle for temperatures of 25°C and 1000°C respectively. For both temperatures, the data are compared with a theoretical treatment of brittle erosion that will be discussed below (Section 4). As can be seen in Figure 10a, erosion rate follows the theoretical curve for angles as low as 30°. However, at 15° impingement, the rate of erosion is approximately 5 times that predicted by theory. At 1000°C, Figure 10b, this same type of comparison also indicates a large difference between measured and predicted erosion rate but now this difference is observed over a wider range of impingement angles. At 15°, the measured erosion rate is approximately 7 times the predicted rate, while at 30° impingement the measured erosion rate is approximately 2 times the predicted rate. At 45° impingement, the measured erosion rate is still 1.6 times the predicted rate. Similar differences between measured and predicted wear were also found in the data for both types of alumina. These deviations between measured and predicted behavior suggest that wear of the materials studied occurs by a mixed mode of erosion, with the ductile mode becoming important at low angles of impingement. On this basis,

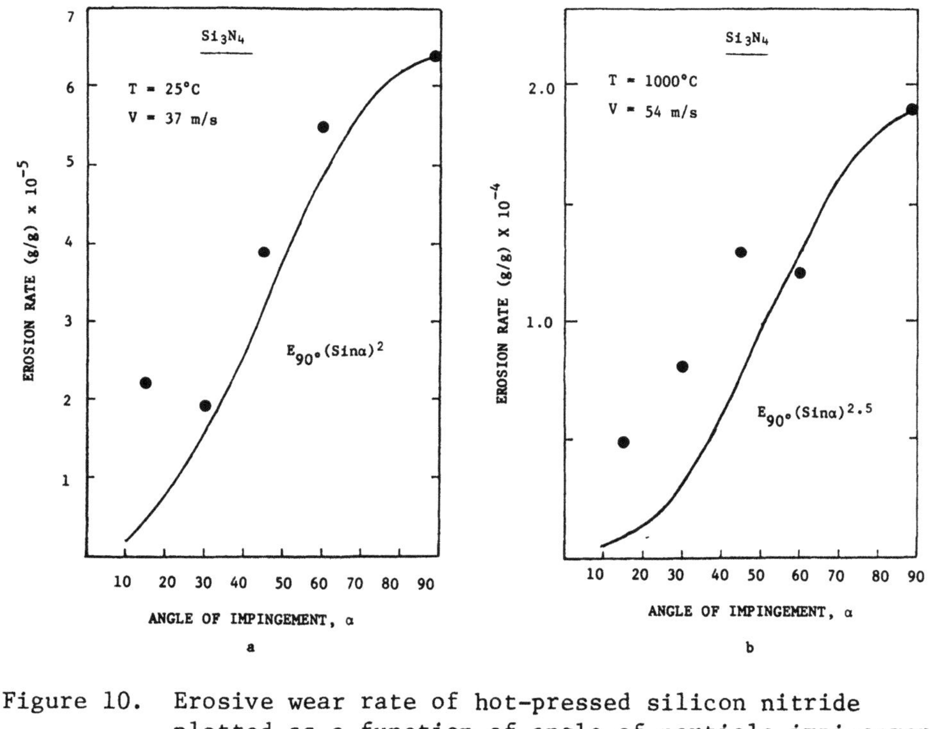

Figure 10. Erosive wear rate of hot-pressed silicon nitride plotted as a function of angle of particle impingement α. Full-line curves represent predicted dependence for purely brittle behavior (see text).

the data also indicates that as the temperature is increased, the role of plastic flow becomes increasingly more important in the erosion process.

The above conclusions are supported by optical and scanning electron microscopy studies of the erosion surfaces. Under erosion conditions where there is good agreement between measured and predicted wear, (i.e. $\alpha = 90°$ to $30°$ at 25°C, and $\alpha = 90°$ to $\sim 45°$ at 1000°C, Figure 10) the resultant surfaces are similar and clearly indicate that material has been removed primarily by a brittle chipping process, Figure 11a. By contrast, erosive wear at lower angles of impingement (i.e. $\alpha = 15°$ at 25°C and $\alpha = 15°$ and 30° at 1000°C), results in relatively smooth surfaces, Figure 11b. These surfaces, moreover, are characterized by numerous smooth furrows or wear scars indicative of material removal by a ductile ploughing process. While lateral venting is also observed along the edges of the wear scars, and obviously contributes to material removal, it is nevertheless clear that, for these conditions, the mechanism of erosion more closely resembles the ductile mode of wear, normally attributed only to metals. These observations, moreover, are consistent with the single particle impact damage results discussed in Section 2.

4. DISCUSSION

The results presented in this paper all point to the importance of plastic deformation in the erosion of materials that otherwise exhibit brittle behavior. Plastic flow appears to enter the erosion process in two distinct ways: by subsurface deformation that enhances surface chipping; and by shear deformation wherein fracture is minimal and material is removed primarily by a plastic cutting process.

In the chipping mode of erosion, material removal occurs primarily by the extension of lateral vent type cracks (as in Figure 3). The nature of these cracks and the way in which they grow during quasi-static indentation has been described by Lawn and Swain (14) for glass, and by Hockey and Lawn (6) and Evans and Wilshaw (7) for crystalline materials. These studies have shown that lateral vent cracks are peculiar to "sharp indenter" contact situations and that they fully develop only after removal of the applied load. Previous studies by transmission electron microscopy (5,6) have shown that the indentation of brittle materials (e.g. Al_2O_3 and SiC) also results in intense deformation, which is generally constrained within the region of contact. As a consequence, large residual stresses are produced. These stresses are believed to provide the driving force for lateral-vent crack formation. Based on results obtained in the present study, a similar explanation is advanced for the formation of lateral vent cracks during solid particle impact events. Accordingly, the

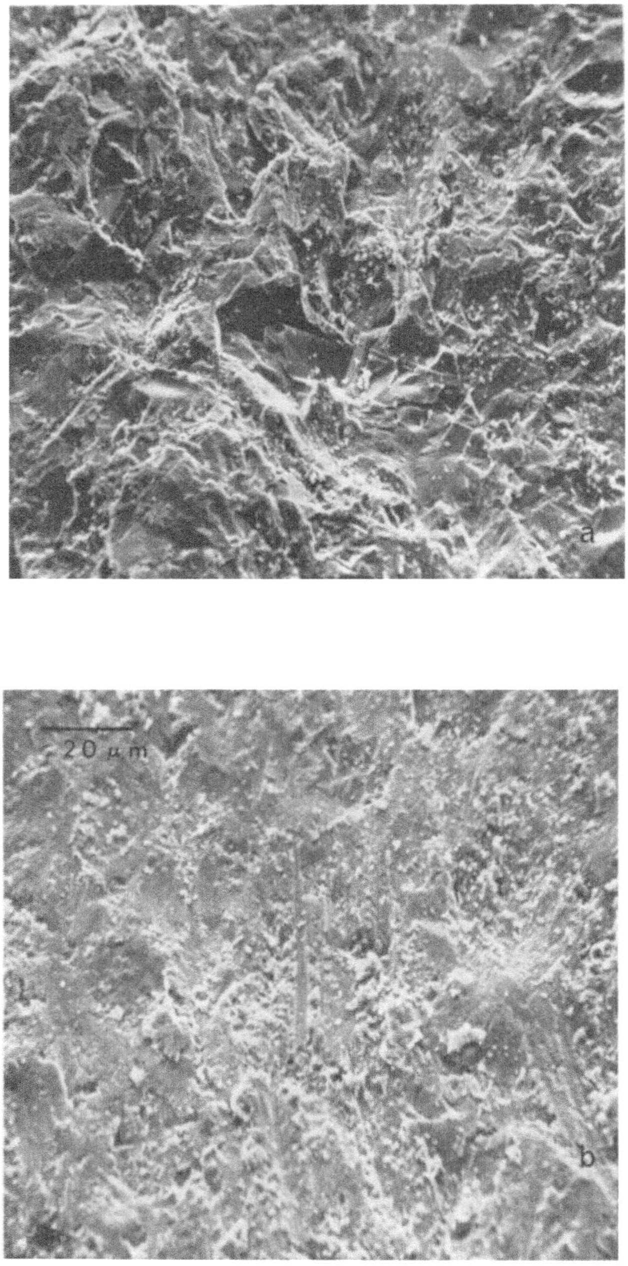

Figure 11. Scanning electron micrographs showing surface morphology of sintered alumina test samples after erosion at 1000°C: (a) 90° impingement angle and (b) 15° impingement angle.

erosion of brittle materials by surface chipping should depend both on resistance to crack extension (fracture toughness) and on resistance to plastic flow (hardness).

Recently, Evans et al.(8,9) have presented a theoretical treatment for the erosion of brittle materials by "sharp" particle impact. They estimate the erosion rate from the depth of penetration of the incoming particle and the size of the lateral cracks that form during the impact event. Accordingly, this treatment incorporates both the hardness, H, and the critical stress intensity factor, K_c, in the equation that describes the amount of material removed per impact $\bar{v}_i$:

$$\bar{v}_i \overset{\sim}{\propto} V_o^{2.5} R_p^{4} \rho_p^{0.3} H^{-0.3} K_c^{-1.5} \tag{1}$$

where V_o is the incoming particle velocity; R_p is the particle radius; and ρ_p is the particle density. The exponents in this equation are obtained in part through a mathematical analysis of the impact problem, and in part through an experimental investigation of the extent of damage that occurs in materials during impact. In applying this expression to multiple impact situations, it is assumed that interactions between adjacent impacts are negligible and that each impact event is similar. The theory is the first to describe the erosion of brittle materials in terms of an elastic-plastic phenomenon and provides a good basis for discussing the erosion data presented in this paper.

Several aspects of the data in this paper can be described in terms of the theory developed by Evans and his colleagues. First and foremost, the theory assumes that plastic flow occurs in the region beneath the contact area. The results of Section 2, provide direct evidence to support this assumption. Furthermore, as discussed earlier in this Section, the formation of lateral vent cracks appears to be directly related to the occurrence of plastic deformation and the development of large residual stresses.

A second experimental parameter that can be compared with the theory developed by Evans et al. is the exponent of the impact particle velocity. In the present study, velocity exponents ranging from 1.7 to 2.7 were obtained; the lower values being representative of room temperature data at low impingement angles, the high values being representative of the high temperature data. The theory developed by Evans et al. predicts an exponent of 2.5, which differs slightly but significantly from the values obtained in this study. The difference between the predicted and experimental values of the velocity exponent can be traced to the assumption used to predict the amount of material removed during multiple impacts. The erosion rate per particle impacting the surface was assumed to be proportional to the depth of the plastic zone and the square of the size of the lateral cracks. Minor variations in this assumed proportionality with velocity would lead to a velocity exponent that differed from the one given by Evans et al. Thus, a dependence on velocity of lateral crack shape or the depth from which these cracks originate could significantly alter the value of the velocity exponent. Furthermore, the theory assumes that all particles hitting the surface cause the same amount and type of damage. Because of a spread in the velocity of particles in an erosion experiment, the arbitrary orientation of particles as they contact the surface, and the surface roughness that develops during the erosion process, this assumption can not be satisfied rigorously. As a consequence, the velocity exponent might be expected to differ from the value of 2.5 predicted from the theory. Considering the difficulty of incorporating these various parameters into the theory, modification of the theory to handle these parameters may not be possible.

A final comparison between theory and experiment can be made by considering the relative erosion rates of alumina and silicon nitride at normal angles of impact. Based on volume loss, the erosion rate of the sintered aluminum oxide is approximately four times that of the silicon nitride at 25°C. Since the room temperature hardness of these two materials are about the same, relative differences in the erosive wear of the two materials should depend only on the critical stress intensity factors of the two materials. Specifically, the ratio of erosion rates should be given by ratio of the two K_c values raised to the 1.5 power, i.e.,

$$\bar{v}_i(Al_2O_3)/\bar{v}_i(Si_3N_4) \propto [K_c(Al_2O_3)/K_c(Si_3N_4)]^{-1.5} \qquad (2)$$

Given that K_c for Al_2O_3 is 2.5 $MN/m^{3/2}$ (15) and K_c for Si_3N_4 is 5 $MN/m^{3/2}$ (16), the predicted ratio of erosion wear is ∿2.9*. Considering the approximations involved in the theory, and the uncertainties of the experimental values of K_c, the predicted ratio is considered to be in good agreement with the experimental value. As predicted by the theory, normal-impingement erosion seems to be controlled by the resistance of the material to crack propagation.

A similar prediction of the erosion rate can be made from the high temperature erosion data. However, since the hardness of

*Because 30 μm gain Al_2O_3 was used in these studies, the contact area during impact was usually confined to a single grain. Therefore, single crystal values of K_c were used in this comparison.

Si_3N_4 has not been measured at elevated temperatures, the prediction must be considered more approximate than the room temperature prediction. Assuming the hardness of aluminum oxide and silicon nitride are equal at 1000°C, and using measured values of the critical stress intensity factors for the two materials (∿1.8 $MN/m^{3/2}$ for single crystal Al_2O_3 at 800°C; ∿5 $MN/m^{3/2}$ for Si_3N_4 at 1000°C), the erosion rate of the aluminum oxide is expected to be ∿4.6 times the erosion rate of silicon nitride. The ratio, ∿3, of the two erosion rates obtained experimentally is in qualitative agreement with the predicted value.

A similar qualitative comparison of data between the hot-pressed aluminum oxide and the other two materials cannot be made because the critical stress intensity factor for this material was not measured. However, because of the small grain size of hot-pressed aluminum oxide, a polycrystalline value of K_c would have to be used to describe the erosion rate. Since polycrystalline values of K_c are typically higher than single crystal values, the erosive wear of the hot-pressed aluminum oxide would be expected to be less than that of the sintered aluminum oxide. This expectation is confirmed by experimental observation.

Considering all of the data presented in this paper on erosion at normal incidence of impact, the elastic/plastic treatment developed by Evans et al.(8) seems to provide a good semi-quantitative description of the erosion process. Plastic flow is always observed beneath the contact area; the dependence of erosion rate on particle velocity agrees qualitatively with the predicted dependence; and the relative erosion resistance of the three materials studied is roughly that expected from the theory. However, consideration of the observed variation in erosive wear with impingement angle indicates other factors influence the erosion process. By modifying the theory presented by Evans et al. to include oblique angles of impact, it can be shown that erosion at low angles of impact involves shear deformation as well as brittle fracture.

Although the theory of Evans et al. was developed for normal particle incidence, it can be extended to oblique angles of incidence by assuming that lateral crack extension and penetration is determined only by the normal component of the particle velocity. Based on this assumption, the erosion equation of Evans et al. is modified by substituting the normal component for the particle velocity, $V_o = V_o \sin \alpha$, into equation 1. Thus, the dependence of erosion rate on angle of impingement should be given by:

$$\bar{v}_i \propto (V_o \sin \alpha)^n , \tag{3}$$

where V_o is the particle velocity, α is the angle of impingement, and where the experimental velocity exponent, n, has been substituted for the value of 2.5 given in equation 1.

Equation 3 is compared in Figures 10a and b with erosion data for hot-pressed Si_3N_4. At 25°C, n was found to have a value of ~ 2 for normal-impingement erosion. The data is, therefore compared with a plot of the rate of erosion for normal impact times $\sin^2\alpha$ in Figure 10a. The comparison of the theoretical and experimental values of the erosion rate is satisfactory for impingement angles greater than 30°. However at 15° impingement, the measured rate of erosion is approximately five times the predicted value. A similar comparison of experimental with predicted erosion rates is given in Figure 10b, for data obtained on hot-pressed silicon nitride at 1000°C. At this temperature, the value of n obtained for normal impact was ~ 2.5. A plot of $\sin^{2.5}\alpha$ times the erosion rate for normal impact is found to agree with the experimental data only for 60° impact. At lower impact angles, the erosion rate is larger than expected.

The deviation of experiment from theory described above can be attributed to the increased importance of a shear mechanism of erosion at low impact angles. Particles impacting upon the target surface remove material primarily by ploughing in a manner described in much of the erosion literature on metals. In view of the microscopic evidence for plastic flow underneath the impact sites, it is reasonable to suggest that the mechanism of material removal in the plowed area is by ductile flow. However considering the fact that some lateral venting also occurs, a contribution of fracture to erosion at oblique angles cannot be dismissed. If plastic ploughing is the primary mechanism of erosion, then the erosion rate should depend more on the resistance to deformation (i.e. hardness), than on fracture toughness. Therefore, materials that have the same hardness, but different fracture toughness are expected to have the same erosion rate for similar erosion conditions. This conclusion is supported by the results shown in Figure 9b, which compares the 15° impingement data obtained at 1000°C. Little difference is seen between the data for the three materials. Had the erosion data been presented in terms of the volume loss rather than the weight lost per weight of impacting particles, the comparison would have been even closer. In contrast, distinct differences in the room temperature erosion behavior for oblique angles of impact, Figure 9a, suggests that at room temperature fracture still plays a role in the erosion process.

5. SUMMARY

In this paper, results are presented which show that plastic flow plays an important role in the erosion of brittle materials.

At high impact angles, the occurrence of plastic flow provides an explanation for the growth of cracks that result in material removal from the target surface. At oblique angles of impingement, impact damage is often free of cracks suggesting that material removal occurs by a shear deformation process. Erosive wear data obtained in this study further support the importance of plastic deformation to the erosion process. The experimental data presented in this paper also provides support for the elastic-plastic erosion theory proposed recently by Evans et al. (8), which describes material removal in terms of the penetration and extension of lateral vent-type cracks. At oblique angles of impingement, however, material removal also occurs by a shear deformation process of the type usually used to describe the erosion of metals. By increasing the target temperature from 25°C to 1000°C, deformation of the target materials during impact is enhanced, and material removal by shear deformation occurs at higher impact angles. Since both brittle and ductile modes of material removal occur, erosive wear is determined by both hardness and fracture toughness.

Acknowledgement

The authors wish to acknowledge the support of the Office of Naval Research.

REFERENCES

1. Lawn, B. R., and Marshall, D. B., this volume.

2. Sheldon, G. L., and Finnie, I., Trans. ASME 88B, 393 (1966).

3. Oh, H. L., Oh, K. P. L., Vaidyanathan, S., and Finnie, I., "The Science of Ceramic Machining and Surface Finishing". NBS Special Publication 348, edited by S. J. Schneider and R. W. Rice, p. 119 (1972).

4. Lawn, B. R., and Wilshaw, T. R., J. Mater. Sci. 10, 1049 (1975).

5. Hockey, B. J., J. Amer. Ceram. Soc., 54, 223 (1971).

6. Hockey, B. J., and Lawn, B. R., J. Mater. Sci., 10, 1275 (1975).

7. Evans, A. G., and Wilshaw, T. R., Acta Met. 24, 939 (1976).

8. Evans, A. G., Gulden, M.E., Rosenblatt, M.; to be published in Proc. Roy. Soc.

9. Evans, A. G., this volume.

10. Wiederhorn, S. M., and Roberts, D. E., Bull. Am. Ceram. Soc. 55, 185 (1976).

11. Ruff, A. W., and Ives, L. K. Wear, 35, 195 (1975).

12. Sheldon, G. L., and Finnie, I., Trans. ASME 89B, 387 (1966).

13. Sheldon, G. L., Trans. ASME 92D, 619 (1970).

14. Lawn, B. R., and Swain, M. V., J. Mater. Sci. 10, 113 (1975).

15. Wiederhorn, S. M., Hockey, B. J., and Roberts, D. E., Phil. mag. 28, 783 (1973).

16. Evans, A. G., and Wiederhorn, S. M., J. Mater. Sci. 9, 276 (1974).

COMPRESSION TESTING OF CERAMICS

George Sines

Materials Dept. School of Engineering, UCLA
Los Angeles, California 90024

Marc Adams

Jet Propulsion Laboratory
Pasadena, California 91103

INTRODUCTION

There is a need to know accurately the strength of brittle materials in compression, both to distinguish between competing statistical, micromechanical theories of strength and to provide data necessary for engineering designs for the optimum use of these materials as load bearing elements. The scope of application of high density, high-strength brittle materials as structural elements is increasingly rapidly. Materials, such as alumina, silicon carbide and silicon nitride, are being used in application where their corrosion resistance, strength at high temperature, light weight and very high compressive strength can be used to advantage. Potential design applications include biaxially prestressed hypersonic airfoils [1,2], tiled domed structures beneath the sea [3], jet engine bearings [4], and gas turbine components [5].

Even at temperatures of 600 to 800°C, no general plastic flow occurs in many brittle materials to relieve local stress concentrations; consequently, mechanical testing of these materials or design with them is very difficult. The extremely high stresses required to break, in compression, materials such as alumina, silicon carbide, and tungsten carbide (greater than 500 ksi) create considerable problems in the design of testing devices and specimens to ensure failure of the material in the desired compressive stress state [6] and not failure of the test apparatus. For these reasons and because only recently has there been interest in high-strength, high-performance ceramics, there is a lack of reliable compressive strength data for such materials and only a small body of knowledge on testing techniques and apparatus. Some data is suspect because the unusual difficulty of designing the specimen and loading device to attain the desired stress state may not have been recognized. Since many engineering applications of these materials involve biaxial compressive stress states, there is a need for the accurate determination of the failure envelope for high-strength brittle materials in this biaxial stress quadrant, as well as the uniaxial strength.

Testing techniques which have been successful on weaker brittle materials such as plaster [7], weak ceramics [8,9] mortar [10] concrete [11] and rocks [12] may not be able to be adapted to materials 10 to 20 times stronger. We direct our attention here to the techniques for determining the compressive strengths of this rapidly developing family of super-strong, but brittle, materials.

THE UNIAXIAL COLUMN COMPRESSION TEST

The testing of brittle materials is difficult because an inadvertant localized high stress can cause fracture of the specimen, thereby giving a false low value for the strength. In the uniaxial column compression test, problems may arise at the interface between the loading blocks and the specimen ends, in the test section from bending stresses induced by eccentric loading, and from stress concentrations at the transition from the butt section and the test section

1.1 Loading Block Size Missmatch

The face of the loading block contacting the end of the specimen should be the same size as the end of the specimen in order to have a uniform distribution of compressive load across the interface. An extreme example of non-uniform loading results on the end of a rigid cylinder pressing on the surface on an elastic, infinite half-space. The compressive stress at the interface [13] varies with the radius according to the expression

$$\sigma = P/2\pi R(R^2-r^2)^{1/2}.$$

From which it can be seen that at the center the stress is only half the nominal and that it becomes infinite at the edge of the cylinder where $r = R$. If the loading block is a cylinder having a radius of the same order of magnitude as that of the specimen instead of being an infinite half-space and if the specimen is also elastic, the stress will be more uniformly distributed over the contacting surface but the infinite stress at the edge of the cylinder can not be avoided when the loading block has a face larger than the specimen end.

This can be seen from the stress concentration that exists at the transition from one diameter to another in a round bar, subjected to an axial load. [14]. The radius of the transition of the fillet is zero for the block-specimen assembly which gives an infinite stress concentration factor. Thus either plastic deformation must occur in the block or in the specimen at the edge of the contacting area or local fracture must occur. Local fracture and spalling of the edge of the specimen can result in a re-entrant fillet radius which can reduce the concentrated stress to the level of the compressive strength. If the specimen does not have a sufficiently reduced cross-section for the test section, this damage at the edge could cause compressive failure to occur at a lower stress than its true strength.

Even the very careful tests on dense alumina by Ryshkewitch may have experienced some of this deleterious effect [15][16]. The specimen was a 6 mm cube but the face of the loading block was 10 mm square; however, the edge effect was less in extent than it might have been because of the high modulus of the tungsten carbide block. He measured a uniaxial compressive strength of 427 ksi on a dense alumina which is much higher than the 221 Ksi of Broutman and Cornish [17] on a similar material but lower than the 448 ksi measured by Sedlacek [18] with his ring test. Bieniawski [19] in his tests on rocks, took care to have the diameter of the loading blocks the same as that of the test specimen.

1.2 Loading Block Compliance Missmatch

A source of error in the measurement of the compressive strength of brittle materials is the premature failure that may be caused by the transverse tensile stress that may be induced in the end of the specimen by the missmatch of the transverse compliances of the loading block and the specimen end. This can be caused by an incorrect choice of the material of the loading block and/or the shape of the loading block. First consider only the effect of the material by treating a loading block of constant cross-section and of the same cross-sectional area as the specimen end. The lateral extension strain ϵ_{11} under a compressive stress σ_{33} is $\epsilon_{11} = (\nu/E)\sigma_{33}$ for isotropic materials, where ν is Poisson's ratio and E is Young's modulus. For anisotropic materials $\epsilon_{11} = S_{1133}\sigma_{33}$ where S_{1133} is the compliance. No transverse stress is induced when the transverse strain of the block ${}^{B}\epsilon_{11}$ equals that of the specimen ${}^{S}\epsilon_{11}$, which for this case occurs when their respective transverse compliances are equal ${}^{B}\nu/{}^{B}E = {}^{S}\nu/{}^{S}E$ or ${}^{B}S_{1133} = {}^{S}S_{1133}$. If the compliances of the block is greater than that of the specimen, tensile stresses are induced in the end of the specimen and premature failure may occur by its splitting.

It may be desirable to have some transverse compressive stress in the specimen in the region near the interface in order to cancel unwanted tensile stresses which might be induced by an imperfect match between the surface of the specimen and that of the loading block. This can be attained by having ${}^{B}S_{1133} < {}^{S}S_{1133}$. Of course, if a transverse compressive stress is induced in the specimen, an equal transverse tensile stress will be induced in the block but the material for the block can be chosen to have sufficient tensile strength to avoid failure.

If the loading block should yield under the compressive load, the combined transverse plastic strain and elastic strain would probably induce enough tensile stress in the specimen to split it; therefore, it is necessary to choose a material with a compressive yield strength high enough and have the cross sectional area of the butt of the specimen over that of the test section large enough to reduce the compressive stress below the yield.

1.3 Load Transmitting Surfaces

Assuming that the transverse compliance of the loading block matches that of the specimen,there still remains the problem of local high stresses from disparities at the interface and from the two surfaces not exactly matching. A tempting solution to this problem is to insert a soft layer of material between the two surfaces so that its plastic and elastic deformation remove these two sources of local stress. Attempts to do this have met with mixed success. Nadai [20] reporting on unpublished work of Lambert, Manjoine and Arnold:

> "When a very thin lead foil was used... the compressive strength dropped to quite low values".

Nadai attributes this drop in strength to the lead intruding into surface flaws in the ceramic and thereby causing local transverse tensile stresses. This has also been our experience with soft aluminum sheets with dense alumina.

There are two other mechanisms by which a soft ductile layer can cause unfavorable stresses. Near the free edge of the layer the shear stresses are high and the lead extrudes from between the block and specimen, but in the interior the stresses are almost equi-triaxial and no general flow takes place. Thus the compressive load is removed from the edge region and concentrated in the interior. Also the sharp gradient of compressive stress might cause local shear and tensile stresses. Another source of unfavorable stress is caused by the radial shear traction on the surface of the ceramic from the viscous flow of the soft metals. The same investigator found that a soft sheet of towel paper gave the highest results in some of their tests.

They also got high values when the ceramic was "perfectly"

fitted to the hardened steel end plates. We assume that "perfectly fitted" means lapped together. In our work [21] the cylindrical ends of the plates were lapped to the specimen. During this lapping, the compliance tubes and specimen were placed on a 2-in. diameter steel mandrel and rotated against each other so that the conical surfaces would remain conical and concentric with the test section. The amount of contact between the two mating surfaces in each of these steps was determined by applying radial lines with a 6H hard lead pencil to each of the surfaces before lapping and then observing how the lines disappeared as the lapping proceeded. Lapping was continued with each grade of compound until the lines drawn on both surfaces were removed uniformly. Final lapping with the 1 μm diamond compound yielded a very uniform, fine matte finish. Unsucessful initial tests indicated the quality of figure of the contacting surfaces was extremely critical for successful testing. To confirm the importance of complete contact, one test was made with only about 80% contact and the stress at fracture was reduced by 40% and it must be emphasized that the area of the contacting surface was three times the cross-sectional area of the test section.

Shanley et al. [22] investigated the effect of 15 interface materials on the measured uniaxial compressive strength of an alumina based ceramic having a modulus of rupture of 5.2 ksi. The compressive strength measured with direct contact to 5" x 5" hardened steel loading blocks on a 1" x 1" x 2" specimen was 42.7 ksi. A 0.044" thick asbestos sheet at the interface gave the highest value - a 25% increase, 0.030" copper+16%, 0.005" brass+13% while a 0.067" rubber sheet gave the lowest value - a 42% decrease in measured strength. The Young's modulus of the ceramic was 6.9×10^6 psi. Subsequent studies by the group favored 0.067 thick Garlock "900" gasket material.

The effect of material at the contacting surfaces also has been studied by Budnikov et al. [23]. Their test specimen was a solid cylinder 8 mm in diameter and 9.6 mm tall. When they tested 97% dense alumina using hardened steel disks they measured a compressive strength of 199 ksi but this was reduced by 29% when 0.1 mm thick soft copper foil was put at the interface. Similarily, beryllium oxide with a compressive strength of 213 ksi was reduced by 40%. Filter paper soaked in paraffin reduced the strength by 30%. In contrast, electric porcelain of 23 ksi and magnesium oxide of 33 ksi had about the same strength whether hard steel, paper or copper foil was at the interface.

Since all three oxide ceramics have high moduli of elasticity, but only magnesium oxide was insensitive to the interface material, no correlation appears with the moduli. Perhaps correlation exist with the intensity of the compressive stress transmitted across the interface because the two weaker materials showed no sensitivity to the choice of interface material while the stronger did.

1.4 Eccentricity of Loading

Care must be taken that bending stresses are not inadvertently introduced [24]. When the loading blocks are rigidly guided to prevent rotation, their surfaces must be parallel and the ends of the specimens equally so. A very small deviation from being parallel can cause bending so that premature failure is caused by the added compressive stress. The bending can easily be great enough to cause a net tensile stress and premature failure from it.

One method used to prevent the application of bending moments is to abandon the precisely guided parallel surfaces and use a flexible load train. This can be accomplished by a spherical joint; however, such a joint is usually made completely ineffective by friction between the mating surfaces. Therefore, the surfaces must be separated by a pressurized fluid layer [25]. An air bearing has been developed and successfully used [26], but it is a rather complicated device, and it would be expensive to construct one of sufficient capacity to test a specimen reasonably large in size considering the very high compressive strength exhibited by many ceramics. Another approach that removes many of the complications is to use a long load train and to replace the bearing by knife edges [27]. The knife edges could be replaced by flexure plates in order to increase the capacity and still retain the benefits of the flexible load train. Although the load train may be completely flexible, a moment can be applied to the test section if the line of loads is not concentric with the axis of the test section. A surprisingly small eccentricity can cause significant bending stress. Simple calculations superimposing the bending stress from beam theory and the axial stresses give the maximum and minimum stresses at opposite sides of a round specimen of radius r with the line of load application eccentric a distance ϵ from the specimen axis,

$$\sigma_{\substack{max \\ min}} = \sigma_{ave}(-1 \pm 4\epsilon/r),$$

and for a rectangular cross-section, b x h, eccentric a distance ϵ in the direction of the h dimension,

$$\sigma_{\substack{max \\ min}} = \sigma_{ave}(-1 \pm 6\epsilon/h).$$

For these reasons it is advisable, in the initial states of a study, before the testing procedures are refined, to have three strain gages mounted longitudinally, equally spaced, around the circumference on each of the test specimens in order to ensure that the desired stress is being applied and that it is uniform in the test section.

THE UNIAXIAL HOLLOW CYLINDER COMPRESSION TEST

2.1 Unpublished studies on several designs of hollow cylindrical specimen by Lambert, Manjoine and Arnold in 1941 at the Westinghouse Research Laboratories are reported by Nadai.[20] Table I summarizes their results and the specimen shapes are shown in Fig. 1.

TABLE I

Shape	Surface	Compressive Strength (ksi)
Straight solid cylinder (d=1-1/2" h=3" plane ground ends)	unglazed	73
	glazed	83
With reinforced shoulders (a)	unglazed	50
Straight hollow cylinder (b)	unglazed	56
	glazed on outside	70
	glazed on both sides	58
Cylinder with curved outside surface (c)	unglazed	121*
		66†
Cylinder with curved outside surface (d)	unglazed	130*
		117†
	glazed on outside	112
	glazed on both sides	102
Cemented ends (e)	unglazed	64

* Highest stress
† Average stress

The small solid unglazed cylinder reported on the first line of Table I broke at 73 ksi while the same material when tested by the cylindrical specimen with 12-deg. conical ends and a reduced cross-section illustrated in Fig. 1d had an average strength of 117 ksi and a high value of 130 ksi. These good results led Babel and Sines [28]to design a specimen similar in shape to that in Fig. 1d for biaxial tests under combined tension and compression which included uniaxial compression.

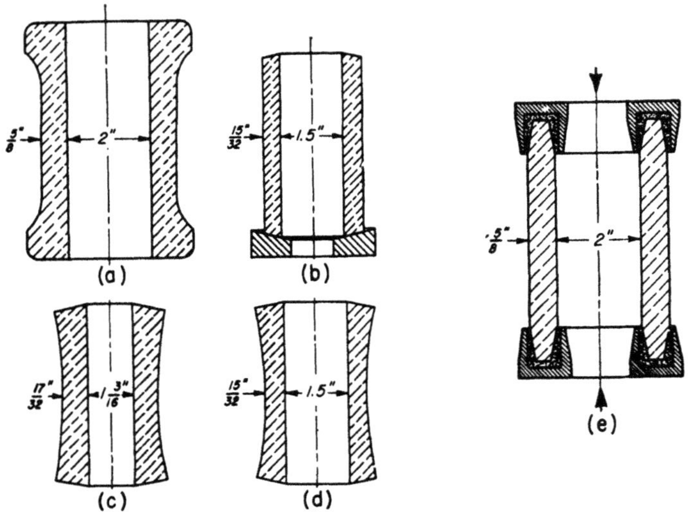

Figure 1. Porcelain Compression Specimen of Lambert, Manjoine and Arnold [20] (From "Theory of Flow and Fracture" by Nadai, copyright 1950, McGraw-Hill. Used with permission of McGraw-Hill Book Co.)

The specimen shown in Fig. 2 is more conservative than that of Lambert et al. in that the cross-sectional area of the butt is 4 times that of the test section and it was loaded by a compliance tube of the same material as that of the specimen and of the same cross-sectional area as the butt to eliminate the many difficulties cited above. Photoelastic and strain gage measurements showed that the stress in the test section varied less than the 5% accuracy of the measurements.

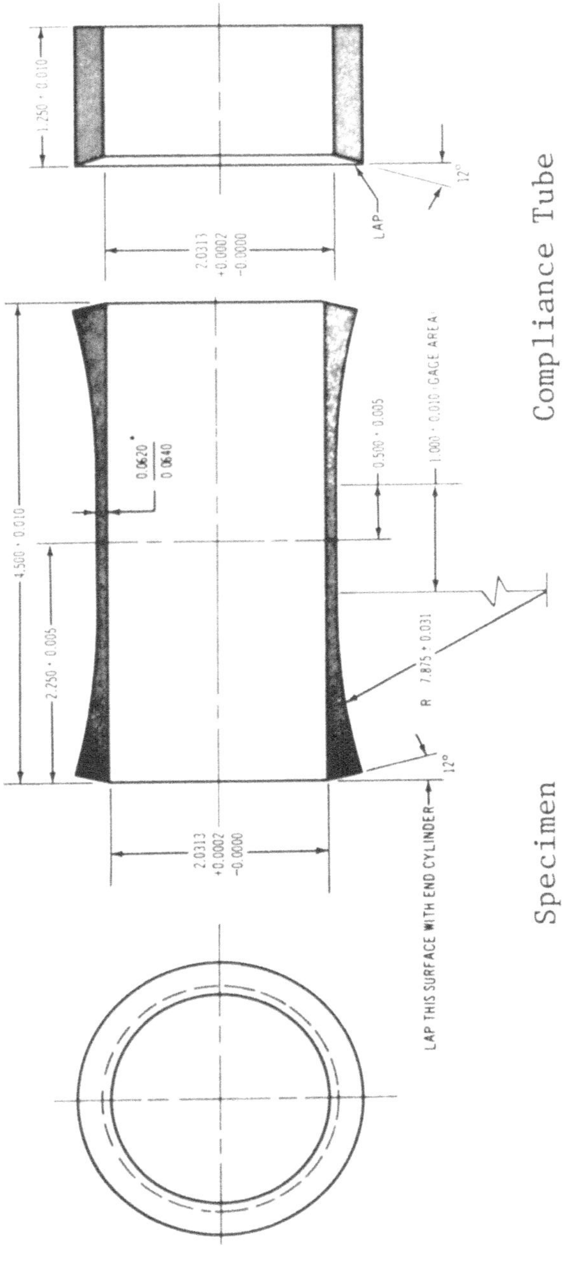

Figure 2. Tension-compression specimen for porous zirconia.

Test results on 82% dense zirconia given in Table II show little scatter for tests on 4 different batches.

TABLE II

Test	Batch Number			
	2	3	4	5
Uniaxial hoop tension:				
n_{1j}	3	7	7	7
$\bar{\sigma}_{1j}$ (psi)..................	3,423	3,215	3,040	3,359
s_{1j} (psi)..................	141	138	133	272
Biaxial[c] tension-compression:				
n_{2j}	2	6	6	2
$\bar{\sigma}_{2j}$ (psi)..................	2,948	2,803	2,593	2,867
s_{2j} (psi)..................	361	293	132	184
Uniaxial compression:				
n_{3j}	3	6	7	5
$\bar{\sigma}_{3j}$ (psi)..................	24,763	25,517	22,178	23,535
s_{3j} (psi)..................	1,986	1.148	1.589	319

n_{ij} = number of tests, $\bar{\sigma}_{ij}$ - average fracture strength, and s_{ij} = standard deviation.

The favorable results of little scatter as well as failure consistently originating in the test section led to a similar design for biaxial compression tests [21] shown in Fig. 3. The wall thickness is thicker and the test section shorter in order to avoid out-of-round buckling from external fluid pressure.

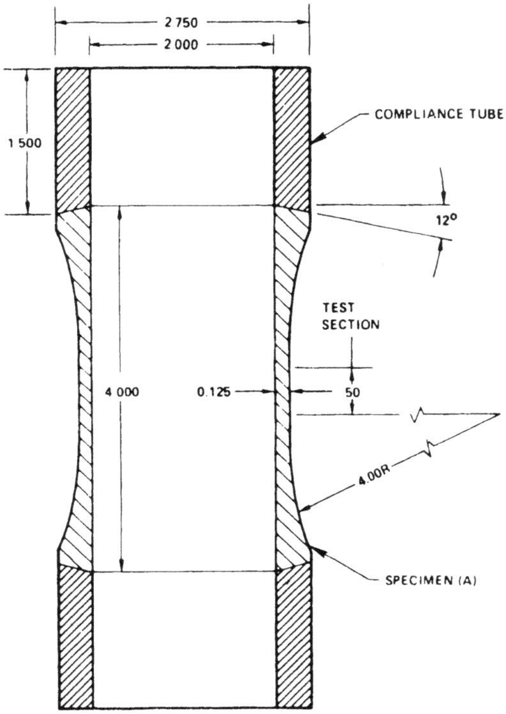

Figure 3. Compression-compression specimen for dense alumina.

2.2 Premature failure from compliance tube splitting

The specimen shown in Fig. 3 was satisfactory for uniaxial circumferential compression and several biaxial compressive stress states described below; however, it was not satisfactory for uniaxial column loading which was the preloading for the 1:1 compression test discussed below in Section 4.3. The compliance tube split logitudinally causing cracks to porpagate into the specimen butt at an axial compressive stress in the test section of only 0.3 of the compressive strength. The butt area was about 3 times the net test area so the splitting occurred at a compressive stress in the butt and ring of only about 0.1 of the compressive strength.

The premature failure of the compliance tube is attributed to sliding of the conical surfaces thereby causing circumferential tension in the compliance tube. The success of the earlier axial tests on zirconia may be attributed to the higher tensile strength to compressive strength ratio, to higher friction between the surfaces, and to the higher ratio of butt area to test area. It is thought that reducing the angle to 6-deg might solve the problem.

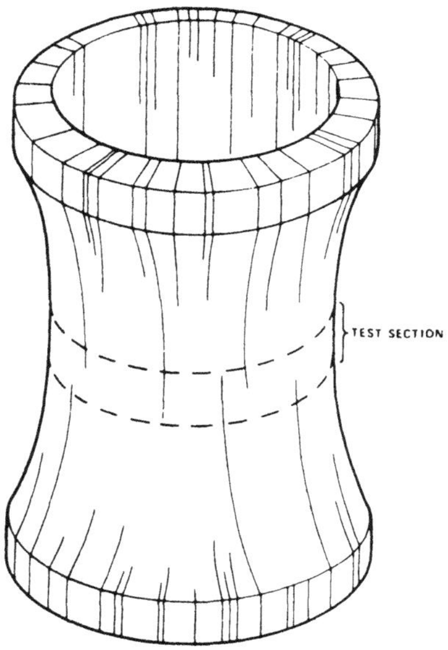

Figure 4. Specimen cracked from compliance missmatch.

The ceramic compliance tubes were replaced by hardened steel ones, but this resulted in splitting of the specimen as shown in Fig. 4.

2.3 Stresses from compliance missmatch of loading block.

The problem of the hardened steel compliance tubes splitting the alumina test specimen tube under axial load is an example of a general design problem which will be encountered when trying to use the full compressive strength of such brittle ceramics. When the structural element which applies the compressive load to the brittle ceramic has a lateral stiffness different from the ceramic element, high local stresses can be produced because of the mismatch in the lateral displacements under load. The static coefficient of friction between alumina and steel under high normal forces is probably quite large because the much harder alumina surface will tend to imbed in the steel and produce mechanical interlocking of the surfaces. For this reason, the alumina and steel will tend to attain the same displacements at their interface. If the transverse stiffnesses of the two elements are not matched, the steel will put the alumina into tension or compression in the vicinity of the interface, depending on whether the stiffness of the steel part is less than or greater than the ceramic, respectively. A simple model was used to develop the equation of the transverse stresses which can occur from the differences in Young's moduli and Poisson's ratios [29] when the two have the same cross-sectional area and are long.

The equation with abbreviated development is:

$$\varepsilon_{xx}^{(s)} = \varepsilon_{xx}^{(c)}$$

where:

$\varepsilon_{xx}^{(s)}$ = strain in steel element

$\varepsilon_{xx}^{(c)}$ = strain in ceramic element

then:

$$\frac{\sigma_{xx}^{(s)}}{E_s} - \frac{\nu_s \sigma_{yy}^{(s)}}{E_s} - \frac{\nu_s \sigma_{zz}^{(s)}}{E_s} = \frac{\sigma_{xx}^{(c)}}{E_c} - \frac{\nu_c \sigma_{yy}^{(c)}}{E_c} - \frac{\nu_c \sigma_{zz}^{(c)}}{E_c}$$

assuming:

$$\sigma_{zz}^{(s)} = \sigma_{zz}^{(c)} = \sigma_{zz}$$

$$|\sigma_{xx}^{(s)}| = |\sigma_{yy}^{(s)}| = |\sigma_{xx}^{(c)}| = |\sigma_{yy}^{(c)}|$$

then:

$$\sigma_{xx}^{(c)} = \sigma_{yy}^{(c)} = -\sigma_{zz}\ \frac{E_c\nu_s - E_s\nu_c}{E_c - E_c\nu_s + E_s - E_s\nu_c}$$

where:

E_c = modulus of ceramic

E_s = modulus of steel

ν_c = Poisson's ratio for ceramic

ν_s = Poisson's ratio for steel

σ_{zz} = normal stress across interface.

2.4 Premature failure from local rotations.

Even though the magnitude of these σ_{xx} and σ_{yy} transverse tensile stresses in the 1:1 compression test were only one-third to one-half the tensile strength of the alumina, the specimen tube cracked. A second phenomenon was probably occurring which induced this cracking. Because of the geometry and stress condition, rotations were being produced in the alumina caused by the displacements necessary to match the steel tube displacements. This rotation produced high concentrations of stress in the alumina in the regions indicated in Fig. 5.

Of course, it is unlikely that the rotation would be so great that it would cause separation of the surfaces near the inner diameter as shown in Figure5. The exaggeration was made to illustrate the phenomenon.

Our 1:1 compression strength measurements are questionable because of these effects, and several other reported values of compressive strengths for brittle ceramics may be too low for similar

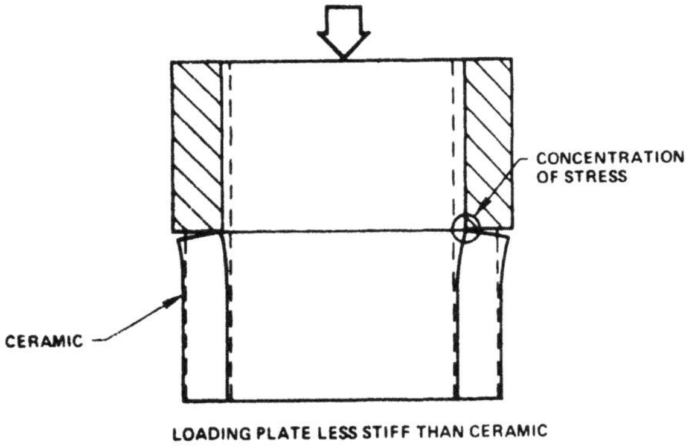

Figure 5. Concentrated stresses from local rotations.

reasons. The use of finite elements could be employed to avoid such situations; however, the amount of effort required to accomplish such an analysis was prohibitively large for our testing program. If a radial pressure is applied to the loading element and the ceramic element, as in our 1:0.5 test described in Section 4.1 below. Where external fluid pressure creates compressive σ_{rr} and $\sigma_{\theta\theta}$ stresses, the mismatch in transverse stiffness is not nearly as severe a problem since the magnitudes of the radial displacements in the two elements are much less. Under most loading conditions where a surrounding external pressure is present, all three principal stresses in both elements will remain compressive. The magnitude of the transverse displacements, which induce the rotations, will be smaller and the ε_{xx} and ε_{yy} strains will approach zero or become negative. This is undoubtedly one of the reasons why the highest compressive strength values are measured in tests utilizing external pressure acting on a closed end tube. One of the major causes of lower strength values measured in uniaxial compression, as compared to biaxial compression, is that the uniaxial compressive strengths were measured by applying an axial load with no provision for eliminating the loading block-specimen stiffness mismatch effect.

There is an additional phenomenon in brittle ceramic materials which may be important when designing structural elements to transmit the load to the ceramic in a fashion that allows full utilization of the compressive strength. Poisson's ratio in these materials is not constant to failure. Although Young's modulus is substantially constant, Poisson's ratio increases with increasing applied compression, due to a greater than linear elastic increase in transverse displacement. Sedlacek's compression test data on Al-995 alumina [32] shows this increase to be 10 percent at a stress level 80 percent of failure. Although data is not available

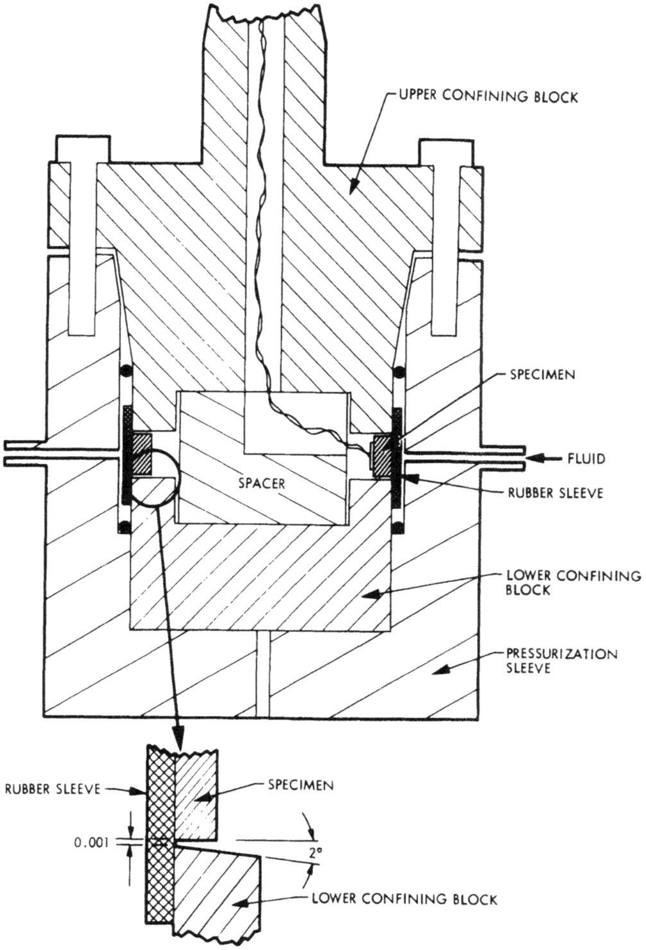

Figure 6. Sedlacek's Ring Compression Test.

above this stress level for high density, high strength brittle ceramic, extensive testing with hard rock material [33, 34] indicates that the greater than linear increase in transverse strain with applied stress may be exponential above 0.8 of the failure stress.

THE RING COMPRESSION TEST

3.1 Description of Apparatus

Sedlacek developed a uniaxial compressive test [18, 35] in which no misalignment is possible and parasitic stresses are eliminated so long as the specimen geometry is ideal. The test device is shown in Fig. 6. The end faces of the ring specimen experience no constraint since the specimen holder does not contact the specimen. External fluid pressure is applied through a rubber sleeve to a thin, ring-shaped specimen which is 2.200 inches outside diameter, 0.450 long and has a wall thickness of 0.075. The gap between the specimen ends and the upper and lower confining blocks is carefully controlled to be approximately 0.001 inches so that the specimen is unconstrained and free to displace with pressure increase. The elongation of the specimen due to the Poisson effect is accomodated by sloping the confining block end faces on a 2° angle towards the center of the test device. Uniformity of load is assured by careful machining of the specimen ring to prevent ellipticity and lack of concentricity of the inside and outside diameters.

The state of stress produced in the specimen is similar to that shown, in Fig. 10 below in Section 4.0, except no axial, σ_{zz}, stress is present so long as the confining block ends do not contact the specimen. The maximum stress, $\sigma_{\theta\theta}$, occurs on the inside surface and the radial stress, σ_{rr}, is zero there. Both stresses vary slightly through the wall thickness.

3.2 Potential Difficulties of Ring Test.

The pressure required to produce circumferential (out-of-round) buckling in this specimen is quite low because the ends are not constrained to be circular and because of the large diameter to wall thickness ratio. The theory of elastic instability indicates the critical buckling pressure for the unconstrained specimen is approximately 5 ksi although the specimens failed at a pressure of 25 to 35 ksi. Partial constraint of the specimen ends would increase considerably the critical buckling pressure and may be the reason why buckling did not occur at the lower pressures. Such constraint could occur by the specimen bearing against the confining blocks or by extrusion of the rubber into the 0.001 gap. Contact with the specimen ends from the displacements of incipient buckling could produce normal and shear tractions on these surfaces which would inhibit buckling failure.

In a visit to Sedlacek's laboratory, we were impressed by the extremely close tolerances to which he personally fitted his specimens and apparatus. Incipient out-of-round buckling might have been suppressed by the ring deflecting to contact the end closures. Since the critical buckling stress is below the measured strength, it is not certain whether the specimens failed in direct compression or by buckling. The average strength measured by Sedlacek for the Al-995 was 448 ksi (standard deviation 36 ksi) which is 85% of the strength measured by other investigators on nominally the same material. However, later unreported results mentioned by Sedlacek to us were considerably higher, approaching 600 ksi.

THE HOLLOW CYLINDER BIAXIAL COMPRESSION TEST

Three kinds of test devices have been developed by the authors [21,29,31,38,39] which can produce any desired biaxial compressive stress state in a specimen of common geometry. The three test devices are shown in Figs. 7, 8 and 9. The loading schemes and stress states developed are shown in Fig. 10. The

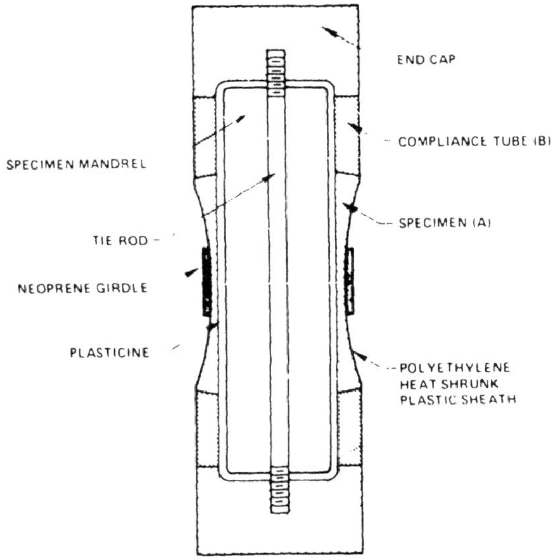

Fig. 7. 1:0.5 biaxial compression test device.

general features of the common specimen are described in Section 2.1, Fig. 3. The three devices utilize external fluid pressure acting on a closed-end tube to produce the circumferential compressive stress which eliminates localized high stresses. The axial load is applied to the ends of the specimen or compliance tubes which have a wall thickness three times greater than that of the test section, thereby eliminating end cracking if the loading blocks are properly designed. The axial load is varied by two methods: an axial prestressing bolt is used to obtain stress ratios greater than 1:0.5 ($\sigma_{\theta\theta}:\sigma_{zz}$) and a center post is used to obtain ratios less than 1:0.5. In all cases there exists a small radial stress equal to the pressure on the outer surface that decreases to zero on the inside diameter which is the location of maximum circumferential stress. A description of the three test devices is given below.

4.1 Biaxial Compression Test, 1:0.5

The simplest test in concept is the 1:0.5 biaxial compression test shown in Fig. 7 in which a closed end, hollow cylinder is

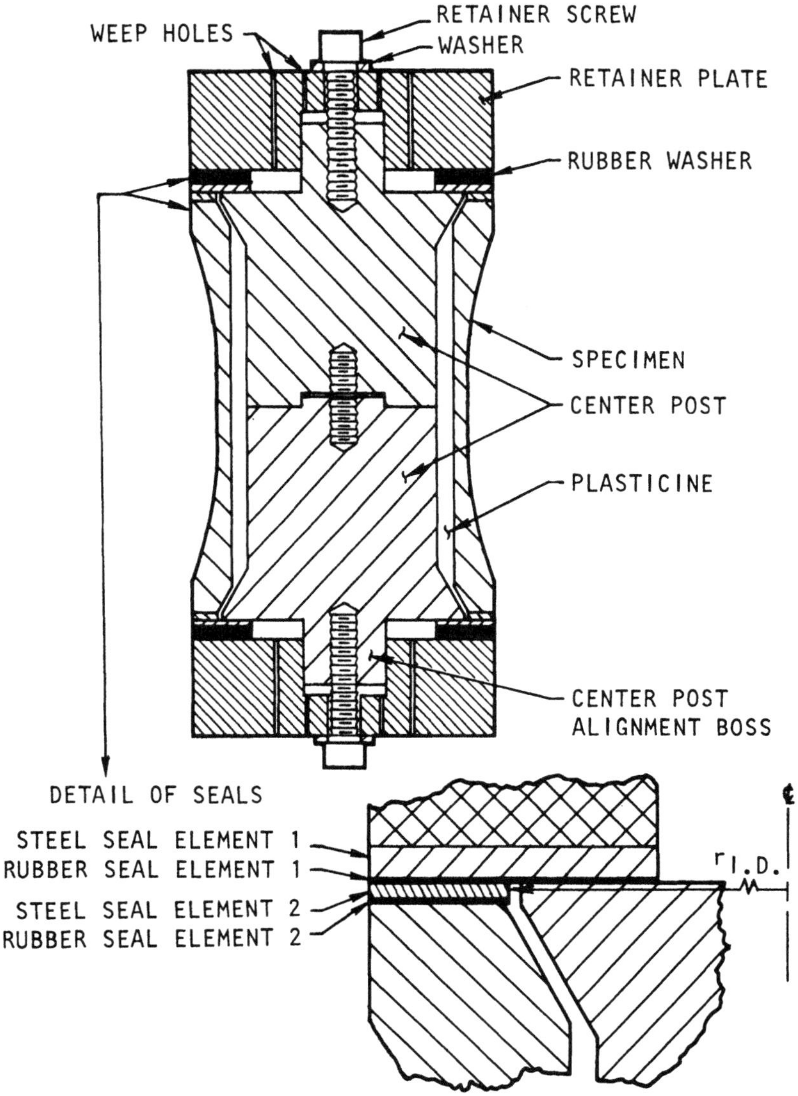

Fig. 8 1:0.01 biaxial compression test device.

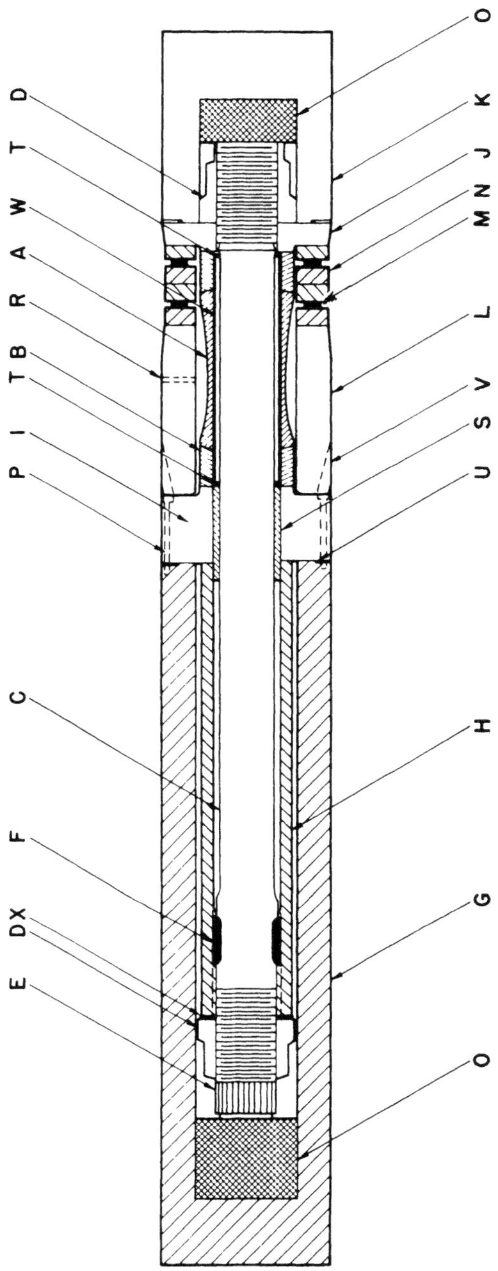

Fig. 9 1:1 biaxial compression test device.

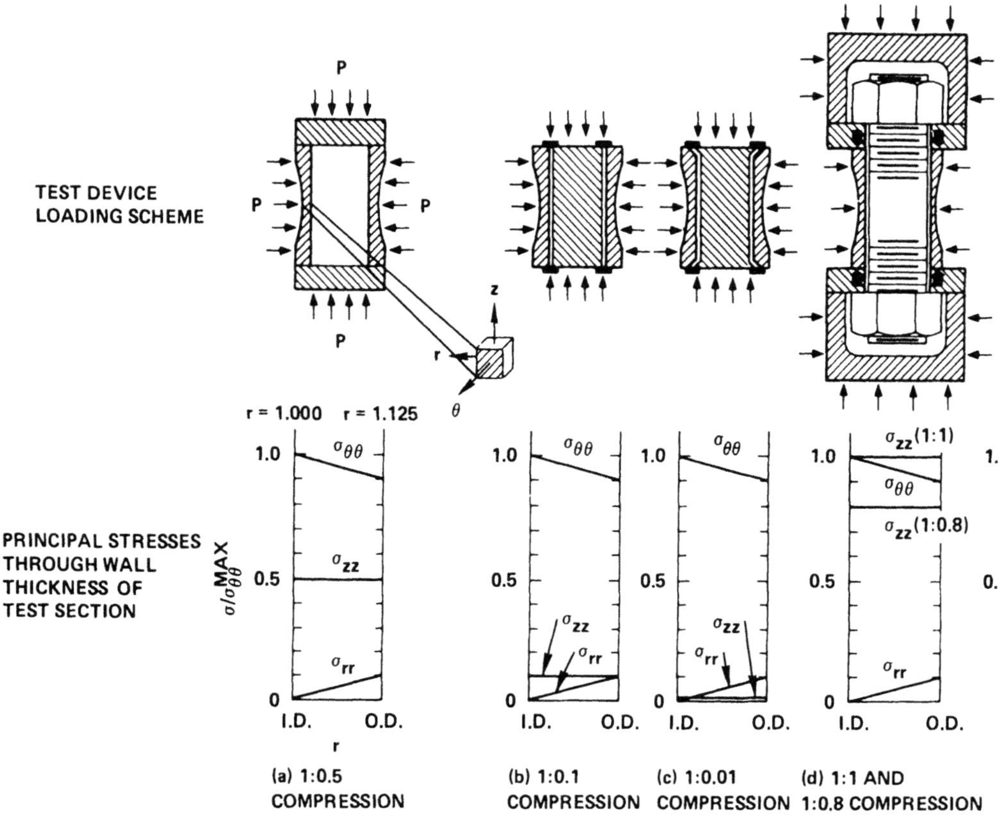

Fig. 10 Loading schemes and stress states developed in test specimens.

subjected to external fluid pressure. The ends of the specimen-compliance tube assembly are closed with two hardened steel end caps; the assembly is held aligned by a soft brass tie rod. External fluid pressure is applied in an isostatic press to the specimen which is covered with a heat-shrunk, plastic sleeve. A steel specimen mandrel is placed in the bore of the specimen in order to prevent the end caps from moving inward a distance large enough to completely destroy the test section at failure. Most of the remaining volume in the bore is filled with Plasticine R to prevent a destructive release of energy on specimen failure; however, a controlled vacant space is provided to prevent internal pressure from arising when the inner specimen volume is reduced by elastic deformation during testing.

The stresses in the specimen are described by the equations of elasticity for a thick-walled pressure vessel. They are as follows:

Circumferential stress $\sigma_{\theta\theta} = - p \frac{r_o^2(r_1^2+r^2)}{r^2(r_o^2-r_1^2)}$

At inner surface $\sigma_{\theta\theta(max)} = - p \frac{2r_o^2}{(r_o^2-r_1^2)}$

Axial stress $\sigma_{zz} = - p \frac{r_o^2}{(r_o^2-r_1^2)}$ (constant through wall)

Radial stress $\sigma_{rr} = - p \frac{r_o^2(r^2-r_1^2)}{r^2(r_o^2-r_1^2)}$

At outer surface $\sigma_{rr(max)} = - p$

where p = applied external pressure,

r_o = 1.125 in. = outside radius of test section,

r_1 = 1.000 in. = inside radius of test section, and

r = radial position within the wall ($1.000<r<1.125$ in.)

The circumferential stress $\sigma_{\theta\theta}$ attains its maximum value at the inside diameter and decreases to a value 90% of this maximum on the outside diameter in an approximately linear fashion. The intermediate principal stress σ_{zz} has a value about one-half that of $\sigma_{\theta\theta}$ and is of constant magnitude through the wall thickness.

The stress state in the 1:0.5 compression test specimen (and for the other two compression tests) is triaxial compression for all radii except the inside diameter. The largest amount of triaxial compression occurs on the outside diameter where the stress state $\sigma_{\theta\theta} = 0.9\ \sigma_{\theta\theta}^{max}$, $\sigma_{zz} \simeq 0.5\ \sigma_{\theta\theta}^{max}$, and $\sigma_{rr} = 0.1\ \sigma_{\theta\theta}^{max}$ prevails. It is believed [31,36] that fracture initiates on the inside diameter where $\sigma_{rr} = 0$ and $\sigma_{\theta\theta}$ is at its maximum. Strength values are reported as the maximum $\sigma_{\theta\theta}$ stress present in the specimen at failure.

All mating surfaces were carefully lapped to ensure nearly perfect contact under load. Imperfect contact at these surfaces was found to significantly reduce the measured strengths.

The 1:0.5 compressive strengths were measured for two batches of WESGO AL-995 alumina using this device. The measured values were some of the highest obtained for this material and the scatter of the individual values was quite small (See Table) which indicates that parasitic tensile stresses and bending were minimal.

TABLE III

	No. of specimens	Average Strength $\sigma_{\theta\theta}$(ksi)	Standard deviation (ksi)
Batch #1	4	572	8.7
Batch #2	4	514	22

4.2 Biaxial Tests near Uniaxial Compression

Successful compressive tests in biaxial stress states near uniaxial compression are difficult to achieve because inadvertant tensile stresses which can develop are not cancelled by the intermediate compressive stress. Since the tensile strengths of dense ceramics are only 1/15 to 1/20 the compressive strength [37] such inadvertant tensile stress greatly reduces the measured compressive strength. More reliable tests are those which maintain a magnitude of transverse compressive stress capable of cancelling such tensile stresses. The lower strength measured by the authors in 1:0.01 compression [37] may have been affected slightly by such tensile stresses.

The device for measuring biaxial compressive strengths near uniaxial compression is similar to the 1:0.5 test device as can be seen in Fig. 8. A center post is placed in the specimen equal to the length of the specimen, and an annular pressure seal is made between the specimen ends and the center post. The axial load produced by the fluid pressure is carried by both the center post and specimen wall. The end diameter of the center post and the end, inside diameter of the specimen establish the proportions of the axial load carried by the center post and the specimen. The load carried through the center post decreases the axial, end load which must be carried by the specimen. By appropriately sizing the seal area and its position, stress states may be obtained in which σ_{zz} is any value from 0.5 $\sigma_{\theta\theta}$ to near zero. The failure stress ($\sigma_{\theta\theta}$) is calculated as for the 1:0.5 test and σ_{zz} is determined from the constant proportion established by the geometry. In the device shown in Fig. 8, $\sigma_{zz} = 0.01\ \sigma_{\theta\theta}$.

The end diameter of the specimen bore was over-cut with a 30-deg chamfer, and steel seal element No. 2 was used to establish the area over which the fluid pressure acted to apply the axial load to the specimen tube. The inside radius of this seal element r_{id} was sized to produce a slight biaxial compression stress state, 1:0.01. A worst-case calculation was made which assumed maximum eccentricity of device parts and maximum possible misalignment that could occur during assembly. This calculation verified that under the worst conditions of eccentricity and misalignment no tensile stress components would be produced in the test section, that is, the stress state in the test section could vary only between 1:0.02 and 1:0.00.

Alignment of the center post axis with the axis of the specimen tube is critical in this device. This was accomplished by very accurate machining and careful inspection of all parts. The center post is aligned to the specimen axis by alignment bosses on the ends of the center post and by an alignment sleeve. Surrounding fluid pressure is applied which acts directly on the center post through the weep holes. The No. 1 seal elements create the seal between the center post and specimen. The steel seal element No. 1 is designed to yield assuring that the axial pressure force is distributed uniformly between the specimen tube and the center post across the seal gap. Steel seal element No. 2 does not yield since it established the centerline of the seal gap force and hence the stress state.

The problem of matching the displacements in the specimen tube side and center post side of the seal area, to prevent a step from forming due to differences in displacements under pressurization, is quite complex in this device. Calculations show that the specimen tube becomes slightly longer, whereas the center post becomes slightly shorter. The difference in the displacements between the two is approximately 0.005 in. at the failure pressure. The second effect to be considered is that which arises from the rubber seal element No. 2. The low modulus of this material results in a much greater axial displacement than for an equivalent thickness of steel. A further complication is that the thin film form of the rubber seal element raises its effective modulus under the compressive axial load. A straightforward calculation of the axial compliance of the specimen side of the seal components was not possible. Instead, several trial pressurizations were made with the device using different thicknesses of No. 2 rubber seal element until a thickness was obtained which matched the displacements of the specimen side and center post side of the seal gap. This was revealed by careful measurement of the permanent deformation produced in the steel seal element No. 1

A simpler test device (shown in Fig. 11 and described in References 21 and 31) was designed and used for compression testing in the 1:0.1 stress state. This device requires no chamfering of the specimen ends and a simple, single seal element effects the end closure.

4.3 Biaxial Compression Testing in Stress States between 1:0.5 and 1:1

To achieve stress states in which $\sigma_{zz} > 0.5\sigma_{\theta\theta}$, the stresses developed by an axial compressive load can be superimposed on

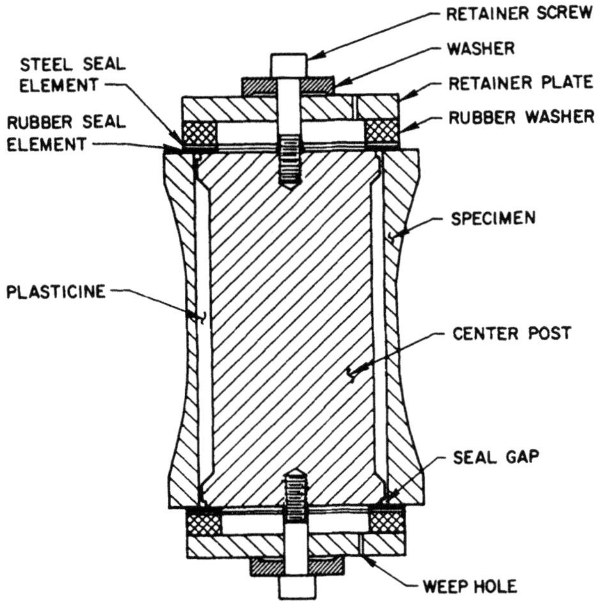

Fig. 11 1:0.1 biaxial compression test device.

those caused by a surrounding external fluid pressure. The axial preload is produced by tightening the nut on the threaded shaft which passes through the specimen. After sufficient axial preload is applied, the specimen is subjected to a surrounding fluid pressure with the ends of the threaded shaft protected from the fluid pressure. By varying the amount of axial preload introduced, stress states between 1:0.5 and 1:1 can be produced in the specimen. The $\sigma_{\theta\theta}$ stress at failure is calculated as in the preceding tests. The axial stress (σ_{zz}) is the sum of the σ_{zz} produced by the fluid pressure ($\sigma_{\theta\theta}/2$) plus the σ_{zz} developed by the axial preload.

A detailed drawing of this test device is shown in Fig. 9. A bolt (C) passes through the assembly of loading washer (J), specimen and compliance tubes (A) and (B), loading ring (I), loading sleeve (H) and thrust washer (X). There is a large 12-point wrenching nut (D) on each end. Integral with one end of the bolt is a spline (E). The spline end of the bolt is enclosed by the prestress pressure case (G) which seals to the loading washer (I) by means of the "O" ring (U). This seal is set by prestressing the "O" ring with bolts (P). The opposite end of the bolt is protected from fluid pressure by pressure cap (K) which seals to the loading washer (J). Parts (O) are lead blocks positioned at each end of the bolt to absorb any impact. A safety sleeve (L) along with extrusion washers (N) and lead washers (M) perform protection and energy dissipating functions in the design. The keys (F) lock the loading sleeve to the bolt. They permit the sleeve to slide longitudinally on the bolt but prevent rotation. Sleeves and rings, (W), (S), and (T) are for alignment purposes.

The specimen is axially preloaded in a large lathe. The spline is grasped in the chuck of the lathe and the opposite end is allowed to slide freely in a modified tail stock. A wrench is fixed over the spline-end nut, and its force is reacted through a column to the ground. The lathe turns the entire assembly and the wrench holds the spline-end nut stationary. The keys (F) prevent the bolt or loading sleeve from being elastically would up in torsion and possibly transmitting torsional stresses to the specimen. The torque applied to the spline is reacted a short distance away at the nut. When a sufficient preload has been applied to the specimen, the device is removed from the lathe and the prestress pressure case (G) and the pressure cap (K) are installed. The device is put into a pressure vessel and fluid pressure is applied until the specimen breaks.

The fluid pressure acts over the cylindrical surface of the specimen through weep hole R and axially through the pressure cap and loading washer. The prestress pressure case G prevents the fluid pressure from being applied to the spline end of the bolt; otherwise, most of the axial load from the fluid pressure would be assumed by the bolt and not by the specimen. The resultant stress state in the specimen is $\sigma_{\theta\theta} = 2\ \sigma_{zz}$ from the external pressurization plus the additional axial (σ_{zz}) compressive stress from the preload tension of the bolt. As the pressure is applied, some of the axial compressive prestress in the specimen, produced by the tension in the bolt, is lost because of the longitudinal elastic shortening of the specimen and consequently the bolt. The bolt and loading sleeve were designed to be elastically soft compared to the specimen to reduce this loss in preload on pressurization. [40]

Strength measurements were made in two stress states with this device, 1:1 and 1:0.8, biaxial compression. As in other tests where steel loading blocks bear on the specimen, excellent contact between mating surfaces must be produced or lower strength values are obtained. A difficulty was encountered in the 1:1 compression tests. During application of the axial preload, the ceramic compliance tubes cracked at a value of approximately

90% of the desired preload. A study was made of this cracking problem [29] and it was concluded that the cracking was probably induced by the tensile hoop stresses caused by slippage of the conical faying surface of the compliance tube over the specimen tube as the axial load was increased. The ceramic compliance tubes were replaced by hardened steel ones, but this resulted in the specimen splitting as shown in Fig. 4. Nevertheless, the tests using the split specimen tubes were completed and, in spite of the splitting, the tubes had remarkable strength (average strength measured on three specimens was 465 ksi). If a smaller conical angle had been used between the ceramic compliance tubes and specimen tube, the splitting might have been avoided; however, the effect that changing this angle would have on the parasitic stresses in the test section is not known. By reducing the preload stress level in the 1:1 test device to yield a stress at failure of 1:0.8, the compliance tube cracking problem was eliminated. Successful 1:0.8 compression tests could be performed with the ceramic compliance tubes without damage occurring to the specimen or compliance tubes during the preloading. These tests measured an average 1:0.8 compressive strength of 528 ksi with a standard deviation of 13.5 ksi. This scatter in individual measurements is comparable to that obtained in the much simpler 1:0.5 tests and the strengths in the two stress states appear to be the same.

4.4 Effect of End Constraints

If the specimen is a straight hollow cylinder with constant wall thickness which is tested by using massive end caps and subjecting the assembly to external fluid pressure to obtain the 1:0.5 stress state, care must be taken that the specimen is long enough that the constraint of the end caps does not reduce the circumferential stress from the one calculated by the thick-wall formula for long cylinders. An approximation of the error can be calculated from shell theory [41]. Simple shell theory can only give an approximation because the wall is rather thick and it must be so in order to avoid out-of-round buckling from the external fluid pressure [Reference 21, Appendix]; a typical value of thickness/radius is 0.12. For a cylinder with its ends constrained from radial displacement and rotation, for an outer radius = 0.90", wall thickness = 0.100", and length 1.42", the calculated circumferential stress at a distance midway between the ends is only 91% of that for a long cylinder and, of course, at the ends it is zero.

The only tests of externally pressurized cylinders of constant wall thickness known to the writers are those of Broutman and Cornish [17] in which they used the radius and thickness considered above, but used specimens 1" and 2" long. The values they reported for the 1:0.5 compressive strength for WESGO Al 995 alumina are:

Specimen Length (inches)	Strength(ksi)	End Cap Material
2	550	Steel
2	530	Steel w/Alumina
2	640	Tungsten Carbide
2	606	" "
1	630	" "

These tests may give values somewhat high for the compressive strength because the effect of the end constraints were not considered in the calculations. Note also that higher values were obtained with the tungsten carbide caps than with the steel ones. Tungsten carbide gives greater constraint than the steel because of its higher elastic modulus- 10×10^7 compared to 3×10^7 psi. It should also be noted that the circumferential stress varies from almost zero at the ends to the peak at the middle, thus only a small fraction of the total volume of a short specimen is near the peak stress and a "size effect" may contribute also to a higher measured value for the strength.

The high values also indicated that the ends of the specimens were properly lapped to the end caps and that in general, careful techniques were used because, except for the constraint, the slightest carelessness results in low measured values. In this test premature failure from tensile stresses induced by bending from the end constraints was avoided because the tensile stresses would be cancelled by the axial compressive load. These tests show that the expense added by using a reduced thickness for the test section may not be necessary for this 1:0.5 compression test. However, with a constant wall thickness, a longer length is needed to attain the desired circumferential stress and to have it constant over a reasonable fraction of the test volume.

A second difficulty arises if the specimen is long enough to provide an adequately large central region which is free from end constraint. In Reference 21, Appendix, it is shown that for the strengths and Young's moduli typical of high density alumina, failure by out-of-round buckling from the external fluid pressure may occur at circumferential stresses below the compressive strength, if the distance between the end-caps, which constrain the test tube from out-of-round displacements, is too long.

A careful, experimental and theoretical study of the effect of the end caps on stress distribution and buckling is needed.

COMMENTS

It is hoped that the above presentation of some of the problems inherent in the compressive testing of very strong, but brittle, materials will aid others in the design of tests to fit their individual requirements. It is premature to propose standardized tests when there is still so little experience; however, some generalizations can be made: it is desirable to have the cross sectional area of test section sufficiently less than the loading area so that the failure can be confined to the test section to ensure failure did not originate at the loading block. It is desirable to use special energy absorbing material and/or provide a parallel path that can assume the load when the specimen fails in order to prevent its pulverization and loss of the details of the failure. It is desirable to have strain gages on all specimens to ensure uniformity of the stress as well as the level of stress. Not all of these goals were met in all of our own tests.

We have cited work which we feel illustrates the problems inherent in testing of this kind of material. The literature cited is not intended to be exhaustive. We welcome information on tests where the investigators feel that the techniques have ensured accurate measurement of the strength.

ACKNOWLEDGEMENTS

This work was supported by the Naval Air Systems Command N00019-73-C-0294, Charles Bersch, monitor, and the Office of Naval Research ONR 76-C-0445, Nicholas Perrone, monitor. We gratefully acknowledge the help of R. Sedlacek of Stanford Research Institute and K. Gray of the Naval Civil Engineering Laboratories for sharing their experience with us during the initial stages of the program.

REFERENCES

1. Shanley, F.R. and W.J. Knapp, "Ceramics as structural materials", Journal of the Structures Division of the Proceedings of the American Society for Civil Engineers, Volume 91, 1965, pp. 47-55.

2. Knapp, W.J. and F.R. Shanley, "Ceramic materials-properties for structural applications", Aero/Space Engineering, Volume 17, 1958, pp. 34-38.

3. Stachiw, J.D., "Glass and ceramics for underwater vehicle structures", Ceramic Age, Volume 80, 1964, pp. 20-23.

4. VanWyk, J.W., "Ceramic airframe bearings", The Boeing Company, Seattle, Washington, Final Report, Contract N00019-73-0230, Naval Air Systems Command, February 1974, 110 pages.

5. van Reuth, E.C., A.F. McLean and R.J. Bratton, "Ceramic gas turbines for improved fuel consumption", Naval Engineers Journal, April 1975, pp. 109-114.

6. Sines, G. and M. Adams, "Measurement of biaxial compressive strength of ceramics", Proceedings of the 1971 International Conference on Mechanical Behavior of Materials, Japan, Volume 5, 1972, pp. 295-303.

7. Salmassey, O.K., W.H. Duckworth, and A.D. Schwope, "Behavior of brittle-state materials", WADC Technical Report, 53-50, Parts I and II, June 1955, 145 pages.

8. Babel, H.W. and G. Sines, "A biaxial fracture criterion for porous brittle materials", Journal of Basic Engineering Transactions of the American Society of Metallurgical Engineers, June 1968, pp. 285-291.

9. Ely, R.E., "Strength of titania and aluminum silicate under combined stresses", Journal of the American Ceramic Society, Volume 55, Number 7, July 1972, pp. 347-350.

10. Vile, G.W.D., "Behavior of concrete under simple and combined stresses", Ph.D. Thesis, University of London, Imperial College, September 1965.

11. Liu, T.C., "Stress-strain response and fracture of concrete in biaxial compression", Department of Structural Engineering, School of Civil Engineering, Cornell University Report Number 339, NSF Grant GK-10214, February 1971, 196 pages.

12. Mazanti, B.B. and G.F. Sowers, "Laboratory testing of rock strength", Testing Techniques for Rock Mechanics, ASTM STP 402, 1966, pp. 207-231.

13. Roark, R.J. and W.C. Young, Formulas for stress and strain , 5th Edition, McGraw-Hill, 1975, p. 519.

14. Peterson, R.E., Stress concentration factors, Wiley-Interscience, 1974, p. 89.

15. Ryshkewitch, E., Oxide ceramics, Academic Press, New York, 1960, pp. 139-140.

16. Ryshkewitch, E., Berichte deutsch keramischen gesellschaft, Volume 22, 1941, pp. 54-65.

17. Broutman, L.J. and R.H. Cornish, "Effect of polyaxial stress states on failure strength of alumina ceramics", Journal of the American Ceramic Society, Volume 48, October 1965, pp.519-524.

18. Sedlacek, R., "Investigation of elasticity and strength of ceramics subjected to tensile and compressive loads", Stanford Research Institute, Menlo Park, California, Technical Report AFML-TR-68-231, August 1968.

19. Bieniawski, Z.T., "Determination of rock properties", Council for Scientific and Industrial Research, CSIR Report MEG 518, Pretoria, January 1967, 50 pages.

20. Nadai, A., Theory of flow and fracture of solids, Volume 1, McGraw-Hill, New York 1950, pp. 343-344.

21. Adams, Marc and George Sines, "Methods for determining the strength of brittle materials in compressive stress states", Journal of Testing and Evaluation, Volume 4, Number 6, November 1976, pp. 383-396.

22. Shanley, F.R., W.J. Knapp, J.F. Brahtz, F.A. Needham, R.D. Johnston, R.B. Simonson, R.D. Chipman, and J.S. Mizushima, "Prestressed ceramic structures", WADC Technical Report 54-75, January 1954, 65 pages.

23. Budnikov, P.P., N.A. Marakueva and V.A. Vasil'ev, "Effect of test conditions on compressive strength tests on refractory ceramics", Glass and Ceramics, (English translation of Steklo i Keramika -Russian-) Volume 25, Number 12, December 1968, pp. 759-761.

24. Rudnick, A., C.W. Marschall, W.H. Duckworth and B.R. Emrich, "The evaluation and interpretation of mechanical properties of brittle materials", Technical Report AFML-TR-67-316 and DCIC 68-3, April 1968, 191 pages.

25. Schwaninger, O, "A 2000-ton compression testing machine", Proceedings of the Institute of Mechanical Engineers, Volume 180, Part 3A, 1965-1966, pp. 380-387.

26. Peters, C.D. and F.S. Digesu, "Gas-bearing facilities for determining axial stress-strain and lateral strain of brittle materials to 5500°F", ASTM Proceedings, Volume 65, 1965, pp. 855-873.

27. Owen, J.B.B., "Cheap strength-testing machines with freely hinged ends and a known line of loading", Proceedings of the Institution of Mechanical Engineers, Volume 180, Part 3A, 1965-1966, pp. 411-418.

28. Babel, H.W. and G. Sines, "An improved method for uniaxial and biaxial testing of brittle materials", Journal of Materials, Volume 3, Number 1, March 1968, pp. 134-152.

29. Adams, M. and G. Sines, "An experimental study on the Compressive biaxial strength of ceramics", Third Annual Report, UCLA-ENG-7395, February 1974, 57 pages.

30. Adams, M. and G. Sines, "Compressive biaxial strength of ceramics", Technical Report UCLA-ENG-7537, August 1975, 290 pages.

31. Adams, M.A., "Strength of brittle ceramics in compressive stress states", Ph.D. Theses, University of California, Los Angeles, August 1975, 290 pages.

32. Sedlacek, R., "Investigation of elasticity and strength of ceramics subjected to tensile and compressive loads", Technical Report, Stanford Research Institute, Menlo Park, California, AFML-TR-68-231, August 1968.

33. Brace, W.F., B.W. Paulding and C. Scholz, "Dilatancy in the Fracture of Crystalline Rocks", Journal of Geophysical Research, Volume 71, Number 16, August 1966, pp. 3939-3953.

34. Bieniawski, Z.T., "Mechanisms for brittle fracture of rock", National Mechanical Engineering Research Institute Council for Science and Industrial Research, Pretoria, South Africa, CSIR Report, MEG 580, August 1967, 226 pages.

35. Sedlacek, R., "Comparison of elastic properties of ceramics under compressive and tensile stress", SRI Project 184531-122, May 1965.

36. Adams, M. and G. Sines, "Spalling and cracking in alumina by compression, Journal Amer. Ceramic Society, Volume 60, No. 5-6, May-June, 1977, pp. 221-226.

37. Adams, M. and G. Sines, "Determination of biaxial compressive strength of a sintered alumina ceramic," Journal Amer. Ceramic Society, Volume 59, No. 7-8, July-August, 1976, pp. 300-304.

38. Adams, M. and G. Sines, "An experimental study on the compressive biaxial strength of ceramics," Second Annual Rept., UCLA-ENG-7297, December 1972 41 pages.

39. Adams, M. and G. Sines, "An experimental study on the compressive biaxial strength of ceramics," Summary Rept., UCLA-ENG-7524, February 1975, 65 pages.

40. Adams, M.A., "An experimental investigation of the biaxial compressive strength of ceramics," M.S. thesis in Engineering, Materials Department, University of California, Los Angeles, March 1972, pp. 46-55.

41. Flugge, W., Stresses in Shells, Springer, 1960, p. 285.

THE STRENGTH OF CERAMICS UNDER BIAXIAL STRESSES

G. Tappin, R.W. Davidge, J.R. McLaren

Materials Development Division, Building 552

A.E.R.E. Harwell, OX11 ORA, U.K.

An improved apparatus is described for studying the effects of biaxial stress states on the strength of ceramics. This uses a specimen in the form of a tube. Three pressures can be applied to the tube: inside, outside, and to the end caps on the tube. By selection of a ratio of two of these pressures it is possible to induce selected biaxial stress states. A key feature of the system is that the specimen geometry is kept constant for all stress states. REFEL* silicon carbide has been used as a model material and tubes in three surface conditions have been tested in stress states ranging through, longitudinal tension, equibiaxial tension, hoop tension, to longitudinal compression/hoop tension. Surface conditions comprised: the original shot blasted surface; a single longitudinal notch; an indentation-induced single sharp longitudinal crack. In each case the compression/tension strength is similar to the unidirectional tensile strength but the equibiaxial tensile strength is significantly lower (by $\sim$15%). A quantitative evaluation of the data in terms of existing theories is not possible because these are based on assumptions that are too simple for normal ceramics. Highly non-idealised fracture occurs due to interactions between the fracture process and the material microstructure. Attention must therefore be given to effects of the non-planar fracture surface formed by a discontinuous fracture process during which the crack front undergoes numerous and localised changes in direction. This means for example that, contrary to the simple theories, the strength of specimens with single large flaws does vary with biaxial stress state.

---

*Registered trademark of United Kingdom Atomic Energy Authority

1. INTRODUCTION

Most experimental and theoretical data for the strength of ceramics relate to a simple condition of uniaxial tensile stress. Many engineering ceramic components however are not subjected to this simple stress state and a multi-axial stress situation is much more common. It is thus essential to have reliable information and understanding of the effects of multi-axial stresses. Unfortunately this is a confusing field in that there are a number of theories and hypotheses matched by only a modest amount of reliable experimental data. The biaxial stress state is the simplest to consider and also of widespread relevance. We shall restrict our discussion to the tensile/tensile and tensile/compressive states for fracture in REFEL silicon carbide.

The approach used incorporates two novel features. An improved testing technique is described which enables a wide range of biaxial stress ratios to be used on specimens with a constant volume of material under maximum stress, and with the stress sensibly uniform throughout this volume. Secondly, the stresses to propagate large single flaws are compared with those to propagate the randomly distributed inherent flaws.

2. EXPERIMENTAL

2.1 Material and Specimens

REFEL silicon carbide is manufactured by a reaction sintering route (1) involving a reaction between molten silicon and a body comprising silicon carbide and graphite which results in an impervious self-bonded SiC containing 8-10% Si. The microstructure near an as-reacted surface is shown in fig. 1. After fabrication components are normally covered with small nodules of silicon which exude from the body of the material during cooling; these are readily removed by grit blasting. The material contains surface layers which are not typical of the bulk material: immediately at the surface is a layer (25-50 μm) of dense carbide containing little silicon and below this is a second layer (150-250 μm) of silicon rich material containing 30-40% Si.

Material was supplied in the form of grit blasted tubes 300 mm long, 21 mm outside diameter, and 18 mm inside diameter.

The strength of REFEL SiC depends on both the method of mechanical treatment of the surface and the microstructure of the material at the surface (2). Specimens with grit blasted surfaces have a strength 200-300 MN m^{-2} whereas ground surfaces give higher strengths, 400-500 MN m^{-2}. The strengths associated with the inner and outer

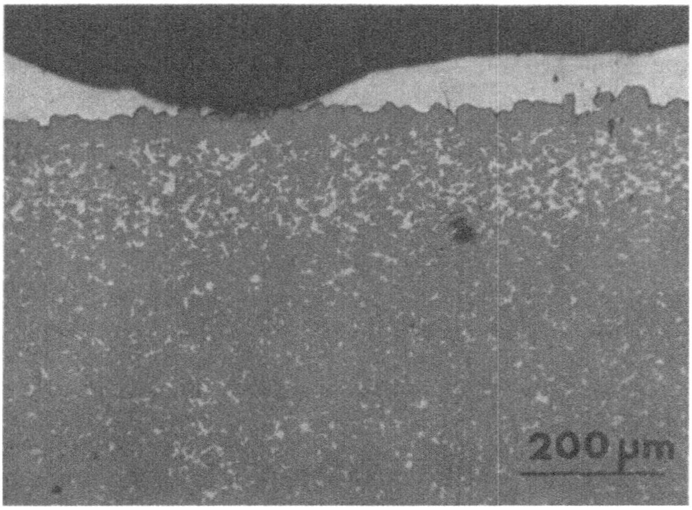

Figure 1. Microstructure of REFEL SiC near as-fabricated surface.

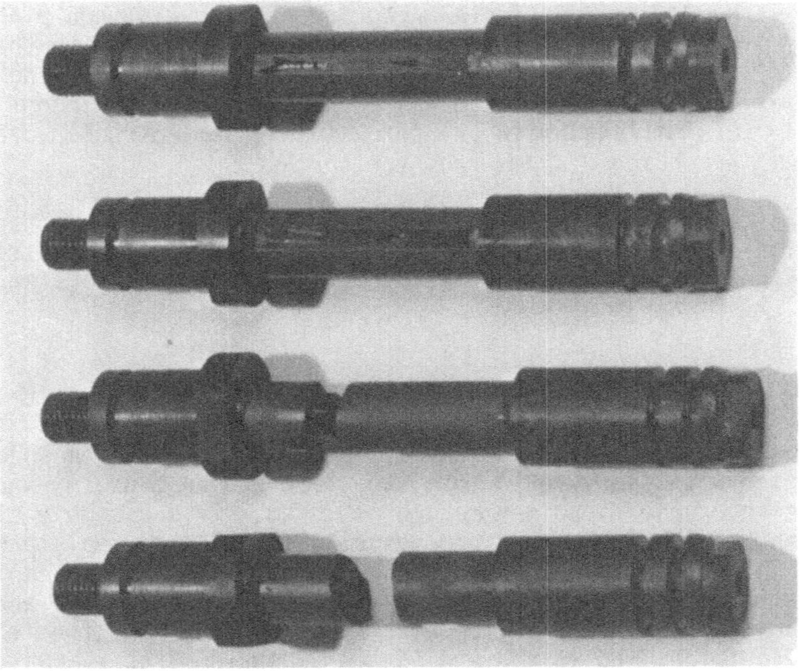

Figure 2. Fractured specimens. Above, under hoop stress: below, under axial stress.

surfaces of the current tubes were determined by cutting thin rings from the tubes and then compressing diametrally either whole or half rings to failure. With whole rings the maximum tensile stress is on the inside surface and with half rings on the outside surface. The strength associated with the inside surface was 510 MN m^{-2} and that with the outside surface 350 MN m^{-2}. The higher strength is thus associated with the inside surface; this presumably reflects the difference in grit blasting procedure which would give tangential impact on the inside surface but radial impact on the outside surface.

For the biaxial strength measurements specimens were cut to lengths of 120 mm. Three sets of specimens were tested.

(a) As-received, which would be expected to fracture from the outside surface.

(b) With a 0.5 mm deep secant shaped longitudinal notch introduced with a narrow diamond slitting wheel near the mid-length of the tube.

(c) With a narrow flat along the length of the tube so that the wall thickness was reduced in the centre of the flat by 0.5 mm. A single Knoop indentation was then placed near the mid-length of the tube on the flat with the long axis of the indentation parallel to the axis of the tube. The indentation load was 10 kg which gave an indent 0.5 mm long, with an approximately semi-circular crack extending 0.25 mm from the surface.

Tests on these specimens thus permitted examination of specimens with a random flaw distribution, a restricted flaw distribution and a single flaw. Mounted specimens, after fracture, are shown in fig. 2.

2.2 Biaxial Strength Apparatus

The apparatus is shown schematically in fig. 3. The specimen S is mounted with epoxy resin into two steel end bobbins, A_1 and A_2, to give an exposed specimen length of 70 mm. This assembly slides into the body B of the pressure vessel with pressure seals provided by two sets of O-rings R. End caps C_1 and C_2 seal the assembly in the chamber. The bobbin A_2 is attached to the end cap C_2, whereas the bobbin A_1 is free to slide within the body B. The free volumes, inside, outside, and at the end of the tube are first evacuated and then filled with oil, the medium through which the pressure is transmitted to the specimen. Two pressure lines are provided: one to the inside of the tube (P_2); the other, either to the outside of the tube (P_3) or to the end of the tube (P_1).

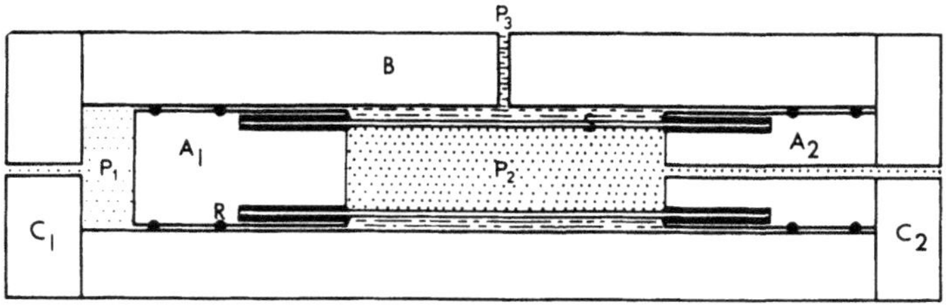

Figure 3. Principle of apparatus for measuring biaxial strength of ceramic tubes

The geometry of the apparatus is designed such that the cross sectional area of the tube is about half that of the bore in B. Pressure (P) combinations used are related approximately to the tensile (T) and compressive (C) stresses in the hoop and axial directions, as indicated in Table I.

Pressure			Stress	
End P_1	Inside P_2	Outside P_3	Hoop σ_h	Axial σ_a
0	P	P	0	T
0	3P/4	P/4	T/2	T
0	3P/2	P/2	T	T
0	P	0	T	T/2
P	2P	0	T	0
P	P	0	T	C/2
3P/2	P	0	T	C
5P/2	P	0	T	2C

Table I. Ratios of hoop (σ_h) and axial (σ_a) stresses for various ratios of applied pressure.

During a test the pressures, in the chosen ratio, were increased steadily until fracture occurred, typically at pressures $\sim$20 MN m^{-2} and in times $\sim$ 1 min.

2.3 Calculations of Stresses and Calibration of Apparatus

Stresses were calculated from the pressures according to the standard equations for pressure tubes as in Table II. Deviations

	Axial σ_a	Hoop σ_h	Radial σ_r
End P_1	$\frac{-P_1 r_b^2}{(r_o^2 - r_i^2)}$	0	0
Inside P_2	$\frac{P_2 r_i^2}{(r_o^2 - r_1^2)}$	$\frac{P_2 r_i^2}{(r_o^2 - r_i^2)} (1 + \frac{r_o^2}{r^2})$	$\frac{P_2 r_i^2}{(r_o^2 - r_1^2)} (1 - \frac{r_o^2}{r^2})$
Outside P_3	$\frac{P_3 (r_b^2 - r_o^2)}{(r_o^2 - r_i^2)}$	$\frac{-P_3 r_o^2}{(r_o^2 - r_i^2)} (1 + \frac{r_i^2}{r^2})$	$\frac{-P_3 r_o^2}{(r_o^2 - r_i^2)} (1 - \frac{r_i^2}{r^2})$

Table II. Equations relating stress to pressure. r_o and r_i are the outer and inner radii of the specimen, r_b, the radius of the test vessel body and r a co-ordinate.

from an ideal uniform biaxial stress state thus exist but these are small: typically, the hoop stress varies by ∿15% throughout the wall thickness and the radial stress is small, varying from zero to ∿one sixth of the hoop stress.

The apparatus was calibrated by means of strain gauges attached to the outside of tubes of silicon carbide and of steel. Connections were provided in the pressure vessel which permitted recordings to be made from up to 6 strain gauges on a particular specimen. In the central regions of the specimens the observed strains were within a few percent of the calculated strain, thus giving confidence in the use of the equations in Table II. The ends of the specimen were constrained by the method of attachment to the bobbins, and strains increased from low values near the bobbins to the expected values in a distance ∿15 mm, giving an effective specimen length of ∿40 mm. The main observational inaccuracy is in the measurement of the bursting pressures which could be read from the pressure dial gauges within ± 2%.

3. RESULTS

A total of ∿150 tubes was broken. At least 6 specimens were broken under each experimental condition. As-received tubes were fractured under the 8 approximate pressure ratios indicated in

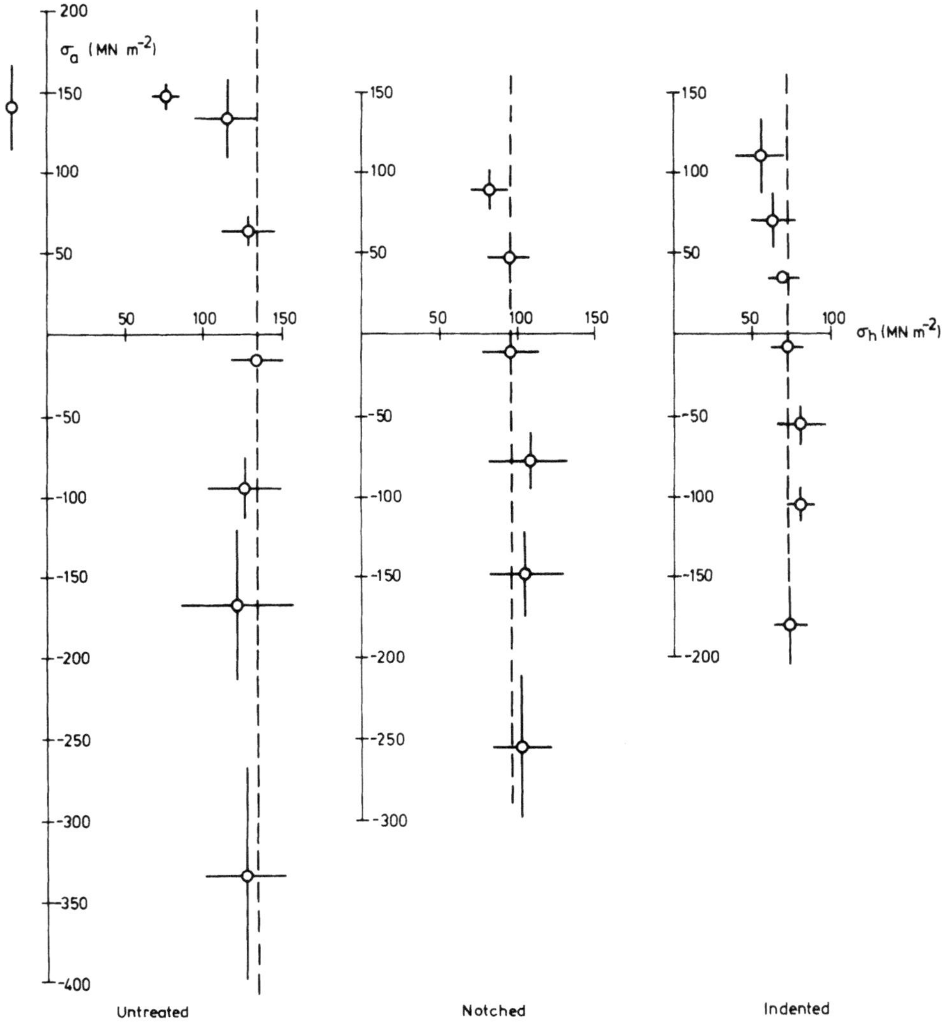

Figure 4. Fracture stresses as a function of biaxial stress state for three sets of tubes of REFEL SiC.

Table 1. The notched, and the indented specimens were tested under the 7 pressure ratios having a positive hoop stress. Results are shown in fig. 4, where the error bars represent one standard deviation for an individual set of results.

The as-received specimens fractured with a hoop fracture crack when tested under pure axial tensile stress, and with axial cracking when tested under a pure hoop stress. The latter fracture mode was also observed when the axial tensile stress was lower than the hoop stress or was compressive. All the notched, and the indented specimens fractured with an axial crack passing through either the notch or the indentation. Examples are shown in fig. 2. In many cases the tube fractured into a number of pieces.

4. UNIAXIAL STRENGTH AND FLAW SIZES

The strengths obtained by diametral compression of rings cut from the tubes, where the maximum stress is in a hoop direction, are 350 MN m^{-2} when fracture originates from the external surface and 510 MN m^{-2} when fracture originates from the inside surface. The difference between these strengths is presumably associated with differences in the grit blasting procedure. The implication is that grit blasting in an axial direction produces more surface damage than does that in a tangential direction. These values compare with strengths of the bulk material (3) of $\sim$400 MN m^{-2} obtained from bend specimens, of similar size to the rings, with machined surfaces. The fracture origin of the complete tubes in the pressure apparatus should therefore lie on the external surfaces.

The mean recorded strengths for tubes in uniaxial tension in the axial and hoop directions are 140 and 134 MN m^{-2} but this difference is small and the material may be regarded as isotropic in its mechanical properties. The much lower strengths associated with complete tubes compared with small specimens can be understood in terms of the distribution of flaws in the material. Previous data (2) on the statistical distribution of strengths of REFEL tubes 150 mm long are available but these tubes differed from the current ones in that the outer surface had been centreless ground to leave a wall thickness of 1 mm; the inside surface was the same as that for the current tubes. The mean strength of the tubes was 180 MN m^{-2} which is slightly greater than the current tubes indicating that grinding of the external surface has a beneficial effect on strength. The strength data were analysed using Weibull statistics and this indicated a Weibull modulus m = 6.5. The fracture origin in these specimens however was not ascertained.

The ratio of the strengths of the current rings and tubes (σ_o/σ_t) can be estimated approximately from the formula

$$\sigma_o/\sigma_t = (A_t/A_o)^{1/m} \qquad (1)$$

where $A_{o,t}$ refer to the areas of material under maximum stress. This assumes that Weibull fracture statistics apply and that fracture originates at the external surface. A_t = 2640 mm^2: A_o is not known exactly but is estimated at $\sim$10 mm^2. Assuming m = 6.5 gives σ_o/σ_t = 2.36 which compares with the experimental value of 2.55. Although this calculation is only approximate it suggests that the difference between the small and large specimens can be accounted for by standard statistical arguments.

It should be noted that REFEL SiC is not generally used in the surface condition of the current experiment for engineering components. When all surfaces are machined much higher strengths are obtained, and the scatter in strength is reduced to give a Weibull modulus of $m \sim 10$.

The size of the flaws from which fracture originates can be calculated from the standard fracture mechanics equation (4)

$$\sigma_f = \frac{Z}{Y} \left(\frac{2E\gamma_i}{C}\right)^{\frac{1}{2}} \quad (2)$$

where σ_f is the fracture stress, E is Young's modulus, γ_i is the effective surface energy for the initiation of fracture, C is the flaw size, Y is a constant relating to the test geometry and Z a constant determined by the shape of the flaw. For a crack of length 2C through the thickness of the wall of the tube and aligned parallel to the tube axis a fracture mechanics analysis (5) shows that $Z = 1$ and $Y = \pi^{\frac{1}{2}}$. This is accurate to within a few percent for small cracks, say < 1 mm. Using the current value of $\sigma_f = 138$ MN m^{-2} and $E = 406$ GN m^{-2} (6), $\gamma_i = 18$ J m^{-2} (3) in equn. (2) gives a flaw size $2C = 0.49$ mm. The flaws in the current tubes are unlikely to go through the thickness of the tube wall otherwise removal of the surface layers would not result in an increase in strength. Assuming that the flaw lies on the surface and is semi-circular in shape then the appropriate value for $Z = 1.57$. In this case the flaw would have a length 1.22 mm and a depth 0.61 mm.

The presence of such large flaws on the external surface of the tubes is in agreement with two experimental observations. Firstly, the breaking pressure of the tube is not decreased until an external surface layer > 0.5 mm has been removed by grinding. In a homogeneous material such a reduction of the wall thickness from 1.5 to 1.0 mm would result in a bursting pressure decrease of one third. Secondly, 10 kg Knoop indentations on the as-received tubes did not result in a decrease in busting pressure.

The flaws produced by indentation are of semicircular shape 0.5 mm length and 0.25 mm depth in a region of reduced wall thickness of 1.0 mm. Substitution into eqn. (2) indicates a strength of 144 MN m^{-2} compared with an observed value of 108 MN m^{-2}. Similar discrepancies have been noted previously (7) in fracture mechanics calculations using flaw sizes observed under Knoop indentations; it is believed that the residual stresses near the indentation enable the flaw to propagate at a lower stress than expected. The flaw produced by indentation thus has an effect larger than that expected from its observed size.

5. BIAXIAL STRENGTH

The key observation from the current experiment is that for all three sets of specimens the strength under a biaxial tensile stress is less than that under uniaxial tension. Comparing equi-biaxial tensile and uniaxial tensile stress states this difference is $\sim$15% which is statistically significant at the 99.9% level. The difference for the indented specimens is slightly greater than this and for the other specimens slighly less; these differences between types of specimen however are not statistically significant.

Several theories and hypotheses exist from which predictions can be made about the variation in strength with the data of biaxial stress. Care must be taken to distinguish between the situations where the specimen contains a single flaw and that where there is a random distribution of flaws. A serious limitation of the underlying theories is that they generally assume that the material is isotropic and that the crack propagates in a single plane. In practice the crack surface is far from flat and there are marked variations in surface energy requirements over a scale related to microstructural features. This means that the crack proceeds through a number of discrete jumps as indicated by acoustic emission evidence. We shall see that further theoretical developments are required before a satisfactory explanation of the current results can be formulated. A consideration of the current theories however will indicate the areas where developments are required.

5.1 Specimens with Single Flaws

The original and classical theory is due to Griffith (8) who considered a single sharp flaw through the thickness of a thin sheet of material under a biaxial stress. Griffith made the hypothesis that the general condition for rupture should be the attainment of a specific stress at the tip of the crack. Under the conditions where there is a tensile stress T applied normal to the flaw and a stress parallel to the flaw ranging from T in tension to compressive values, fracture should occur at a stress normal to the flaw independent of the value parallel to the flaw. For values of the compressive stress > 3T fracture should occur at a lower stress provided that the flaw is in a specific orientation but different from that above; this however is outside the range of the current experiments and according to the Griffith hypothesis the strength of present specimens containing a single flaw should be independent of the biaxial stress ratio.

The basic Griffith calculations have been re-examined and there is now doubt as to whether the failure criterion is independent of the stress parallel to the crack. Swedlow (9), for example, presents calculations based on strain energy considerations which indicate that in a biaxial stress situation the stress applied parallel to the crack does have an effect on the failure criterion and the tensile/tensile strength differs from that prediced by Griffith depending on the value of Poisson's ratio. For ν <0.25 the biaxial strength should be greater than the Griffith criterion and for ν >0.25 less than the Griffith criterion. The results of Swedlow however have been disputed by a number of workers (10), who have questioned the validity of the mathematical analyses, but accepted by others (11). A concensus view on this question would be extremely valuable.

There appears to be little data in the literature concerning the biaxial strength of ceramics with single flaws, apart from Griffith's original limited data on just 10 specimens of glass which were in agreement with his hypothesis.

The current results for material with a single flaw are not in accordance with existing theories. The Griffith criterion predicts no variation in strength with stress state whereas the calculations of Swedlow indicate that the equibiaxial strength should be greater than the uniaxial tensile strength (Poisson's ratio for REFEL SiC is 0.17). Theory however relates to the propagation of a plane crack and this is far from the actual situation for normal ceramic materials. The path of the crack under uniaxial tension for example deviates quite markedly from the plane normal to the stress direction, even though the average direction of crack propagation is in this plane. Figure 5 gives data obtained from analyses of

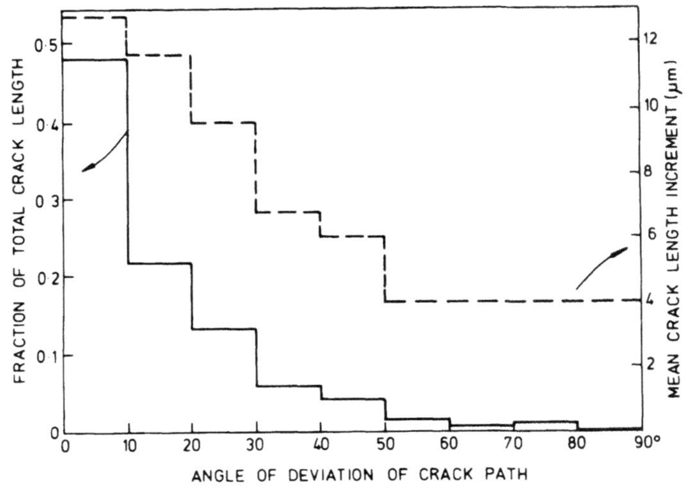

Figure 5. Variation of incremental crack length and proportion of crack length as a function of crack orientation.

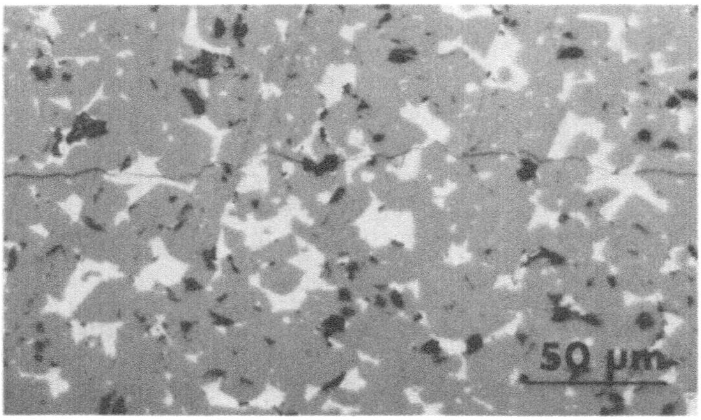

Figure 6. Crack produced under uniaxial tensile stress.

crack paths in material fractured under uniaxial tension. Photographs of planes cut normal to the fracture plane, as in fig. 6, were analysed by assuming that the line of intersection of the fracture plane with the examined surface could be split up into a number of approximately linear sections. The angle that each section made with the average direction of crack propagation was measured, as was the length of the section. Sections <2 µm were arbitrarily ignored, and longer sections were allocated to crack lengths in 4 µm increments. The analysis in fig. 5 shows that ~50% of the crack path is within ± 10° of the average direction of propagation and ~70% within ± 20°; ~15% of the crack path however deviates by >30° from the average direction. The length of crack propagation in a given orientation range decreases as the angle of deviation increases. For cracks within ± 20° of the average direction the mean crack length observed is ~12 µm. At angles >50° this has fallen to 4 µm, the shortest length increment selected.

Calculation of the stress to propagate a crack moving in such a complex manner is a difficult interpretive challenge although there is understanding of some features of the process such as the stress for a plane crack to deviate into a new plane. For a uniaxial applied tensile stress normal to the crack, the tensile stress σ_θ near the tip of a crack in a direction normal to a plane making an angle θ with the plane of the crack is (12)

$$\sigma_\theta = \frac{K_I}{(2\pi r)^{\frac{1}{2}}} \cos^3(\theta/2) \qquad (3)$$

K_I is the tensile opening mode stress intensity factor and r is the distance from the tip of the crack. For $\theta = 45^0$, σ_θ is reduced to 0.79 of the value for $\theta = 0$. Thus for a plane crack to deviate through 45^0 under a tensile stress normal to the original crack requires an applied stress increase of ~25%. Noting that the crack increments shown in fig. 5 are very much smaller than the original crack size (1-2% of this) the chance that a crack will make a significant deviation from its original direction in a relatively short distance is quite large. The stress to propagate such a real crack should therefore be rather greater than the value estimated from that for a simple planar crack. In the case of a biaxial tensile stress the stress enhancement required to produce a deviation in crack direction will be less than for uniaxial tension. It should thus be easier for a complex crack to propagate under a biaxial tensile stress compared with a uniaxial tensile stress. The experimental data for the specimens with single flaws are in sympathy with this qualitative argument.

In the real situation the mode of crack propagation is even more complex. Small areas of fracture will propagate from the crack front at random positions and in orientations determined by the local microstructure. Adjacent regions of crack front will thus contain localised fractures in different orientations. For these localised cracks to progress further then they must link together and this will almost certainly involve a tearing or shearing mechanism. Some progress towards a theoretical understanding of this incremental crack growth mechanism has been obtained for the case of a simple two-dimensional microstructure where a flat crack is considered to pass through a number of regions having different localised fracture energy requirement (13). It seems unlikely however that such a model could be extended realistically to three dimenions to include considerations of the linking process by tearing between adjacent regions of crack front and also to estimate the effects of multiaxial stressing. On the experimental side further progress could be made by investigating the biaxial stress behaviour of a simple isotropic material like glass, as did Griffith, containing single large flaws. In this case further agreement with the simpler theories might be expected.

5.2 Specimens with Random Flaws

In most practical cases we are not concerned with the stress to fracture a sample with a single flaw but with that to propagate whichever flaw, from the statistical distribution available in the material, is subjected to the maximum stress. In uniaxial tension only those flaws which are oriented approximately normal to the stress direction are likely to propagate and thus the flaws sampled by a unidirectional stress are just a small fraction of the total within the specimen. If however the specimen is subjected to an equibiaxial tensile stress then all flaws normal to the plane of the stress are subjected to the same stress. The number of flaws sampled by a biaxial tensile stress is thus much greater than that in the unidirectional case and thus one expects the strength to be lower in the former case. Similarly, the effect of a tensile/compressive stress is also likely to be different from the unidirectional tensile case.

The Weibull statistical theory of fracture has been extended to cover biaxial stresses (14), for a body of constant volume subjected to a uniform biaxial stress. Results for Weibull moduli, of 5, 10 and 20 are shown in fig. 7. The theory predicts that tensile/tensile stresses reduce the fracture strength compared with uniaxial tension, and tensile/compressive stresses the reverse. These effects increase as m decreases, and are more pronounced in the tensile/tensile quadrant. With m = 10, for example, the equibiaxial strength is 0.8 times the unidirectional strength. Although more sophisticated statistical treatments are now available (15) the general conclusions are qualitatively similar.

However, in view of the lack of precise understanding of the situation for material containing a single flaw, any comments on

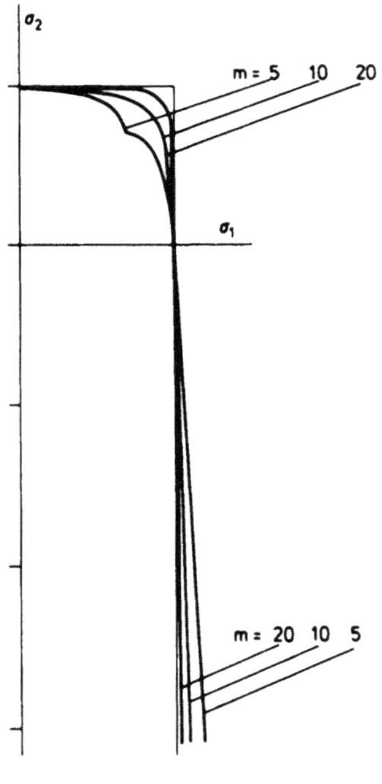

Figure 7. Predicted variation of strength with biaxial stress state from Weibull statistical theory (14).

the current as-received specimens containing a random distribution of flaws must be highly speculative. Although the existing statistical theories are in agreement with the data for such specimens it should be noted that the effects of biaxial stress are similar for all sets of specimens examined. It appears therefore that in REFEL SiC the dominant factor affecting behaviour under biaxial stress is the detailed and complex manner in which real cracks propagate, and that the existing statistical theories are inappropriate.

ACKNOWLEDGEMENT

Thanks are due to Mr. A.T. Slater for his skilled assistance on the design and commissioning of the experimental apparatus.

REFERENCES

1. C.W. Forrest, P. Kennedy and J.V. Shennan, Special Ceramics 5 (1972), 99.
2. P. Kennedy, J.V. Shennan, P.M. Braiden, J.R. McLaren and R.W. Davidge, Proc. Brit. Ceram. Soc. 22 (1973), 67.
3. J.R. McLaren, G. Tappin and R.W. Davidge, Proc. Brit. Ceram. Soc. 20 (1972) 259.
4. R.W. Davidge, Mechanical Properties of Ceramics, Cambridge University Press (1978).
5. F. Erdogan and J.J. Kibler, Int. J. Frac. Mech. 5 (1969), 229.
6. H.E. Rosinger, I.G. Ritchie and A.J. Shillinglaw, Mat. Sci. Eng. 16 (1974) 143.
7. R.R. Wills and J.M. Wimmer, J. Amer. Ceram. Soc. 59 (1976) 437.
8. A.A. Griffith, Phil. Trans. Roy. Soc. Lond. A221 (1920), 163; Proc. First Int. Congress Appl. Mech. (ed. C.B. Buzeno and J.M. Burgers), Delft (1924), p. 55.
9. J.L. Swedlow, Int. J. Frac. Mech. 1 (1965), 210.
10. P.L. Key, Int. J. Frac. Mech. 5 (1969), 287.
11. A.G. Evans and T.G. Langdon, Prog. Mat. Sci. 21 (1976) 171.
12. B.R. Lawn and T.R. Wilshaw, Fracture of Brittle Solids, Cambridge University Press (1975).
13. A.G. Evans and L.J. Graham, Acta. Met. 23 (1975), 1303.
14. O.K. Salmassy, W.H. Duckworth and A.D. Schwope, WADC Technical Rep. 53-50 Part 1, (1955).
15. A.G. Evans, this Volume.

A FRACTURE SPECIMEN FOR HIGH-TEMPERATURE TESTING

A.S. Kobayashi, L.I. Staley, A.F. Emery and W.J. Love

University of Washington, Department of Mechanical Engineering, Seattle, Washington 98195

A wedge-loaded tapered DCB specimen is proposed for fracture toughness and stable crack growth measurements of ceramic materials at elevated temperatures. The specimen is wedge-loaded by a compression pin seated in the starter notch. By appropriate selection of the specimen's taper and starter notch angle, the stress intensity factor is maintained relatively constant through 50 percent of the test section. Room temperature fracture toughness testings of commercial grade polycrystalline alumina specimens indicate that the crack will run straight for approximately 50 percent of the length of the test section without side grooving. A moderate 10 percent side grooving was sufficient to maintain straight crack extension throughout the entire specimen length.

INTRODUCTION

Because of their inherent brittleness, ceramic materials can be readily analyzed by linear fracture mechanics. Despite the extensive use of fracture mechanics in metal structures, the use of fracture mechanics in failure prediction and analysis of ceramic structures is relatively new.[1,2] Perhaps the extreme brittleness, which in some instances could lower the critical crack length derived from fracture mechanics considerations to a dimension comparable to the microstructural grain size of the ceramic material, was responsible for the reluctance of ceramic engineers to apply the continuum theory of fracture mechanics to the heterogeneous grainy ceramics. Recent developments of tougher ceramic materials together with readily available finite element codes for detailed structural analysis, however, have stimulated the use of fracture mechanics in prototype design of ceramic structures. Fracture mechanics parameters developed for metallic structures, such as fracture toughness and stable crack growth rate, are now viable parameters in the failure analysis of ceramic structures and thus the many measurements techniques developed for characterizing fracture of metals can also be used in ceramics. In particular, the inherent brittleness of ceramics relaxes the thickness requirements of metal fracture specimens and thus allows the use of relatively thin ceramic fracture specimens. A comprehensive review of typical ceramic fracture specimens is given in Reference 1. Special purpose fracture test specimens, such as the notched-ring fracture toughness specimen[3] and the short rod fracture toughness specimen[4], are also useful specimen configurations for testing ceramic materials.

A distinct advantage of some ceramic materials is the relatively small reduction in mechanical properties at the extremely high temperatures of 1000°C to 1400°C where most metals fail as a structural material. Fracture testing at such high temperatures requires simple test procedures where much of the test fixture is fabricated from the same high temperature ceramic which is undergoing testing. The most commonly used fracture test specimen for such high temperature testing is the double-torsion specimen which yields essentially a constant stress intensity factor under constant load regardless of the crack length.[5,6] The experimental convenience of the double-torsion specimen is, however, overshadowed by the lack of rigorous analysis of the stress intensity factor for the curved crack front which is inherent in this specimen. Although a recent three-dimensional finite element analysis[7] of the curved crack front in the double-torsion specimen confirmed the results of a simple one-dimensional analysis[8] of this specimen, this simplified FEM analysis was not sophisticated enough to reveal the expected complex variations in stress intensity factors in three-dimensional fracture mechanics.[9]

The purpose of this paper is to propose a fracture specimen, a wedge-loaded tapered DCB specimen, which can be conveniently used in high temperature testing and for which an analytical background can be provided.

WEDGE-LOADED TAPERED DCB SPECIMEN

Wedge-loaded DCB specimens have been used extensively in determining the dynamic fracture toughness and crack arrest intensity factors in steel.[10] Straight-sided DCB specimens loaded under essentially fixed grip conditions were used at the Battelle Columbus Laboratories to achieve crack arrest in the field of decreasing static stress intensity factor. For static and quasi-static fracture testing at high temperature, however, a field of constant static stress intensity factor is desirable regardless of the crack length since instantaneous crack length are difficult to measure in this environment.[5] A tapered DCB specimen[11], with appropriate taper, loaded under essentially a fixed grip condition could be more appropriate for such application.

Although wedge-loading by the use of two pins, as used by Battelle Columbus Laboratory (BCL specimen) would allow better control because of the defined point of specimen load application, this loading mechanism was considered impractical for high temperature testing where the rigidity of the loading fixture would be compromised. A pin-loaded wedge as shown in Figure 1 appeared to offer the minimum load fixture compliance for this wedge-loaded DCB specimen.

The taper angle of 45° was established through a trial-and-error search for a static stress intensity factor which is independent of the crack length at a given load. A standard finite element method was used for this analysis with the stress intensity factor being determined through the use of the methods of the strain energy release rate and crack opening displacement.* Figure 2 shows the finite element idealization and a typical finite element

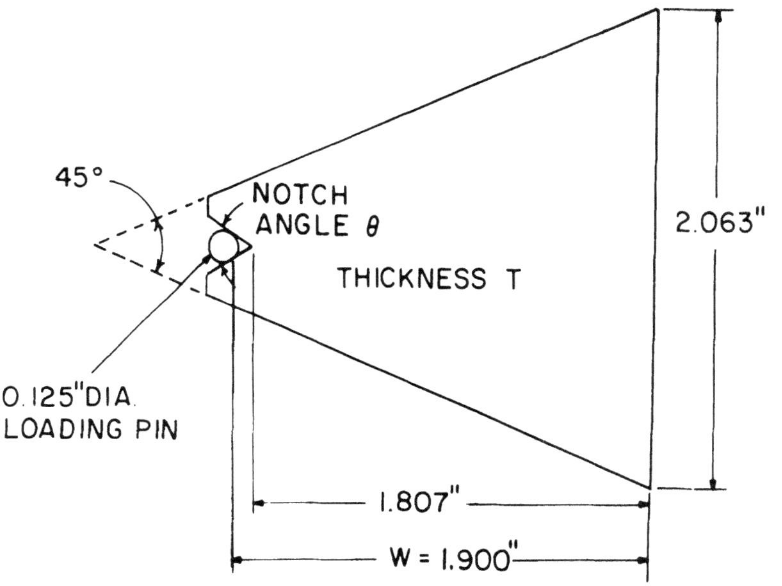

Figure 1. Wedge-Loaded Tapered DCB Specimen

* An abbreviated description of these computational procedures can be found in example Reference 12.

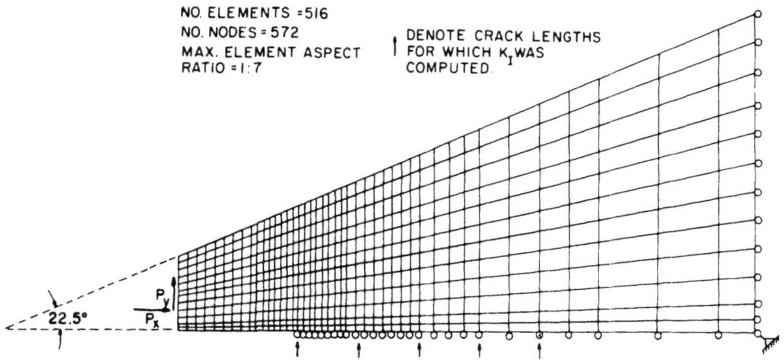

Figure 2. Finite Element Mesh for Tapered DCB Specimen

breakdown used in this series of analysis. Analyses were conducted for unit forces applied in directions parallel and perpendicular to the crack with the point of force application at the contact point between the pin and notch surface. A modulus of elasticity of $E = 56 \times 10^6$ psi and Poisson's ratio of $\nu = 0.22$ were used in the numerical calculation to simulate commercial grade alumina. The resultant stress intensity factor, however, is not strongly dependent upon the material properties in a simply connected body under this essentially force-prescribed boundary condition.

Figures 3 through 5 show the variations in stress intensity factors, K_I, with crack length, a, coefficient of friction, μ, and notch angle, θ. The variations in stress intensity factors with friction coefficients between the loading pin and notch surface show the sensitivity of this fracture specimen to the exact contact condition at the pin-notch interface. In general, an acute notch angle will increase the sensitivity of stress intensity factor to coefficient of friction between the pin-notch surface but a smaller applied load will be required to generate the necessary stress intensity factor for fracture. Similar reductions in the crack tip stress intensity factors were found in the BCL type of wedge-loaded DCB specimen. This dependence of stress intensity factor upon the coefficient of friction can be tolerated under ordinary testing environment where the wedge-load measurement can be supplemented with crack opening displacement or wedge-pin displacement measurements. For high temperature testing of ceramic materials, however, such displacement measurements are impractical if not impossible and thus the actual coefficient of friction under the test environment must be determined as accurately as possible.

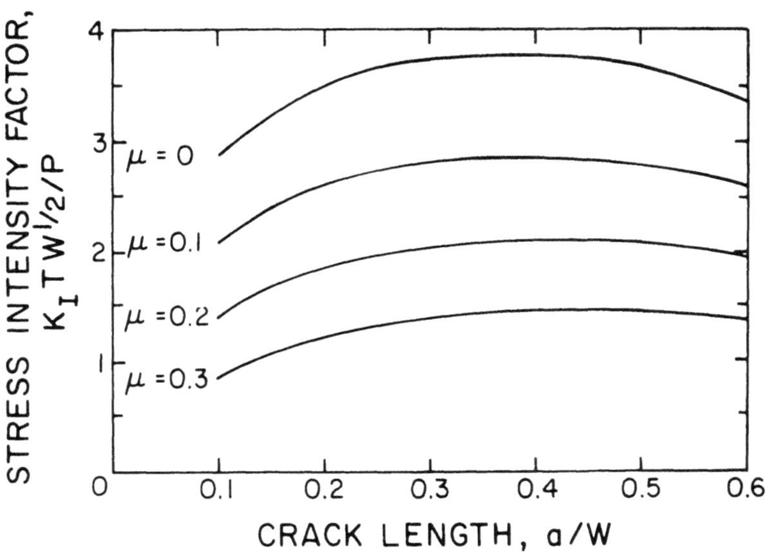

Figure 3. Stress Intensity Factor of a Wedge Loaded DCB Specimen, Notch Angle θ = 90°.

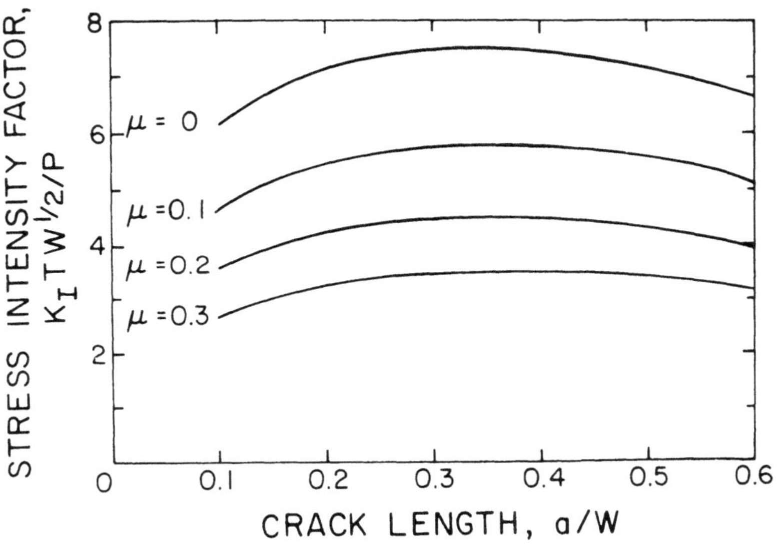

Figure 4. Stress Intensity Factor of a Wedge Loaded DCB Specimen, Notch Angle θ = 60°.

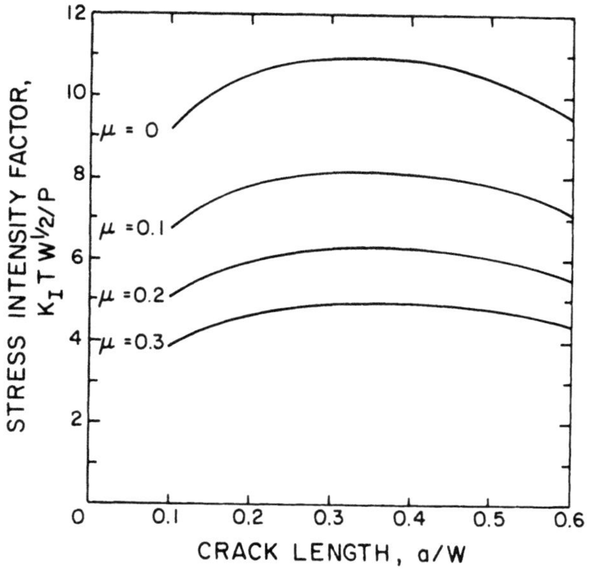

Figure 5. Stress Intensity Factor of a Wedge Loaded DCB Specimen, Notch Angle θ = 45°.

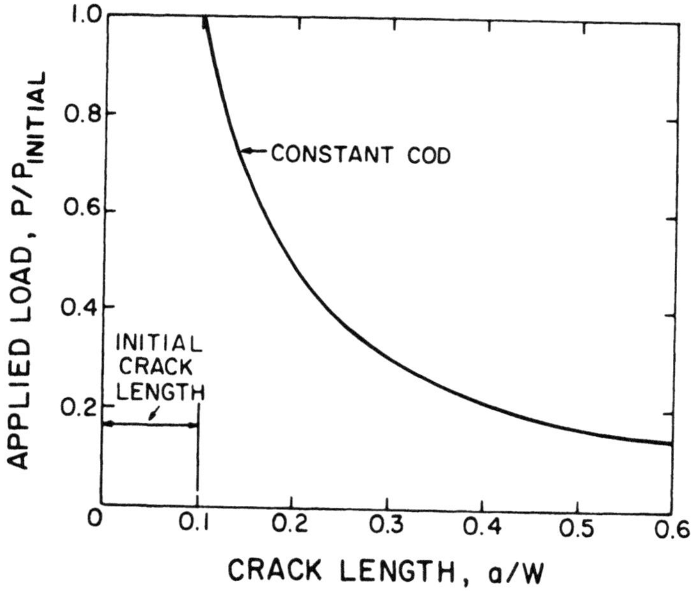

Figure 6. Load Relaxation Due to Crack Extension

On the other hand, the relaxing applied wedge load versus crack length relation under a fixed wedge displacement, as shown in Figure 6, is independent of the coefficient of friction. Barring other experimental errors involved in high temperature testing, the rate of load relaxation can thus be used to monitor the instantaneous crack length in this wedge-loaded DCB specimen undergoing stable crack growth in a state of constant stress intensity factor.

EXPERIMENTAL PRELIMINARIES

Upon completion of the analysis of the wedge-loaded DCB specimen, preliminary fracture testing was performed at room temperature on a series of twenty-seven alumina specimens with dimensions shown in Figure 1. The specimens were fabricated from a commercial grade hot-sintered alumina, COORS AD-998 ground to a nominal thickness of 0.25 in.

As shown in Figures 3 through 5, the stress intensity factor depends upon the coefficient of friction between the loading pin and the notch surface of the alumina specimen. The coefficient of friction was experimentally measured by pushing a hardened steel pin (not permitted to rotate) between two parallel alumina plates with a compressive load applied normal to the plate surface. A similar procedure may be easily adapted for high temperature testing. Coefficient of friction tests were also performed using a graphite lubricant (Graffoil) applied between the loading pin and alumina surface. For the series of preliminary tests, the graphite lubricant was used because the lower coefficient of friction significantly reduced the applied load necessary to initiate fracture and also eliminated the variability in coefficient of friction due to varying surface hardness of the steel loading pin. For a given critical stress intensity factor, it can be seen from Figures 3-5 that a lower applied load will result from a lower coefficient of friction. It is noted that the efficacy of the wedge-loading in elevated temperature testing will be determined by the consistency of the data on the coefficient of friction obtained in the actual high temperature environment. Measured variations in the coefficient of friction at room temperature testing were on the order of 10 percent for graphite lubricated contact surface.

An important feature of this fracture testing procedure is the precracking of the tapered DCB specimen. This precracking was accomplished by preloading the notch root of the tapered DCB specimen through the steel loading pin with the two edges of the specimen compressed in the fixture shown in Figure 7. This fixture effectively limits the possible crack opening displacement near the point of load application and consequently results in easily controlled crack extension. Little difficulty in crack curving as

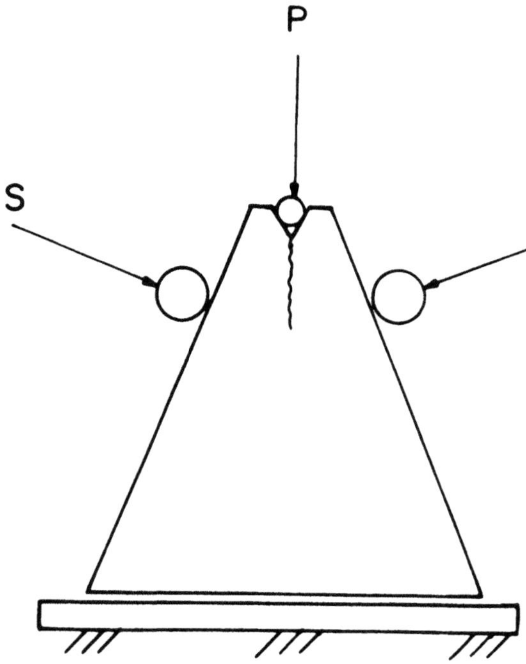

Figure 7. Precracking Fixture

reported in Reference 1 was found to occur in this precracking process for precracks of less than one-half inch.

FRACTURE TOUGHNESS

Seventeen specimens with precracks of length approximately .2∿.3' beyond the notch tip and ten specimens with approximately 10 percent side grooving and with similar precracks were tested in an Instron testing machine. Figures 8 and 9 show the measured fracture toughness for these two types of specimens. Insignificant differences between the average fracture toughness obtained from specimens with and without side grooving are noted. The side grooving, however, appears to decrease the amount of data scatter and also eliminates the crack curving which existed in the specimens without side-grooving.

CONCLUSION

The above preliminary test results show that a side-grooved, precracked, tapered DCB specimen can be used for fracture testing

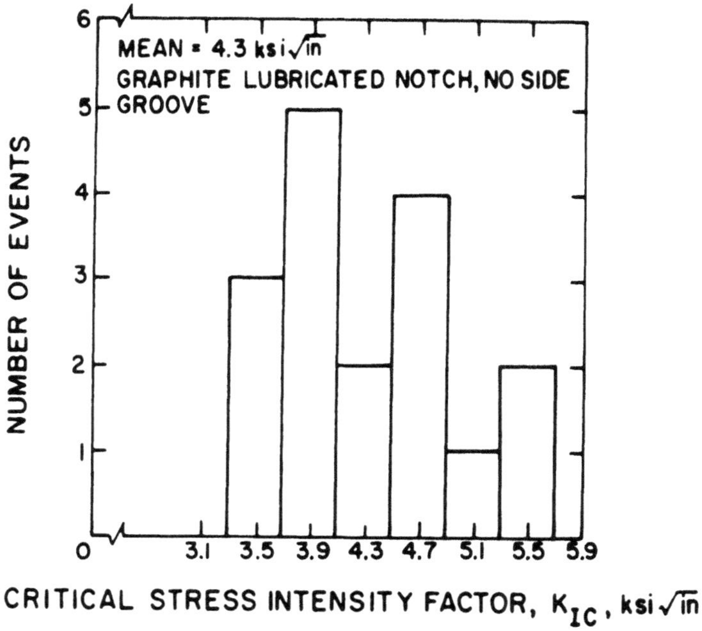

Figure 8. Fracture Toughness of Alumina

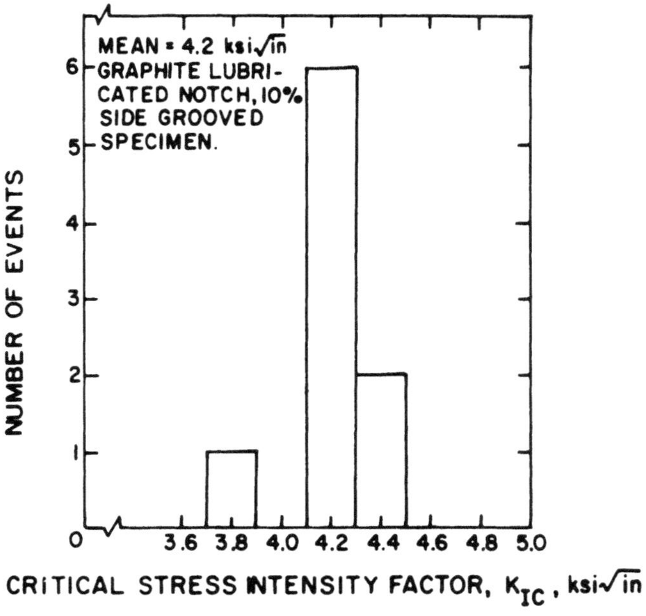

Figure 9. Fracture Toughness of Alumina

of ceramic materials with a relatively simple test setup which can be readily adapted for high temperature testing. Also load relaxation under fixed wedge-displacement loading can be used to monitor stable crack growth in this specimen. The test procedure is, however, sensitive to the coefficient of friction and thus the success of this wedge-loaded tapered DCB specimen will hinge on the accuracy of determining the coefficient of friction in the testing environment.

ACKNOWLEDGEMENT

This preliminary investigation was supported under NASA grant NGL 48-002-004. The authors wish to thank Professor J. I. Mueller, University of Washington, Department of Ceramic Engineering, for his encouragement and many helpful discussions throughout the course of this investigation.

REFERENCES

1. A.G. Evans, "Fracture Mechanics Determination", Fracture Mechanics of Ceramics, Vol. 1, edited by R.C. Bradt, D.P.H. Hasselman and F.F. Lange, Plenum Press, 1974, pp. 17-48.

2. S.M. Wiederhorn, A.G. Evans and D.E. Roberts,"A Fracture Mechanics Study of the Skylab Windows", Fracture Mechanics of Ceramics, Vol. 2, edited by R.C. Bradt, D.P.H. Hasselman and F.F. Lange, Plenum Press, 1974, pp. 829-842.

3. D.J. Rowcliffe and P.T. Lum, "A Notched-Ring Fracture Toughness Test for Ceramics", a paper presented at this symposium.

4. L.M. Barker, "Short Rod K_{Ic} Measurements of Al_2O_3", a paper presented at this symposium.

5. A.G. Evans, "A Method for Evaluating the Time Dependent Failure Characteristics of Brittle Materials - and Its Application to Polycrystalline Alumina", Journal of Material Science, I, 1972, pp. 1137-1146.

6. A.G. Evans, "High-Temperature Slow Crack Growth in Ceramic Materials", Ceramics for High Performance Applications, Proceedings of the 2nd Army Materials Technology Conference, 1973, pp. 373-395.

7. G.G. Trantina, "Stress Analysis of the Double Torsion Specimen", to be published in the Journal of American Ceramic Society.

8. D.P. Williams and A.G. Evans, "Method for Studying Slow Crack Growth", Journal of Testing and Evaluation, 1, (4), 1973, pp. 264-270.

9. "Three-Dimensional Fracture Analysis", edited by L.E. Hulbert, Proc. of a Workshop Held at Battelle's Columbus Laboratory, April 26-28, 1976, Battelle Columbus Laboratories Research Report.

10. R.G. Hoagland, R.C. Gehlen, A.R. Rosenfield and G.T. Hahn, "The Application of D.C.B. Specimens for Measuring the Crack Arrest Properties of A533 B and Other Steels", to be published in Fast Fracture and Crack Arrest, ASTM STP 627, 1977.

11. J.E. Srawley and B. Gross, "Stress-Intensity Factors for Crack-line-loaded Edge-crack Specimens", NASA TN D-3820, February 1967.

12. A.S. Kobayashi, "Experimental Techniques in Fracture Mechanics", Iowa State University Press, 1973, pp. 14-24.

APPLICATION OF FRACTURE MECHANICS TO THE ADHERENCE OF THICK FILMS AND CERAMIC BRAZE JOINTS

P.F.Becher, W.L.Newell[†] and S.A.Halen[††]

Naval Research Laboratory

Washington, D. C. 20375

ABSTRACT

Ceramic-to-ceramic and ceramic-to-metal adherence plays an important role in the use of thick films for both microelectronic and protective coating applications, as well as in joining bulk materials. Test techniques have been developed to measure adherence in terms of the fracture resistance of such systems. This approach can be combined with a simple fracture mechanics model of the systems based on bulk material properties to examine the role of materials parameters in the adherence of such systems. This offers a means of designing for improved adherence and of examining the effects of processing conditions and materials employed on the fracture toughness of the system.

BACKGROUND

The characterization and analysis of the adherence between dissimilar materials are important factors , from the stand-point of both the design and reliability, in such technologies as printed microelectronic circuitry

[†]Currently enrolled in the Graduate School, Virginia Polytechnic Institute and State Univ., Blacksburg, VA.

[††]Currently enrolled in the Graduate School, Rutgers University, Piscataway, N.J.

ceramic to metal seals, protective coatings (e.g. ceramic coatings for metal turbines) and non-mechanical joining of components. Adherence is often determined via the stress (or load) required to cause separation which is quite useful in both 1) ascertaining a relative ranking of the adherence between systems and, 2) establishing quality during production. However, because the failure stress of brittle systems is a function of both the flaw size (e.g. an unbonded interfacial area or void) and the true adherence, a strength test by itself cannot determine the cause of failure. NDE or post failure analysis may allow the flaw size to be determined thus give a more comprehensive evaluation, but this in such systems is difficult at best.

On the other hand, fracture mechanics can be used to determine adherence in a fashion similar to that used to describe fracture in bulk brittle materials. Either the stress intensity factor (K) or strain energy release rate ($\mathcal{G}$) can be used to determine the ease of crack propagation either along an interface (adhesive failure) or in an adjacent phase (cohesive failure). Adherence is thus described in terms of the ability of the total system to resist fracture, which serves not only as a tool to examine adherence mechanisms, but also as an important design aid.

The application of fracture mechanics to adherence problems to date has centered on the areas of structural adhesives and composites (1,2). Here, not only the analysis but also the measurements of crack tip stress intensity (K) can become quite complex even for crack motion along a flat interface between two materials as the stresses acting on a crack are not symmetric and combined loading modes exist (e.g. mode I plus II). However, when the elastic properties E (Young's modulus) and ν (Poisson's ratio) of the two materials are comparable [in theory they must be identical (3) but in practice, need only approach each other (4)], then the solution is the same as for the crack in a single phase body. This is the case for the analysis of adherence of many ceramic-to-ceramic and ceramic-to-metal systems. The result is that one can describe adherence in terms of K_I, and not combined K_I-K_{II}, and a much simpler description of crack propagation is obtained.

Experimentally, determination of the adherence of the system is simplified by use of the energy balance concept equating the change in the system's strain energy [the strain energy release rate ($\mathcal{G}_I$)] with the total energy for crack propagation ($2\gamma_I$). Because the system acts as a homogeneous body, $\mathcal{G}_I$ is related to K_I, i.e.,

$$K_I = \frac{\sigma c^{\frac{1}{2}}}{A} = (\mathcal{G}_I E)^{\frac{1}{2}}$$

where c is the flaw size and A is a geometric factor thus either $\mathcal{G}_I$ or K_I can be determined experimentally. Several experimental techniques have been evolved to determine the adherence of not only adhesives to metal (1,2), but most recently to ceramics to metals or to ceramics (5-7).

EXPERIMENTAL CONSIDERATIONS

The experimental techniques used to examine the adherence must consider several aspects. First, the measurement of crack length during or after testing is generally difficult at best due to the opacity and the complex structure of the specimens. As a result, the applied moment and the constant compliance double cantilever beam specimen have generally been used, as crack length data is not required to calculate $\mathcal{G}_I$ of K_I.

The second feature concerns the effects of test specimen geometry where one such effect is best illustrated by examining the applied moment double cantilever beam geometry used for testing the adherence of thick film circuitry to ceramic substrates. For this test, the test specimen, Fig. 1(a), involves a brass

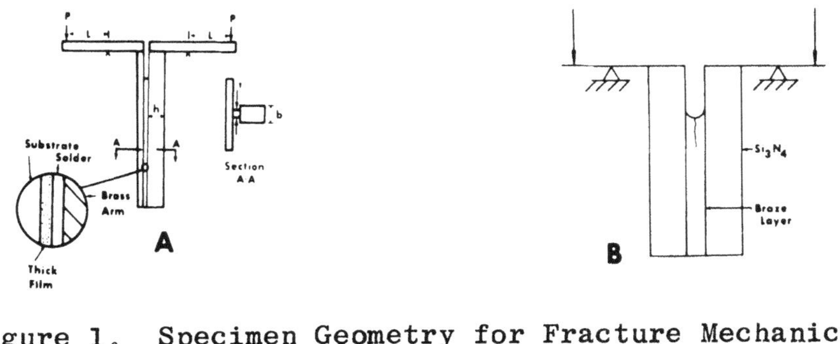

Figure 1. Specimen Geometry for Fracture Mechanics Tests. The applied moment [load (P) x fulcrum (L)] cantilever geometry for A) thick film adherence and B) brazing samples.

loading arm attached via a 2 to 5 mil solder layer joining it to a thick film strip which is itself attached to an alumina substrate. For such a non-symmetric sample, calculation of $\mathcal{G}_I$ must consider the strain energy in both arms (i.e., the brass and alumina arm):

$$\mathcal{G}_I = \frac{M^2}{EtI} = \frac{M^2}{2t}\left(\frac{1}{E_1I_1} + \frac{1}{E_2I_2}\right)$$

where M is the applied bending moment, t is the crack plane width and I is the moment of inertia ($bh^3/12$) with subscripts 1 and 2 referring to the brass and alumina arms respectively. Unless the strain energies in each arm are comparable (i.e., $E_1I_1 = E_2I_2$), the crack propagation will be artificially forced toward the arm with the highest strain energy. In the case of the thick film adherence, this non-symmetry of the strain energy appears to be responsible for obtaining both high and low $\mathcal{G}_{IC}$ values via the constant compliance(6b) specimen which was more asymetric than the applied moment specimen where only the low $\mathcal{G}_{IC}$ values were obtained. The greater assymmetry of the strain energy in the constant compliance test then often resulted in the crack being forced by the higher strain energy half of the specimen into a region requiring higher $\mathcal{G}_{IC}$ for crack propagation. Similar side-ways motion of a crack due to asymmetry of the strain energy is seen in bulk ceramic specimens when the crack plane is not in the center of the specimen.

As noted in the thick film applied moment specimen, $\mathcal{G}_{IC}$, is obtained by calculating the strain energy loss using the properties and dimensions of only the cantilever arms and not the thick film-solder layer values as well. The reason is that the strain energy contributed by this latter region to crack propagation is negligible. A similar case involves the calculations of $\mathcal{G}_{IC}$ for Si_3N_4 braze samples where a thin (≤ 0.010") layer of braze material (e.g., ZrO_2) is used to bond two pieces of Si_3N_4. Here only values based on the Si_3N_4 "arms" are considered in the $\mathcal{G}_{IC}$ equation. This can be used as long as the thickness of the layer is small in comparison to the Si_3N_4 arm width (h). For example, with the Si_3N_4 braze samples where E of the arm material and the center layer (e.g., ZrO_2) are nearly equal, , the error on the calculated $\mathcal{G}_I$ based only on the Si_3N_4 values is ≤ 5% as long as the ZrO_2 center layer thickness is ≤ 5% of the width (h) of the Si_3N_4 loading arm. Here as in the case of the thick film samples as long as the volume of this "center" layer is small, the product EI of this layer should be insignificant as compared to that of the cantilever arms. Thus, the strain energy contribution of this "center" layer can be ignored in calculating $\mathcal{G}_{IC}$.

Finally, in examining the use of the fracture mechanics test for adherence studies, the effects of several parameters may need to be examined (e.g., loading rates, effect of attachment schemes and other specimen preparation processes). For example, in the work on the adherence of the printed thick film to alumina substrates soldering was used to attach brass load arm. In this case, one had to show that neither strains developed during cooling after the soldering step nor plastic deformation of the solder could occur during testing which would affect the $\mathcal{G}_{IC}$ result (6a).

FRACTURE MECHANICS TESTING OF ADHERENCE

Si_3N_4 Solid State Braze Joints

Use of brazing, welding or other non-mechanical techniques to join ceramics offer expanded opportunities, however, such joining requires that sufficient bonding or adherence be achieved to support minimal design loads. The adherence of Si_3N_4 braze joints has been examined by determining both the strength and K_{IC} of the joint region (7). Using the applied moment specimen Fig. 1(b), it was shown when a ZrO_2 layer was used to braze higher purity Si_3N_4 pieces (HS 132) the K_{IC} value for crack propagation within a joint was comparable to that of bulk partically stabilized ZrO_2 (PSZ) (~ 5 MN/m$^{3/2}$). Indeed, fracture was found to occur totally within the PSZ braze layer such that crack motion involves cohesive failure of the ZrO_2. The adherence of such a brazed system is reduced ($K_{IC} \sim 2.7$ MN/m$^{3/2}$), when low purity Si_3N_4 (i.e., HS 110) is brazed due to the formation of ZrN at the Si_3N_4/ZrO_2 interface which results in interfacial failure.

The incorporation of fracture mechanics in the characterization of these braze joints or similar systems is a very useful tool. As the value of K_{IC} for the braze samples where failure occurs in the ZrO_2 layer is comparable to that measured for partially stabilized ZrO_2) it locates the system's strength controlling region. Interfaces are thus not necessarily the weak link in the system; whether cohesive or adhesive failure is involved in the fracture process is controlled via composition and/or microstructures.

FRITTED THICK FILM CONDUCTOR ADHERENCE TO ALUMINA SUBSTRATES

Initial application of fracture mechanics tests to the adherence of a glass (frit) bonded thick film conductor to an alumina substrate showed that adherence is primarily a function of interface between the metal conductor layer and the glass layer which bonds it to the substrate. By varying the temperature, time, or atmosphere used to fire the thick film onto the substrate or the substrate used, one can alter the metal/glass interface microstructure; thus change the thick film adherence (8-10). In the case of Pt-Au conductor thick films, the maximum adherence occurs when the metal/glass interface has a highly interlocking structure. Fracture then occurs within the glass layer (Fig. 2a) and the K_{IC} (or $_{IC}$) value obtained equals that in bulk samples of similar glasses (8). However, when this interlocking nature of the metal/glass interface is removed failure occurs at extremely low K_{IC} values due to interface separation (adhesive failure)

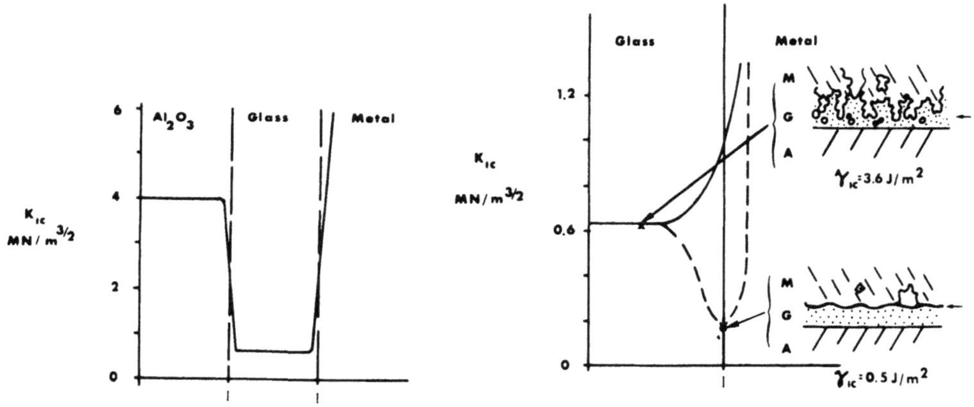

Figure 2. Model of Stress Intensity Profile Through Thick Film-Substrate. Overall K_{IC} profile (2A) is based on bulk materials parameters which assumed a linear profile across interfaces. For optimum adherence failure in glass layer is predicted (note measured value x) while interfacial failure (o) occurs at a much lower K_{IC}.

These results suggested that the adherence of a glass (frit) bonded thick film conductor-substrate system might be modeled via fracture mechanics to locate the mechanical weak link of a system with optimum adherences and thus examine means of improving adherence. In systems involving the adherence of multiple layers like that of the thick film/substrate sample, the locus of failure can be predicted by the use of a fracture diagram based on bulk material fracture mechanics parameters (either K_{IC} or $\mathcal{G}_{IC}$). Thus, the K_{IC} values for the major phases present in each layer (i.e., conductor metal/glass/alumina) must be examined for progressive positions through the thick film into the substrate (Fig. 2a).

The K_{IC} diagram in Fig. 2a for the present thick film-substrate system (based on measured bulk K_{IC} values) assumes strong interfaces between both the conductor metal-glass and glass-alumina. The thick film is then predicted to have optimum or maximum adherence when the glass layer has the least resistance to fracture, not the interfaces. When optimum adherence is achieved, the analysis predicts failure will occur within the glass layer, and the glass K_{IC} value controls the upper limit of the thick film adherence. In fact, in Fig. 2b, the data point (x) on the upper limit K_{IC} curve (i.e., interpenetrating glass metal interface-adequate interfacial adherence)was obtained and failure occurred in the glass layer. The worst situation that can occur is when a minimum in the K_{IC} occurs at the interface. Such is shown in Fig. 2b when failure occurred along a smooth glass-metal interface in the present thick film substrate system.

Even such a simplified analysis points out what might be done to improve the optimum adherence of a glass bonded thick film conductor. As the present glass layer limits the optimum adherence, one would look to improvements in the K_{IC} (or $\mathcal{G}_{IC}$) of this layer. The toughness of glass can be improved by subsequent partial crystallization (11) or by initially adding second phase particles (12). Recently improved fracture toughness of glass to metal seal systems has been achieved by use of post sealing heat treatments to obtain glass crystallization (13).

SUMMARY

It has been shown that fracture mechanics testing

can be applied to examine the adherence between dissimilar materials as long as the effects of specimen geometry and test conditions are accounted for. The adherence can thus be measured in terms of K_{IC} (or $\mathcal{G}_{IC}$) and the role of processing can be accounted for in terms of locus of fracture. Also, by considering bulk material properties, it is possible to describe the adherence of the system.

ACKNOWLEDGEMENTS

It is with pleasure that we acknowledge the assistance of our NRL colleagues, W. Bascom, T. Donahue, S. W. Freiman, L. A. Mann, S. Morey, J. S. Murday, D. Mulville, and R. W. Rice during various portions of this work. The original fracture mechanics-adherence analysis evolved in the thick film studies was supported by NAVAIR (J. Willis, Contract Monitor). The use of this same approach to evaluate ceramic brazing was supported by both ARPA/NAVSEA (C. Miller, Contract Monitor) and the Office of Naval Research.

REFERENCES

1. H. T. Corten, pp. 676-769 in Fracture, Vol. VII, H. Liebowitz (Ed.), Academic Press, N.Y. (1972).

2. D. R. Mulville and R. H. Vaishnav, pp. 45-58 in Proc. Symp. on Solid Mechanics, Army Materials and Mechanics Research Center, AMMRC MS-74-81, Watertown, MA. (1974).

3. J. R. Rice and G. C. Sih, J. Appl. Mech., Ser E 32 (2), pp. 418-423 (1965).

4. E. Wu and R. L. Thomas, pp. 575-587 in Interfacial Fracture Phenomena, Proc. 5th Int. Congress Rheol. Vol. 1, S. Onogi (Ed.), University Park Press, Baltimore, MD (1969).

5a. G. Elssner and R. Pabst, pp. 179-186 in Mechanical Properties of Ceramics, Vol. II, R. W. Davidge (Ed.) Proc. Brit. Ceram. Soc., No. 25 (1975).

5b. G. Elssner and R. Pabst, High Temperature-High Pressures 6, 321-327 (1974).

6a. P. F. Becher, W. D. Bascom, J. L. Bitner and J. S. Murday, pp. 279-286 in Proc. 1975 Int. Microelectronics Symp., ISHM, Mongomery, Ala. (1975).

6b. W. D. Bascom, P. F. Becher, J. L. Bitner and J. S. Murday in Proc. ASTM Symp. on Adhesion Measurement of Thin Films, Thick Films and Bulk Coatings, ASTM, Phila., PA, in press (1976).

7. P. F. Becher and S. A. Halen, in Ceramics for High Performance Applications-II, Proc. 5th Army Materials Technology Conf., in press (1977).

8. P. F. Becher and W. L. Newell, J. Mater. Sci. 12, 90-96 (1977).

9. P. F. Becher, submitted to J. Mater. Sci. (1977).

10. P. F. Becher and J. S. Murday, J. Mater. Sci. 12 1088-1094 (1977).

11. S. W. Freiman and L. L. Hench, J. Am. Ceram. Soc. 55, 86 (1972).

12. D. R. Biswas and R. M. Fulrath, J. Am. Ceram. Soc. 58, 526-527 (1975).

13. J. C. Swearengen and R. J. Eagen, J. Mater. Sci. 11 1857-1866 (1976).

A NOTCHED RING FRACTURE TOUGHNESS TEST FOR CERAMICS

D. J. Rowcliffe, R. L. Jones, and J. K. Gran

SRI International

Menlo Park, California 94025

ABSTRACT

This paper describes a new method of measuring the fracture toughness of brittle materials that uses internal pressurization of an externally edge-notched ring specimen to generate an accurately known stress state. Stress intensity calibrations derived by finite element analysis are presented for three ring geometries. The results obtained when one geometry was investigated experimentally are shown to be consistent with the analytical solution. Values of K_{IC} obtained using the new test method are given for quartz and zinc sulfide.

INTRODUCTION

Many specimen geometries have been developed to determine fracture mechanics parameters.[1] Stress intensity factor calibrations for a given specimen geometry usually are derived either by finite element analysis or through the equivalent route of compliance testing. In both approaches, the cracked body is treated as an elastic continuum loaded under prescribed stress conditions. In metal systems, the desired state of stress can be easily attained, because the specimen can deform plastically, so that stresses arising from local misalignment of the loading system, for example, parasitic bending stresses set up during uniaxial tensile loading, are reduced by small amounts of yielding. This is not so for materials behaving in a completely elastic manner, for example, many ceramics at low temperatures, and the presence of unrelaxed parasitic stresses leads to the well-known difficulties in measuring accurately the strengths of ceramics in uniaxial tension.

Most specimen geometries used to determine both conventional mechanical properties and fracture mechanics parameters rely on uniaxially applied loads and thus are not ideally suited for ceramic testing. The expanded ring test was developed by Sedlacek and Halden to overcome these difficulties with regard to strength measurements in ceramics.[2] We have recently extended this test to determine fracture mechanics parameters from notched ceramic rings while retaining the unique loading conditions inherent in the method. This paper describes the test system, presents the finite element analysis solutions for stress intensity factors in internally pressurized rings containing external notches, and reports some preliminary fracture toughness measurements on two ceramic materials.

THE EXPANDED RING TEST

The method has undergone a number of refinements since its original development.[2] A block diagram of the current testing system and a schematic of the pressure head is shown in Figure 1a and b, respectively. The basis of the test is to develop a pure hoop stress by internally pressurizing a thin-walled ring. The hydrostatic pressure is transferred from a pressurized fluid to the inner surface of the ring through a thin rubber toroid. Spacers $\sim 40\ \mu m$ taller than the test ring are set between the upper and lower support rings to ensure that no parasitic stresses are developed in the ring due to end constraints, while adequately supporting the toroid. Under these circumstances, the hoop stress, σ_θ, at a radius R in the wall of the ring is given by the relation

$$\sigma_\theta = \frac{P_i\, R_I^{\,2} \left(1+R_O^{\,2}/R\right)}{R_O^{\,2} - R_I^{\,2}} \tag{1}$$

where R_I and R_O are the internal and external radii, respectively, and P_i is the internal pressure. For thin-walled test specimens, that is, rings having $R_O/R_I \leq 10/9$, the tensile strength, σ, is

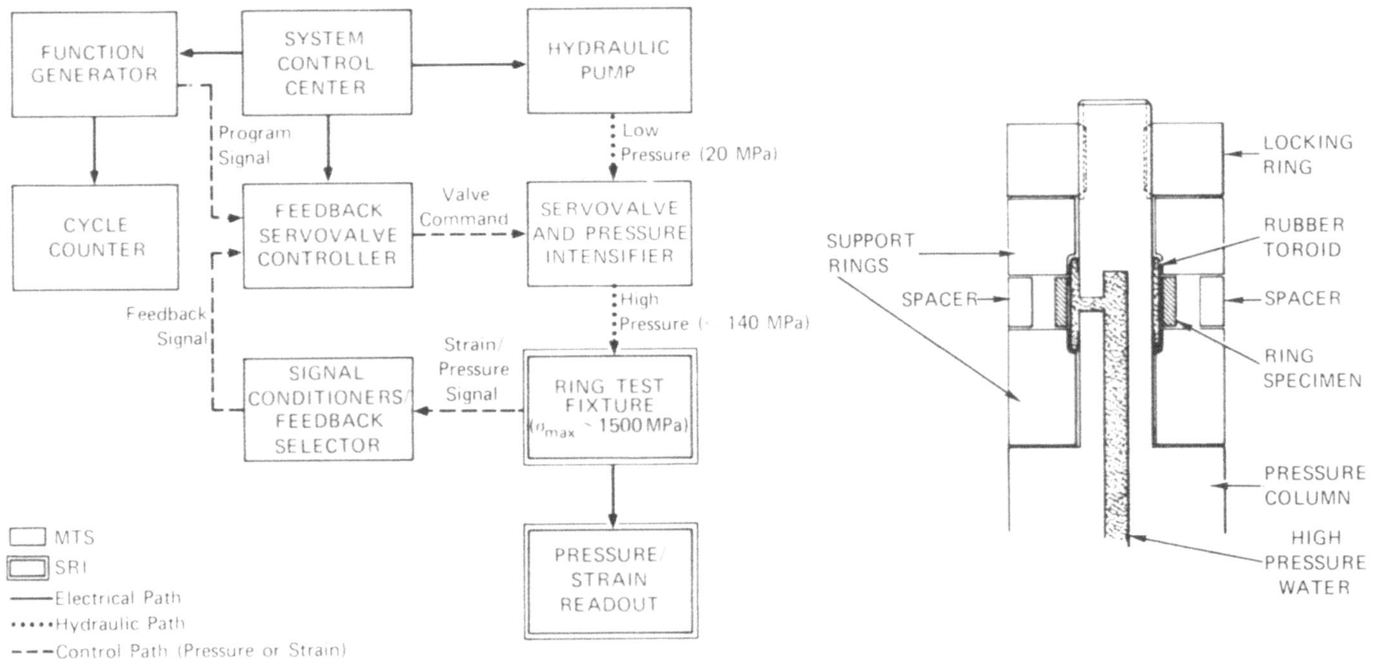

(a) (b)

Figure 1 SRI/MTS Ring Test System (a) Block Diagram of Control System (b) Schematic of Pressure Head.

given to a close approximation by

$$\sigma = \left(\frac{R_O^{\ 2} + R_I^{\ 2}}{R_O^{\ 2} - R_I^{\ 2}} \right) P_F \tag{2}$$

where P_F is the internal pressure required to fracture the ring.

The reliability of the expanded ring test has been established through extensive measurements of the tensile strength of many ceramics, including alumina,[3] PZT,[4] silicon carbide,[5] and silicon nitride.[5] The tensile strength data obtained from the expanded ring tests on silicon nitride are in excellent agreement with those derived from precision uniaxial tensile tests using strain-gaged specimens.[6]

As well as having the advantage of avoiding parasitic stresses, the expanded ring test uses a relatively simple specimen geometry, permits testing of large volumes of material, and offers an extremel simple and rapid test procedure. In engineering ceramics, a number of important component shapes either are rings (such as seals and ball races) or are based on ring shapes (such as the rotors and stators of small turbines). Thus, the expanded ring test frequently offers a method for directly evaluating the properties of full size ceramic components, which is important because subsize test specimens can often provide misleading indications of the mechanical properties of engineering ceramics.[7]

When the expanded ring test is used to measure fracture toughness, an axial notch is cut into the wall of the ring. In principle the notch could be cut into either the internal or the external surface, but it is preferable and experimentally much simpler to cut into the external surface. For an externally notched ring, the stress intensity K_I can be related to the stress σ_O at the external surface of the ring by an equation of the form

$$K_I = f(a/w)\ \sigma_O \sqrt{\pi a} \tag{3}$$

where w is the wall thickness ($R_O - R_I$), a is the length of the notch, and f(a/w) is a function that depends on the ring geometry. Equation 3 defines the fracture toughness K_{IC} when

$$\sigma_O = \left(\frac{2R_I^{\ 2}}{R_O^{\ 2} - R_I^{\ 2}} \right) P_F \tag{4}$$

where P_F is the internal pressure required to fracture the notched ring. Equations 3 and 4 apply to both thick-walled and thin-walled rings. Thus, to measure K_{IC} using the expanded ring test, it is necessary to evaluate f(a/w) for the ring geometry of interest. We have used finite element analysis to evaluate f(a/w) for three ring geometries, as described below

FINITE ELEMENT CALCULATIONS

Kobayashi et al.[8] have calculated normalized stress intensity factors for internally pressurized two-dimensional cylinders (rings) containing internal edge cracks for various ring geometries. The analysis used a total of 320 elements containing 338 radial points. In the region containing the crack, a fine net was used with 20 elements through the ring thickness and an included angle spacing of 1 degree. Similar K-calibrations were calculated[9] by the same procedure for rings with radius ratios R_O/R_I = 5/4 and 10/9 containing external edge cracks.

In our work, we calculated K-calibration curves for externally edge-cracked rings of radius ratio R_O/R_I = 5/4 and 3/2, using a much smaller number of elements with different properties from those used by Kobayashi et al.[8] Full details will be published elsewhere but the main features of the procedure are presented here.

In the analytical solution of planar problems, the distribution of stress near the tip of a crack has been shown to be[10]

$$\left.\begin{aligned} \sigma_y &= K_I \frac{1}{\sqrt{2\pi r}} \operatorname{Cos} \frac{\theta}{2} \left(1 + \sin\frac{\theta}{2} \sin\frac{3\theta}{2}\right) + \sigma_{Oy} \\ \sigma_x &= K_I \frac{1}{\sqrt{2\pi r}} \operatorname{Cos} \frac{\theta}{2} \left(1 - \sin\frac{\theta}{2} \sin\frac{3\theta}{2}\right) + \sigma_{Ox} \end{aligned}\right\} \qquad (5)$$

where K_I is the stress intensity factor for mode I, and r and θ are polar coordinates originating at the crack tip. Furthermore, K_I is defined in terms of the strain energy release rate associated with crack extension under fixed-grip conditions.[10] The modified Griffith theory, in the case of plane stress, gives

$$K_I = E\sqrt{-\left(\frac{dU}{da}\right)} \qquad (6)$$

where U is the strain energy, and E is Young's modulus.

The $\frac{1}{\sqrt{r}}$ singularity in the stresses represented by Equation 5 is obtained in an 8-node isoparametric quadrilateral element when two of the midside nodes are placed at the "quarter" position. Although the values of stress and displacement at internal points are not exact, this "singularity element" provides a reasonably accurate stiffness matrix.[11]

The approach used in this work takes advantage of the more accurate stiffness of the singularity element at the crack tip, avoiding an extremely fine element mesh. The application of the modified Griffith theory strictly adheres to the fixed-grip requirement in the definition of dU/da. For example, the element mesh shown in Figure 2 was used for $R_O/R_I = 3/2$. Values of dU/da were

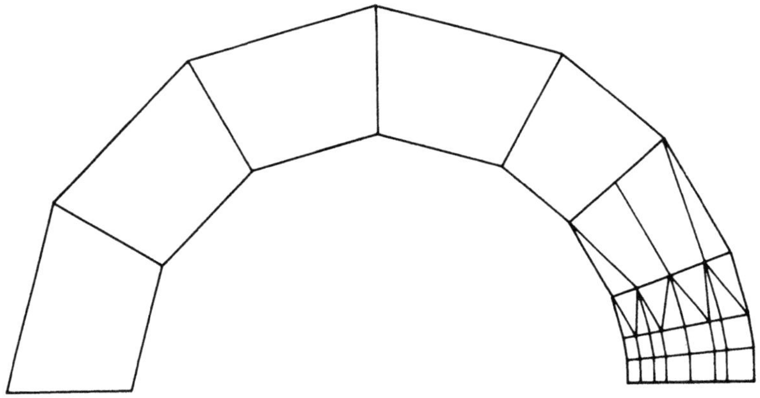

Figure 2 Finite Element Net for an Internally Pressurized Ring Containing an External Edge Crack

calculated for a/w = 0.2, 0.3, 0.5, 0.7, and 0.8. Each of these cases required three computer runs. First, an internal pressure load was applied. The resulting displacements on the inside perimeter were then prescribed with the crack length incremented by plus and minus one element width. The three values of strain energy from these computations were then used to calculate dU/da from which K_I was obtained using Equation 6. A similar element mesh was used for $R_O/R_I = 5/4$. Table 1 shows the calculated values of $K_I/\sigma_O\sqrt{\pi a}$ for both ring geometries, together with Kobayashi's values[9] for $R_O/R_I = 5/4$. For this ring geometry, the data from the two finite element methods are in excellent agreement. The present method was also used to calculate the stress intensity solution for an edge-cracked plate in tension, and the values were within 3% of those provided by Sih.[1]

Table 1 $K_I/\sigma_O\sqrt{a\pi}$ for Externally Cracked Rings

a/w	$R_O/R_I = 3/2$	$R_O/R_I = 5/4$	$R_O/R_I = 5/4$*
0.2	1.16	1.19	-
0.3	1.34	1.25	1.24
0.5	1.89	1.89	1.89
0.7	2.74	2.82	2.93
0.8	3.31	-	-
0.9	-	4.30	4.52

*Data from A. S. Kobayashi, Reference 9.

K calibrations for three ring geometries containing external edge cracks are presented in Figure 3a. The close proximity of the curves indicates that the stress intensity calibration is relatively insensitive to ring geometry. The best-fit polynomial expressions are as follows:

$R_O/R_I = 10/9, \quad 0.3 \leq a/w \leq 0.9$

$$K_I/\sigma_O\sqrt{\pi a} = 0.619 + 0.15(a/w) - 0.875(a/w)^2 + 5.0(a/w)^3 \tag{7}$$

$R_O/R_I = 5/4 \quad 0.3 \leq a/w \leq 0.9$

$$K_I/\sigma_O\sqrt{\pi a} = 0.265 + 4.15(a/w) - 4.5(a/w)^2 + 5.42(a/w)^3 \tag{8}$$

$R_O/R_I = 3/2 \quad 0.2 \leq a/w \leq 0.8$

$$K_I/\sigma_O\sqrt{\pi a} = 0.925 + 0.836(a/w) + 1.35(a/w)^2 + 1.67(a/w)^3 \tag{9}$$

CRITICAL STRESS INTENSITY FACTOR MEASUREMENTS

As a preliminary assessment of the test, edge-notches were machined with a diamond saw to several lengths into the external surfaces of quartz rings of radius ratio 5/4, internal diameter 25.4 mm, and height 12.5 mm, and the rings were pressurized to failure. The value of a/w = 0.3 was arbitrarily chosen to derive K_{IC} using Equation 8. The derived value of K_{IC} was then used to calculate new values of the dimensionless parameter $K_I/\sigma_O\sqrt{\pi a}$ for each of the other crack lengths that had been tested. The results are plotted in Figure 3b and show excellent agreement with the analytical curve (Eq. 8) over the whole range of crack lengths

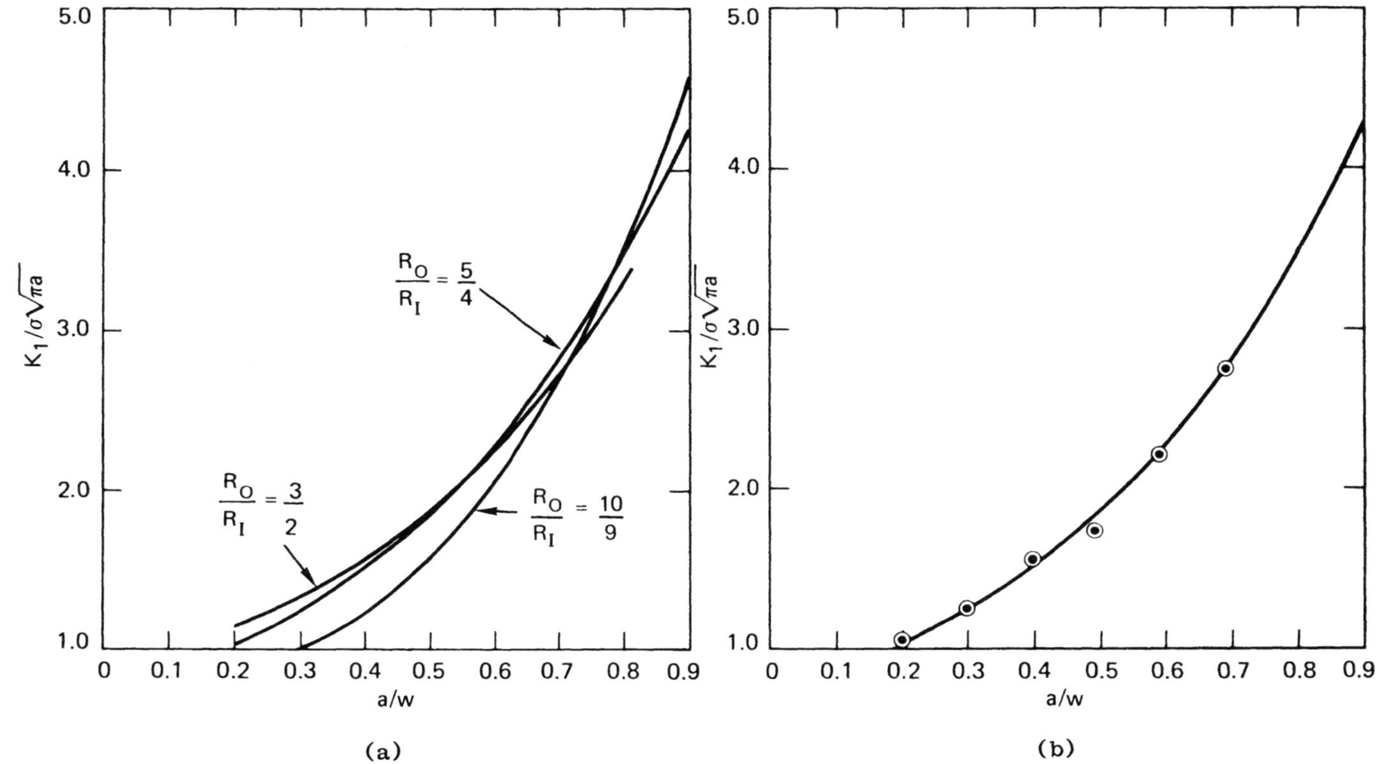

Figure 3 Stress Intensity Calibrations for Externally Edge Cracked Rings. (a) Analytical Curves for Three Ring Geometries, (b) Analytical Curve and Experimentally Determined Data for $R_O/R_I = 5/4$.

studied. The mean value of K_{IC} for quartz calculated from the six data points is 0.92 ± 0.02 $MPa \cdot m^{1/2}$. More extensive testing clearly is required to compare results from the expanded ring test with those from other methods of determining K_{IC}.

K_{IC} was also measured for ZnS rings having the same radii as the quartz specimens but with a height of 6.4 mm. The results for four specimens having the same nominal notch length ($a/w \approx 0.5$) were 0.77, 0.75, 0.71, and 0.75 $MPa \cdot m^{1/2}$, giving a mean value of 0.74 ± 0.03 $MPa \cdot m^{1/2}$. The larger scatter of these values, compared with those for the quartz rings, is believed to be associated with the precision with which the notches were positioned. To test the sensitivity to notch position, nonradial notches were cut in some ZnS rings. These rings showed significantly lower values of K_{IC}. Therefore, we devised a method for ensuring that the notches would be exactly radial and used it for the subsequent tests on quartz.

CONCLUSIONS

1. A test method based on the expanded ring tensile test offers significant advantages for precision K_{IC} determinations in brittle materials.

2. Finite element analysis has been used to derive stress intensity calibrations for externally edge-cracked rings of a range of useful geometries. The K-calibrations obtained are relatively insensitive to specimen geometry.

3. For the one ring geometry that was investigated experimentally, the experimental data were in extremely close agreement with the computed K-calibration.

4. Values of K_{IC} for quartz and ZnS, using the method, are 0.92 ± 0.02 and 0.74 ± 0.03 $MPa \cdot m^{1/2}$, respectively.

ACKNOWLEDGMENT

The authors are extremely grateful to Albert S. Kobayashi for a number of helpful discussions and for making available unpublished data. This research has been sponsored by Army Materials and Mechanics Research Center, with Roger Lamothe as project monitor.

REFERENCES

1. G. C. Sih, "Handbook of Stress-Intensity Factors," Lehigh University, Bethlehem, Pa. 1973.

2. R. Sedlacek and F. A. Halden, Rev. Sci. Inst. 33, 298 (1962).

3. R. Sedlacek,"Processing of Ceramics--Surface Finishing Studies,' Final Technical Report Under Contract N00019-68-C-0388, May 1969.

4. R. Sedlacek, "Strength Characterization of Ceramic Sonar Rings,' Final Technical Report Under Contract N00024-70-C-1224. April 1973.

5. R. Sedlacek and R. L. Jones, "Mechanical Testing of Ceramic Rings," Final Report Under Contract DAAG 46-72-C-0112, May 1974.

6. A. F. McClean, E. A. Fisher, and R. J. Bratton, "Brittle Materials Design, High Temperature Gas Turbine," AMMRC CTR 73-32, Sept. 1973.

7. A. G. Evans and R. L. Jones, to be published. J. Amer. Ceram. Soc.

8. A. S. Kobayashi, N. Polvanich, A. F. Emery and W. J. Love, in "Computational Fracture Mechanics," eds. E. R. Rybicki and S. E. Benzley, ASME, 1975.

9. A. S. Kobayashi, private communication

10. G. R. Irwin, J. Appl. Mech. 24, 361, 1957.

11. R. D. Henshell and K. G. Shaw, I.J.N.M.E., 9, 495 (1975).

SHORT ROD K_{Ic} MEASUREMENTS OF $A\ell_2O_3$

L. M. BARKER

Terra Tek, Inc.

University Research Park, 420 Wakara Way, SLC, Utah

INTRODUCTION

Fracture toughness measurements of ceramics have traditionally been made with side-grooved double cantilever beam (DCB) specimens (1-7), single edge notch bend (SENB) specimens (6-9), double torsion specimens (7,10,11), and work of fracture specimens (12-15). Each of these specimens suffer from at least some of the following problems: expense and complexity of the test, initiation and arrest of a natural crack, crack length control and measurement, crack angle and straightness, theoretical justification of the technique, and lack of consistent correlation with other techniques.

Most of these drawbacks seem to be overcome by the recently introduced short rod method of measuring the fracture toughness of both brittle and ductile materials (16-18). In this paper the short rod technique is reviewed, and short rod test results on several commercial aluminas are presented and discussed. In addition, a new device (named a "Fractometer") which greatly simplifies fracture toughness testing is described.

SHORT ROD TEST DESCRIPTION

The short rod specimen configuration is illustrated in Figure 1. It consists of a rod with a diameter 2/3 of its length, and with longitudinal slots which form an internal V-shaped region. An opening load is applied at the front of the specimen or within the slot, causing a crack to initiate at the point of the V. The crack

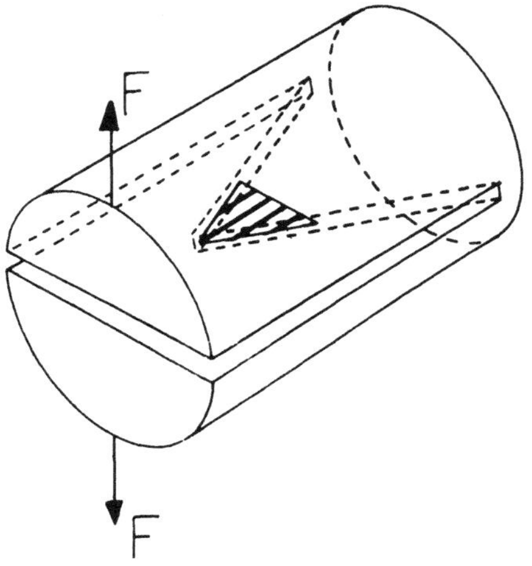

Figure 1. Short rod specimen. The shaded area denotes the crack.

is initially stable, however, and requires an increasing load for continued crack advance until the crack reaches a critical length, a_c, where the load goes through a smooth maximum. Thereafter, the crack-advancing load decreases. It can be shown (16) that a_c is essentially independent of the specimen material, provided only that the plastic zone size at the crack tip is sufficiently small compared to the specimen dimensions, i.e., provided linear elastic fracture mechanics is applicable. Thus, once a_c has been determined for a particular short rod specimen configuration, one always knows where the crack is at the time of the peak load, regardless of the specimen material. K_{Ic} can therefore be calculated from the peak load alone. No separate measurement of crack location is required in each test. Further, because of the initial crack growth stability arising from the constantly widening crack front, the required pre-crack initiation and arrest are automatically achieved in the same single loading sequence that tests for K_{Ic}. The testing simplifications resulting from the use of a specimen with an initial region of crack growth stability (viz. no fatigue precracking, no crack length measurements, and calculation of K_{Ic} from the peak load alone) were recognized by Pook (19), who suggested using a modified Tattersall-Tappin three point bend specimen. Pook's suggestion was apparently never tested, however (20).

Two methods have been used to load short rod specimens. In the first, shallow grip grooves are machined in the face of the specimen, and grips provide the loading as indicated in Figure 2. The peak load is used to calculate K_{Ic} in the equation (16, 18)

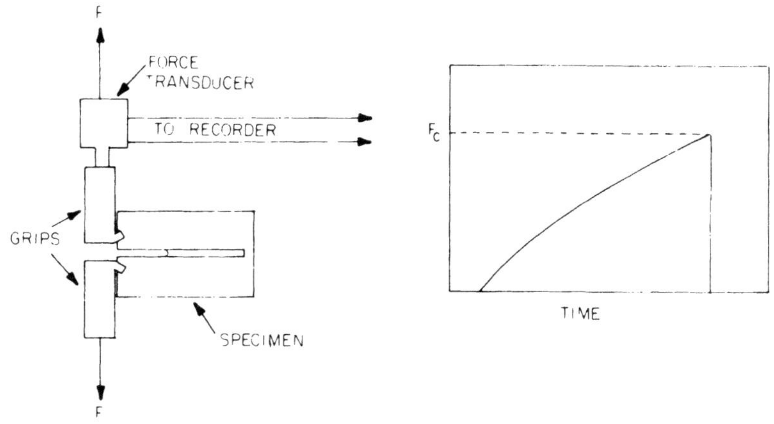

Figure 2. Grip method of loading the short rod specimen.

$$K_{Ic} = \frac{AF_c}{B^{3/2}\,(1-\nu^2)^{1/2}}\,, \qquad (1)$$

where A is a dimensionless constant (equal to 20.8 ± 1.0), F_c is the peak load (force) applied to the specimen in the test, B is the specimen diameter, and ν is Poisson's ratio.

The second and more recent method of loading short rod specimens is illustrated in Figure 3. A thin inflatable bladder called a flatjack is inserted in the slot opening at the end of the speciment until it fills the slot almost to the point of the V. A fluid is then pumped into the flatjack, causing it to "jack" the slot open as the fluid volume is increased. The peak load (pressure in the flatjack, in this case) during the experiment is used to calculate K_{Ic} using equation (9) of the next section. The flatjack, the fluid pressurizing system, the pressure transducer, and the electronics necessary to provide a digital readout of K_{Ic} immediately following the test have been combined into a test machine named a Fractometer.* The Fractometer method of short rod specimen testing has several additional advantages. For example, no grip grooves need be machined in the specimen face. Specimen alignment relative to the loading fixture is more reproducible and more easily accomplished, and extremely stiff machine loading (desirable in certain applications) is easily attained. With the Fractometer, a specimen can be installed, tested, and the K_{Ic} recorded in less than a minute. A derivation of the Fractometer equation for K_{Ic} is given in the next section.

---

*Patent Pending

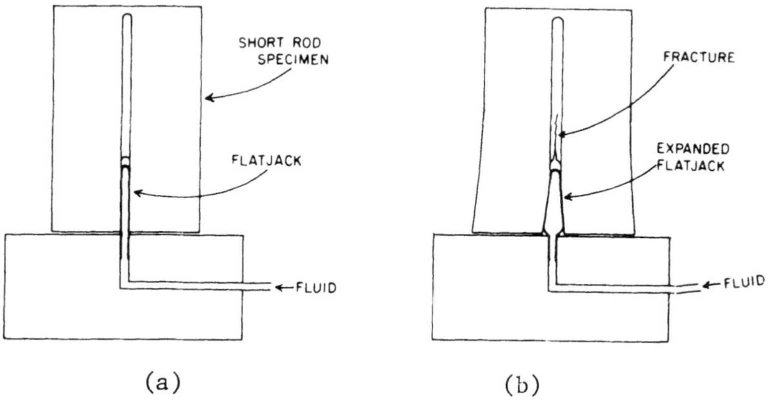

Figure 3. Fractometer method of loading short rod specimens. In (a), the specimen is seated with the flatjack in the specimen slot. In (b), fluid pressure in the flatjack is producing fracture of the specimen. The specimen deflection is greatly exaggerated.

DERIVATION OF THE FRACTOMETER EQUATION

The derivation of the K_{Ic} equation for flatjack loading of the short rod specimen (Figure 3) is similar to the grip loading derivation (18) in which the principles of linear elastic fracture mechanics are used. It is assumed that the energy per unit area of new crack surface created in plane strain is a material constant γ. Thus, the energy required to advance the crack a small distance, da (Figure 4a) is

$$dW_1 = 2\gamma\ b\ da, \qquad (2)$$

where b is the width of the crack front. The energy to create the crack surfaces, dW_1, comes from the irrecoverable work done on the specimen during the test.

Figure 4b shows a plot of the pressure-volume curve for the pressurized region of the specimen slot. During the initial loading a linear elastic path is obtained whose slope can be related to the elastic modulus of the specimen material. The onset of nonlinearity occurs when the crack initiates at the point of the V. Suppose that the volume in the slot is increased until point A on the P-V curve is reached, corresponding to a crack length, a, in Figure 4a. If the fluid volume were then retracted from the specimen slot, the unloading path would be a straight line to the origin, since no crack growth occurs on unloading and since the crack would close completely under the assumption of elasticity. Such behavior is in fact observed in ceramics.

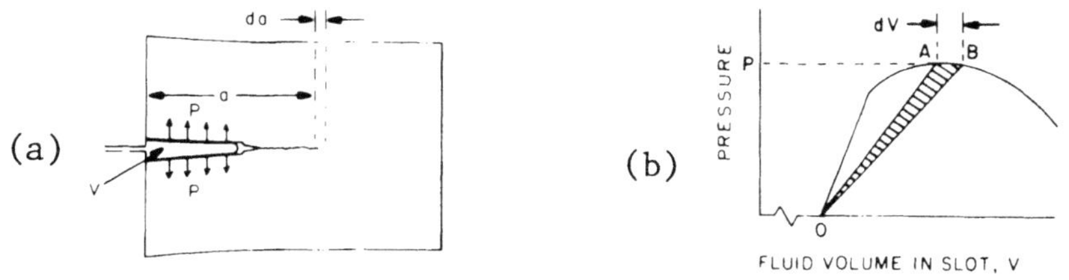

Figure 4. In (a), the flatjack's loading of the specimen, and incremental crack growth da is depicted. In (b), a plot of the pressure-volume loading curve is shown, including the straight line unloading and reloading paths AO and BO.

Now, let the specimen be loaded again. The loading path will retrace the unloading straight line OA, but when point A is reached again, further crack growth begins. Let the volume be increased beyond point A by an amount dV, taking the loading path to point B, and causing the crack to grow an additional increment, da. Then let the fluid again be retracted from the specimen slot, resulting in the straight line unloading path BO. It is clear that the irrecoverable work, dW_2, done on the specimen during the OABO loading-unloading cycle is given by the shaded area of the triangle OAB, i.e.,

$$dW_2 = 1/2\ PdV. \tag{3}$$

Letting $c_V = V/P$ be the volume compliance of the specimen, and letting P be the average pressure (i.e., a constant) during the crack advance, da, we have $dV = Pdc_V$, or

$$dW_2 = 1/2\ P^2 dc_V. \tag{4}$$

According to the Griffith-Irwin-Orowan criterion, dW_2 represents a decrease of strain energy in the specimen which is used up in the creation of the new crack area. Thus, equating dW_1 (equation 2) and dW_2, we obtain

$$2\gamma b = \frac{P^2}{2}\left[\frac{dc_V}{da}\right]. \tag{5}$$

Here, b, P, and dc_V/da are evaluated at the crack length, a, at which the incremental crack advance took place. Equation (5) is qualitatively the same equation published by Irwin and Kies (21) in 1954. Its derivation here shows its applicability to the short rod specimen configuration and to the fractometer method of loading the specimen. Now γ is related to K_{Ic} by the equation

$$2\gamma E = K_{Ic}^2\ (1 - \nu^2), \tag{6}$$

where E is the elastic modulus and ν is Poisson's ratio. Thus, equation (5) becomes, after some manipulation,

$$K_{Ic} = \frac{P\sqrt{B}}{(1-\nu^2)^{1/2}} \; f(a/B), \tag{7}$$

where B is the specimen diameter, and where

$$f(a/B) = \left[\frac{B}{2b}\frac{d(c_v E/B^3)}{d(a/B)}\right]^{1/2}, \tag{8}$$

The quantity in brackets is dimensionless, and is a function only of the ratio a/B. It is independent of the specimen material, as long as the scaled specimen configuration remains constant. Furthermore, it has been shown (16) that the value of a/B at the time of the peak load in the experiment is a constant, a_c/B. Thus, if we always use the peak load, P_c, in equation (7), then f(a/B) becomes the dimensionless constant $A_F \equiv f(a_c/B)$. Equation (7) then reduces to

$$K_{Ic} = A_F P_c \sqrt{B} \Big/ (1-\nu^2)^{1/2}, \tag{9}$$

or simply

$$K_{Ic} = A_F P_c \sqrt{B} \tag{10}$$

if we follow custom by replacing $(1-\nu^2)^{1/2}$ by unity. Calibration experiments, described below, have evaluated A_F for the short rod specimen configuration in use. They give

$$A_F = 7.2 \pm 0.4. \tag{11}$$

Finally, when specimens are restricted not only to a given scaled configuration, but also to a given size, B becomes a constant in equation (10), and K_{Ic} is then equal to a constant times the peak load in the experiment. This has been used to allow the Fractometer to provide a direct digital readout of K_{Ic} immediately following the test. The output of a transducer which senses the pressure in the flatjack is appropriately amplified such that its peak reading during the test is equal to K_{Ic}. An electronic memory circuit retains the peak output, such that K_{Ic} can be recalled at any time after the test.

SHORT ROD SPECIMEN CALIBRATIONS

The initial calibration of the short rod specimen configuration (16) was done by testing 2014 T651 aluminum according to ASTM E 399 to establish its K_{Ic} value. Tests on the same billet of material using the same crack orientation were then done using short rod specimens, and the calibration constant, A, in equation (1) (the short rod grip equation) was solved for, inasmuch as K_{Ic} was already known for that material. The value of A was found to be 20.8 ± 1.4. Since those experiments were performed, an additional inter-laboratory check of the short rod's grip calibration has been made. Professor D. L. Jones of George Washington University kindly supplied compact tension halves of two ASTM-valid tests of 2048 T6 aluminum of two different crack orientations. Short rod specimens were machined from the CT halves and tested. Without prior knowledge of the material type, crack orientation relative to rolling direction, or the ASTM-valid values of K_{Ic}, the short rod results compared 2% low on one specimen and 4% high on the other (22). Based on this excellent agreement, the uncertainty of the value of A in equation (1) is reduced:

$$A = 20.8 \pm 1.0 \text{ (grip tests)} \qquad (12)$$

The value of A_F in equation (11) has been determined in a manner similar to that described above, i.e., the value of K_{Ic} was measured, in this case using short rod grip tests, and subsequent Fractometer tests of the same material allowed A_F to be determined. A large number of tungsten carbide and alumina specimens have been tested to establish the best value of A_F; c.f. equation (11).

K_{Ic} MEASUREMENTS OF $A\ell_2O_3$

The Fractometer was used to test six grades of alumina kindly furnished by Coors Porcelain Company. Table I provides some physical properties of the various grades (23), along with the measured values of the plane strain critical stress intensity factor, K_{Ic}. The short rod specimen dimensions were 12.70 ± .025 mm (.500 ± .001 in) diameter, and 19.05 ± .075 mm (.750 ± .003 in) long. The slot thickness was .35 mm, the depth to the point of the V was 6.35 ± .075 mm (.250 ± .003 in), and the chord angle of the V was 58 ± 1/2 degree.

The Fractometer K_{Ic} tests were done for the most part in air at room temperature and pressure. However, the presence of stress corrosion cracking in air could easily be seen by loading the specimen to a given slot opening, and then observing the decrease in pressure as stress corrosion cracking progressed. Evans (11) has measured stress corrosion crack growth rates in $A\ell_2O_3$ in both 50% humid air and in toluene. He found that SCC effects could cause crack growth rates in 50% humid air of at least 1 mm/sec, but that SCC effects in toluene were several orders of magnitude smaller. Accordingly, since the crack growth rates in some of the initial Fractometer K_{Ic} tests may have been as low as 1 mm/sec, a few of the later tests were conducted after submerging the specimen and the flatjack portion of the Fractometer in toluene. It was found that the toluene did indeed greatly inhibit SCC effects, but that the

K_{Ic} values obtained from the tests in toluene were the same as those done in air to well within experimental uncertainties. Thus, it was shown that all of the K_{Ic} measurements were free of errors due to stress corrosion cracking.

DISCUSSION

The K_{Ic} data of Table I appear to be self-consistent in that both higher $A\ell_2O_3$ percentages and larger grain sizes tend to increase K_{Ic}. The dependence on grain size is consistent with previous DCB test results (5,24-26), but disagrees with SENB tests (9,25,27) which show little grain size dependence. Pratt (27) seemed to obtain DCB test results which gave smaller K_{Ic} values for continuous "rapid fracture" than for "slow crack growth" in which the specimen was unloaded following each finite advance of the crack. No such behavior was observed in the present study, however, although some specimens, including the 20μm grain size VISTAL material, were subjected to unloading and re-loading when the crack was at the critical location, while other specimens were rapidly fractured through the critical crack location.

Some comparisons of the measured K_{Ic} values of the present study with those for the same or similar materials of previously published research are possible. Freiman, et al (28), and Swanson (26) obtained K_{Ic}'s of 3.0 and 3.14 $MPa\sqrt{m}$, respectively, for Coors AD-85, which compares very well with the 2.98 $MPa\sqrt{m}$ measured here. However, for the AD-999 material, Swanson (26) obtained 4.44 $MPa\sqrt{m}$

TABLE I. Physical Properties and K_{Ic} Test Results for Six Commercial Aluminas.

COORS DESIGNATION		AD-85	AD-90	AD-96	AD-995	AD-999	VISTAL
PERCENT $A\ell_2O_3$		85	90	96	99.5	99.9	99.9
SPECIFIC GRAVITY		3.39	3.60	3.72	3.89	3.96	3.99
GRAIN SIZE (μm)	RANGE	2-12	2-10	2-20	5-50	1-6	15-45
	AVERAGE	6	4	11	17	3	20
ELASTIC MODULUS (GPa)		221	276	303	372	386	393
FRACTURE TOUGHNESS TESTS							
NO. OF K_{Ic} TESTS		5	5	6	5	3	4
AVERAGE K_{Ic} ($MPa\sqrt{m}$)		2.98	3.06	3.31	4.08	3.12	4.41
STANDARD DEVIATION		0.8%	2.9%	1.9%	0.8%	1.1%	0.8%

and Simpson (25) measured 4.00 and 3.93 MPa$\sqrt{m}$. These values are substantially larger than 3.12 MPa$\sqrt{m}$ of this study. On the other hand, Freiman, et al (28) measured 3.7 MPa$\sqrt{m}$ for a 99.9% alumina with an 8μm grain size. This correlates very well with the AD-999 and the VISTAL data here, which had grain sizes of 3 and 20μm, and K_{Ic}'s of 3.12 and 4.41 MPa$\sqrt{m}$, respectively.

The AD-96 K_{Ic} of 3.31 MPa$\sqrt{m}$ (11μm grain size) compares with the 3.2 MPa$\sqrt{m}$ measured by Freiman, et al (28) for a 96% alumina with the same elastic modulus and a 4μm grain size. Bertolotti (29) obtained 3.09 MPa$\sqrt{m}$ for a 94% alumina of the same density and grain size as AD-96, and Swanson and Gross (24) have measured 3.38 MPa$\sqrt{m}$ for another 10μm grain size alumina. These comparisons with existing data show good agreement of the short rod specimen, as tested by the Fractometer, with previous measurements by other techniques. In addition, the small standard deviations obtained here (Table I) indicate both a good uniformity of the specimens tested and a good repeatability of the Fractometer test method.

CONCLUSIONS

The short rod method of K_{Ic} measurement has been applied to six commercial grades of alumina. The test is quite simple, in that (a) a real pre-crack is obtained without any separate pre-cracking operation or precautions; (b) no crack length measurement is needed; and (c) the K_{Ic} value is obtained from the peak load alone. Further, the specimen configuration is short and small. The Fractometer further simplifies both the specimen preparation and the measurement, providing a direct digital read-out of K_{Ic} immediately after the test.

Standard deviations of 3% or less were obtained for the carefully controlled specimen materials, and the measured K_{Ic} values compare favorably with previously published data. The test results show a consistent increase in K_{Ic} with increasing grain size, a finding which agrees with previous DCB tests but disagrees with data from SENB tests. The K_{Ic} of the tested materials also tends to increase with increasing purity of $A\ell_2O_3$.

ACKNOWLEDGEMENTS

The author is grateful to Doug Jones of George Washington University for the interlaboratory comparison of K_{Ic} values, and to Steve Freiman, NRL, for helpful consultations. Short rod specimen blanks for Fractometer calibration were kindly supplied by Larry Pope of Fansteel, Rob Housman of Tungsten Carbide Mfg., Dave Wirth of Coors Porcelain, and Claude Brunswic of Carmet. Joe Farb and Al Rogers prepared specimens and assited in experiments.

LIST OF REFERENCES

1. Berry, J. P., "Determination of Fracture Surface Energies by the Cleavage Technique," J. Appl. Phys., 34 [1] 62 (1963).

2. Wiederhorn, S. M., A. M. Shorb, and R. L. Moses, "Critical Analysis of the Theory of the Double Cantilever Method of Measuring Fracture Surface Energies," J. Appl. Phys., 39 [3] 1569 (1968).

3. Swanson, G. D. and G. E. Gross, "Range of Validity for Use of the Double Cantilever Fracture Energy Analysis," J. Appl. Phys., 40 [11] 4684 (1969).

4. Freiman, S. W., D. R. Mulville, and P. W. Mast, "Crack Propagation Studies in Brittle Materials," J. Mater. Sci. 8 [11] 1527 (1973).

5. Gutshall, P. L. and G. E. Gross, "Observations and Mechanisms of Fracture in Polycrystalline Alumina," Eng. Frac. Mech. 1, 463 (1969).

6. Simpson, L. A., "Microstructural Considerations for the Application of Fracture Mechanics Techniques," in Fracture Mechanics of Ceramics, Vol. 2, Ed. by R. C. Bradt, D. P. H. Hasselman, and F. F. Lange, Plenum Press, N.Y., p. 567 (1974).

7. Evans, A. G., "Fracture Mechanics Determinations," in Fracture Mechanics of Ceramics, Vol. 1, Ed. by R. C. Bradt, D. P. H. Hasselman, and F. F. Lange, Plenum Press, N.Y., p. 17 (1974).

8. Simpson, L. A., "Use of the Notched-Beam Test for Evaluation of Fracture Energies of Ceramics," J. Am. Ceramic Soc., 57 [4] 151 (1974).

9. Calussen, N., R. Pabst, and C. P. Lahmann, "Influence of Microstructure of $A\ell_2O_3$ and ZrO_2 on K_{Ic}, " Proc. British Ceramic Soc., No. 25, 139 (1975).

10. Kies, J. A. and A. B. J. Clark, in Proceedings of the Second International Conference on Fracture, Brighton 1969, paper 41.

11. Evans, A. G., "A Method for Evaluating the Time-Dependent Failure Characteristics of Brittle Materials - And its Application to Polycrystalline Alumina," J. Mater. Sci., 7, 1137 (1972).

12. Tattersall, H. G. and G. Tappin, "Work of Fracture and its Measurement in Metals, Ceramics, and Other Materials," J. Mater. Sci., 1 [3] 296 (1966).

13. Clarke, F. J. P., H. G. Tattersall and G. Tappin, "Toughness of Ceramics and Their Work of Fracture," Proc. British Ceramic Soc., No. 6, 163 (1966).

14. Simpson, L. A., "Effect of Microstructure on Measurements of Fracture Energy of $A\ell_2O_3$," J. Am. Ceramic Soc., 56 [1] 7 (1973).

15. Bluhm, J. I., "Slice Synthesis of a Three Dimensional 'Work of Fracture' Specimen," Eng. Frac. Mech. 7, 593 (1975).

16. Barker, L. M., "A Simplified Method for Measuring Plane Strain Fracture Toughness," Eng. Frac. Mech. 9, 361 (1977).

17. Barker, L. M.. "Evaluation of a Simple Method for Measuring Fracture Toughness in both Brittle and Ductile Materials," Proceedings of the ICM-II Conference, Boston, p. 1547 (1976).

18. Barker, L. M. and W. C. Leslie, "Short Rod K_{Ic} Tests of Several Steels at Temperatures to 700K," Fracture 1977, Vol. 2, ICF-4, Waterloo, Canada, p. 305 (1977).

19. Pook, L. P., "An Approach to a Quality Control K_{Ic} Testpiece," Int. Journ. of Fracture Mech., 8, 103 (1972).

20. Pook, L. P., Private Communication.

21. Irwin, G. R. and J. E. Kies, "Critical Energy Rate Analysis of Fracture Strength," Weld. J. 33, 193 (1954).

22. Jones, D. L., Private Communication.

23. Bulletin 952, Coors Porcelain Co., Golden, Colorado (1974).

24. Swanson, G. D. and G. E. Gross, "Factor Analysis of Fracture-Toughness Test Parameters for $A\ell_2O_3$," J. Am. Ceramic Soc., 54, 382 (1971).

25. Simpson, L. A., "Discrepancy Arising from Measurement of Grain-Size Dependence of Fracture Energy of $A\ell_2O_3$," J. Am. Ceramic Soc., 56 [11] 610 (1973).

26. Swanson, G. D., "Fracture Energies of Ceramics," J. Am. Ceramic Soc., 55 [1] 48 (1972).

27. Pratt, P. L., "Grain Size and Fracture Toughness of Alumina," in Fracture 1977, Vol. 3, ICF-4, Waterloo, Canada, p. 909 (1977).

28. Freiman, S. W., K. R. McKinney, and H. L. Smith, "Slow Crack Growth in Polycrystalline Ceramics," in Fracture Mechanics of Ceramics, Vol. 2, Ed. by R. C. Bradt, D. P. H. Hasselman, and F. F. Lange, Plenum Press, N. Y., p. 659 (1974).

29. Bertolotti, R. L., "Fracture Toughness of Polycyrstalline $A\ell_2O_3$," J. Am. Ceramic Soc., 56, 107 (1973).

IMPACT FRACTURE OF CERAMICS AT HIGH TEMPERATURE

Stephen T. Gonczy and D. Lynn Johnson

Department of Materials Science and Engineering
Northwestern University
Evanston, Illinois 60201

ABSTRACT

Samples of hot-pressed silicon nitride were tested in 3-point bend impact loading at ambient temperature, 1100°C, 1250°C, and 1400°C under isothermal conditions. Dynamic stress intensity factors (K_{IC}) were calculated from the fracture load and the dimensions of an induced surface flaw, produced by a Knoop hardness indenter. Dynamic stress intensities of about 5.25 $MN/m^{3/2}$ were determined for all temperatures up to 1250°C. A 20% drop in K_{IC} was found at 1400°C. This drop in K_{IC} coincides with a change in the fracture surface at the same temperature, indicating a change in the micromechanics of fracture at 1400°C.

I. INTRODUCTION

The use of ceramics as structural components requires a knowledge of their impact properties at ambient and high temperatures. Impact studies on alumina[1], silicon carbide[2,3,4], and silicon nitride[4,5] have been done with instrumented Charpy or drop-weight tests. The Charpy-type tests at high temperatures have been limited by the fact that the specimens have been fractured under transient thermal conditions, or at least with thermal gradients along and through the specimens. The free-fall drop-weight test fractures the specimen under isothermal conditions; the instrumentation measures the deflection of the specimen, rather than the load exerted on the specimen.[4,5]

Recent studies by several investigators[6,7,8] have analyzed the use of a controlled semi-elliptical surface flaw induced on the tensile side of a ceramic slow bend specimen by a Knoop hardness indenter with a 2½ kg load. The shape and dimensions of the flaw produced by a given load are reproducible. In accordance with fracture theory, the flaw reduces the bend strength of the specimen. The use of the induced flaw reduces the scatter of the bend strengths, normally found among unflawed specimens, caused by the statistical distribution of natural flaws.[6] Additionally, the critical stress intensity factor, K_{IC}, can be evaluated by using the induced flaw dimensions.

The present investigation was designed to take advantage of the controlled surface flaws and to break the specimens at controlled impact rates under isothermal conditions.

II. EXPERIMENTAL PROCEDURE

Hot-pressed silicon nitride* was cut into test bars 34.5 mm long with a cross section of 3.0 by 3.0 mm. These bars were polished on all sides with 600 grit SiC powder on a two-surface lapping machine. Polishing continued until all visible surface imperfections were eliminated and until the cross-sectional dimensions were constant along the length to within .03 mm. A surface finish of better than 20 μinches RMS was achieved.

Using a microhardness tester with a Knoop diamond indenter, an indentation was centered on the tensile side of each test bar. The long axis of the indentation was perpendicular to the tensile stress direction; its length was measured with an optical microscope.

Three-point bend impact loading of the test bars was done at room temperature, 1100°C, 1250°C, and 1400°C in air in a test system shown schematically in Fig. 1. The tensile stresses were in a plane perpendicular to the hot-pressing direction of the material. The test beams were held in a notched alumina tube with a bending span of 29.0 mm in a furnace heated by molybdenum disilicide resistance heating elements. The temperature was monitored by a thermocouple placed $\approx$10 mm directly above the center of the beam. Test bars were taken from room temperature to test temperature over a period of 1½ hours.

The velocity of the impact head was controlled by a pulley system and pneumatic damper, consisting of a metal piston traveling inside a metal cylinder closed at one end. Valves in the closed end were adjusted to control the amount of air flowing in behind the moving piston, and thus controlling the drop velocity.

*NC-132, Norton Co., Worcester, MA.

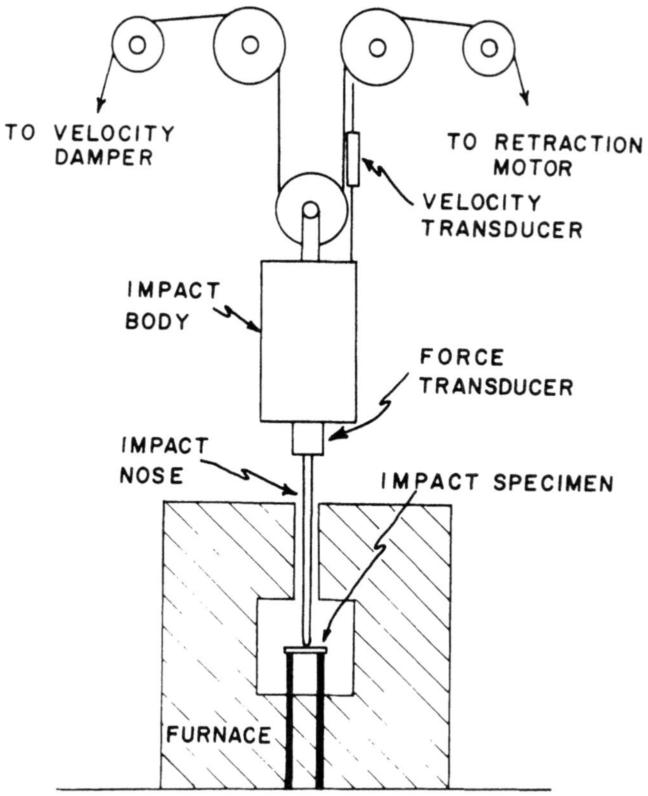

Figure 1. Schematic Diagram of Impact System and Furance.

A retraction motor controlled by position microswitches withdraws the impact nose from the hot zone of the furnace after fracture occurs. The impact nose enters and leaves the furnace over a period of less than one second. The impact body is kept aligned with the specimen by guides at the top and bottom of the drop path.

A quartz load transducer*, located between the impact nose and the impact body, monitors the force applied to the test beam. The velocity of the impact body is monitored by a velocity transducer.†

*Kistler Instruments, Sundstrand Data Corp., Redmond, WA.

†Trans-Tek Inc., Ellington, CT.

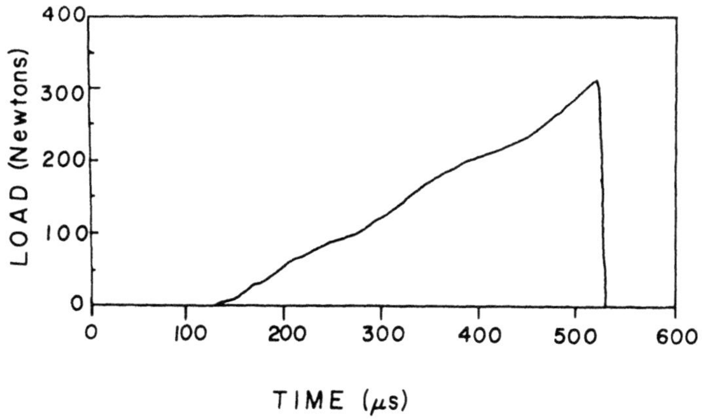

Figure 2. Typical Load-Time Trace for Impact Velocity of 45 cm/s.

The mass of the impact body is 2 kg and is made of brass. The impact nose is made of a titanium alloy and is 7.5 mm in diameter.

A charge amplifier* is used to condition the force signal. Two high speed digital waveform recorders** store the two signals. The recorder used for the load signal has 2048 units of 8-bit memory to record events as short as 409 μs. The recorder used for the velocity signal has 256 units of 6-bit memory to record events as short as 25.6 μs. Both recorders are triggered by the rise of the load signal to a specified trigger level. The recorders are operated in a pretrigger mode, which inusres the recording of that portion of the signal prior to the point at which the recorder is triggered.

The recorders have both digital and analog outputs. The stored signals can be observed on an oscilloscope, transferred to a strip chart recorder for hard analog copy, and transferred to a Teletype punch unit for hard digital copy.

A typical graph of the load signal for an impact fracture is shown in Fig. 2. A graph of the corresponding velocity signal is shown in Fig. 3.

Fracture surfaces were studied by optical microscope using both polarized and unpolarized light. The surfaces were examined

*Kistler Instruments, Sundstrand Data Corp., Redmond, WA.

**Biomation Corp., Cupertino, CA.

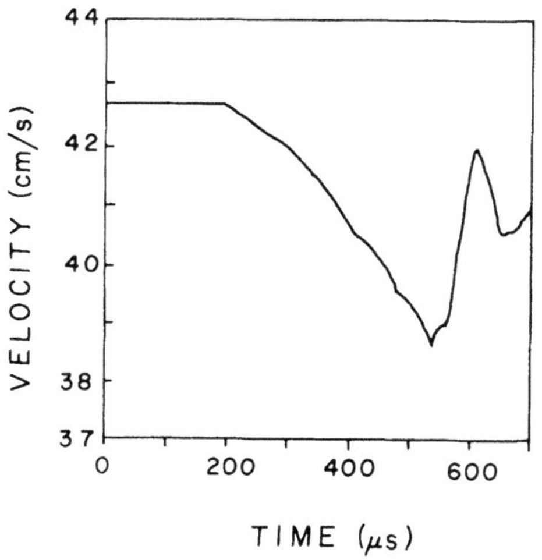

Figure 3. Typical Velocity-Time Trace for Impact Velocity of 45 cm/s.

primarily for characterization of the fracture surface and for measurement of the depth of the induced flaw.

III. CALCULATION OF K_{IC}

The Knoop hardness indenter introduces a semi-elliptical surface crack on the tensile side of the bend specimen. This surface flaw will grow catastrophically when the stress intensity at the flaw reaches a critical value. According to fracture mechanics theory, the stress intensity factor for a semi-elliptical surface flaw in bending is given by:[6,9]

$$K_I = \sigma M (\pi a/Q)^{\frac{1}{2}} \tag{1}$$

where σ is the maximum outer fiber tensile stress, M is a numerical factor related to flaw and beam geometry, a is the flaw depth, and Q is given by:[6,9]

$$Q = \Phi^2 - 0.212 (\sigma/\sigma_{ys})^2 \tag{2}$$

where σ_{ys} is the tensile yield stress and $0.212 (\sigma/\sigma_{ys})^2$ is a plastic zone correction factor. At the loading rates used in this work, this plastic correction factor is considered negligible. The value Φ is the elliptical integral[9]

$$\Phi = \int_0^{\pi/2} \left[1 - \frac{c^2 - a^2}{c^2} \sin^2\theta\right]^{\frac{1}{2}} d\theta \qquad (3)$$

which is tabulated in standard mathematical tables; c is the major axis of the flaw.

The value for M was taken as 1.03 based upon bend tests by Petrovic et al[6] of silicon nitride with similar beam and flaw geometry.

IV. RESULTS AND DISCUSSION

Analysis of the Load-Time and Velocity-Time Traces

A study of the load-time traces shows that the average time between the points of initial loading and fracture was approximately 325 μs for all specimens broken with an impact velocity in the 40-45 cm/s range. The magnitude of the fracture load for all the specimens was in the range of 300 N and depended upon the cross-sectional geometry of individual beams.

The shape of the load-time traces was generally of the form shown in Fig. 2. Small scale oscillations were seen in some cases. These oscillations, superimposed on the linear load trace, had a wavelength of 30-40 μs and a magnitude of less than 10% of the final fracture load. These oscillations are most likely caused by vibration of the test beam itself under inertial loading. A second explanation is stress wave reflections between the tip of the impact nose and the junction of the impact nose with the load transducer.

There was no sign of non-catastrophic crack growth for any specimens broken at any temperature. The load trace dropped sharply to zero within 10 μs after fracture initiation.

The velocity curves were of the form seen in Fig. 3. Impact velocities were not precisely reproducible between tests, but ranged between 41 and 45 cm/s. The drops in velocity that occurred with impact were generally linear, and the duration of the linear portion agreed within ±10% of the duration of the corresponding load trace. The fact that the velocity transducer is attached to the upper part of the impact body explains the discrepancies. The transducer monitors the velocity of the impact body, rather than the precise velocity of the impact nose.

Fracture Surfaces and Flaw Size Analysis

The fracture surfaces of all the broken specimens had regions described by Kirchner[10] as mirror, mist, and hackle. The mirror is a region of smooth flat surface. It is surrounded by mist, which is less reflective and stippled. The mist is bordered by hackle, which is characterized by a rough region of hills and valleys. For all specimens the location of the mirror region proved conclusively that the Knoop flaw was the point at which fracture started.

The fracture surfaces of specimens broken at room temperature, 1100°C, and 1250°C were all similar, showing mirror, mist and hackle. The relative amounts of each region were not markedly different between the different temperatures. However, on the specimens broken at 1400°C virtually no hackle was seen, and the mist region was smoother and more homogeneous. This change in fracture surface at 1400°C is a possible indication of a change in the mode of fracture propagation; it coincides with the 20% decrease in critical stress intensity observed at 1400°C.

A polarizing microscope was used to measure the depth of the crack beneath the Knoop flaw. The general location of the flaw on the fracture surface was known from both the original positioning of the diamond indenter and the location of the mirror region on the fracture surface. However, in only 45% of the 27 specimens broken could the flaw depth be identified and measured with certainty. In 33% of the specimens the flaw could be observed and measured with only a fair degree of accuracy. In 22% the flaw could not be identified at all. This difficulty in identification and measurement was partially caused by oxidation of the fracture surface after fracture. There was also pick-up of alumina fiber furnace insulation, when the specimen struck the furnace walls while rebounding from the impact nose. In some cases the oxidation and alumina totally obscured the portion of the test beam on which the Knoop flaw was located.

The flaw itself, even when the fracture surface was clean, was not always identifiable under either polarized light or unpolarized light. On other specimens the flaw was easily located and measured; all identifiable flaws were clearly elliptical in shape. Figure 4 shows the average flaw dimensions for Knoop indentation loads of 2 kg and 2½ kg. Similar data from Petrovic et al[6] for NC-132 silicon nitride is in good agreement.

For those specimens broken at high temperature, there was no sign of slow crack growth. All identifiable Knoop flaws were of

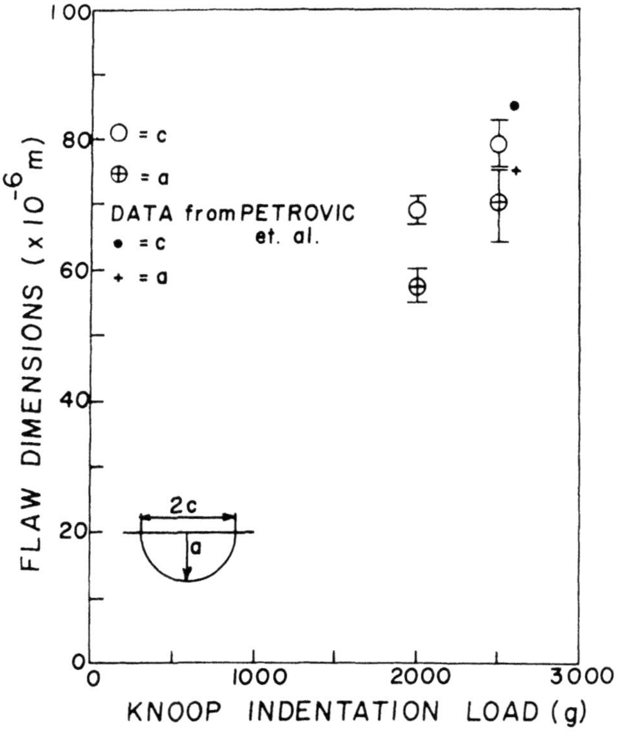

Figure 4. Flaw Dimensions vs Knoop Load for NC-132 Silicon Nitride (Error Bars Indicate Standard Deviation)

the same general dimensions as the flaws on the specimens broken at room temperature. Petrovic observed in his bend tests of HS-130 silicon nitride that the flaw grew at all tests above 1200°C using a cross-head speed of .01 cm/min. On tests at 1300°C the flaw grew approximately 10 times its initial size before catastrophic failure. They also noted that at 1300°C the amount of slow crack growth decreased with increasing cross-head speed. The fastest cross-head speed used by Petrovic et al was .5 cm/min. With the present impact velocity of about 2500 cm/min, the lack of any signs of slow crack growth at high temperatures is consistent with theory.

The fracture surfaces showed no sign of Hertzian-type damage. In many cases, however, there was secondary cracking on the flawed side of the specimen. No study was made of the character or causes of these secondary cracks.

Calculation of K_{IC} Factors

The bending stress at fracture and the measured flaw depth were used in K_{IC} calculations. When the flaw could not be identified or accurately measured, an expected value for the flaw depth was interpolated from the measured flaw length on the surface and the average flaw depth/flaw length ratio determined among all specimens. This interpolated value was then used to determine K_{IC}.

Room Temperature K_{IC}

Petrovic et al[6] showed that the room temperature strength and K_{IC} values for flawed HS-130 Si_3N_4 specimens could be increased 30% by annealing the flawed specimens at over 1100°C in air for 6 hours. They attributed this increase to the elimination of residual stresses at the flaw site; the plastic deformation caused by the diamond indenter leaves the immediate area in a compressive state. This plastically strained region acts as a wedge in the crack, placing the crack tip in a tensile stress state. This residual tensile stress acts with any applied stress, and thus reduces the amount of applied stress needed to reach a critical stress level. Annealing the specimen eliminates this residual stress.

Specimens of NC-132 with flaws were tested at room temperature in two conditions. One set of specimens was tested in an unannealed condition, while a second set was annealed in air for 4 hours at 1100°C. The K_{IC} values for both sets are shown in Fig. 5. The 30% increase in K_{IC} values for the annealed specimens is in agreement with Petrovic et al. The K_{IC} values at room temperature for the annealed specimens are also in agreement with dynamic stress intensity measurements of HS-130 silicon nitride conducted at impact velocities of 1.3×10^4 cm/min by Mendiratta, Wimmer, and Bransky.[5]

High Temperature K_{IC}

This study tested the NC-132 specimens at 1100°C, 1250°C, and 1400°C. The computed critical stress intensity values are shown in Fig. 5. The K_{IC} values for the annealed specimens tested at room temperature and the values for unannealed specimens tested at 1100°C and 1250°C are constant near 5.25 $MN/m^{3/2}$. These values are slightly higher than the K_{IC} values determined at similar temperatures by Evans and Wiederhorn[11] in double torsion and by Petrovic et al[6] in 4-point bend for HS-130 Si_3N_4.

The data in the room temperature to 1250°C range also agrees with impact K_{IC} data by Mendiratta et al.[5] They tested HS-130 with induced surface flaws. The impact velocity in that study was an order of magnitude larger than this study; but the difference in

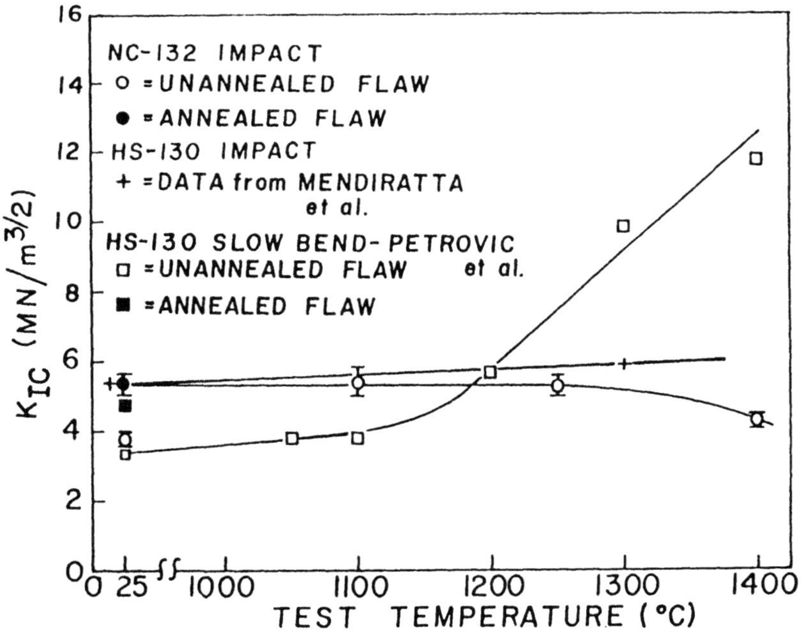

Figure 5. K_{IC} versus Temperature for Silicon Nitride (Error Bars Indicate Standard Deviation)

velocities is not apparent in the K_{IC} values.

In tests at 1400°C the K_{IC} value dropped 20% to 4.25 $MN/m^{3/2}$. This is in marked contrast to the high temperature double torsion and slow-bend data for HS-130. In those studies the K_{IC} value increased by a factor of 3. But this increase was caused by slow crack growth. The researchers determined K_{IC} using the flaw depth marked by the onset of catastrophic fracture. As Petrovic et al pointed out, the depth dimension increased at 1300°C by a factor of 10 by the slow crack growth mechanism.

At the faster loading rates characteristic of impact, this slow crack growth does not have time to occur. The flaw does not grow, and the critical stress intensity does not increase. Mendiratta et al tested their samples at a maximum temperature of 1300°C and saw little change in K_{IC} values compared to room temperature values.

Thus the drop in K_{IC} values for Si_3N_4 at 1400°C under impact loading has not been observed prior to this study. However, a drop in K_{IC} at high temperatures under slow-bend testing was observed by Petrovic et al[7] in studies of hot-pressed silicon carbide. In that study silicon carbide, unlike silicon nitride, showed no signs of slow crack growth, even at the highest test temperature of 1400°C. Petrovic and Jacobson theorized that the drop in K_{IC} for silicon carbide at high temperatures was caused by a change in the micromechanism of the fracture process. There was a marked difference between SEM micrographs of the fracture surfaces broken at 1400°C in vacuum and at room temperature in air. They also theorized that a grain boundary phase rich in impurities provided an easier path for brittle fracture at high temperatures.

This same theory is a possible explanation for the drop in K_{IC} for silicon nitride at 1400°C under impact loading. The difference in fracture surfaces noticed between samples broken at 1250°C and at 1400°C adds credence to this theory. A grain boundary phase containing the hot-pressing additives is often cited as the source of the grain boundary sliding which contributes to the slow crack growth occurring over 1200°C. If at 1400°C this grain boundary region becomes significantly less viscous and much weaker, it will provide an easier path for brittle fracture; and this would cause a marked decrease in high-temperature strength even under impact load.

V. CONCLUSIONS

The controlled surface flaw technique used for slow bend tests is also applicable to impact loading tests of ceramics. Using the impact fracture load and the measured flaw dimensions, the dynamic stress intensity factors can be determined.

Impact tests of NC-132 silicon nitride at high temperatures show no sign of the slow crack growth seen in 4-point bend studies at similar temperatures.

The fracture surfaces of the NC-132 impact specimens show definable areas of the fracture regions of mirror, mist, and hackle.

Room temperature impact K_{IC} values for the flawed NC-132 can be raised by annealing the specimen in air at 1100°C.

The impact K_{IC} values in the region of ambient temperature up to 1250°C show a consistent value of roughly 5.25 $MN/m^{3/2}$ for NC-132 silicon nitride.

There is a marked decrease of 20% in the impact K_{IC} of NC-132 at 1400°C. This decrease correlates with a change in fracture surface, compared with specimens broken at 1250°C or less. This indicates a change in the micromechanism of fracture at this temperature level. Thus the impact resistance of NC-132 silicon nitride is not constant beyond 1300°C, but drops. This indicates a temperature limitation for NC-132 in its consideration for high-temperature gas turbines.

ACKNOWLEDGEMENT

The research upon which this work is based was supported by the Air Force Office of Scientific Research, Grant #AFOSR-76-2920-A.

REFERENCES

1. R. L. Bertolotti, "Strength and Absorbed Energy in Instrumented Impact Tests of Polycrystalline Al_2O_3", J. Am. Ceram. Soc. 57 [7] 300 (1974).
2. H. Abe, "Instrumented Charpy Impact Testing of Silicon Carbide", PhD Dissertation, Pennsylvania State University, May 1976.
3. H. C. Chandan, H. Abe, and R. C. Bradt, "Elevated Temperature Charpy Impact Test of Silicon Carbide", Abstract, Am. Ceram. Soc. Bul. 56 [3] 289 (1977).
4. J. M. Wimmer, I. Bransky, "Impact Resistance of Structural Ceramics", Am. Ceram. Soc. Bul. 56 [6] 552 (1977).
5. M. G. Mendiratta, J. Wimmer, I. Bransky, "Dynamic K_{IC} & Dynamic Flexural Strength of HS-130 Si_3N_4", J. Mat. Sci. 12 212 (1977).
6. J. J. Petrovic, L. A. Jacobson, P. K. Talty, and A. K. Vasudevan, "Controlled Surface Flaws in Hot-Pressed Si_3N_4", J. Am. Ceram. Soc. 58 [3-4] 113 (1975).
7. J. J. Petrovic, L. A. Jacobson, "Controlled Surface Flaws in Hot-Pressed SiC", J. Am. Ceram. Soc. 59 [1-2] 34 (1976).
8. Y. W. Mai, "Thermal Shock Resistance and Fracture Strength Behavior of Two Tool Carbides", J. Am. Ceram. Soc. 59 [11-12] 491 (1976).
9. G. R. Marrs and C. W. Smith, "A Study of Local Stresses Near Surface Flaws in Bending Fields", p. 23 in Stress Analysis and Growth of Cracks, Am. Soc. of Test. Mat., Spec. Tech Publ. #513, 1972.
10. H. P. Kirchner, R. M. Gruever, and W. A. Sotter, "Use of Fracture Mirrors to Interpret Impact Fractures in Brittle Materials" J. Am. Ceram. Soc. 58 [5-6] 188 (1975).
11. A. G. Evans and S. M. Wiederhorn, "Crack Propagation and Failure Prediction in Silicon Nitride at Elevated Temperatures", J. Mat. Sci. 9 [2] 270 (1974).

CONTRIBUTORS

Co-Chairmen

R. C. Bradt, Professor, Department of Materials Science and Engineering, Ceramic Science and Engineering Section, Pennsylvania State University, University Park, Pennsylvania.

D. P. H. Hasselman, Whittemore Professor of Engineering, Department of Materials Engineering, Virginia Polytechnic Institute and State University, Blacksburg, Virginia.

F. F. Lange, Senior Research Scientists, Structural Ceramics Group, Rockwell International/Science Center, Thousands Oaks, California.

Conference Session Chairmen

E. M. Lenoe, AMMRC; Watertown, Massachusetts

R. W. Davidge, AERE-Harwell; England

V. D. Frechette, Alfred University; Alfred, New York

H. C. Palmour, III, North Carolina State University; Raleigh, North Carolina

M. L. Williams, University of Pittsburgh; Pittsburgh, Pennsylvania

K. R. Kinsman, E.P.R.I., Palo Alto, California

Conference Coordinator

Ronald Avillion, J. Orvis Keller Conference Center, Pennsylvania State University; University Park, Pennsylvania.

AUTHORS

M. Adams, Jet Propulsion Laboratory, Pasadena, California

W. F. Adler, Effects Technology Inc., Santa Barbara, California

C. G. Annis, Pratt and Whitney, West Palm Beach, Florida

K. Arin, General Electric Co., Schenectady, New York

M. Bakioglu, Lehigh University, Bethlehem, Pennsylvania

G. J. Bansal, Battelle Columbus Labs., Columbus, Ohio

L. M. Barker, Terra Tek, Salt Lake City, Utah

S. B. Batdorf, UCLA, Los Angeles, California

E. K. Beauchamp, Sandia Labs., Albuquerque, New Mexico

P. F. Becher, Naval Research Lab., Washington, District of Columbia

R. C. Bennet, University of British Columbia, Vancouver, British Columbia, Canada

D. R. Biwas, University of California, Berkeley, California

W. C. Bourne, The Pennsylvania State University, University Park, Pennsylvania

R. C. Bradt, The Pennsylvania State University, University Park, Pennsylvania

S. D. Brown, University of Illinois, Urbana, Illinois

J. G. Bruce, Honeywell, Minneapolis, Minnesota

F. E. Buresch, Kernforschungsanlage Julich, Julich, West Germany

J. S. Cargill, Pratt and Whitney, West Palm Beach, Florida

H. C. Chandan, Bell Laboratories, Norcross, Georgia

R. J. Charles, General Electric Co., Schenectady, New York

M. M. Chaudhri, Cavendish Laboratory, Cambridge, England

J. L. Chermant, Universite de Caen, Caen, Cedex, France

N. Claussen, Max Planck Institute, Stuttgart, West Germany

P. H. Conley, Koppers Co., Inc., Pittsburgh, Pennsylvania

R. W. Davidge, AERE-Harwell, Harwell, England

A. DesChanvres, University of Caen, Caen, Cedex, France

W. Duckworth, Battelle Columbus, Ohio

R. J. Eagen, Sandia Labs., Albuquerque, New Mexico

A. F. Emery, University of Washington, Seattle, Washington

F. F. Erdogan, Lehigh University, Bethlehem, Pennsylvania

A. G. Evans, Rockwell International, Thousand Oaks, California

H. Fessler, University of Nottingham, Nottingham, England

I. E. Field, University of Cambridge, Cambridge, England

V. D. Frechette, Alfred University, Alfred, New York

S. W. Freiman, U. S. Naval Research Laboratory, Washington, District of Columbia

E. R. Fuller, Jr., National Bureau of Standards, Washington, District of Columbia

R. M. Fullrath, University of California, Berkeley, California

A. Gaddipatti, University of Washington, Seattle, Washington

S. T. Gonczy, Northwestern University, Evanston, Illinois

R. S. Gordon, University of Utah, Salt Lake City, Utah

J. K. Gran, SRI, Palo Alto, California

D. J. Green, Canada Center for Min. & Energy Tech. Ottawa, Canada

R. M. Gruver, Ceramic Finishing Company, State College, Pennsylvania

T. K. Gupta, Westinghouse, Pittsburgh, Pennsylvania

J. T. Hagan, Cambridge University, Cambridge, England

S. A. Halen, USNRL, Washington, District of Columbia

D. P. H. Hasselman, Virginia Polytechnic Institute and State University, Blacksburg, Virginia

N. Hattu, N. V. Philips Gloelampen Fabrieken, Eindhoven, Holland

B. J. Hockey, National Bureau of Standards, Washington, District of Columbia

S. V. Hooker, Effects Technology, Santa Barbara, California

C. O. Hulse, United Technologies, East Hartford, Connecticut

C. A. Johnson, General Electric, Schenectady, New York

D. L. Johnson, Northwestern University, Evanston, Illinois

H. Johnson, National Bureau of Standards, Washington, District of Columbia

B. F. Jones, Admiralty Materials Laboratory, Dorsett, England

R. L. Jones, SRI, Palo Alto, California

H. P. Kirchner, Ceramic Finishing Company, State College Pennsylvania

A. S. Kobayashi, University of Washington, Seattle, Washington

B. G. Koepke, Honeywell International, Bloomington, Minnesota

F. F. Lange, Rockwell International, Thousand Oaks, California

J. Lankford, Southwest Research Institute, San Antonia, Texas

B. R. Lawn, University of New South Wales, Kensington, Australia

M. W. Lindley, Admiralty Materials Laboratory, Dorsett, England

W. J. Love, University of Washington, Seattle, Washington

B. D. Marshall, University of New S. Wales, Kensington, Australia

M. Matsui, NGK, Ltd., Nagoya, Japan

J. R. McClaren, AERE-Harwell, Harwell, England

C. L. McKinnis, Owens-Corning Fiberglas, Granville, Ohio

J. J. Mecholsky, Jr., USNRL, Washington, District of Columbia

T. Michalske, Alfred University, Alfred, New York

J. S. Nadeau, University of British Columbia, Vancouver, Canada

P. S. Nicholson, McMaster University, Hamilton, Ontario, Canada

H. A. Nied, General Electric, Schenectady, New York

W. L. Newell, USNRL, Washington, District of Columbia

I. Oda, NGK, Ltd., Nagoya, Japan

E. Olsen, IBM, San Jose, California

F. Osterstock, University of Caen, Caen, Cedex, France

R. F. Pabst Max-Planck Institute, Stuttgart, West Germany

K. C. Pitman, Admiralty Materials Labs., Dorsett, England

P. Pizzi, Fiat Research Center, Orbassano, Italy

B. J. Pletka, National Bureau of Standards, Wasington, District of Columbia

R. Pohanka, Office of Naval Research/NRL, Arlington, Virginia

R. A. Queeney, The Pennsylvania State University, University Park, Pennsylvania

R. W. Rice, U. S. Naval Research Laboratory, Washington, District of Columbia

J. E. Ritter, Jr., University of Massachusetts, Amherst, Massachusetts

E. M. Rockar, National Bureau of Standards, Washington, District of Columbia

D. J. Rowcliffe, SRI International, Menlo Park, California

N. L. Rupert, Naval Surface Weapons Center, Dahlgren, Virginia

W. D. Scott, University of Washington, Seattle, Washington

D. K. Shetty, University of Utah, Salt Lake City, Utah

G. Sines, UCLA, Los Angeles, California

A. D. Sivil, Lucas Ltd., Birmingham, England

V. Snijders, N. V. Philips Gloeilampen Fabrieken, Eindhoven, Holland

W. D. Snowden, General Electric Company, Nela Park, Ohio

T. Soma, NGK, Ltd., Nagoya, Japan

L. I. Staley, University of Washington, Seattle, Washington

P. Stanley, University of Nottingham, Nottingham, England

J. Steeb, Max-Planck Institute, Stuttgart, West Germany

M. V. Swain, Cambridge University, Cambridge, England

J. C. Swearengen, Sandia, Albuquerque, New Mexico

G. Tappin, AERE-Harwell, Harwell, England

B. K. Tariyal, Bell Laboratories, Norcross, Georgia

R. M. Thomson, National Bureau of Standards, Washington, District of Columbia

G. G. Trantina, General Electric Research & Development, Schenectady, New York

R. E. Tressler, The Pennsylvania State University, University Park, Pennsylvania

J. R. Varner, UCLA, Los Angeles, California

J. D. P. Veldkamp, N. V. Philips Gloeilampen Fabrieken, Eindhoven, Holland

A. V. Virkar, University of Utah, Salt Lake City, Utah

A. M. Walley, Cavendish Laboratory, Cambridge, England

S. M. Wiederhorn, National Bureau of Standards, Washington, District of Columbia

C. Cm. Wu, USNRL, Washington, District of Columbia

INDEX

Pages 1-506 are found in Volume 3, pages 507-987 in Volume 4.